I0053144

Yeasts: From Nature to Bioprocesses

Edited by

Sérgio Luiz Alves Júnior

Laboratory of Biochemistry and Genetics
Federal University of Fronteira Sul
Campus Chapecó - SC
Brazil

Helen Treichel

Laboratory of Microbiology and Bioprocesses
Federal University of Fronteira Sul
Campus Erechim - RS
Brazil

Thiago Olitta Basso

Department of Chemical Engineering
University of São Paulo
São Paulo - SP
Brazil

&

Boris Ugarte Stambuk

Department of Biochemistry
Federal University of Santa Catarina
Florianópolis - SC
Brazil

Mycology: Current and Future Developments

Volume # 2

Editors: Sérgio Luiz Alves Júnior, Helen Treichel, Thiago Olitta Basso and Boris Ugarte Stambuk

Yeasts: From Nature to Bioprocesses

ISSN (Online): 2452-0780

ISSN (Print): 2452-0772

ISBN (Online): 978-981-5051-06-3

ISBN (Print): 978-981-5051-07-0

ISBN (Paperback): 978-981-5051-08-7

© 2022, Bentham Books imprint.

Published by Bentham Science Publishers Pte. Ltd. Singapore. All Rights Reserved.

BENTHAM SCIENCE PUBLISHERS LTD.
End User License Agreement (for non-institutional, personal use)

This is an agreement between you and Bentham Science Publishers Ltd. Please read this License Agreement carefully before using the ebook/echapter/ejournal (**"Work"**). Your use of the Work constitutes your agreement to the terms and conditions set forth in this License Agreement. If you do not agree to these terms and conditions then you should not use the Work.

Bentham Science Publishers agrees to grant you a non-exclusive, non-transferable limited license to use the Work subject to and in accordance with the following terms and conditions. This License Agreement is for non-library, personal use only. For a library / institutional / multi user license in respect of the Work, please contact: permission@benthamscience.net.

Usage Rules:

1. All rights reserved: The Work is the subject of copyright and Bentham Science Publishers either owns the Work (and the copyright in it) or is licensed to distribute the Work. You shall not copy, reproduce, modify, remove, delete, augment, add to, publish, transmit, sell, resell, create derivative works from, or in any way exploit the Work or make the Work available for others to do any of the same, in any form or by any means, in whole or in part, in each case without the prior written permission of Bentham Science Publishers, unless stated otherwise in this License Agreement.
2. You may download a copy of the Work on one occasion to one personal computer (including tablet, laptop, desktop, or other such devices). You may make one back-up copy of the Work to avoid losing it.
3. The unauthorised use or distribution of copyrighted or other proprietary content is illegal and could subject you to liability for substantial money damages. You will be liable for any damage resulting from your misuse of the Work or any violation of this License Agreement, including any infringement by you of copyrights or proprietary rights.

Disclaimer:

Bentham Science Publishers does not guarantee that the information in the Work is error-free, or warrant that it will meet your requirements or that access to the Work will be uninterrupted or error-free. The Work is provided "as is" without warranty of any kind, either express or implied or statutory, including, without limitation, implied warranties of merchantability and fitness for a particular purpose. The entire risk as to the results and performance of the Work is assumed by you. No responsibility is assumed by Bentham Science Publishers, its staff, editors and/or authors for any injury and/or damage to persons or property as a matter of products liability, negligence or otherwise, or from any use or operation of any methods, products instruction, advertisements or ideas contained in the Work.

Limitation of Liability:

In no event will Bentham Science Publishers, its staff, editors and/or authors, be liable for any damages, including, without limitation, special, incidental and/or consequential damages and/or damages for lost data and/or profits arising out of (whether directly or indirectly) the use or inability to use the Work. The entire liability of Bentham Science Publishers shall be limited to the amount actually paid by you for the Work.

General:

1. Any dispute or claim arising out of or in connection with this License Agreement or the Work (including non-contractual disputes or claims) will be governed by and construed in accordance with the laws of Singapore. Each party agrees that the courts of the state of Singapore shall have exclusive jurisdiction to settle any dispute or claim arising out of or in connection with this License Agreement or the Work (including non-contractual disputes or claims).
2. Your rights under this License Agreement will automatically terminate without notice and without the

need for a court order if at any point you breach any terms of this License Agreement. In no event will any delay or failure by Bentham Science Publishers in enforcing your compliance with this License Agreement constitute a waiver of any of its rights.

3. You acknowledge that you have read this License Agreement, and agree to be bound by its terms and conditions. To the extent that any other terms and conditions presented on any website of Bentham Science Publishers conflict with, or are inconsistent with, the terms and conditions set out in this License Agreement, you acknowledge that the terms and conditions set out in this License Agreement shall prevail.

Bentham Science Publishers Pte. Ltd.
80 Robinson Road #02-00
Singapore 068898
Singapore
Email: subscriptions@benthamscience.net

BENTHAM SCIENCE

CONTENTS

PREFACE

Yeasts: from nature to bioprocesses travels back in time through the history of yeasts from the early days up to now, with an evolutionary, taxonomic, and biotechnological approach. Along this journey, its chapters present numerous bioprocesses which use these microorganisms, from the Neolithic revolution to the present.

While the budding yeasts subphylum has been estimated to appear on earth about 400 million years ago, some yeast species known today are certainly more recent, such as the workhorse *Saccharomyces cerevisiae*, which probably diverged from its sister species approximately 5 million years ago. Indeed, yeasts play a fundamental ecological role in nutrient recycling and angiosperm reproduction. Thus, directly and indirectly, they have guaranteed the maintenance of the biodiversity of plants and, consequently, animals that establish an ecological relationship with them. Yeast ecology has a chapter of its own in this book, although other chapters have also punctuated this theme in different contexts.

The main yeast genera are discussed in specific chapters of the book. Likewise, important biotechnological applications of these microorganisms are also addressed in different chapters. It should be noted that industrial sectors dependent on yeasts comprise a trillion-dollar annual market value. Therefore, yeasts stand out as the most profitable microorganisms in industrial microbiology.

Although humans appeared on earth much more recently, several yeast species have been widely domesticated by them, aiming for yeast-based bioprocesses. Given the benefits that yeasts provide to humanity, either as the leading figures in various bioprocesses or indirectly through their ecological role, the book ends up bringing up a question that has already been asked other times before: would yeasts be the best friends of humans? Although the question does not need to be categorically answered, the reading of *Yeasts: from nature to bioprocesses* will undoubtedly convince the reader of the importance of these microorganisms for the development of civilization, economy, and science.

We wish everybody an excellent reading.

Beyond grateful,

Sérgio Luiz Alves Júnior
Laboratory of Biochemistry and Genetics
Federal University of Fronteira Sul
Campus Chapecó - SC
Brazil

Helen Treichel
Laboratory of Microbiology and Bioprocesses
Federal University of Fronteira Sul
Campus Erechim - RS
Brazil

Thiago Olitta Basso
Department of Chemical Engineering
University of São Paulo
São Paulo - SP
Brazil

&

Boris Ugarte Stambuk
Department of Biochemistry
Federal University of Santa Catarina
Florianópolis - SC
Brazil

List of Contributors

Alan Rempel

Graduate Program in Environmental and Civil Engineering, University of Passo Fundo (UPF), Passo Fundo/RS, Brazil

Albertyn-Pohl Carolina

SARChI Research Chair in Pathogenic Yeast, Department of Microbiology and Biochemistr, University of the Free State, PO Box 339 Bloemfontein, 9300, South Africa

Aline F. Camargo

Laboratory of Microbiology and Bioprocess, Federal University of Fronteira Sul (UFFS), Erechim/RS, Brazil

Andrea Origone

Instituto de Investigación y Desarrollo en Ingeniería de Procesos, Biotecnología y Energías Alternativas (PROBIEN, CONICET-UNCo), Biotecnología y Energías Alternativas (PROBIEN, CONICET-UNCo), Argentina

Andressa Warken

Laboratory of Microbiology and Bioprocesses, Federal University of Fronteira Sul, Erechim/RS, Brazil

Atrayee Chattopadhyay

Department of Biotechnology, Indian Institute of Technology Kharagpur, Kharagpur-721302, India

Barbara Dunn

Department of Genome Sciences, University of Washington, Seattle, WA, United States of America

Benjamas Cheirsilp

Center of Excellence in Innovative Biotechnology for Sustainable Utilization of Bioresources, Faculty of Agro-Industry, Prince of Songkla University, Hat Yai, 90110, Thailand

Boris U. Stambuk

Department of Biochemistry, Federal University of Santa Catarina, Florianópolis/SC, Brazil

Bruno Venturin

Center for Exact and Technological Sciences, Graduate Program in Agricultural Engineering, Western Paraná State University (UNIOESTE), Cascavel/PR, Brazil

Caiti Smukowski Heil

Department of Biological Sciences, North Carolina State University, Raleigh, NC, USA

Camila E. Hollas

Center for Exact and Technological Sciences, Graduate Program in Agricultural Engineering, Western Paraná State University (UNIOESTE), Cascavel/PR, Brazil

Charline Bonatto

Laboratory of Microbiology and Bioprocess, Federal University of Fronteira Sul (UFFS),)Erechim/RS, Brazil

Christian A. Lopes

Instituto de Investigación y Desarrollo en Ingeniería de Procesos, Biotecnología y Energías Alternativas (PROBIEN, CONICET-UNCo), Neuquén, Argentina

César Hernández-Rodríguez

Instituto Politécnico Nacional, Escuela Nacional de Ciencias Biológicas, Departamento de Microbiología, Ciudad de México C.P. 11340, México

Delphine Sicard

SPO, Univ Montpellier, INRAE, Institut Agro, Montpellier, France

Diego Libkind

Centro de Referencia en Levaduras y Tecnología Cervecera (CRELTEC), Instituto Andino Patagónico de Tecnologías Biológicas y Geoambientales (IPATEC), CONICET / Universidad Nacional del Comahue, Quintral 1250 (8400), Bariloche, Rio Negro, Argentina

Dielle Pierotti Procópio

Department of Chemical Engineering, University of São Paulo, São Paulo/SP, Brazil

Esaú De-la-Vega-Camarillo

Instituto Politécnico Nacional, Escuela Nacional de Ciencias Biológicas, Departamento de Microbiología, Ciudad de México C.P. 11340, México

Eunice Valduga

Universidade Regional Integrada do Alto Uruguai e das Missões, Campus Erechim, Avenida Sete de Setembro, 1621, Erechim/RS, Brazil

Gamero Amparo

Dep. Preventive Medicine and Public Health, Food Science, Toxicology and Forensic Medicine, Faculty of Pharmacy, University of Valencia, Valencia, Spain

Geciane Toniazzo Backes

Universidade Regional Integrada do Alto Uruguai e das Missões, Campus Erechim, Avenida Sete de Setembro, 1621, Erechim/RS, Brazil

Guilherme Hassemer

Universidade Regional Integrada do Alto Uruguai e das Missões, Campus Erechim, Avenida Sete de Setembro, 1621, Erechim/RS, Brazil

Helen Treichel

Laboratory of Microbiology and Bioprocess, Federal University of Fronteira Sul (UFFS), Erechim/RS, Brazil

J. Alfredo Hernández-García

Universidad Autónoma de Nuevo León, Facultad de Ciencias Forestales, Departamento de Silvicultura, Carretera Nacional No. 85, Km. 145, Linares, Nuevo León C.P. 67700, México

Jamile Zeni

Universidade Regional Integrada do Alto Uruguai e das Missões, Campus Erechim, Avenida Sete de Setembro, 1621, Erechim/RS, Brazil

Jéssica Mulinari

Laboratory of Membrane Processes, Department of Chemical Engineering and Food Engineering, Federal University of Santa Catarina (UFSC), Florianópolis/SC, Brazil

Jolly Neil

Post-Harvest and Agro-Processing Technologies, ARC Infruitec-Nietvoorbij, Agricultural Research Council, Private Bag X5026, Stellenbosch, 7600, South Africa

Julieta A. Burini

Centro de Referencia en Levaduras y Tecnología Cervecera (CRELTEC), Instituto Andino Patagónico de Tecnologías Biológicas y Geoambientales (IPATEC), CONICET / Universidad Nacional del Comahue, Quintral 1250 (8400), Bariloche, Rio Negro, Argentina

Kate Howell

School of Agriculture and Food, University of Melbourne Victoria 3010, Australia

Lourdes Villa-Tanaca

Instituto Politécnico Nacional, Escuela Nacional de Ciencias Biológicas, Departamento de Microbiología, Ciudad de México C.P. 11340, México

Luiz Carlos Basso

Department of Biological Sciences, Escola Superior de Agricultura Luiz de Queiroz, University of São Paulo, Piracicaba/SP, Brazil

Manoela Martins

Bioprocess and Metabolic Engineering Laboratory (LEMEB), Department of Food Engineering, Faculty of Food Engineering, University of Campinas (UNICAMP), Brazil

Marcus Bruno Soares Forte — Bioprocess and Metabolic Engineering Laboratory (LEMEB), Department of Food Engineering, Faculty of Food Engineering, University of Campinas (UNICAMP), Brazil

Maria Paula Jiménez Castro — Bioprocess and Metabolic Engineering Laboratory (LEMEB), Department of Food Engineering, Faculty of Food Engineering, University of Campinas (UNICAMP), Brazil

María C. Bruzone — Centro de Referencia en Levaduras y Tecnología Cervecera (CRELTEC), Instituto Andino Patagónico de Tecnologías Biológicas y Geoambientales (IPATEC), CONICET / Universidad Nacional del Comahue, Quintral 1250 (8400), Bariloche, Rio Negro, Argentina

María E. Rodríguez — Instituto de Investigación y Desarrollo en Ingeniería de Procesos, Biotecnología y Energías Alternativas (PROBIEN, CONICET-UNCo), Neuquén, Argentina

Mehlomakulu Ngwekazi Nwabisa — Department of Consumer and Food Sciences, University of Pretoria - Hatfield Campus, Cnr Lynnwood Road and Roper Street, Pretoria

Melisa González Flores — Instituto de Investigación y Desarrollo en Ingeniería de Procesos, Biotecnología y Energías Alternativas (PROBIEN, CONICET-UNCo), Neuquén, Argentina

Motlhalamme Thato Yoliswa — South African Grape and Wine Research Institute, Department of Viticulture and Oenology, Stellenbosch University, P/Bag X1 Matieland, South Africa

Mrinal K. Maiti — Department of Biotechnology, Indian Institute of Technology Kharagpur, Kharagpur-721302, India

Natalia Klanovicz — Laboratory of Microbiology and Bioprocesses, Federal University of Fronteira Sul, Erechim/RS, Brazil
Research Group in Advanced Oxidation Processes (AdOx), Department of Chemical Engineering, Escola Politécnica, University of São Paulo, São Paulo/SP, Brazil

Natalia Paroul — Universidade Regional Integrada do Alto Uruguai e das Missões, Campus Erechim, Avenida Sete de Setembro, 1621, Erechim/RS, Brazil

Rogerio Luis Cansian — Universidade Regional Integrada do Alto Uruguai e das Missões, Campus Erechim, Avenida Sete de Setembro, 1621, Erechim/RS, Brazil

Rosana Goldbeck — Bioprocess and Metabolic Engineering Laboratory (LEMEB), Department of Food Engineering, Faculty of Food Engineering, University of Campinas (UNICAMP), Brazil

Rosicler Colet — Universidade Regional Integrada do Alto Uruguai e das Missões, Campus Erechim, Avenida Sete de Setembro, 1621, Erechim/RS, Brazil

Setati Mathabatha Evodia — South African Grape and Wine Research Institute, Department of Viticulture and Oenology, Stellenbosch University, P/Bag X1 Matieland, 7600 South Africa

Sergio Álvarez-Pérez — Department of Animal Health, Faculty of Veterinary Medicine, Complutense University of Madrid, Madrid, Spain

Sérgio Luiz Alves Júnior — Laboratory of Biochemistry and Genetics, Federal University of Fronteira Sul, Chapecó/SC, Brazil

Thalita Peixoto Basso	Department of Genetics, Escola Superior de Agricultura Luiz de Queiroz, University of São Paulo, Piracicaba/SP, Brazil
Thamarys Scapini	Laboratory of Microbiology and Bioprocesses, Federal University of Fronteira Sul, Erechim/RS, Brazil
Thiago Olitta Basso	Department of Chemical Engineering, University of São Paulo, São Paulo/SP, Brazil
Thamarys Scapini	Laboratory of Microbiology and Bioprocess, Federal University of Fronteira Sul (UFFS), Erechim/RS, Brazil
Viviani Tadioto	Laboratory of Biochemistry and Genetics, Federal University of Fronteira Sul, Chapecó/SC, Brazil
Yasmi Louhasakul	Faculty of Science Technology and Agriculture, Yala Rajabhat University, Yala, 95000, Thailand
Zhou Nerve	Department of Biological Sciences and Biotechnology, Botswana International University of Science and Technology, P/Bag 16, Palapye, Botswana

CHAPTER 1

Origin and Evolution of Yeasts

Thato Yoliswa Motlhalamme[1], Nerve Zhou[2], Amparo Gamero[3], Ngwekazi Nwabisa Mehlomakulu[4], Neil Jolly[5], Carolina Albertyn-Pohl[6] and Mathabatha Evodia Setati[1,*]

[1] *South African Grape and Wine Research Institute, Department of Viticulture and Oenology, Stellenbosch University, P/Bag X1 Matieland 7600, South Africa*

[2] *Department of Biological Sciences and Biotechnology, Botswana International University of Science and Technology, P/Bag 16, Palapye, Botswana*

[3] *Dep. Preventive Medicine and Public Health, Food Science, Toxicology and Forensic Medicine, Faculty of Pharmacy, University of Valencia, Valencia, Spain*

[4] *Department of Consumer and Food Sciences, University of Pretoria - Hatfield Campus, Hatfield, Pretoria 0002, South Africa*

[5] *Post-Harvest and Agro-Processing Technologies, ARC Infruitec-Nietvoorbij, Agricultural Research Council, Private Bag X5026, Stellenbosch 7600, South Africa*

[6] *SARChI Research Chair in Pathogenic Yeasts, Department of Microbiology and Biochemistry, University of the Free State, PO Box 339 Bloemfontein 9300, South Africa*

Abstract: Yeasts are generally unicellular fungi that evolved from multicellular ancestors in distinct lineages. They have existed in this form for millennia in various habitats on the planet, where they are exposed to numerous stressful conditions. Some species have become an essential component of human civilization either in the food industry as drivers of fermentative processes or health sector as pathogenic organisms. These various conditions triggered adaptive differentiation between lineages of the same species, resulting in genetically and phenotypically distinct strains. Recently genomic studies have expanded our knowledge of the biodiversity, population structure, phylogeography and evolutionary history of some yeast species, especially in the context of domesticated yeasts. Studies have shown that a variety of mechanisms, including whole-genome duplication, heterozygosity, nucleotide, and structural variations, introgressions, horizontal gene transfer, and hybridization, contribute to this genetic and phenotypic diversity. This chapter discusses the origins of yeasts and the drivers of the evolutionary changes that took place as organisms developed niche specializations in nature and man-made environments. The key phenotypic traits that are pivotal to the dominance of several yeast species in anthropic environments are highlighted.

* **Corresponding author Mathabatha Evodia Setati:** South African Grape and Wine Research Institute, Department of Viticulture and Oenology, Stellenbosch University, P/Bag X1 Matieland 7600, South Africa; Tel: +27 21 808 9203; E-mails: setati@sun.ac.za

Sérgio Luiz Alves Júnior, Helen Treichel, Thiago Olitta Basso and Boris Ugarte Stambuk (Eds.)
All rights reserved-© 2022 Bentham Science Publishers

Keywords: Adaptation, Abiotic stressors, Aneuploidy, *Brettanomyces*, Crabtree effect, Copy number variations, Domestication, Dimorphism, Fermentation, Fructophilic, Glucophilic, Horizontal gene transfer, Pathogenicity, Saccharomy cotina, *Starmerella, Saccharomyces cerevisiae*, Whole-genome duplication, *Wickerhamiella*, 4-vinylguiaiacol.

INTRODUCTION

The term "yeast" generally refers to a polyphyletic group of unicellular or dimorphic fungi that maintain a unicellular cell structure through most of their life cycle, divide asexually through budding or fission, and have a sexual structure not enclosed in fruiting bodies [1]. As members of the Kingdom Fungi, yeasts share a common ancestor with other opisthokonts, all of which are believed to have transitioned from unicellular to multicellular organisms. However, the yeasts seem to have subsequently "de-evolved" back to unicellularity from multicellular filamentous ancestors in distinct lineages of Ascomycota, Basidiomycota and certain Mucoromycota, containing more complex forms of fungi [2] and have lost most of the genes associated with multicellularity. This "de-evolution" was accompanied by convergent changes in regulatory networks, reduction and compaction of the genome marked by extensive gene losses [1 - 3]. Evidently, 3000 – 5000 genes, including those encoding plant cell wall degrading enzymes, fungal cell wall synthesis and modification, hydrophobins and fungal lysozymes, were dispensed, while genes required for essential cellular processes such as DNA replication, sequence recognition, chromatin binding and chromosome segregation were retained [4]. Moreover, it is hypothesized that the transcription factors regulating the Zn-cluster gene family, which contributes to the suppression of filamentous forms throughout the life cycle and under different conditions, were expanded [4]. Yeasts have evolved at least five times independently within the Kingdom Fungi. Today, yeasts are mainly distributed in two phyla, the Ascomycota and Basidiomycota. Within the Ascomycota, they are distributed between two subphyla, the Saccharomycotina (representing almost two-thirds of all known yeast), the Taphrinomycotina (representing ~ 3% of the total of members of the Ascomycota) [5].

Interestingly, even in their unicellular life forms, some yeasts can display multicellular growth under specific environmental conditions. For instance, dimorphic yeasts can switch from yeast to multicellular hyphae or pseudohyphae. These include pathogenic fungi of mammals, such as *Candida* spp. (*e.g., C. albicans, C. parapsilosis, C. dubliensis, C. guilliermondii* and *C. lusitaniae*), *Exophiala dermatidis* and *Trichosporon cutaneum*, as well as phytopathogens such as *Taphrina deformans, Ustilago maydis, Ophiostoma ulmi* and saprophytic biotechnologically important yeasts such as *Saccharomyces cerevisiae, Yarrowia lipolytica* and *Debaryomyces hansenii*. In pathogenic fungi, the yeast-mycelial switch is involved in virulence; however, in other yeasts, this switch is induced in response to environmental stimuli, *e.g.*, nutrient limitation, pH, oxygen availability, ethanol concentrations, *etc* [6, 7]. Pseudohyphal or hyphal growth leads to clonal multicellularity, where daughter cells "stay together" after mitotic divisions. Alternatively, individual single cells can form multicellular aggregates generally referred to as flocs. In *S. cerevisiae*, where such aggregates have been extensively studied, a group of proteins called flocculins is responsible for the phenotype [8]. While most ascomycetous yeasts are distributed in the subphylum Saccharomycotina, a few unicellular or dimorphic fungi in which the unicellular form is restricted to specific environmental conditions also exist in the subphylum Pezizomycotina [5].

Multicellularity improves yeast access to complex substrates, allows for efficient nutrient uptake, and enhances the stress and toxin resistance [3, 8]. Despite these benefits, most yeasts maintain a long-term single-celled lifestyle. With this morphology and limited dispersal, most yeasts have evolved adaptive mechanisms that allow them to thrive in liquid environments containing concentrated simple sugars (*e.g.*, plant-derived liquids, such as fruit juices, honeydew, and nectar), where they have a fitness advantage over prokaryotes [1, 2]. Such adaptations are explained by many genetic features that have undergone multiple rounds of modifications to endow different species with traits that allow for niche specialization. These genetic signatures and their associated phenotypes will be discussed in detail in subsequent sections.

MOLECULAR DRIVERS OF EVOLUTION

Gene losses, expansions and concomitant fine-tuning were the important drivers in the switch of yeast from their multicellular origins to single-celled lifestyle; these and additional modifications have also contributed significantly to yeast evolution and species diversification. Mainly, these modifications include Whole-Genome Duplication (WGD), Large Scale Genome Rearrangements (LSGR), Horizontal Gene Transfer (HGT), Copy Number Variations (CNV). WGD is a process by which additional copies of the genome are generated due to nondisjunction during meiosis. Through this process, organisms can acquire more than two complete sets of chromosomes, leading to a change in ploidy. Acquisition of genome copies can arise through interspecies hybridization, resulting in allopolyploids or intraspecies hybridization, leading to autopolyploidization. WGD is typically followed by inter-chromosomal rearrangements and the loss of one of the gene duplicates [9]. Large-scale genome rearrangements may occur through chromosome duplications or aneuploidy, thus creating copy number variations that may change gene dosage [10]. CNVs refer to duplication or deletion of a 50 bp fragment to a whole chromosome that results in a change in the copy number of a respective genetic locus across individuals in a population [11]. CNVs can change gene dosage, interrupt coding sequences, contribute to population genetic and phenotypic diversity such as virulence, growth rate, growth on various substrates, and stress tolerance [11].

In addition to WGD and CNV, increasingly available genomic data reveal that HGT has had an extensive impact on yeast evolution. HGT is defined as the exchange of genetic material between different strains or species. In yeast, HGT has evidently occurred through eukaryote-to-eukaryote and prokaryote-t--eukaryote [12, 13] transfers. Ecological proximity together with stressful environmental conditions, have been highlighted as important factors that trigger and facilitate HGT events. While eukaryote-to-eukaryote HGT can occur through introgression (interspecific hybridization), bacterial genes can be acquired through various means, including virus-aided transmission, environmental stress-induced DNA damage, and repair, a phagocytosis-based ratchet [13]. In addition to events such as WGD, HGT, and CNVs, where large gene fragments can be altered simultaneously, small changes engle genes or small-scale nucleotide changes (SSNC) also contribute to yeast adaptation to various environments. SSNC can occur from single nucleotide and frameshift mutations, insertions or deletions and loss of function mutations. These changes may alter the structure or function of the encoding protein or gene expression [14, 15]. The molecular mechanisms described here have driven many adaptive evolutionary events in many yeasts, allowing them to thrive in different niches. Notably, dispersal of yeasts by insects, humans, and animals across different ecosystems and induction of adaptive evolution events by strong selection pressure for specific niches has ultimately led to changes in phenotypic traits. In the transition to a unicellular life form, yeast developed traits to efficiently grow on simple sugars.

EVOLUTION OF CARBON METABOLISM IN YEASTS: PREFERENCE FOR GLUCOSE AND FRUCTOSE

The utilisation of disaccharides, such as sucrose and maltose, hexoses, such as glucose, galactose, fructose and mannose, as main carbon substrates is well conserved in yeasts. However, there is a huge diversity in sugar metabolism in yeasts, most probably due to evolutionary history based on niche specializations in nature. Respiratory yeasts could have evolved in environments with a low amount of carbon sources, where efficiency and ATP yield would warrant a competitive advantage in the face of a limited carbon source [16]. MacLean and Gudelj [17] suggested that respiratory yeasts completely oxidise glucose to limit the accumulation of ethanol and organic acids, probably as a strategy to reduce toxicity, which subsequently increases their chances of survival and reproduction.

The ancestral yeast that lived before the appearance of angiosperms about 125 million years ago was probably a respiratory yeast, as suggested by a carbon-limited niche [18 - 20]. On the other hand, the emergence of fruit trees coincides with the split between the *Kluyveromyces* and *Saccharomyces* lineages, suggesting the presence of a glut of sugars responsible for the emergence of a new lifestyle, the make-accumulate and consume strategy (MAC), exhibited by yeasts in the *Sacharomycetaceae* family [21]. This strategy is characterised by the fermentation of excess glucose into ethanol irrespective of the presence or absence of oxygen. This trait is not unique to this family because the *Dekkera/Brettanomyces* and *Schizosaccharomyces pombe* lineages, as distant relatives of the *Saccharomyces* yeasts, also independently evolved the respiro-fermentative lifestyle [22, 23].

PREFERENCE FOR GLUCOSE AND EVOLUTION OF ETHANOL PRODUCTION

There is an evolutionarily conserved preference for specific carbon sources, with glucose and fructose as the most common among yeasts [24], despite glucose and fructose having the same empirical formula and being dependent on the same hexose transporters [25, 26]. The preferential consumption of glucose until depletion before switching to an available alternative carbon source is well studied in *S. cerevisiae* [27, 28]. The proposed justifications of this preference are the lower metabolic costs associated with a direct entrance into the glycolytic pathway [29] and five times higher affinity for glucose over fructose of the hexose transporters [26, 30]. A respiro-fermentative lifestyle in yeasts, where respiratory and fermentative pathways are run concurrently in the presence of abundant sugar and oxygen, has also been described [31]. A hallmark for this group of yeasts is their preference for and rapid consumption of glucose, followed by the production of pyruvate and a subsequent exclusive dissimilation of ethanol either for redox balancing or for ecological advantages. $NAD(P)^+$, an essential oxidoreductase cofactor required for ATP production during substrate-level phosphorylation, is regenerated at a faster rate during the ethanol production pathway to allow the re-run of glycolysis [32, 33]. Production of ATP in the absence of oxygen is energetically inefficient but is thought to have enabled glucophilic yeasts to utilise anaerobic niches [34, 35]. These yeasts were designated as Crabtree positive yeasts [31, 34, 35]. This trait circumscribes *Saccharomyces* lineage yeasts separated from *Saccharomyces-Lachancea* and *Kluyveromyces-Eremothecium* lineages about 125 – 150 million years ago [36]. The Crabtree effect (CE) is more pronounced in yeasts that underwent a WGD about 100 million years ago [21]. The presence of glucose in these lineages represses the expression of genes encoding enzymes required for the utilization of alternative carbon sources [27].

This phenomenon, also known as glucose repression, is similar to glucophily in some way, and the two words are interchangeable.

MOLECULAR EVENTS OF ETHANOL PRODUCTION AMONG GLUCOPHILES

The extensive studies of yeasts belonging to the *Saccharomycetaceae* family have highlighted the origins of the CE. The advent of a powerful field of comparative genomics has made it possible to point out several molecular events responsible for this effect. These mechanisms, which include loss of the respiratory complex I [37], HGT of *URA1* gene [38], WGD [39], gene duplications [40], and, possibly, rewiring of the rapid growth elements (RGE)/transcriptional networks [41], have been described (Fig. **1**).

The HGT of *URA1* from *Lactococcus lactis* encoding a dihydroorotate dehydrogenase (enzyme for *de novo* synthesis of pyrimidines independent of the respiratory chain) is thought to have allowed yeasts to grow under anaerobic conditions [12, 38]. This trait may have evolved after the split of *Kluyveromyces* and the *Lachancea-Saccharomyces* lineages (Fig. **1**). The timing is concordant with the inability of anaerobic growth in *Kluyveromyces* and *Eremothecium* lineages [16, 32, 35, 36, 38, 42]. Supplementation of the growth medium with pyrimidines together with other factors required for anaerobic growth allows resumption of growth [16, 43, 44]. Contrastingly, yeasts from the *Dekkera/Brettanomyces* clade can grow under anaerobic conditions without anaerobic factors, despite the absence of *URA1* gene [45, 46]. It is speculated that this novel gene could have been crucial for the exploration of anoxic environments as new or novel niches [16, 35, 36, 38, 42]. This invention was not outright beneficial due to the absence of genes responsible for the consumption of the accumulated ethanol, which meant a metabolic dead end [38, 40]. Therefore, the duplication of an ancestral alcohol dehydrogenase gene *ADHA*, required for ethanol production under anaerobic conditions with the sole purpose of recycling NADH during glycolysis, giving rise to *ADH1* and *ADH2* [40], led to the accumulation of ethanol as well as its consumption [40, 47]. This duplication event predates the WGD, as suggested by the ethanol metabolism in pre-WGD yeasts [48].

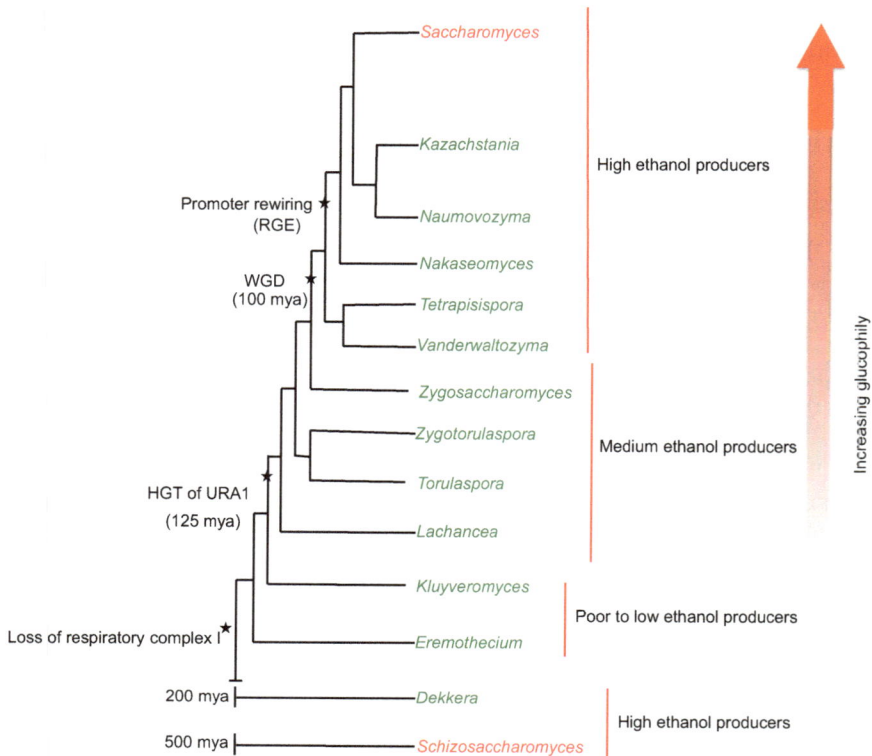

Fig. (1). Molecular mechanisms explaining the evolution of ethanol production among *Saccharomycetaceae* yeasts based on Kurtzman *et al.* [49]. Specific evolutionary events such as the loss of respiratory complex I [37], the horizontal gene transfer (HGT) of *URA1* gene [38], the WGD [39], and the loss of RGE associated with the rewiring of promoters associated with respiration [41]. The discrepancies and organised fermentative capacity, as described by Hagman and co-workers [21] is also shown.

The WGD event in six clades that diverged from the fructophilic *Zygosaccharomyces* lineages increased the CE as the genetic reservoir for increased glycolytic flux [39, 50, 51]. This flux increase [52] could be explained by the presence of duplicate genes. Out of the 10 genes required to run glycolysis, 6 were retained as duplicates [39, 52]. The duplicate genes could lead to dosage imbalance, fitness decrease or even being lethal [53, 54]; however, studies on post-WGD yeasts suggest that increased glycolytic enzymes increased the growth rate by a factor of 2 [52]. An increase in glycolytic flux, unfortunately, led to an overflow metabolism where the cellular demand for cell biomass production and respiration was above normal requirements [55], also described by Hagman and co-workers [21] as a short-term CE. It is speculated that this overflow metabolism could be the basis of the evolution of aerobic fermentation in yeasts that diverged after the WGD [35, 36].

The glucophilic lifestyle in post-WGD or the glucose repression phenotype could have been an invention to enhance ethanol production within a short period of time. However, aerobic ethanol production coupled with an increased glycolytic flux in pre-WGD lineages, such as *Sc. pombe* and *Dekkera/Brettanomyces* yeasts, suggests that the WGD event was not a crucial trigger for the CE but a perfection of the trait. Another molecular mechanism which could have led to the perfection and probably increased glucophilic phenotype is the loss of RGE associated with respiratory genes in post-WGD yeasts [56]. This rewired the transcriptional networks and negated the use of fully functional mitochondria in generating energy for metabolic processes [16, 41]. *Dekkera/Brettanomyces* lineages, whose Crabtree positive phenotype is comparable to WGD yeasts, also lack the RGE elements [35]. Recently, Ata and colleagues [57] reported that a single Gal4-like transcription factor activates the CE in *Komagataella phaffii,* suggesting that the molecular basis of the evolution of respiro-fermentative metabolism in yeast remains unclear.

ECOLOGICAL BASIS SUPPORTING GLUCOPHILY AND EVOLUTION OF ETHANOL PRODUCTION

The preference for glucose is characteristic among *Saccharomycetaceae* family yeasts that accumulate ethanol when excess glucose and oxygen are available. This strategy undermines the principles of cellular energetics because it is energetically inefficient, yielding 15 times less ATP than conventional oxidative respiration [31, 35, 40, 58]. However, this trait was selected in nature, suggesting that it is a winning trait in glucose-rich environments [36, 41, 51, 59]. In fact, it provides a net fitness advantage of about 7% [60]. To date, there are two hypotheses that have been brought forward: (1) the MAC strategy, which ascertains that organisms do so to "starve off" and annihilate competitors by the fast depletion of glucose and production of ethanol, CO_2 and heat [36, 16, 61], (2) that ascertains that the trait arose as a rate/yield trade-off (RYT) for ATP production, which compensates for the inefficiency of the ATP production rate during alcoholic fermentation [59, 62]. RYT is supported by the coexistence of energetically inefficient and efficient cells where cooperation rather than competition (ascertained by MAC) could be a preferred outcome of resource conflicts, where common resources are used efficiently [17]. MAC fails to account for the fitness advantage endowed by ethanol toxicity among competing microorganisms [48] but offers solid speculation of an ecosystem engineering strategy where products of alcoholic fermentation "kill" off alcohol-sensitive microorganisms as a niche defense solution.

EVOLUTION OF FRUCTOPHILY, A NON-ETHANOL PRODUCING SUGAR UTILISATION STRATEGY AMONG SOME YEASTS

Fructophilic yeasts prefer fructose to other carbon sources, including glucose [63 - 65]. Fructophily is a rare trait patchily distributed among Ascomycetous yeasts of the Saccharomycotina lineage, specifically in the *Zygosaccharomyces* and *Wickerhamiella/Starmerella* lineages (Fig. **2**) [66]. Fructophily is more pronounced in the Basidiomycetous ancestral yeasts [67]. However, some fructophilic yeasts grow very well in glucose in the absence of fructose [65]. The most likely explanation of preference for fructose in the presence of glucose is that fructose metabolism is important as a source of carbon as well as for regeneration of NAD $(P)^+$, a co-factor required to run the glycolytic pathway [68]. The fate of fructose in fructophilic yeasts is the production of mannitol thought to be important in redox balancing in the absence of or inefficient alcoholic fermentation pathways [69]. In addition to redox balancing, the production of mannitol could have later been pertinent for stress protection [69].

Molecular Mechanisms of Fructophily

A low affinity, high capacity and uniquely specific fructose transporter, Ffz1 (fructose facilitator *Zygosaccharomyces*), was reported as a genetic requirement for fructophily in yeasts [67, 70]. Initial research suggested that the Ffz1 was a prerequisite for the trait [71]. However, more genetic studies and related comparative genomics suggested that there was another transporter known as the Ffz2 transporter family only found in the *Zygosaccharomyces* genus. This family was shown to transport both glucose and fructose contrary to the Ffz1 fructose-only transporter [70 - 72]. The trait is present only in Dikarya (Ascomycota and Basidiomycota) and absent in all Basal fungi [67]. Comparative genomic analyses of the *Ffz* genes from the Saccharomycotina and Pezizomycotina suggest that the *FFZ1* gene was not present in the most recent common ancestor of the Saccharomycotina [73]. Yeasts associated with fructose-rich niches are thought to have acquired the trait through HGT from the filamentous Ascomycetes and, Pezizomycotina [67]. The existence of this gene in the Dikarya and patchy distribution in the Saccharomycotina (Fig. **2**) suggest that there was a loss of the gene in Saccharomycotina ancestor followed by acquisition from a species close to *Monascus,* as described by Goncalves and co-workers [74]. This is evident as the *FFZ1* gene homologs clustered with those from Pezizomycotina. *Zygosaccharomyces* fructophily was described as a second HGT from *Wickerhamiella/Starmerella* clades [67].

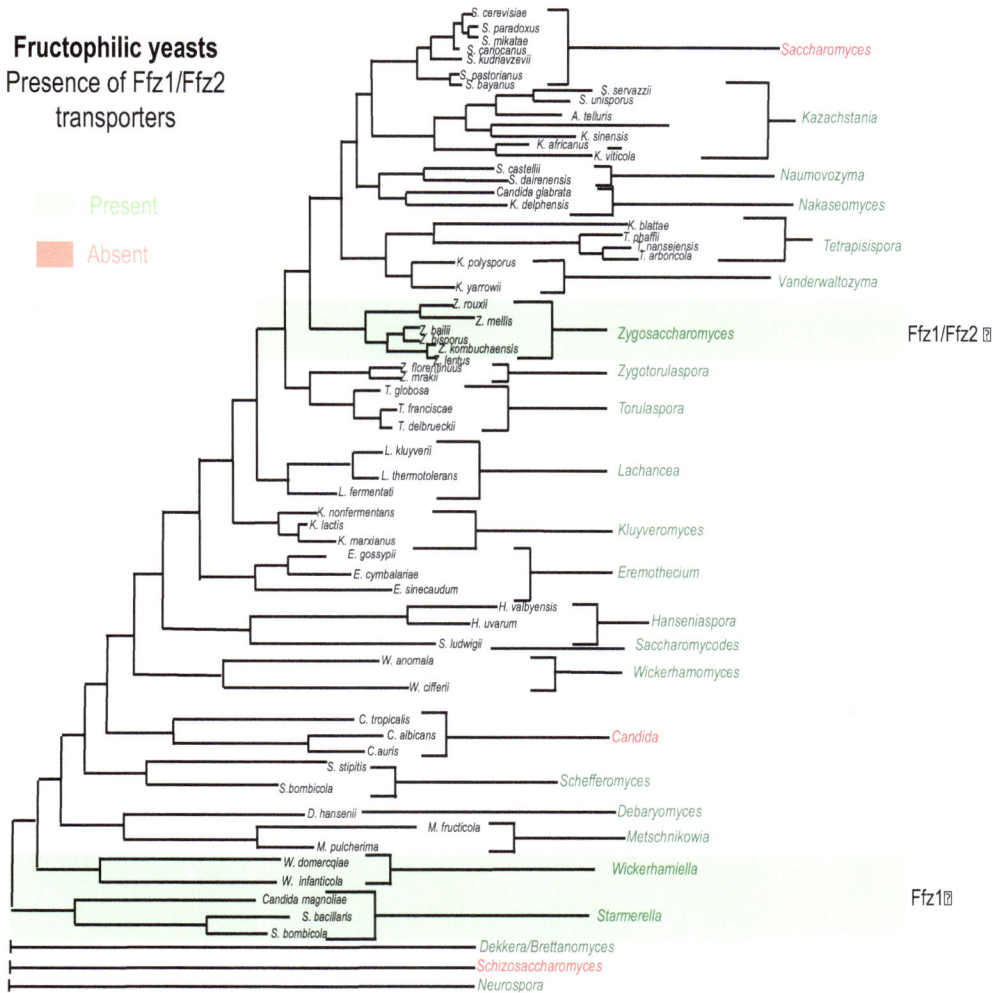

Fig. (2). Evolution of Ffz-like fructose transporter family among chosen Ascomycetous yeasts. This figure was drawn based on results presented by Goncalves and co-workers [67].

FFZ1-like genes are not a prerequisite for fructophily based on the finding of Cabral and co-workers [63]. This suggests that there could be an FFZ-like independent pathway responsible for fructophilic behaviour. In agreement with this hypothesis, a recent genome sequencing study revealed that fructophilic *W. bombicola* and *W. occidentalis* lacked the *FFZ1* gene [74]. Yeasts did not only evolve to grow efficiently on simple sugars in nature, but they have also been contemporary with human civilization and are key drivers of many fermentation processes. They evolved to express varying niche-specific traits. Invariably, the early fermentation processes occurred spontaneously; however, human interventions have promoted the adaption of microbes to man-made environments, thus leading to the development of wild and domesticated microbial lineages.

YEAST DOMESTICATION

Domestication is the result of co-evolutionary mutualisms that develop in the context of active niche construction by both humans and their plant/animal partners [75]. This niche construction, whether intentional or not, has shaped the evolution of several yeast species. For instance, the evolution of *Saccharomyces* spp., *Lachancea thermotolerans, Torulaspora delbrueckii, Brettanomyces* spp. as well as *Kluyveromyces* spp., has been shaped by anthropisation, geographic origin and flux between ecosystems often mediated by humans, birds, animals and insects [15, 76 - 78]. The vast majority of microbial domestications seem to have occurred through a commensal pathway in which the organisms first started to habituate to a human niche but through increasing degrees of well-considered human actions and continuous cultivation evolved and acquired traits that expedite niche specialization [79]. Backslopping is one such ecosystem engineering practice. In backslopping, brewers re-used the yeast sediment to inoculate the next batch [79, 80]. Such transfers allow for new generations of species and strains that would have adapted to the changing environment of the fermentation process and are, therefore, fit to be selected over time. Consequently, further diversification driven by the ecology of specific niches is evident within the domesticated populations of some yeast species [15, 76]. Several domestication signatures have been described in various yeast species and the drivers of these signatures will be discussed in the sections below (Fig. **3**).

The genetic differentiation of wild and domesticated strains is also reflected at the phenotypic level, with domesticated strains often largely displaying industry-specific traits for stress tolerance, sugar consumption and flavour production. For instance, natural isolates of species such as *S. cerevisiae, L. thermotolerans* and *T. delbrueckii* display inferior fermentation performances. Furthermore, within the domesticated populations, sub-specialization for specific niches can be observed. This is seen within *Saccharomyces* spp., where strains are specialized for beer, bread, sake and wine [15], while in *T. delbrueckii*, strains sub-specialized for dairy products have been identified [76].

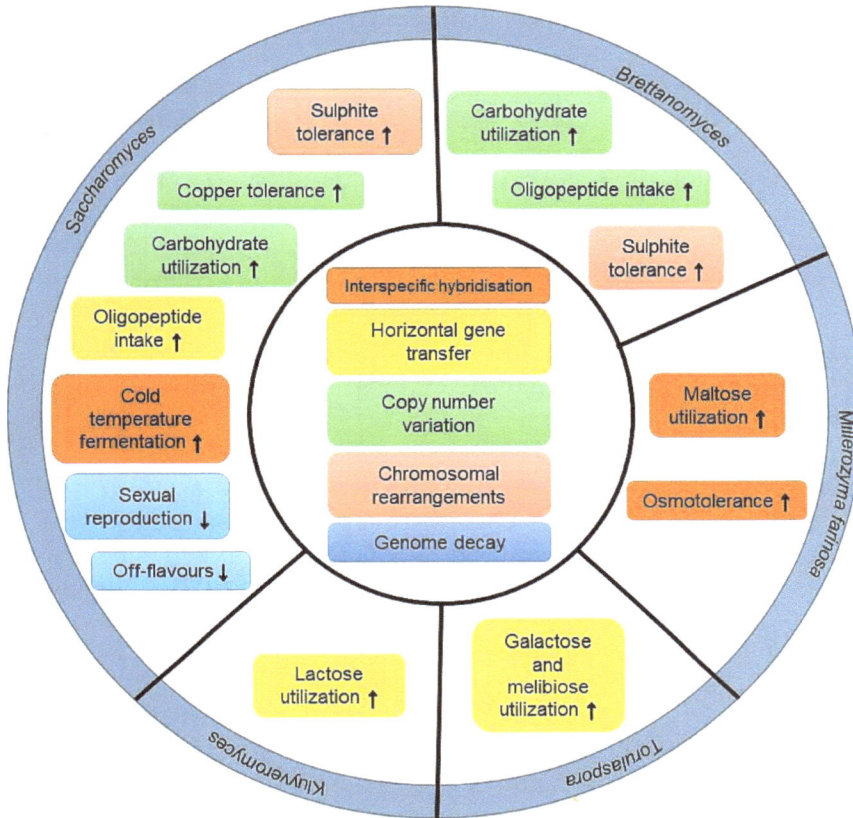

Fig. (3). Major domestication phenotypes in various yeast species. Phenotypes are coloured according to the genetic driver of that phenotype. Orange = interspecific hybridization; yellow = horizontal gene transfer; green = copy number variation; blue = genome decay. Arrows indicate increase (up) or decrease (down) of specific phenotypes in domesticated strains. Adapted from Steensels *et al.*'s study [79].

Utilization of Carbon Substrates in Domesticated Yeast

Expansions of genes encoding enzymes responsible for the utilization of various sugars is one of the hallmarks of domestication of beer and wine *S. cerevisiae* strains. For instance, beer strains exhibit a considerable expansion of the *MAL3* locus, which includes *MAL31* (encoding a permease), *MAL32* (encoding a maltase) and *MAL33* (a transcription factor), with most strains containing 6 or more copies. Similarly, bread strains were enriched in copies of these genes, while sake strains were not, and wine strains showed variations between 2-6 copies [81]. Moreover, SNVs of a particular allele of *MAL11* (sugar transporter gene) in beer strains enhanced the utilisation of maltotriose, a carbon source found in beer medium [82]. *S. cerevisiae* strains isolated from low glucose environments had an increased number of hexose transporter genes, leading to higher expression and increased glucose transport into the cell [83]. Transportation of glucose is carried out by the hexose transporter (*HXT*) gene family highly CN variable in wine yeast strains. In this group of strains, *HXT13*, *HXT15*, and *HXT17* exhibited CN variation, whereas *HXT1, HXT6, HXT7,* and *HXT16* are more commonly duplicated, and *HXT9* and *HXT11* are more commonly deleted [11, 84].

Cheese-derived strains of *S. cerevisiae* were found to contain a unique region, Region T, which carries *GAL* orthologues believed to have been acquired from an unknown donor through trans-species introgression [85, 86]. These orthologues replaced the *GAL* gene cluster (*GAL1, GAL7* and *GAL10*) present in most *S. cerevisiae* strains by recombination. Furthermore, the cheese strains also harbor a high-affinity transporter (Gal2) and specific alleles of *GAL4* and *GAL80* that allow the strains to grow on galactose [85]. Recently, strains in the genus *Torulaspora* were shown to harbour larger *GAL* clusters, which in addition to the *GAL1-GAL10-GAL7* genes, include genes for melibiose (*MEL1*), phosphoglucomutase (*PGM1*) and the transcription factor (*GAL4*). Together, these genes confer an ability to catabolize extracellular melibiose [87]. This cluster is thought to have been acquired by HGT from *Torulaspora franciscae* to *T. delbrueckii*, and from *Torulaspora maleeae* to strains of *Torulaspora globosa*. However, the *MEL1* gene is in most strains a pseudogene, with only one strain of *T. delbrueckii* (CBS1146[T]), having a functional *MEL1* [87].

In *Brettanomyces* species, such as *B. bruxellensis* and *B. nanus*, copy number expansions of ORFs predicted to encode several glycosidases involved in the metabolism of fermentation substrates such as starch, galactose and sugars from complex polysaccharides have been reported [88]. In addition, *B. anomalus* and *B. bruxellensis* seem to have an invertase of bacterial origin through HGT that allows them to utilize sucrose as the sole carbon source [89].

Kluyveromyces species (*K. marxianus* and *K. lactis*) are the only yeast species that can ferment lactose. This trait is associated with the acquisition of *LAC12* (lactose permease) and *LAC4* (β-galactosidase) genes. *K. marxianus* is thought to have acquired the *LAC4* gene *via* HGT from bacteria [90]. These genes, together with a flocculin encoding gene (*FLO*), were then later acquired from a dairy strain of *K. marxianus* into *K. lactis* through introgression (*i.e.*, interspecies mating). The *FLO* gene, which subsequently underwent frameshift mutations, is now a pseudogene [86]. Within *K. lactis*, two varieties exist, *i.e.*, *K. lactis var drosophilarum* is lactose negative (found in plant and invertebrates) and *K. lactis var lactis* (dairy products) is lactose positive [90].

ADAPTATION TO NITROGEN UPTAKE

Nitrogen acquisition is pivotal to the outcome of fermentation. Genes involved in the utilization of amino acids and nitrogen, such as *VBA3* and *VBA5* (amino acid permeases), and *PUT1* (a gene that aids in the recycling or utilization of proline), are often duplicated in wine yeast [11]. One of the three genomic regions in wine yeast (region C) was acquired through HGT from *Torulaspora microellipsoides*. This region contains the *FOT1-2* encoding oligopeptide transporters, which preferentially assimilate glutathione and oligopeptides rich in glutamate/glutamine. These are some of the most abundant amino acids in grape berry cultivars [91], suggesting that the *FOT* transporters may give yeasts a competitive edge during fermentation of musts from different cultivars or towards the end of fermentation where nitrogen sources are scarce [92]. Other genes with putative functions associated with nitrogen metabolisms such as asparaginase, oxoprolinase, ammonium and allantoate transporters, as well as lysine and proline transcription factors were also present in the genomic regions of wine yeasts [14, 93]. *Brettanomyces custersianus* and *B. anomalus* displayed large expansions of a sarcosine oxidase/L-pipecolate oxidase (PIPOX) encoding gene, which also occurred in multiple copies in the other *Brettanomyces* species. PIPOX is a broad substrate enzyme that acts on several N-methyl amino acids and D-proline, an abundant amino acid in winemaking [88].

Modifications in Thiamine Metabolism

Thiamine, commonly known as vitamin B1, is essential for all living organisms because its active form, thiamine pyrophosphate (TPP), is an indispensable cofactor of enzymes participating in amino acid and carbohydrate metabolism. While some yeasts can synthesize this vitamin de novo, others cannot; but they acquire it from the environment through the thiamine salvage pathway. With respect to vitamins, the *THI* family of genes involved in thiamine or vitamin B1 metabolism are CN variables. *THI13* is commonly duplicated, whereas *THI5* and *THI12* were deleted in wine yeast strains. *THI5* is associated with an undesirable rotten egg smell and taste in wine [11]. Although this gene is deleted in most wine strains, it is duplicated in other strains of *S. cerevisiae, Saccharomyces paradoxus* and the hybrid species *Saccharomyces pastorianus*.

ADAPTATION TO ABIOTIC STRESSORS

Yeast living in association with human habitats are constantly exposed to antimicrobial agents such as sulphites (in the winery), copper sulphate (in the vineyard), and antifungal drugs in clinical settings. These microorganisms have developed strategies to withstand these agents [79]. For example, the use of copper sulfate as a fungicide in the vineyards since the 1880s has resulted in strains of *S. cerevisiae* that display increased resistance to $CuSO_4$. This resistance phenotype is driven by high copy numbers of *CUP1* encoding the copper-binding metallothionein [14]. Sulphur dioxide (SO_2) is added to grape must at various stages of fermentation. In *S. cerevisiae* wine strains, the reciprocal translocation between chromosome VIII and XVI generated a dominant allele of the sulfite pump, *SSU1-R1,* which is expressed at much higher levels than *SSU1* and confers a high level of sulfite resistance [14]. Another translocation between chromosome XV and XVI allows for a short lag phase during the alcoholic fermentation of grape juice. Together, these two translocations confer a selective advantage by shortening the lag phase in a medium containing SO_2 [14].

Yeasts in clinical settings evolve resistance to antimicrobials through various mechanisms. In the yeast *Candida albicans*, drug resistance is facilitated by the acquisition of aneuploidies, in particular, the duplication of the left arm of chromosome 5, resulting in the formation of an isochromosome i(5L), which harbours the azole target gene *ERG11* and a transcriptional activator Tac1, which regulates the efflux pumps Cdr1 and Cdr2 [94]. Moreover, numerous mutations in the *ERG11* and Upc2, the transcriptional regulator that causes the overexpression of *ERG11*, have been reported in response to azole exposure. The formation of i(5L) is often followed by a loss of heterozygosity, rendering the acquired mutations homozygous, thereby conferring higher levels of azole resistance [95]. Similarly, mutations in the echinocandins target gene *FKS1* are commonly followed by a loss of heterozygosity [94].

Fermentative conditions are stressful environments associated with nutrient depletion and increases in ethanol, and wine and beer yeasts have developed strategies that favour their survival. One of these strategies is flocculation, which is controlled by the *FLO* family of genes. Analysis of patterns of CNV in this gene family shows frequent duplications in *FLO11* as well as numerous duplications and deletions in *FLO1*, *FLO5*, *FLO9*, and *FLO10*. A partial duplication in the serine/threonine-rich hydrophobic region of *FLO11* is associated with the adaptive phenotype of floating to the air-liquid interface to access oxygen among "flor" or "sherry" yeasts. Another strategy is the hybridization of strains lacking a beneficial trait with those that have a trait that will confer competitive fitness in a specific environment. In the *Saccharomyces* clade, hybrids of *S. cerevisiae* and other species have been reported, especially in the wine fermentation and brewing environments. Hybrids thriving in brewing mostly display the acquisition of cold tolerance from non-*cerevisiae* strains and the ability to use maltotriose from *S. cerevisiae* strains [80]. For instance, the hybridization of *Saccharomyces eubayanus* and *S. cerevisiae* generating *S. pastorianus,* a partial allotetraploid, has enabled cold fermentation and lager brewing [90, 95].

Millerozyma farinosa is a hybrid osmotolerant yeast derived through interspecific hybridization. This yeast has acquired specific stress resistance genes that allow it to thrive in high solute environments from both parents, albeit through unequal contributions of its parents. Having been isolated from a 70% (w/v) concentrated sorbitol solution, *M. farinosa* boasts a collection of genes that make up an osmoregulatory system that allows for the production and intracellular maintenance of glycerol and other osmolytes. Amongst the genes involved are two potassium transporters, *HAK1* (high-affinity K transporter) and *TRK1* (Transport of K); the P-type ATPase *ACU1* mainly mediating efficient H^+ uptake in high NaCl environments, as well as the NHA1-2 Na^+/H^+ antiporter (involved in Na^+ and also K^+ efflux and *TOK1* (a permeable channel for K^+ efflux) are all involved in K^+ homeostasis. Furthermore, this yeast has H^+/glycerol symport activity and lacks the glycerol permease responsible for glycerol leakage (aquaglyceroporin *FPS1*), thereby retaining the osmolyte in the cells. Through the uniparental acquisition of *MAL* genes (*MALX1*, *MALX2* and *MALX3*), this yeast strain acquired the ability to hydrolyze maltose [96].

FLAVOUR PRODUCTION SPECIALISATIONS

Adaptation of industrial yeasts to specific niches has resulted in the accentuation of the traits that are desirable for humans but would be a disadvantage for the organisms in natural settings. An example of such domestication trait can be seen in beer yeast strains. SNVs have led to the loss of function of genes in *S. cerevisiae* that result in the production of undesirable compounds, enhancing the fitness of the yeast for beer production. An example of this is the loss of function of genes related to ferulic acid decarboxylation, which leads to the production of 4-vinylguaiacol, a phenolic compound with a distinct clove-like aroma. This phenolic compound is considered an off-flavour in most beer styles. *PAD1* (phenylacrylicacid decarboxylase) and *FDC1* (ferulic acid decarboxylase) regulate the decarboxylation of ferulic acid to 4-vinylguaiacol [15]. In response to human selection against the production of off-flavours, different strains have acquired different mutations. In many industrial brewing strains, the *PAD1* and *FDC1* seem to be inactive and acquired a frameshift mutation or a premature stop codon in the gene sequence [97, 98].

ELIMINATING SEXUAL REPRODUCTION

Domesticated yeasts have not only acquired traits that make them suitable for the man-made niche environment they inhabit, but they have also relaxed the selection of traits that are not advantageous or too costly in these environments. This results in gene loss or pseudogenisation of genes that are not needed for survival, which is referred to as genome decay [80]. One of these traits is sexual reproduction which helps yeasts adapt to new, harsh niches but plays a lesser role in more favorable environments. A genotypic and phenotypic study of *S.*

cerevisiae found that beer yeast strains have adapted to living in a nutrient-rich environment and have become obligate asexual. Additionally, beer yeast lineages had high levels of heterozygosity and lacked genetic admixture. This suggests that heterozygosity was acquired during long periods of asexual reproduction rather than through outbreeding [84].

EVOLUTION OF PATHOGENIC YEASTS

There are currently nearly 1500 described yeast species. While most of these species are non-pathogenic, a few species in the phylum Ascomycota and Basidiomycota are opportunistic pathogens of humans and animals. Overall, the ascomycetous yeasts, mainly members of the genus *Candida*, comprise the largest group of pathogenic fungi. Amongst the basidiomycetous yeasts, the major pathogenic genera are *Cryptococcus* and *Malassezia*. Most fungal pathogens of humans are opportunistic pathogens and acquire several virulence and virulence-associated factors through several mechanisms. These include gene duplication and subsequent expansion of specific gene families and clusters, telomeric expansion, gene loss and pseudogenisation, as well as HGT [99]. In the genus *Candida,* tandem duplication and expansion of gene encoding proteins that facilitate host recognition and adhesion has been reported. These include genes encoding Als adhesins and Epa family in *Candida albicans* and *Candida glabrata*, respectively [99, 100]. Moreover, gene families such as *TLO* involved in morphogenesis and virulence and the *IFF* gene family, which confers neutrophil resistance, have been expanded in *C. albicans,* while in *Candida dubliniensis,* which is undergoing reductive evolution, they have already been lost or are in the process of being lost through pseudogenization [99, 101]. Like *Candida* spp., pathogenicity in *Cryptococcus* can be attributed to various virulence factors, *e.g.,* adherence to host tissues, biofilm formation, and secretion of extracellular enzymes such as proteases, ureases and phospholipases [102, 103]. However, the most prominent feature shared by many pathogenic fungi is dimorphism. Morphogenesis promotes host invasion and evasion by dimorphic fungi. The widely characterized human pathogens *Candida albicans* and *Cryptococcus neoformans* are trimorphic, showing the ability to transition between yeast morphology, pseudohyphae and hyphae. In *C. albicans,* the yeast phase is important for dissemination within the host while the hyphal growth is essential for infection and colonization of host tissues and for biofilm formation on catheter and mucosal surfaces, while in *Cryptococcus* spp., the yeast form is responsible for human infections [104]. *Cryptococcus neoformans* and member of the *Cryptococcus gattii* species complex are encapsulated, and genes directly or indirectly associated with capsule formation are crucial for virulence and have been shown to play a role in resistance to oxidative stress, antimicrobial peptides and phagocytosis [102, 103].

The *C. neoformans/C. gattii* pathogenic species complex has not been as extensively studied as members of the genus *Candida*; nevertheless, phylogenetic studies revealed that the pathogenic lineages originated from non-pathogenic saprobic species. Their divergence is largely attributed to chromosomal alterations in the MAT loci [103]. The two lineages differ in certain biochemical, ecological, and pathological features; however, they display several evolutionarily conserved signaling pathways crucial for the pathobiology of both species. These include the cAMP/PKA pathway which is involved in the production of the capsule and melanin, as well as the calmodulin/calcineurin pathway, which plays a role in thermotolerance, virulence and cell wall/membrane integrity in both species [105].

While yeast belonging to the genera *Candida* and *Cryptococcus* are regarded as the most important pathogens, there are other yeasts such as *Malassezia* and *Coccidioides* spp. that can cause severe diseases in humans. These yeasts also display morphogenesis as a key trait associated with virulence. For instance, under certain conditions, *Malassezia* populations can switch between yeasts and hyphyae or pseudophyphae both of which express different virulence factors. Similarly, *Coccidioides* species such as *Coccidioides immitis* and *Coccidioides posadasii* can produce spherules that release endospores into host tissues. The endospores would subsequently germinate to produce hyphal growth or more spherules [106].

Overall, dimorphism is a widespread trait amongst pathogenic fungal species of plants, insects, humans and other mammalian hosts [2]. Most of the fungal pathogens can primarily proliferate either as budding yeasts, pseudohyphae or hyphae. These morphological switches aid pathogens in adhesion to host tissues, dissemination through the body, and manipulation of the host immune responses. Here, we have highlighted mainly those fungi that predominantly exist in their unicellular form in nature and not those that thrive mainly as saprotrophic moulds but can convert to yeast phase upon tissue invasion.

CONCLUDING REMARKS

In the past two decades, advances in molecular techniques, as well as the accessibility of omics technologies and associated bioinformatics tools, have revolutionized population genetic studies. Indeed, genome sequencing has improved our understanding of the evolutionary divergence of yeast species and strains. However, studies into the evolutionary history of yeast adaptation to various niche environments are still in the early stages, and only a few industrially relevant and pathogenic yeasts have received research attention. Indeed, the adaptation of *S. cerevisiae* to various man-made environments has been widely described. Similarly, a lot of insight has been gained regarding the genus *Candida* and the pathogenicity of species in this genus. This chapter has detailed the origin of yeasts and the mechanisms underpinning the evolution of a few widely researched species. However, these are less than a drop in the ocean of thousands of yeast species known to man. Numerous yeast species have been isolated in extreme environments such as deserts, volcanoes, deep oceans, glaciers, stratosphere, *etc*. These extremophilic/extremotolerant yeasts have evolved numerous adaptation strategies to overcome the negative effects that characterise their extreme environments; however, the adaptive evolutionary history of these organisms requires further investigations.

CONSENT FOR PUBLICATION

Not applicable.

CONFLICT OF INTEREST

The author declares no conflict of interest, financial or otherwise.

ACKNOWLEDGEMENTS

Motlhalamme TY was supported by the National Research Foundation through the Competitive Programme for Rated Researchers Grant No. 118505

REFERENCES

[1] Naranjo-Ortiz MA, Gabaldón T. Fungal evolution: major ecological adaptations and evolutionary transitions. Biol Rev Camb Philos Soc 2019; 94(4): 1443-76.
[http://dx.doi.org/10.1111/brv.12510] [PMID: 31021528]

[2] Nagy LG, Tóth R, Kiss E, *et al.* Six key traits of fungi: their evolutionary origins and genetic bases 2017.
[http://dx.doi.org/10.1128/9781555819583.ch2]

[3] Nagy LG, Varga T, Csernetics Á, *et al.* Fungi took a unique evolutionary route to multicellularity: Seven key challenges for fungal multicellular life. Fungal Biol Rev 2020; 34: 151-69.
[http://dx.doi.org/10.1016/j.fbr.2020.07.002]

[4] Nagy LG, Ohm RA, Kovács GM, *et al.* Latent homology and convergent regulatory evolution underlies the repeated emergence of yeasts. Nat Commun 2014; 5: 4471.
[http://dx.doi.org/10.1038/ncomms5471] [PMID: 25034666]

[5] Dujon BA, Louis EJ. Genome diversity and evolution in the budding yeasts (Saccharomycotina). Genetics 2017; 206(2): 717-50.
[http://dx.doi.org/10.1534/genetics.116.199216] [PMID: 28592505]

[6] Barth G, Gaillardin C. Physiology and genetics of the dimorphic fungus *Yarrowia lipolytica*. FEMS Microbiol Rev 1997; 19(4): 219-37.
[http://dx.doi.org/10.1111/j.1574-6976.1997.tb00299.x] [PMID: 9167256]

[7] Cruz JM, Domínguez H, Parajo JC. Dimorphic behaviour of *Debaryomyces hansenii* grown on barley bran acid hydrolyzates. Biotechnol Lett 2000; 22: 605-10.
[http://dx.doi.org/10.1023/A:1005677618040]

[8]　Opalek M, Wloch-Salamon D. Aspects of multicellularity in *Saccharomyces cerevisiae* yeast: A review of evolutionary and physiological mechanisms. Genes (Basel) 2020; 11(6): 690.
[http://dx.doi.org/10.3390/genes11060690] [PMID: 32599749]

[9]　Wolfe KH. Origin of the yeast whole-genome duplication. PLoS Biol 2015; 13(8): e1002221.
[http://dx.doi.org/10.1371/journal.pbio.1002221] [PMID: 26252643]

[10]　Chang S-L, Lai H-Y, Tung S-Y, Leu JY. Dynamic large-scale chromosomal rearrangements fuel rapid adaptation in yeast populations. PLoS Genet 2013; 9(1): e1003232.
[http://dx.doi.org/10.1371/journal.pgen.1003232] [PMID: 23358723]

[11]　Steenwyk JL, Rokas A. Copy number variation in fungi and its implications for wine yeast genetic diversity and adaptation. Front Microbiol 2018; 9: 288.
[http://dx.doi.org/10.3389/fmicb.2018.00288] [PMID: 29520259]

[12]　Hall C, Brachat S, Dietrich FS. Contribution of horizontal gene transfer to the evolution of *Saccharomyces cerevisiae*. Eukaryot Cell 2005; 4(6): 1102-15.
[http://dx.doi.org/10.1128/EC.4.6.1102-1115.2005] [PMID: 15947202]

[13]　Kominek J, Doering DT, Opulente DA, *et al.* Eukaryotic acquisition of a bacterial operon. Cell 2019; 176(6): 1356-1366.e10.
[http://dx.doi.org/10.1016/j.cell.2019.01.034] [PMID: 30799038]

[14]　Marsit S, Dequin S. Diversity and adaptive evolution of *Saccharomyces* wine yeast: a review. FEMS Yeast Res 2015; 15(7): fov067.
[http://dx.doi.org/10.1093/femsyr/fov067] [PMID: 26205244]

[15]　Giannakou K, Cotterrell M, Delneri D. Genomic adaptation of *Saccharomyces* species to industrial environments. Front Genet 2020; 11: 916.
[http://dx.doi.org/10.3389/fgene.2020.00916] [PMID: 33193572]

[16]　Rozpędowska E, Hellborg L, Ishchuk OP, *et al.* Parallel evolution of the make-accumulate-consume strategy in *Saccharomyces* and *Dekkera* yeasts. Nat Commun 2011; 2: 302.
[http://dx.doi.org/10.1038/ncomms1305] [PMID: 21556056]

[17]　MacLean RC, Gudelj I. Resource competition and social conflict in experimental populations of yeast. Nature 2006; 441(7092): 498-501.
[http://dx.doi.org/10.1038/nature04624] [PMID: 16724064]

[18]　Friis EM, Pedersen KR, Crane PR. Cretaceous angiosperm flowers: Innovation and evolution in plant reproduction. Palaeogeogr Palaeoclimatol Palaeoecol 2006; 232: 251-93.
[http://dx.doi.org/10.1016/j.palaeo.2005.07.006]

[19]　Sun G, Dilcher DL, Wang H, Chen Z. A eudicot from the Early Cretaceous of China. Nature 2011; 471(7340): 625-8.
[http://dx.doi.org/10.1038/nature09811] [PMID: 21455178]

[20]　Sun G, Ji Q, Dilcher DL, Zheng S, Nixon KC, Wang X. Archaefructaceae, a new basal angiosperm family. Science 2002; 296(5569): 899-904.
[http://dx.doi.org/10.1126/science.1069439] [PMID: 11988572]

[21]　Hagman A, Säll T, Compagno C, Piskur J. Yeast "make-accumulate-consume" life strategy evolved as a multi-step process that predates the whole genome duplication. PLoS One 2013; 8(7): e68734.
[http://dx.doi.org/10.1371/journal.pone.0068734] [PMID: 23869229]

[22]　Hellborg L, Piškur J. Complex nature of the genome in a wine spoilage yeast, *Dekkera bruxellensis*. Eukaryot Cell 2009; 8(11): 1739-49.
[http://dx.doi.org/10.1128/EC.00115-09] [PMID: 19717738]

[23]　Uribelarrea JL, De Queiroz H, Goma G, Pareilleux A. Carbon and energy balances in cell-recycle cultures of *Schizosaccharomyces pombe*. Biotechnol Bioeng 1993; 42(6): 729-36.
[http://dx.doi.org/10.1002/bit.260420608] [PMID: 18613106]

[24]   Barnett JA. The utilization of sugars by yeasts. Adv Carbohydr Chem Biochem 1976; 32: 125-234.
       [http://dx.doi.org/10.1016/S0065-2318(08)60337-6] [PMID: 782183]

[25]   Berthels NJ, Cordero Otero RR, Bauer FF, Thevelein JM, Pretorius IS. Discrepancy in glucose and
       fructose utilisation during fermentation by *Saccharomyces cerevisiae* wine yeast strains. FEMS Yeast
       Res 2004; 4(7): 683-9.
       [http://dx.doi.org/10.1016/j.femsyr.2004.02.005] [PMID: 15093771]

[26]   Cirillo VP. Galactose transport in *Saccharomyces cerevisiae*. I. Nonmetabolized sugars as substrates
       and inducers of the galactose transport system. J Bacteriol 1968; 95(5): 1727-31.
       [http://dx.doi.org/10.1128/jb.95.5.1727-1731.1968] [PMID: 5650080]

[27]   Gancedo JM. Yeast carbon catabolite repression. Microbiol Mol Biol Rev 1998; 62(2): 334-61.
       [http://dx.doi.org/10.1128/MMBR.62.2.334-361.1998] [PMID: 9618445]

[28]   Carlson M. Glucose repression in yeast. Curr Opin Microbiol 1999; 2(2): 202-7.
       [http://dx.doi.org/10.1016/S1369-5274(99)80035-6] [PMID: 10322167]

[29]   Guillaume C, Delobel P, Sablayrolles J-M, Blondin B. Molecular basis of fructose utilization by the
       wine yeast *Saccharomyces cerevisiae*: a mutated *HXT3* allele enhances fructose fermentation. Appl
       Environ Microbiol 2007; 73(8): 2432-9.
       [http://dx.doi.org/10.1128/AEM.02269-06] [PMID: 17308189]

[30]   Kotyk A. Properties of the sugar carrier in baker's yeast. II. Specificity of transport. Folia Microbiol
       (Praha) 1967; 12(2): 121-31.
       [http://dx.doi.org/10.1007/BF02896872] [PMID: 6047678]

[31]   De Deken RH. The Crabtree effect: a regulatory system in yeast. J Gen Microbiol 1966; 44(2): 149-56.
       [http://dx.doi.org/10.1099/00221287-44-2-149] [PMID: 5969497]

[32]   Compagno C, Dashko S, Piškur J. Introduction to Carbon Metabolism in Yeast.Molecular Mechanisms
       in Yeast Carbon Metabolism. Springer Berlin Heidelberg 2014; pp. 1-19.
       [http://dx.doi.org/10.1007/978-3-662-45782-5_1]

[33]   Pronk JT, van Dijken JP, Pronk JT. Regulation of fermentative capacity and levels of glycolytic
       enzymes in chemostat cultures of *Saccharomyces cerevisiae*. Enzyme Microb Technol 2000; 26(9-10):
       724-36.
       [http://dx.doi.org/10.1016/S0141-0229(00)00164-2] [PMID: 10862878]

[34]   Pronk JT, Yde Steensma H, Van Dijken JP. Pyruvate metabolism in *Saccharomyces cerevisiae*. Yeast
       1996; 12(16): 1607-33.
       [http://dx.doi.org/10.1002/(SICI)1097-0061(199612)12:16<1607::AID-YEA70>3.0.CO;2-4] [PMID:
       9123965]

[35]   Piskur J, Rozpedowska E, Polakova S, Merico A, Compagno C. How did *Saccharomyces* evolve to
       become a good brewer? Trends Genet 2006; 22(4): 183-6.
       [http://dx.doi.org/10.1016/j.tig.2006.02.002] [PMID: 16499989]

[36]   Dashko S, Zhou N, Compagno C, Piškur J. Why, when, and how did yeast evolve alcoholic
       fermentation? FEMS Yeast Res 2014; 14(6): 826-32.
       [http://dx.doi.org/10.1111/1567-1364.12161] [PMID: 24824836]

[37]   Dujon B. Yeast evolutionary genomics. Nat Rev Genet 2010; 11(7): 512-24.
       [http://dx.doi.org/10.1038/nrg2811] [PMID: 20559329]

[38]   Gojković Z, Knecht W, Zameitat E, *et al.* Horizontal gene transfer promoted evolution of the ability to
       propagate under anaerobic conditions in yeasts. Mol Genet Genomics 2004; 271(4): 387-93.
       [http://dx.doi.org/10.1007/s00438-004-0995-7] [PMID: 15014982]

[39]   Wolfe KH, Shields DC. Molecular evidence for an ancient duplication of the entire yeast genome.
       Nature 1997; 387(6634): 708-13.
       [http://dx.doi.org/10.1038/42711] [PMID: 9192896]

[40]   Thomson JM, Gaucher EA, Burgan MF, *et al.* Resurrecting ancestral alcohol dehydrogenases from yeast. Nat Genet 2005; 37(6): 630-5.
[http://dx.doi.org/10.1038/ng1553] [PMID: 15864308]

[41]   Ihmels J, Bergmann S, Gerami-Nejad M, *et al.* Rewiring of the yeast transcriptional network through the evolution of motif usage. Science 2005; 309(5736): 938-40.
[http://dx.doi.org/10.1126/science.1113833] [PMID: 16081737]

[42]   Piskur J. Origin of the duplicated regions in the yeast genomes. Trends Genet 2001; 17(6): 302-3.
[http://dx.doi.org/10.1016/S0168-9525(01)02308-3] [PMID: 11377778]

[43]   Verduyn C, Stouthamer AH, Scheffers WA, van Dijken JP. A theoretical evaluation of growth yields of yeasts. Antonie van Leeuwenhoek 1991; 59(1): 49-63.
[http://dx.doi.org/10.1007/BF00582119] [PMID: 2059011]

[44]   Ishtar Snoek IS, Yde Steensma H. Factors involved in anaerobic growth of *Saccharomyces cerevisiae.* Yeast 2007; 24(1): 1-10.
[http://dx.doi.org/10.1002/yea.1430] [PMID: 17192845]

[45]   Piškur J, Ling Z, Marcet-Houben M, *et al.* The genome of wine yeast *Dekkera bruxellensis* provides a tool to explore its food-related properties. Int J Food Microbiol 2012; 157(2): 202-9.
[http://dx.doi.org/10.1016/j.ijfoodmicro.2012.05.008] [PMID: 22663979]

[46]   Tiukova IA, Petterson ME, Tellgren-Roth C, *et al.* Transcriptome of the alternative ethanol production strain *Dekkera bruxellensis* CBS 11270 in sugar limited, low oxygen cultivation. PLoS One 2013; 8(3): e58455.
[http://dx.doi.org/10.1371/journal.pone.0058455] [PMID: 23516483]

[47]   Boulton R, Singleton V, Bisson L, *et al.* Yeast and Biochemistry of Ethanol Fermentation.Principles and Practices of Winemaking. Boston, MA: Springer 1999; pp. 102-92.
[http://dx.doi.org/10.1007/978-1-4757-6255-6_4]

[48]   Zhou N. Carbon metabolism in non-conventional yeasts: biodiversity, origins of aerobic fermentation and industrial applications. Department of Biology, Lund University 2015.

[49]   Kurtzman CP. Phylogenetic circumscription of *Saccharomyces, Kluyveromyces* and other members of the Saccharomycetaceae, and the proposal of the new genera *Lachancea, Nakaseomyces, Naumovia, Vanderwaltozyma* and *Zygotorulaspora.* FEMS Yeast Res 2003; 4(3): 233-45.
[http://dx.doi.org/10.1016/S1567-1356(03)00175-2] [PMID: 14654427]

[50]   Gu Z, Nicolae D, Lu HH, Li WH. Rapid divergence in expression between duplicate genes inferred from microarray data. Trends Genet 2002; 18(12): 609-13.
[http://dx.doi.org/10.1016/S0168-9525(02)02837-8] [PMID: 12446139]

[51]   Kellis M, Birren BW, Lander ES. Proof and evolutionary analysis of ancient genome duplication in the yeast *Saccharomyces cerevisiae.* Nature 2004; 428(6983): 617-24.
[http://dx.doi.org/10.1038/nature02424] [PMID: 15004568]

[52]   Conant GC, Wolfe KH. Increased glycolytic flux as an outcome of whole-genome duplication in yeast. Mol Syst Biol 2007; 3: 129.
[http://dx.doi.org/10.1038/msb4100170] [PMID: 17667951]

[53]   Papp B, Pál C, Hurst LD. Dosage sensitivity and the evolution of gene families in yeast. Nature 2003; 424(6945): 194-7.
[http://dx.doi.org/10.1038/nature01771] [PMID: 12853957]

[54]   Birchler JA, Bhadra U, Bhadra MP, Auger DL. Dosage-dependent gene regulation in multicellular eukaryotes: implications for dosage compensation, aneuploid syndromes, and quantitative traits. Dev Biol 2001; 234(2): 275-88.
[http://dx.doi.org/10.1006/dbio.2001.0262] [PMID: 11396999]

[55]   Ohno S. Evolution by gene duplication. Berlin: Springer 1970.

[http://dx.doi.org/10.1007/978-3-642-86659-3]

[56]   Zeevi D, Sharon E, Lotan-Pompan M, *et al.* Compensation for differences in gene copy number among yeast ribosomal proteins is encoded within their promoters. Genome Res 2011; 21(12): 2114-28.
        [http://dx.doi.org/10.1101/gr.119669.110] [PMID: 22009988]

[57]   Ata Ö, Rebnegger C, Tatto NE, *et al.* A single Gal4-like transcription factor activates the Crabtree effect in *Komagataella phaffii.* Nat Commun 2018; 9(1): 4911.
        [http://dx.doi.org/10.1038/s41467-018-07430-4] [PMID: 30464212]

[58]   Johnston M. Feasting, fasting and fermenting. Glucose sensing in yeast and other cells. Trends Genet 1999; 15(1): 29-33.
        [http://dx.doi.org/10.1016/S0168-9525(98)01637-0] [PMID: 10087931]

[59]   Pfeiffer T, Schuster S, Bonhoeffer S. Cooperation and competition in the evolution of ATP-producing pathways. Science 2001; 292(5516): 504-7.
        [http://dx.doi.org/10.1126/science.1058079] [PMID: 11283355]

[60]   Goddard MR. Quantifying the complexities of *Saccharomyces cerevisiae*'s ecosystem engineering *via* fermentation. Ecology 2008; 89(8): 2077-82.
        [http://dx.doi.org/10.1890/07-2060.1] [PMID: 18724717]

[61]   Molenaar D, van Berlo R, de Ridder D, Teusink B. Shifts in growth strategies reflect tradeoffs in cellular economics. Mol Syst Biol 2009; 5: 323.
        [http://dx.doi.org/10.1038/msb.2009.82] [PMID: 19888218]

[62]   Pfeiffer T, Morley A. An evolutionary perspective on the Crabtree effect. Front Mol Biosci 2014; 1: 17.
        [http://dx.doi.org/10.3389/fmolb.2014.00017] [PMID: 25988158]

[63]   Cabral S, Prista C, Loureiro-Dias MC, Leandro MJ. Occurrence of *FFZ* genes in yeasts and correlation with fructophilic behaviour. Microbiology 2015; 161(10): 2008-18.
        [http://dx.doi.org/10.1099/mic.0.000154] [PMID: 26253443]

[64]   Sols A. Selective fermentation and phosphorylation of sugars by sauternes yeast. Biochim Biophys Acta 1956; 20(1): 62-8.
        [http://dx.doi.org/10.1016/0006-3002(56)90263-3] [PMID: 13315350]

[65]   Sousa-Dias S, Gonçalves T, Leyva JS, *et al.* Kinetics and regulation of fructose and glucose transport systems are responsible for fructophily in *Zygosaccharomyces bailii.* Microbiology 1996; 142: 1733-8.
        [http://dx.doi.org/10.1099/13500872-142-7-1733]

[66]   Gonçalves C, Wisecaver JH, Kominek J, *et al.* Evidence for loss and reacquisition of alcoholic fermentation in a fructophilic yeast lineage. eLife 2018; 7: e33034.
        [http://dx.doi.org/10.7554/eLife.33034] [PMID: 29648535]

[67]   Gonçalves C, Coelho MA, Salema-Oom M, Gonçalves P. Stepwise functional evolution in a fungal sugar transporter family. Mol Biol Evol 2016; 33(2): 352-66.
        [http://dx.doi.org/10.1093/molbev/msv220] [PMID: 26474848]

[68]   Zaunmüller T, Eichert M, Richter H, Unden G. Variations in the energy metabolism of biotechnologically relevant heterofermentative lactic acid bacteria during growth on sugars and organic acids. Appl Microbiol Biotechnol 2006; 72(3): 421-9.
        [http://dx.doi.org/10.1007/s00253-006-0514-3] [PMID: 16826375]

[69]   Gonçalves C, Ferreira C, Gonçalves LG, *et al.* A new pathway for mannitol metabolism in yeasts suggests a link to the evolution of alcoholic fermentation. Front Microbiol 2019; 10: 2510.
        [http://dx.doi.org/10.3389/fmicb.2019.02510] [PMID: 31736930]

[70]   Pina C, Gonçalves P, Prista C, Loureiro-Dias MC. Ffz1, a new transporter specific for fructose from *Zygosaccharomyces bailii.* Microbiology 2004; 150(Pt 7): 2429-33.
        [http://dx.doi.org/10.1099/mic.0.26979-0] [PMID: 15256584]

[71]   Leandro MJ, Sychrová H, Prista C, Loureiro-Dias MC. The osmotolerant fructophilic yeast *Zygosaccharomyces rouxii* employs two plasma-membrane fructose uptake systems belonging to a new family of yeast sugar transporters. Microbiology 2011; 157(Pt 2): 601-8.
[http://dx.doi.org/10.1099/mic.0.044446-0] [PMID: 21051487]

[72]   Leandro MJ, Cabral S, Prista C, Loureiro-Dias MC, Sychrová H. The high-capacity specific fructose facilitator ZrFfz1 is essential for the fructophilic behavior *of Zygosaccharomyces rouxii* CBS 732^T. Eukaryot Cell 2014; 13(11): 1371-9.
[http://dx.doi.org/10.1128/EC.00137-14] [PMID: 25172765]

[73]   González-Ramos D, Gorter de Vries AR, Grijseels SS, *et al.* A new laboratory evolution approach to select for constitutive acetic acid tolerance in *Saccharomyces cerevisiae* and identification of causal mutations. Biotechnol Biofuels 2016; 9: 173.
[http://dx.doi.org/10.1186/s13068-016-0583-1] [PMID: 27525042]

[74]   Gonçalves P, Gonçalves C, Brito PH, Sampaio JP. The *Wickerhamiella/Starmerella* clade-A treasure trove for the study of the evolution of yeast metabolism. Yeast 2020; 37(4): 313-20.
[http://dx.doi.org/10.1002/yea.3463] [PMID: 32061177]

[75]   Zeder MA. Domestication as a model system for niche construction theory. Evol Ecol 2016; 30: 325-48.
[http://dx.doi.org/10.1007/s10682-015-9801-8]

[76]   Albertin W, Chasseriaud L, Comte G, *et al.* Winemaking and bioprocesses strongly shaped the genetic diversity of the ubiquitous yeast *Torulaspora delbrueckii*. PLoS One 2014; 9(4): e94246.
[http://dx.doi.org/10.1371/journal.pone.0094246] [PMID: 24718638]

[77]   Hranilovic A, Bely M, Masneuf-Pomarede I, Jiranek V, Albertin W. The evolution of *Lachancea thermotolerans* is driven by geographical determination, anthropisation and flux between different ecosystems. PLoS One 2017; 12(9): e0184652.
[http://dx.doi.org/10.1371/journal.pone.0184652] [PMID: 28910346]

[78]   Hranilovic A, Gambetta JM, Schmidtke L, *et al.* Oenological traits of *Lachancea thermotolerans* show signs of domestication and allopatric differentiation. Sci Rep 2018; 8(1): 14812.
[http://dx.doi.org/10.1038/s41598-018-33105-7] [PMID: 30287912]

[79]   Steensels J, Gallone B, Voordeckers K, Verstrepen KJ. Domestication of industrial yeasts. Curr Biol 2019; 29(10): R381-93.
[http://dx.doi.org/10.1016/j.cub.2019.04.025] [PMID: 31112692]

[80]   Gallone B, Mertens S, Gordon JL, Maere S, Verstrepen KJ, Steensels J. Origins, evolution, domestication and diversity of *Saccharomyces* beer yeasts. Curr Opin Biotechnol 2018; 49: 148-55.
[http://dx.doi.org/10.1016/j.copbio.2017.08.005] [PMID: 28869826]

[81]   Gonçalves M, Pontes A, Almeida P, *et al.* Distinct domestication trajectories in top-fermenting beer yeasts and wine yeasts. Curr Biol 2016; 26(20): 2750-61.
[http://dx.doi.org/10.1016/j.cub.2016.08.040] [PMID: 27720622]

[82]   Marsit S, Leducq JB, Durand É, Marchant A, Filteau M, Landry CR. Evolutionary biology through the lens of budding yeast comparative genomics. Nat Rev Genet 2017; 18(10): 581-98.
[http://dx.doi.org/10.1038/nrg.2017.49] [PMID: 28714481]

[83]   Gibbons JG, Rinker DC. The genomics of microbial domestication in the fermented food environment. Curr Opin Genet Dev 2015; 35: 1-8.
[http://dx.doi.org/10.1016/j.gde.2015.07.003] [PMID: 26338497]

[84]   Gallone B, Steensels J, Prahl T, *et al.* Domestication and divergence of *Saccharomyces cerevisiae* beer yeasts. Cell 2016; 166(6): 1397-1410.e16.
[http://dx.doi.org/10.1016/j.cell.2016.08.020] [PMID: 27610566]

[85]   Legras JL, Galeote V, Bigey F, *et al.* Adaptation of *S. cerevisiae* to fermented food environments reveals remarkable genome plasticity and the footprints of domestication. Mol Biol Evol 2018; 35(7):

1712-27.
[http://dx.doi.org/10.1093/molbev/msy066] [PMID: 29746697]

[86]   Varela JA, Puricelli M, Ortiz-Merino RA, *et al.* Origin of lactose fermentation in *Kluyveromyces lactis* by interspecies transfer of a neo-functionalized gene cluster during domestication. Curr Biol 2019; 29(24): 4284-4290.e2.
[http://dx.doi.org/10.1016/j.cub.2019.10.044] [PMID: 31813610]

[87]   Venkatesh A, Murray AL, Coughlan AY, Wolfe KH. Giant *GAL* gene clusters for the melibiose-galactose pathway in *Torulaspora.* Yeast 2021; 38(1): 117-26.
[http://dx.doi.org/10.1002/yea.3532] [PMID: 33141945]

[88]   Roach MJ, Borneman AR. New genome assemblies reveal patterns of domestication and adaptation across *Brettanomyces* (*Dekkera*) species. BMC Genomics 2020; 21(1): 194.
[http://dx.doi.org/10.1186/s12864-020-6595-z] [PMID: 32122298]

[89]   Naumov GI. Domestication of dairy yeast *Kluyveromyces lactis*: transfer of the β-galactosidase (*LAC4*) and lactose permease (*LAC12*) gene cluster? Dokl Biol Sci 2005; 401: 120-2.
[http://dx.doi.org/10.1007/s10630-005-0061-6] [PMID: 16003874]

[90]   Querol A, Belloch C, Fernández-Espinar MT, Barrio E. Molecular evolution in yeast of biotechnological interest. Int Microbiol 2003; 6(3): 201-5.
[http://dx.doi.org/10.1007/s10123-003-0134-z] [PMID: 12898400]

[91]   Yokotsuka K, Fukui M. Changes in nitrogen compounds in berries of six grape cultivars during ripening over two years. Am J Enol Vitic 2002; 53: 69-77.

[92]   Marsit S, Mena A, Bigey F, *et al.* Evolutionary advantage conferred by an eukaryote-to-eukaryote gene transfer event in wine yeasts. Mol Biol Evol 2015; 32(7): 1695-707.
[http://dx.doi.org/10.1093/molbev/msv057] [PMID: 25750179]

[93]   Novo M, Bigey F, Beyne E, *et al.* Eukaryote-to-eukaryote gene transfer events revealed by the genome sequence of the wine yeast *Saccharomyces cerevisiae* EC1118. Proc Natl Acad Sci USA 2009; 106(38): 16333-8.
[http://dx.doi.org/10.1073/pnas.0904673106] [PMID: 19805302]

[94]   Robbins N, Caplan T, Cowen LE. Molecular evolution and antifungal drug resistance. Annu Rev Microbiol 2017; 71: 753-75.
[http://dx.doi.org/10.1146/annurev-micro-030117-020345] [PMID: 28886681]

[95]   Libkind D, Peris D, Cubillos FA, *et al.* Into the wild: new yeast genomes from natural environments and new tools for their analysis. FEMS Yeast Res 2020; 20(2): fo008.
[http://dx.doi.org/10.1093/femsyr/foaa008] [PMID: 32009143]

[96]   Louis VL, Despons L, Friedrich A, *et al.* *Pichia sorbitophila*, an interspecies yeast hybrid, reveals early steps of genome resolution after polyploidization. G3 (Bethesda) 2012; 2(2): 299-311.
[http://dx.doi.org/10.1534/g3.111.000745] [PMID: 22384408]

[97]   Mukai N, Masaki K, Fujii T, Iefuji H. Single nucleotide polymorphisms of *PAD1* and *FDC1* show a positive relationship with ferulic acid decarboxylation ability among industrial yeasts used in alcoholic beverage production. J Biosci Bioeng 2014; 118(1): 50-5.
[http://dx.doi.org/10.1016/j.jbiosc.2013.12.017] [PMID: 24507903]

[98]   Chen P, Dong J, Yin H, *et al.* Single nucleotide polymorphisms and transcription analysis of genes involved in ferulic acid decarboxylation among different beer yeasts. J Inst Brew 2015; 121: 481-9.
[http://dx.doi.org/10.1002/jib.249]

[99]   Moran GP, Coleman DC, Sullivan DJ. Comparative genomics and the evolution of pathogenicity in human pathogenic fungi. Eukaryot Cell 2011; 10(1): 34-42.
[http://dx.doi.org/10.1128/EC.00242-10] [PMID: 21076011]

[100]   Gabaldón T, Naranjo-Ortíz MA, Marcet-Houben M. Evolutionary genomics of yeast pathogens in the Saccharomycotina. FEMS Yeast Res 2016; 16(6): fow064.

[http://dx.doi.org/10.1093/femsyr/fow064] [PMID: 27493146]

[101]   Flanagan PR, Fletcher J, Boyle H, Sulea R, Moran GP, Sullivan DJ. Expansion of the *TLO* gene family enhances the virulence of *Candida* species. PLoS One 2018; 13(7): e0200852.
[http://dx.doi.org/10.1371/journal.pone.0200852] [PMID: 30028853]

[102]   Jong A, Wu C-H, Chen H-M, *et al.* Identification and characterization of *CPS1* as a hyaluronic acid synthase contributing to the pathogenesis of *Cryptococcus neoformans* infection. Eukaryot Cell 2007; 6(8): 1486-96.
[http://dx.doi.org/10.1128/EC.00120-07] [PMID: 17545316]

[103]   Gupta S, Paul K, Kaur S. Diverse species in the genus Cryptococcus: Pathogens and their non-pathogenic ancestors. IUBMB Life 2020; 72(11): 2303-12.
[http://dx.doi.org/10.1002/iub.2377] [PMID: 32897638]

[104]   Bastidas RJ, Heitman J. Trimorphic stepping stones pave the way to fungal virulence. Proc Natl Acad Sci USA 2009; 106(2): 351-2.
[http://dx.doi.org/10.1073/pnas.0811994106] [PMID: 19129500]

[105]   Jung K-W, Lee K-T, Averette AF, *et al.* Evolutionarily conserved and divergent roles of unfolded protein response (UPR) in the pathogenic *Cryptococcus* species complex. Sci Rep 2018; 8(1): 8132.
[http://dx.doi.org/10.1038/s41598-018-26405-5] [PMID: 29802329]

[106]   Fernandes KE, Carter DA. Cellular plasticity of pathogenic fungi during infection. PLoS Pathog 2020; 16(6): e1008571.
[http://dx.doi.org/10.1371/journal.ppat.1008571] [PMID: 32497133]

CHAPTER 2

Ecology: Yeasts on their Natural Environment

Sergio Álvarez-Pérez[1,*]

[1] *Department of Animal Health, Faculty of Veterinary Medicine, Complutense University of Madrid, Madrid, Spain*

Abstract: Yeasts are prevalent in most habitats on Earth, where they often reach high abundance and establish species-rich communities. To date, most research efforts have focused on cataloging the prevalence and diversity (at the phylogenetic and/or physiological level) of yeasts in different habitats and searching for reservoirs of novel yeast taxa. However, little is known regarding the ecological roles that yeasts play in their natural habitats or the relationships that they maintain with other coexisting organisms. This chapter provides a general overview of yeast habitats, with attention to the response of yeasts to diverse abiotic and biotic factors. Furthermore, the chapter presents a detailed description of some relevant systems where yeasts interact with other macro- and microorganisms, namely the insect microbiome, phylloplane, decaying cactus tissues, angiosperm flowers, human microbiome, and industrial processes. Future challenges in the study of yeast ecology are briefly discussed.

Keywords: Anthropogenic environment, Aquatic habitat, Atmosphere, Cactus, Community, Dispersal, Diversity, Ecology, Environmental factor, Evolution, Floral nectar, Flower, Human mycobiome, Industrial process, Insect microbiome, Multipartite interaction, Phylloplane, Soil, Symbiosis, Yeast.

INTRODUCTION

Virtually, all ecosystems on Earth contain yeasts. These taxonomically and phylogenetically diverse unicellular fungi colonize most terrestrial and aquatic habitats, including those most inhospitable, and can also be found in the atmosphere [1 - 5]. Furthermore, many yeast species are integral to human society as they are involved in the production of diverse food products, beverages, and industrial chemicals, and may act as human or animal pathogens. In addition, they provide excellent study models for use in cell biology and other disciplines. However, the ecology of most known yeast species is still poorly understood, and even the natural habitats of renowned model yeasts such as *Saccharomyces cerevisiae* are far from being fully characterized [6, 7].

---
[*] **Corresponding author Sergio Álvarez-Pérez:** Department of Animal Health, Faculty of Veterinary Medicine, Complutense University of Madrid, Madrid, Spain; Tel: +34 91 394 3717; E-mail: sergioaperez@ucm.es

Sérgio Luiz Alves Júnior, Helen Treichel, Thiago Olitta Basso and Boris Ugarte Stambuk (Eds.)
All rights reserved-© 2022 Bentham Science Publishers

An important hurdle in the study of yeast ecology is that, until recently, most studies of yeast presence in natural and anthropogenic environments have utilized culture-based approaches, which tend to be biased toward the most abundant members of a community and often neglect low-abundance and slow-growing species [4, 8, 9]. Additionally, the use of different sampling strategies, culture media, and incubation conditions has made it difficult to compare the results obtained in different studies. Fortunately, recent developments in next-generation sequencing and other DNA-based culturing-independent methods have improved our knowledge regarding the diversity and habitat distribution of yeasts and other fungi in nature [10, 11].

This chapter provides a general overview of yeast habitats, with a special focus on the response of yeasts to diverse environmental factors. Subsequently, it delves into a more detailed description of some systems where yeasts interact with other macro- and microorganisms, often participating in multipartite interactions. Finally, future challenges in the study of yeast ecology are briefly discussed.

YEAST HABITATS

Yeast abundance and diversity in natural and anthropogenic habitats are determined by a variety of abiotic and biotic factors that frequently exhibit spatial heterogeneity and temporal variation (Table 1). In addition, such growth-limiting factors usually come into force together and simultaneously, mutually influencing each other, so that the outcome of these interactions may be difficult to predict [2, 3]. Moreover, there are large-scale phenomena, such as climate and biogeography, which manifest themselves through changes in abiotic and biotic factors (*e.g.*, temperature, humidity, solar radiation, soil composition, vegetation, and animal vectors) [2].

Table 1. Overview of the main environmental factors that influence the metabolic activity, growth, and survival of yeasts [2, 3, 5].

Factors	Short Description
Temperature	Temperature influences yeast growth and generation time. Most yeasts are mesophilic, and grow best between 20 and 30 °C. Some species, mostly pathogens of warm-blooded animals, can grow at 37 °C. The few yeast species capable of growing at 48–50 °C are considered thermotolerant, rather than truly thermophilic. Temperatures >50 °C are usually lethal for vegetative yeast cells. The lower temperature limit of growth for some psychrotolerant species may extend below 0 °C, if water remains fluid (*e.g.*, in salty seawater).
Light and solar radiation	Yeasts are not photosynthetic organisms, so illumination is not a requirement for their existence. However, ultraviolet radiation can be lethal.

(Table 1) cont.....

Factors	Short Description
Pressure	Under natural conditions, the normal atmospheric pressure does not affect yeast growth. However, in the deep sea and some industrial processes, yeast cells must withstand high pressure. The viability of yeast cells decreases with increasing pressures above 100 MPa, and the cells are destroyed between 200 and 300 MPa.
Water activity	Water availability, generally expressed as water activity (a_w), is an important factor affecting yeast growth. Most yeasts can grow well at water activities 0.95–0.90. Only a few yeast species require reduced water activity and are considered truly xerophilic. Nevertheless, many yeast species can grow at high sugar and/or high salt concentrations and are classified as xerotolerant.
Oxygen dependence	Yeasts are basically aerobic organisms. Fermentative yeasts, which represent around half of the species described to date, are only facultative anaerobes.
pH	In general, yeasts prefer a slightly acidic medium and have an optimum pH between 4.5 and 5.5, but most species tolerate a wide range of pH values (generally between 3 and 10). Some species can grow at a strongly acidic pH (≤ 1.5). The tolerance to low pH depends on the type of acidulant, with organic acids possessing a stronger inhibitory effect than inorganic acids. Although acidic conditions are better tolerated than alkaline ones, numerous yeast species can thrive at pH above 10.
Nutrient availability	Yeasts require some sources of carbon, nitrogen, mineral salts, and certain vitamins and growth factors. Differences among yeast species in their ability to assimilate specific nutrients play a major role in habitat specificity. In general, cosmopolitan yeast species are generally the most heterogeneous in their nutritional abilities, whereas yeasts that have specialized habitats exhibit narrower nutritional potentials.
Presence of toxic compounds	Ethanol, which is the main product of alcoholic fermentation, exerts a toxic effect on various yeast species. *Saccharomyces cerevisiae* can tolerate 13–15% ethanol, and some strains even >18%. Carbon dioxide (CO_2), which is the second product of alcoholic fermentation, rarely accumulates at inhibitory concentrations under natural conditions, but yeasts living in the intestinal tract of animals may be subjected to high CO_2 concentrations. CO_2 can dissolve in water and, depending on the pH, form bicarbonate ions that inhibit yeast growth. Acetate, lactate, and other weak organic acids widely used as preservatives in the food industry (*e.g.*, benzoic and sorbic acid) exert specific inhibitory effects on yeasts. Plant and animal tissues contain diverse compounds that may inhibit yeast growth.
Interaction with other organisms	In their natural habitats, yeasts often interact with different macro- and microorganisms. Such interactions can be facultative or obligate, mutual or unidirectional, and they may have a positive (+), negative (-) or neutral (0) effect on the partners involved. The following modalities are possible: mutualism (+/+ interaction), competition (-/-), commensalism (+/0), amensalism (-/0), and predation/parasitism (+/-).

Among all the yeast species found in any habitat, it is important to distinguish those that are essential components of the community from those that are transient members [4]. Moreover, while some yeast species are ubiquitous generalists that occupy a wide geographic range and can dwell in different habitats, other species seem to have a more restricted distribution [5]. Determining whether a given yeast species is an essential or transient member of the community, or is a habitat generalist or specialist, may be challenging, especially when environmental surveys are based on culture-based approaches. Furthermore, advances in yeast taxonomy have shown that some species previously considered habitat generalists actually belong to species complexes, including several morphologically similar sibling species that may differ in their habitat and/or geographical distribution. This is the case, for example, of *Saccharomyces paradoxus*, which was previously regarded as a variety of *S. cerevisiae* [12], and of the '*Sporopachydermia cereana* species complex', which includes at least ten different cryptic species of cactus-inhabiting yeasts [13]. Hybridization and introgression phenomena may result in fuzzy boundaries between sibling species that further complicate the study of their ecology and biogeographic distribution [12]. (note that "complicate" refers to "fuzzy boundaries").

A brief account of the different factors that determine the distribution of yeasts in soils, aquatic habitats, the atmosphere, polar and other terrestrial cold habitats, and human-made environments is provided below.

Yeasts in Soil

A wide diversity of yeasts form part of the microbial communities of natural soils and soils exposed to different degrees of human intervention (*e.g.*, agricultural, orchard, vineyard, and pasture soils) [3, 14 - 18]. In all these cases, yeasts may be found in the bulk soil, the rhizosphere of plants, and/or in association with diverse animals [17, 19].

In general, most soils are highly heterogeneous habitats in which microbial abundance and diversity are unevenly distributed [9, 14, 19]. Yeast abundance in soils depends on different factors acting at the macro- and microscales, including the type of soil, local plant diversity, and the availability of water and different nutrients [9, 15, 17]. Moreover, yeast abundance typically decreases with soil depth, a trend that is explained by the reduced amount of nutrients and oxygen present in the lower soil horizons [9, 19]. The application of fungicides and other toxic agrochemicals, some of which may persist for a long time in the environment, can also determine the abundance and species composition of soil yeast communities [15]. Additionally, some yeast species display spatial and/or temporal restrictions in their distributions, as they have been exclusively or preferentially found in particular soil horizons, localities, and/or seasons [15]. Nonetheless, yeasts are generally less abundant in soil than bacteria and filamentous fungi, which together account for more than 90% of the microbial soil biomass [3, 9, 15, 19]. In fact, yeast abundance rarely exceeds 10^3 cells/g of soil, although counts $>10^6$ cells/g are occasionally found in soils containing abundant organic matter [9, 15, 17].

Extensive lists of yeast taxa found in soil have been published in different comprehensive reviews [15, 16, 19]. Soil yeast communities are often dominated by *Cryptococcus*, *Cystofilobasidium*, *Sporobolomyces*, *Rhodotorula*, and other genera of basidiomycetous yeasts, although ascomycetous yeasts, such as *Candida* spp., *Debaryomyces hansenii*, *Geotrichum* spp., and *Hanseniaspora* spp., are also frequently isolated [3, 15, 16, 19]. However, it must be noted that most currently available knowledge of soil yeasts is biased toward temperate and boreal forests, whereas data from other regions, particularly from the Southern hemisphere, are scarce [16, 17]. In addition, not every yeast species found in soil is an autochthonous inhabitant, as some species may originate from plant litter, decaying fruits, animal frass, or other sources [15 - 17]. Soils have traditionally been regarded as reservoirs of yeasts coming from elsewhere [9].

Despite their polyphyletic nature, most soil yeasts possess common traits that help them thrive in their habitat, including a wide spectrum of metabolic activities enabling them to utilize L-arabinose, D-xylose, cellobiose, and other hydrolytic products of plant materials generated by mycelial fungi and bacteria [14, 19]. Nitrogen oligotrophy is another widespread trait of soil yeasts and confers an advantage when in competition with other microbes [14, 16, 19]. Moreover, some soil yeasts, including members of the genera *Cryptococcus* and *Lipomyces*, have a remarkable ability to produce extracellular polymeric substances (EPS) that sequester and concentrate nutrients, and help them prevent desiccation [3, 14, 16, 19].

An overview of the ecological functions attributed so far to soil yeasts is presented in Fig. (**1**). These functions include the enhancement of plant root growth and protection of plants against root pathogens, the mineralization of organic matter, the solubilization of insoluble phosphates rendering them available for plants and other soil microorganisms, their contribution to soil aggregation and stability and rock weathering, and their role as prey for other soil inhabitants [3, 14 - 17, 19]. A better understanding of these and other unidentified functions may provide insight into the ecological importance of soil yeasts in relation to their prokaryotic counterparts and contribute to the future development of more sustainable agricultural practices [14].

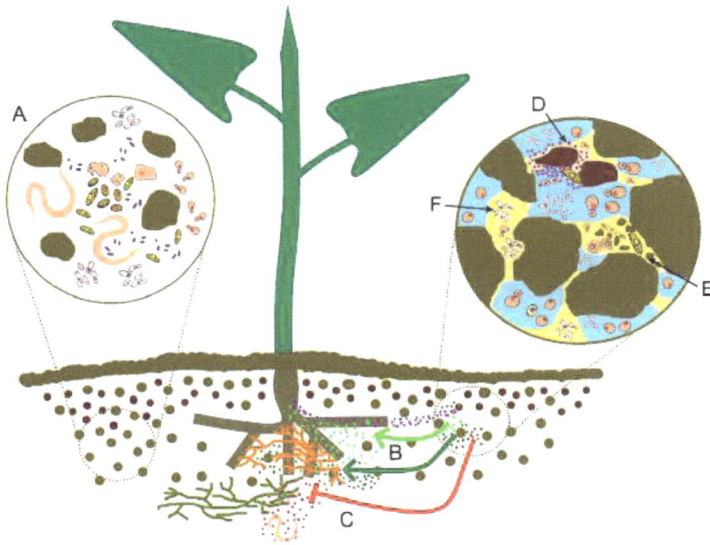

Fig. (1). Overview of the ecological functions attributed to soil yeasts [3, 14 - 17, 19]. Soil-inhabiting yeasts are prey for other organisms, including arthropods, bacteria, nematodes, protists, and other fungi (A). Yeasts can also enhance plant growth, either directly (*e.g.*, by producing diverse plant growth regulators) or by stimulating mycorrhizal colonization of roots (B). Moreover, some yeasts can antagonize diverse fungal root pathogens, nematodes, and protists (C). Additionally, yeasts participate in the mineralization of soil organic matter and the solubilization of insoluble phosphates, rendering them available for plants and other microorganisms (D). Finally, yeasts can contribute to rock weathering and soil formation (E) and soil aggregation and stability via the production of extracellular polymeric substances that bind soil particles together, thus increasing soil porosity and water-holding capacity (F).

Yeasts in Aquatic Habitats

Aquatic ecosystems remain frequently overlooked as fungal habitats, even when phylogenetically and functionally diverse species of mycelial fungi and yeasts are ubiquitous components [20]. In general, it is considered that the unicellular morphology of yeasts makes them better suited than mycelial fungi to fluid systems, and yeasts seem to be the predominant fungal form in some aquatic habitats, such as marine water, the deep open ocean, and hydrothermal vents [4, 20 - 23]. However, there is still limited information regarding the factors affecting the diversity and distribution patterns of aquatic yeasts, and distinguishing any given yeast species as a transient or resident inhabitant of aquatic habitats remains challenging [23, 24].

Most yeasts recovered from water sources are actually associated with terrestrial habitats (*e.g.*, plant sources, soil, effluents from human activity, *etc.*) and arrive in aquatic habitats through runoff phenomena [4, 21, 23, 24]. Nevertheless, the overall abundance of yeasts and the proportion of allochthonous species depend on a variety of factors, including the amount of organic matter, the surrounding vegetation, and the presence of human activity [22 - 24].

Detailed lists of the yeast species found in diverse aquatic habitats are available in the literature [21 - 25]. Although the species most commonly found in freshwater and seawater (*e.g.*, *Aureobasidium pullulans*, *Debaryomyces hansenii*, and *Rhodotorula mucilaginosa*) are considered ubiquitous, the fact that some other species are endemic to specific regions suggests that geographical patterns and local conditions could influence the distribution of aquatic yeast communities [24]. An example of aquatic yeast endemism is *Metschnikowia australis*, a species that has only been found in Antarctica, both in seawater and in association with macroalgae and marine invertebrates [26, 27].

Freshwater Habitats

Yeasts are prevalent in rivers, lakes, lagoons, and other freshwater habitats worldwide [23, 24, 28]. In general, yeast species richness in lakes and rivers is higher in tropical environments than in temperate and cold environments, which is probably due to the occurrence of denser and more diverse plant communities in the former [24, 28]. Yeast communities from tropical lakes are dominated by basidiomycetous yeasts, whereas there seems to be no dominance of either ascomycetous or basidiomycetous taxa in temperate lakes and rivers [24]. Most lakes in temperate regions are characterized by a seasonal pattern of alternate cycles of layering and complete mixing, but the effects of seasonality on the abundance and species diversity of aquatic yeasts are mostly unknown (but see, for example [29]). Some highly acidic rivers and lakes, such as those located in the Iberian Pyrite Belt (southwest of the Iberian Peninsula) and the Rio Agrio and Lake Caviahue system (Argentine Patagonia), have received particular attention as habitats for extremophilic yeasts [30, 31]. In contrast, although yeasts and yeast-like fungi have also been found in groundwater aquifers from different countries [23, 32, 33], groundwater yeast communities have been understudied.

Marine and Oceanic Habitats

Yeasts have been found in all oceans worldwide, in habitats ranging from nearshore environments to oceanic surface waters and deep-sea sediments [21, 23 - 25, 34]. Nearshore waters usually contain $10–10^3$ yeast cells/L, whereas <10 cells/L are typically found in open ocean and deep-sea regions [3, 21, 25]. Traditionally, it has been thought that the yeast species found in marine and oceanic habitats have physiological adaptations to overcome the adverse effects of salinity and high hydrostatic pressure. However, most yeast species found in other habitats can grow in media containing salt concentrations exceeding those normally present in the sea, and few marine yeasts truly have a physiological dependence on sodium chloride or other seawater components [23, 25]. Moreover, except in the proximity of hydrothermal vents, where water temperature can reach 400 °C, the average temperature in most parts of the deep sea ranges from -1 to 4 °C, which means that the yeasts of this habitat should be adapted to cold conditions [35].

In general, marine yeasts may be either terrestrial species that were introduced into coastal waters or endemic marine species whose life histories occur in specific marine habitats [21]. Nearshore areas are subject to terrestrial influx due to natural drainage and human activities [21, 24]. Information regarding the yeast communities inhabiting offshore areas is scarce, as research in the open ocean—particularly, in the deep sea—is expensive and methodologically challenging [21, 24]. The isolation frequency of yeasts usually decreases with increasing distance from the coastline and depth, but phylogenetically diverse yeasts have also been discovered in the deep sea, where they seem to be the predominant fungal representatives [25, 36]. Yeasts of the genera *Candida*, *Cryptococcus*, *Debaryomyces*, *Exophiala*, *Pichia*, *Rhodosporidium*, *Rhodotorula*, and *Trichosporon*, as well as isolates representing undescribed taxa, have been isolated from water and animals surrounding deep-sea hydrothermal vents [37, 38].

Other Aquatic Habitats

Estuaries, salt marshes, and phytotelmata (*i.e.*, small water bodies held by leaves or flowers of plants or in tree holes) often contain high nutrient levels resulting from terrestrial runoff and the degradation of leaf litter and other organic materials, which permits the development of large yeast populations [22, 25]. Moreover, the concentration of nutrients in these areas favors the feeding and breeding sites of many animals that can vector yeasts [22]. Human-associated yeasts such as *Candida glabrata*, *Candida tropicalis*, *Candida parapsilosis*, *Pichia kudriavzevii*, and *Meyerozyma guilliermondii* are frequently found in these ecotone habitats and can serve as pollution indicators [22].

Yeasts in the Atmosphere

The atmosphere has been traditionally regarded as a mere reservoir or dispersal medium for yeasts and other microorganisms, rather than a habitat capable of supporting their growth and reproduction [5, 39]. Although yeasts have been found in air and clouds [40 - 45], there is still limited information regarding their actual abundance and species diversity in this habitat. Furthermore, beyond the possible participation of yeasts as cloud condensation nuclei [45], little is known regarding their role in the atmosphere.

Yeast cells may become airborne because of wind, active ballistoconidia dispersal from phylloplane species, and human activity [3]. Regardless of their origin, yeasts in the atmosphere are exposed to growth-inhibiting factors such as solar radiation, desiccation, low temperatures, low nutrient availability, oxidants, and rapid salinity fluctuation [5, 39, 41]. Furthermore, thunderstorms and other natural meteorological phenomena can carry aerosol particles from the troposphere to the stratosphere, where the action of the aforementioned stressors intensifies [46]. Therefore, the atmosphere may act as an efficient filter for yeast diversity during long-range dispersal. Nevertheless, some extremophilic yeasts such as *Naganishia friedmannii* and other members of the genus *Naganishia* (formerly known as *Cryptococcus albidus* clade) can withstand desiccation, intense UV irradiation, and low pressure and temperature, and seem to be adapted to airborne transportation, which might explain their abundance in high-elevation soils [46, 47]. Other yeasts and yeast-like fungi frequently found in the atmosphere are members of the basidiomycetous genera *Bullera*, *Cryptococcus*, *Dioszegia*, *Rhodotorula*, and *Sporobolomyces*, and the ascomycetous genera *Aureobasidium* and *Exophiala* [3, 40, 44, 45].

Yeasts in Polar and other Terrestrial Cold Habitats

The main terrestrial cold habitats on Earth are Antarctica (~14 million km^2), the Arctic (~4 million km^2), and the subarctic region composed by the continental lands and islands surrounding the ice-covered Arctic Ocean (approximately between 50°N and 70°N in latitude) [48 - 50]. Outside the poles and subarctic region, cold habitats are mostly represented by the glaciers and permafrost soils found in the Himalayas, Andes, and European high mountains [48, 50, 51]. Glacial lakes and lagoons, glacial sediments, cryoconite holes (*i.e.*, holes in a glacier's surface caused by sediment melting), and snowpacks that accumulate for extended periods of time in high altitudes also represent important nonpolar cold habitats [24, 50 - 52].

Although some yeasts found in cold habitats are obligate psychrophiles (*i.e.*, microorganisms with an optimum growth temperature ≤15 °C, an upper growth temperature of about 20 °C, and a minimum growth temperature of 0 °C), most of them are psychrotolerant microorganisms that can grow across a wide range of temperatures and have growth optima of >20 °C [48, 49, 51]. Moreover, in most of cold habitats, low temperature is often associated with other growth limiting factors, such as low water and nutrient availability, high hydrostatic pressure, and increased exposure to ultraviolet (UV) radiation, which render such habitats very inhospitable for life [24, 48 - 51]. Accordingly, the yeasts present in cold habitats display remarkable nutritional plasticity and some other adaptations, such as the production of cold-active enzymes, anti-freezing compounds, and extracellular polymers that protect cells against the damaging effects of subzero temperatures, the possession of an increased proportion of unsaturated fatty acids in their cytoplasm membranes to preserve their fluidity at low temperatures, and the production of different pigments and other photo-protective compounds [24, 48 - 51]. These singular traits make cold-adapted yeasts very attractive for diverse biotechnological applications [48].

In general, yeast diversity in cold habitats is dominated by Basidiomycota taxa, such as the genera *Cryptococcus, Cystobasidium, Dioszegia, Filobasidium, Glaciozyma, Goffeauzyma, Leucosporidium, Mrakia, Naganishia, Papiliotrema, Phenoliferia, Rhodotorula, Solicoccozyma,* and *Vishniacozyma,* although some ascomycetous yeasts such as *Aureobasidium pullulans, Debaryomyces hansenii,* and *Candida* spp. are also frequently found in polar and non-polar locations [48 - 51]. However, there is still very limited information about the actual composition of the yeast communities of most cold habitats on Earth. Similarly, although it has been postulated that yeasts could participate in organic matter decomposition and nutrient cycling inside glaciers and ice sheets [51], the ecological relevance of yeasts in cold habitats remains clearly understudied.

Yeasts in Anthropogenic Habitats

Apart from being found in diverse types of managed soils (see 'Yeasts in soil'), yeasts are present in many other anthropogenic habitats. For instance, wet areas within houses (*e.g.*, kitchens, bathrooms, and domestic saunas) and household appliances are inhabited by diverse yeast and yeast-like species [53 - 55]. In general, these indoor yeasts can form biofilms on synthetic and metal materials, and they have adaptations that allow them to tolerate multiple stress factors, including the production of extracellular polysaccharides, the ability to degrade cleaning agents, and their tolerance to high or low temperatures, as well as high salt concentrations and alkaline pH [54, 55]. Notably, indoor yeast communities include some species that are regarded as potential pathogens that may pose a significant health risk (*e.g.*, if they are present in the houses of immunocompromised individuals), such as members of the genera *Candida, Cryptococcus, Exophiala, Malassezia, Rhodotorula,* and *Trichosporon* [53 - 55].

Other anthropogenic habitats widely analyzed for yeast presence include swimming pools [56, 57], hospitals [58, 59], and the food and beverage industry [60, 61]. In all these cases, published reports have mostly focused on studying the abundance and diversity of pathogenic and/or spoilage yeasts.

INTERACTION WITH OTHER ORGANISMS

In all the habitats mentioned in the previous section, yeasts can be found as free-living microorganisms or in association with different macro- and microorganisms. These interactions can be of different nature and distinctly effect each partner. The ecological typing of interactions distinguishes the following modalities: mutualism (+/+ interaction), competition (-/-), commensalism (+/0), amensalism (-/0), and predation/parasitism (+/-) [62, 63]. Moreover, yeast interactions with other organisms can be facultative or obligate, and they can vary in specificity and symmetry, sometimes depending on the genetic background of the partners and/or the environmental conditions [64]. In addition, the outcome of the interaction may depend on the order and timing of arrival of the different species to the habitat, a process known as priority effects [5].

A thorough review of all the ecological interactions of yeasts with other organisms is beyond the scope of this chapter. Instead, I will focus on some emblematic systems that have been the object of intensive research during the last decades, namely, the insect microbiome, the phylloplane, the decaying tissues of cacti, the flowers of angiosperms, the human microbiome, and diverse industrial processes.

Yeasts in the Insect Microbiome

Insects are the most diverse group of animals on Earth and, by far, they represent the most important vectors of yeasts in nature and one of the most important reservoirs of yeast diversity [3, 9, 65].

Insects engage in a remarkable array of symbiotic interactions with yeasts, which range from parasitic to mutualistic [66, 67]. Moreover, insect–yeast symbioses often include plants as a third partner. In most cases, yeasts are housed in the gastrointestinal tract of insects, and their relationship with the host seems to be facultative [67]. Yeast access to the gastrointestinal tract occurs by ingestion of food sources colonized by the microorganism, or through coprophagy or trophallaxis [67]. The structure and physiology of the insect gut vary among insect orders and within an individual over different life stages, especially in insects with holometabolous life cycles, and such physiological differences are likely to affect yeast growth [67 - 69]. In addition, some insects have specialized body structures to host microbes, such as gastric caeca and mycangia (invaginations of the exoskeleton to carry fungal spores or yeast cells) [66 - 68, 70].

Obligate interactions between insects and microbial symbionts are common, but mostly involve bacterial partners rather than fungi [66]. Nevertheless, intracellular (endosymbiotic) yeast-like symbionts, often located within specialized cells in the fat body known as mycetocytes, have been associated with planthoppers, aphids, and leafhoppers [66, 68]. Such yeast-like symbionts are present in all stages of the insect life history and are vertically transmitted across generations via transovarial infection [66, 68]. Three different hypotheses have been proposed to explain the evolutionary origin of endosymbiotic yeast-insect associations: i) the yeast symbionts may be derived from pathogenic parasites or nonpathogenic commensals; (ii) the yeast symbionts may be the descendants of phytopathogenic or saprophytic fungi; or (iii) the feeding behavior of insects brought them in contact with yeasts occurring endophytically or on the phylloplane [71].

The most important role attributed to insect-associated yeasts is their participation in insect nutrition, as yeasts provide digestive enzymes (*e.g.*, exoproteases, peptidases, and lipases), essential amino acids, vitamins, trace metals, and sterols that insects cannot produce [66 - 69, 71]. In some cases, there seems to be some specialization in nutritional relationships. For example, a high degree of correlation has been found between the occurrence of certain yeasts that process and utilize xylose, which forms the backbone of the hemicellulose component of plant cell walls, and wood-ingesting beetles from distantly related families [72]. Nevertheless, yeasts unable to assimilate and ferment xylose have also been associated with wood-ingesting beetles [72]. Besides their nutritional role, yeasts can also participate in the detoxification of toxic plant metabolites that are present in the insect's diet, suppression of pathogens that might interfere with insect development, and localization of suitable food resources [66 - 68, 71, 73]. In return, yeasts benefit from their dispersal to new habitats and the shelter offered by the insect body [66 - 69, 73]. Additionally, the insect gut seems to represent a suitable environment for the sexual reproduction of some yeast species such as *S. cerevisiae*; this event is rarely observed in other environments [73].

The complementary benefits of yeast-insect interactions are the basis of the 'dispersal–encounter' hypothesis, whereby yeasts are dispersed by insects between different sugar-rich ephemeral patches, and insects obtain the benefits of an honest signal from yeasts for the sugar resources [74]. There is abundant evidence that, in different systems, the communication between yeasts and insects is mediated by volatile organic compounds (VOCs) that result from the fermentation of sugars available in the medium [74]. VOC-mediated communication between plant-associated yeasts and their insect vectors is gaining attention because of its potential applications in the development of new strategies for the biological control of agricultural pests [75]. In addition, the study of insect-microbe symbioses might lead to the discovery of new enzymes and chemical compounds with potential biotechnological applications [76].

Yeasts in the Phylloplane

Plant leaves constitute one of the largest terrestrial habitats for microorganisms [77]. Although bacteria are generally the earliest and the most numerous colonizers of leaf surfaces (also known as 'phylloplane'), diverse mycelial fungi and yeasts can also be found in this habitat, either as resident species or transient inhabitants [77].

Yeast populations are unevenly distributed across individual leaf surfaces and mostly occur in aggregates established on discrete sites (*e.g.*, nearby leaf veins, trichomes, stomata, and wounds of the leaf cuticle) [77]. Yeast abundance is also largely dependent on the type of plant and/or climate. Values of 10^3–10^4 colony forming units (CFU)/cm^2 are common on the leaves of herbs or deciduous trees in temperate climates, but values $>10^6$ CFU/cm^2 have been found in some studies [77]. In contrast, differences in yeast abundance between the lower (abaxial) and upper (adaxial) leaf surfaces or depending on the leaf position in the canopy (higher *vs.* lower leaves) have not been fully demonstrated [77].

The atmosphere has been recognized as a major source and sink for phylloplane yeasts. Immigration to leaves occurs through the impaction of particles onto the leaf surface, gravity settling or sedimentation, or rain-splash dispersal to the leaf surface, whereas emigration from leaves occurs because of active dispersal mechanisms caused by rain, water movement, or wind [77]. Once established in the phylloplane, the dynamics of the yeast populations are determined by a combination of those of individual yeast species and the ontogenetic cycles of plants [78]. Moreover, the phylloplane is often considered an extreme environment where microorganisms must face continuously fluctuating environmental factors, such as temperature, relative humidity, solar radiation, and water and nutrient availability [77, 79]. The nutrients present on the phylloplane, which eventually determine the microbial carrying capacity of the leaf, can have an endogenous or exogenous origin [8, 77, 79]. Endogenous nutrients are the result of the diffusion of compounds out of the leaf tissue via hydathodes, trichomes, or fissures in the leaf cuticle due to injuries or weathering. Exogenous sources of nutrients include pollen, guttation fluids, and nutrients excreted by other organisms, including microbes and insects [77].

In response to these stress factors, most phylloplane yeasts display adaptations such as oligotrophic nutrition, photoprotective compound production (*e.g.*, melanins, carotenoids, and mycosporines), and EPS capsules that act as cellular buffer systems to prevent water loss and contribute to the efficient rehydration of yeast cells after periods of drought [77, 79]. The production of ballistoconidia, which represents an efficient means for yeast dispersal and seems to be stimulated by the nutrient-poor conditions found on leaf surfaces, is also a common trait among phylloplane yeasts. Moreover, given the strong competition for space and nutrients that they face, most phylloplane yeasts have evolved different strategies of intra- and inter-species interactions and communication, including the release of soluble or volatile compounds into the environment [79]. Antagonistic interactions between phylloplane yeasts and other coexisting microorganisms include the production of killer toxins (secreted proteins or glycoproteins that disrupt cell membrane function in susceptible yeasts [80 - 82]), glycolipids, and iron chelators (*e.g.*, the pulcherrimin produced by *Metschnikowia pulcherrima*) [79]. Furthermore, through the production of phytohormones, such as indole-3 - acetic acid and other plant growth promoters, some yeast species might influence the growth and fitness of their plant hosts [79].

In general, the composition of epiphytic yeast communities in temperate regions is nonspecific, as the same yeast species dominate diverse plant species [78]. However, yeast species may exhibit differences in their seasonal dynamics of relative abundance [78]. Most phylloplane yeast communities are dominated by relatively few abundant species, mostly belonging to the Basidiomycota division (*e.g.*, *Cryptococcus*, *Rhodotorula*, and *Sporobolomyces*); however, less abundant species may account for a significant fraction of species richness [3, 77 - 79]. Ascomycetous yeast genera such as *Aureobasidium*, *Candida*, *Debaryomyces*, and *Metschnikowia* are also commonly encountered as phylloplane inhabitants [77 - 79]. Unfortunately, limited information is available regarding the abundance patterns and diversity of phylloplane yeast communities in tropical regions [8]. Similarly, little is known regarding the yeast microbiota colonizing the surface of mosses and other non-vascular plants [79].

Yeasts is Decaying Cactus Tissues

The decaying tissues of cacti provide a habitat for phylogenetically diverse yeasts and their associated vectors [9, 83 - 87]. Among such yeasts, the species that have been found exclusively in this system are often referred to as 'cactophilic yeasts' [83]. Some cactophilic yeasts, such as *Pichia cactophila*, *Candida sonorensis*, *Clavispora opuntiae*, and the *Sporopachydermia cereana* species complex, are prevalent in necrotic cacti worldwide, whereas other species are limited to certain geographic areas or hosts [83 - 87]. The latter is the case of, for example, *Pichia heedii*, an ascomycetous yeast that is found predominantly in the necrotic tissues of saguaro and senita cacti in the Sonoran Desert of the southwestern USA and northwestern Mexico, but it is rarely found in other cactus types and does not occur in other habitats [84, 86, 87]. Host plant identity usually has a larger influence on the diversity and composition of cactus yeast communities than geographic distance such that the communities from the same cactus type are

more similar to one another across different regions than the communities from different cactus types within the same regions [84]. Nevertheless, rare non-cactu-
-specific yeast species found in some regions or cactus types may account for a significant proportion of the diversity of the community [84]. For a complete list of the cactophilic and other cactus-inhabiting yeasts identified to date, the reader can refer to the available comprehensive reviews on this topic [83 - 86].

Cactus rots can be seen as discrete, nutrient-rich patches separated by inhospitable (extremely dry) territory. Yeast dispersal from cactus to cactus is mostly performed by drosophilid flies (Diptera: Drosophilidae) which are recruited to the rot through their attraction to volatiles produced by microbial activity [9, 83, 85 - 88]. Adult drosophilids disperse the yeasts from one cactus rot to another according to their particular host preferences, whereas the larvae spread the yeast within the soft rot [9, 85, 86]. Both adult flies and their larvae feed on the microorganisms present in cactus rot, which supplies them with sterols, concentrated protein, and vitamins [9, 85, 86]. Notably, some *Drosophila* species may select for particular cactus yeasts. This is the case for *Drosophila mojavensis*, which apparently selects for *P. cactophila* in different cactus species [89]. Other cactus-feeding and breeding animals, including insects (*e.g.*, dipterans, beetles, cochineals, and moths), birds, reptiles, and mammals, may also act as dispersal agents of yeasts, but these alternative vectors are generally regarded as occasional visitors of cacti [83, 86]. The relationships between cactophilic yeasts and their drosophilid vectors are not obligate but seem to be strong enough to restrict their distribution to decaying cactus tissues and to maintain yeast communities that are stable over time and space [84].

Establishment of different yeast species within the cactus host depends on the chemical characteristics of the cactus tissues. The flat, fleshy cladodes of *Opuntia* cacti (family Cactaceae, subfamily Opuntioideae) are relatively rich in simple sugars, whereas sugars in most columnar cacti (family Cactaceae, subfamily Cactoideae) are bound to complex carbohydrates or triterpene glycosides [83, 85]. Moreover, most columnar cacti contain alkaloids and other secondary plant metabolites that are toxic to yeasts and/or their dispersal agents [83, 85]. Therefore, *Opuntia* cladode tissues seem to be less restrictive to yeast growth than columnar cactus stem tissues [88]. However, the role of cactus tissue chemistry in determining yeast community ecology is not completely understood [83]. Interactions with other yeasts, including growth facilitation via cross-feeding and interference competition (often achieved by the production of killer toxins), can affect the population size of each particular yeast species and, therefore, determine their likelihood of being vectored to the next habitat [83, 85].

Despite the intense research on cactus yeasts performed during the last six decades, there is still much work to be done in this field. For instance, published surveys follow culture-based approaches, so it remains unknown if there is a significant uncultured component to the biodiversity of cactus yeasts [83]. Additional field and microcosm experiments may contribute to elucidating the overall function of the cactus–insect–yeast system and, in particular, of the microbe–microbe interactions taking place in decaying cactus tissues.

Yeasts in the Floral Microbiome

The flowers of angiosperms offer a wide variety of microhabitats suitable for microbial growth [90 - 92]. In particular, floral nectar provides a habitat for specialized and opportunistic yeasts and bacteria that can withstand high osmotic pressure, scarcity of nitrogen sources, and the presence of diverse secondary compounds of plant origin [93, 94]. Yeast presence in floral parts other than nectaries (*e.g.*, petals, sepals, tepals, stamens, stigmas, styles, ovaries, and pollen) has been studied in much less detail [91]; hence, this section will mostly refer to nectar yeasts.

While floral nectar is assumed to be sterile when the flower opens, this sugary secretion often becomes rapidly colonized by yeasts, reaching densities of 10^4-10^6 cells/mm^3 in different plant species [94 - 96]. Nectar-inhabiting yeasts originate from various sources, including the air, rain drops, the phyllosphere, other floral parts, and the body (generally mouthparts) of pollinators and other flower-visiting animals [97, 98]. Depending on their origin, these yeasts are often categorized into two distinct groups [93, 98]: i) those that originate from different environmental sources and usually show no specific adaptations to nectar conditions; and ii) those species that are mainly dispersed from flower to flower by animal visitors (mostly insects but also hummingbirds and other animals). The yeasts classified in the second group, which are often referred to as 'nectar specialists', show much higher levels of specialization and are highly adapted to survive in nectar [93]. Notably, insects may serve as the overwintering sites of nectar specialists between consecutive flowering seasons [99].

The nectar yeast communities of most plant species surveyed to date are composed of a few species (on average, 1.2 culturable species/sample) and are often dominated by nectar specialists of the genus *Metschnikowia*, including *M. reukaufii* and *M. gruessi* [94, 97]. Other yeast genera that are commonly found in nectar and other floral parts include the ascomycetous genera *Aureobasidium*, *Candida*, *Clavispora*, *Debaryomyces*, *Hanseniaspora*, *Kodamaea*, *Starmerella*, and *Wickerhamiella*, and the basidiomycetous genera *Cryptococcus*, *Papiliotrema*, *Rhodotorula*, and *Sporobolomyces*; however, all these genera are generally less abundant than *Metschnikowia* spp [91, 94]. The aforementioned yeast genera frequently co-occur with bacteria, and some studies have suggested associations between yeasts and bacteria in floral nectar. For example, a survey of nectar microorganisms associated with diverse Mediterranean plants found that culturable bacteria and yeasts co-occurred more often than would be expected by chance, suggesting a positive association between *Metschnikowia* spp. and bacterial species from the genus *Acinetobacter* [100]. Such positive co-occurrence might be facilitated by resource partitioning between *Metschnikowia* and the nectar-inhabiting acinetobacters, with the yeasts depleting glucose and enriching floral nectar in fructose and the bacteria preferentially utilizing fructose [101]. However, laboratory and field experiments performed by other researchers have found that priority effects between *M. reukaufii* and bacteria may result in competitive exclusion between nectar microbes [102 - 104]. Nonetheless, interpretation of these observations is difficult, as mechanisms underlying yeast–bacterium interactions in nectar and other floral microhabitats remain poorly studied [102].

Community assembly in floral nectar depends on the complex interaction between multiple factors, including the nectar secretion pattern of the host plant, the filtering effect of the physical and chemical characteristics of nectar on each particular yeast species, priority effects, and different mechanisms of microbe–microbe interactions (Fig. **2**) [93, 102]. Furthermore, the nectar–yeast system constitutes a metacommunity, in which individual flowers act as island-like ephemeral habitats that are connected by occasional dispersal via animal visitors [93, 102]. Accordingly, nectar yeasts have gained attention as a suitable study system for testing ecological processes affecting community assembly, such as environmental filtering, dispersal, historical contingency, and metacommunity dynamics [91, 105].

There are many unanswered questions regarding the ecology and evolution of flower-inhabiting yeasts and their interactions with animals and plants [91, 92, 102]. Nevertheless, it has been demonstrated that nectar yeasts can alter the chemical composition of their habitat by consuming the available nutrients and/or releasing VOCs, which, in turn, may affect the behavior of pollinators and other flower visitors and have an impact on the reproductive success of the plant [91 - 93, 106, 107]. In addition, flower-inhabiting microbes can affect pollen quantity or quality [92]. Finally, yeast cells can act as a nutritional supplement for nectar- and pollen-feeding insects and suppress the growth of insect pathogens [88, 93, 108, 109]. Therefore, flower-inhabiting yeasts may have diverse effects on the fitness of their host plants and floral visitors [91 - 93, 108, 109].

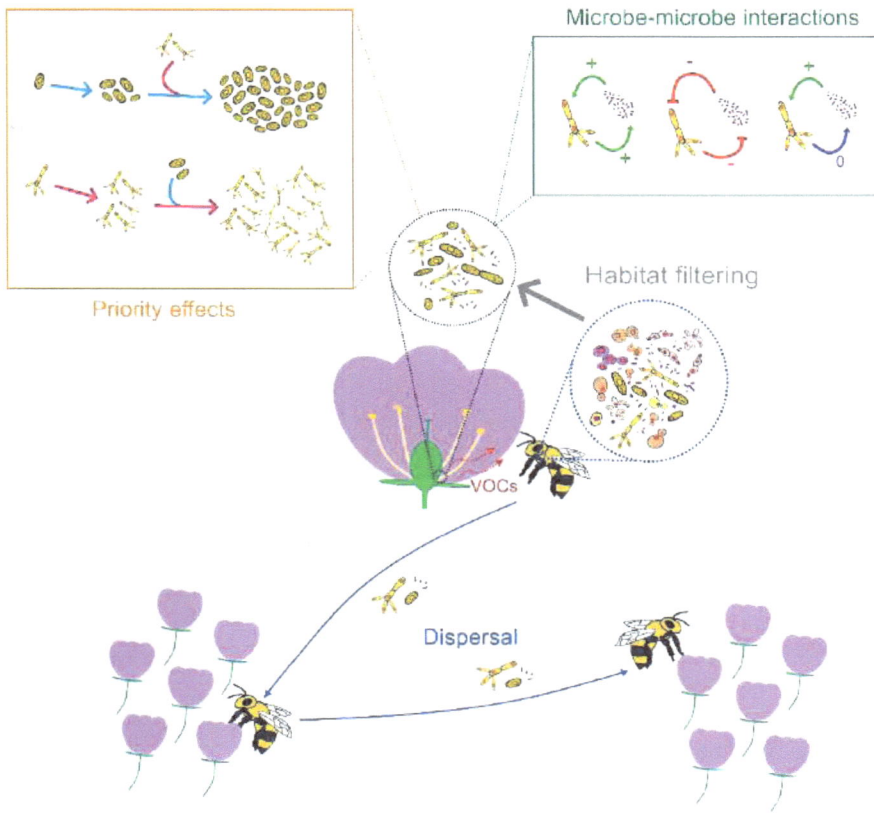

Fig. (2). Overview of the main factors contributing to the assembly of the microbial communities inhabiting floral nectar [93, 102]. Microbes (including yeasts and bacteria) are dispersed from flower to flower by the action of insects and other floral visitors. The harsh conditions of the nectar environment (high osmotic pressure, scarcity of nitrogen sources, and presence of plant toxins) hinder the growth of some yeast and bacterial species, thus acting as a filter of the microbial diversity brought by the floral visitors. Priority effects and microbe–microbe interactions of diverse sign and intensity may determine microbial diversity and abundance in floral nectar. Finally, microbial growth can alter the chemical composition of nectar by consuming the available nutrients and/or releasing volatile organic compounds (VOCs), which, in some cases, may affect the behavior of floral visitors and eventually have an influence on plant–pollinator interactions.

Yeasts in the Human Microbiome

Fungi, and yeasts in particular, are frequent colonizers of the skin and mucosal surfaces of humans and other animals [110 - 115]. However, the study of the fungal microbiome, often referred to as the 'mycobiome' or 'mycome', is a research field that has traditionally lagged behind the study of the bacterial microbiome [110 - 112, 114, 116].

In the absence of reliable estimates of the fungal load in the human body, it is generally assumed that the total number of fungal cells is orders of magnitude smaller than that of the bacterial microbiome [111]. For example, 10^1-10^3 fungal cells/g of stool are typically found in humans, which contrasts with the 10^{11}-10^{12} bacterial cells/g of stool [113]. Regarding species diversity, the members of genera *Candida*, *Cryptococcus*, *Debaryomyces*, *Galactomyces*, *Malassezia*, *Saccharomyces*, and *Trichosporon* are prevalent on the skin and different mucosal surfaces of the human body. Furthermore, these seven yeast genera are included in the list of fungi most commonly detected in gut mycobiome studies [110, 114]. However, nearly half of all fungal taxa found in the human gut have only been observed in a limited number of samples and/or studies, and there is a long list of taxa that have been reported only once [110, 114]. Notably, many of these rare inhabitants are incapable of colonizing or persisting for a long time in the gut [110, 114]. Nonetheless, it is often difficult to determine which yeast species are true residents of the human body and which are transient species originating from foods, the environment, or other sources [110, 112].

Although there is still limited information regarding the functions of the human mycobiome and its interactions with other components of the body microbiota in states of health and disease, it is thought that yeasts may play a role in host immune regulation in different chronic inflammatory diseases (*e.g.*, atopic dermatitis and inflammatory bowel diseases) and metabolic disorders, as well as in other physiological processes [111, 112, 116]. Moreover, human mycobiota has been identified as a source of fungal infections when systemic or local mucosal immune functions are disturbed, or after antibiotic treatments that disrupt bacteria-mediated colonization resistance [111 - 113]. Attention has also been paid to the potential use of yeasts, particularly *Saccharomyces boulardii*, as probiotics for treating gastroenteritis and other diseases [111, 116, 117].

Yeasts in Industrial Processes

The last example of systems where yeast interactions with other organisms can have vital roles are diverse industrial processes, including mixed culture fermentation. Most fermentation processes in the food and beverage industry depend on mixtures of microorganisms that act in concert to produce the desired product characteristics [118]. Substrates for food fermentation usually have a highly heterogeneous physicochemical composition, which allows for the simultaneous occupation of multiple niches by different microorganisms (*e.g.*, through the utilization of different nutrient sources) [119]. When different microbial species (*e.g.*, *Saccharomyces* and non-*Saccharomyces* yeasts, yeasts and bacteria, yeasts and mycelial fungi, *etc.*) are involved in mixed culture fermentation, they often do not coexist passively, but establish direct and indirect interactions of different natures [118, 120]. Mixed cultures can be initiated spontaneously when different microorganisms originate from the initial material or the environment, or by inoculation of the participants as starter cultures, and they can consist of sets of known or unknown species [119, 121]. Such mixed cultures often result in a better utilization of the substrate, as the mixture of microorganisms may possess a wider range of enzymes and, therefore, can attack a greater variety of compounds [121]. Additionally, microbial consortia can perform more complex activities and tolerate more variation in environmental factors than pure cultures (*i.e.*, they are more versatile and robust) [118].

Yeast–bacterium interactions are also essential in the preservation of many food products and beverages [82]. For example, bacteria often excrete organic acids that lower the pH of the medium, which inhibits the growth of undesired pathogens and/or promotes yeast growth [82]. In addition, yeast interactions with other microorganisms may result in more diversified secondary metabolite synthesis, which is of great interest in drug discovery [122].

In all the aforementioned examples, coexisting microorganisms can interact through multiple mechanisms, which may occur through physical contact, by producing signaling molecules, or by modifying the physicochemical conditions of the environment [119]. The use of '-omics' approaches (transcriptomics, proteomics, metabolomics, *etc.*) and modern analytical techniques (*e.g.*, ultrahigh resolution mass spectrometry), which are still poorly developed for non-*Saccharomyces* yeasts, may contribute to the elucidation of the mechanisms involved in microbe–microbe interactions and their effects on industrial processes.

FUTURE CHALLENGES IN YEAST ECOLOGY

Despite recent advances in the study of yeast ecology, many issues still need to be addressed, including the following:

1) Undersampled reservoirs of yeast diversity

Our current knowledge of yeast diversity is heavily biased toward species from some specific regions (mostly Western Europe, Japan, and North America) and isolated from clinical sources, beverages, food products, and a few natural substrates [123]. According to various predictions, a substantial proportion of the diversity of this group of eukaryotic microorganisms remains unknown, which justifies the efforts aimed at studying yeast diversity in different habitats worldwide, especially in megadiverse regions of the planet [124]. It is important to mention that, like other organisms, yeasts are threatened by habitat alterations due to global warming and anthropogenic activities such as agriculture, deforestation, and urbanization. In view of the rapid decline of many natural habitats, the study of yeast diversity in undisturbed or low-managed habitats should be a priority.

2) Biotechnological applications

A better understanding of yeast diversity and ecological interactions in natural habitats may contribute to the development of new agricultural and industrial applications. In particular, detailed metabolic and physiological characterization of environmental yeasts, especially those found in stressful habitats, represents a valuable tool for the detection of biotechnologically relevant traits. Furthermore, yeast culture collections play an essential role in maintaining a rich diversity of yeasts for current and future applications [124].

3) Importance of quorum sensing in yeast ecology

In nature, microbial cells do not live in isolation, but rather they communicate with each other using diverse chemical signals. The term "quorum sensing" refers to a sophisticated mode of cell-to-cell signaling where the behavior of the microbial population is coordinated in a cell-density-dependent manner [125 - 127]. Although quorum sensing was first discovered in bacteria, this phenomenon has also been described in yeasts and other fungi, in which it regulates processes such as morphological differentiation (*e.g.*, from a yeast form to a filamentous form, or vice versa), biofilm formation, secondary metabolite production, and/or pathogenesis [125 - 127]. Furthermore, some yeasts can communicate with bacteria and even their animal or plant hosts through signaling molecules such as farnesol [125]. In any case, most research in this field has focused on a few target species (mostly on *S. cerevisiae*, *Candida albicans*, *Histoplasma capsulatum*, and *Cryptococcus neoformans*), and knowledge about the environmental cues triggering quorum sensing in yeasts and the ecological relevance of this phenomenon is still scarce [125, 126]. The biotechnological potential of yeast quorum sensing is also clearly understudied.

4) Yeast roles in natural habitats

To date, most research in yeast ecology has focused on describing the diversity of yeast habitats. In contrast, much less attention has been paid to the ecological roles of yeasts in nature. In particular, little is known regarding the in-situ response of yeasts to different abiotic factors and their interactions with other organisms, since the majority of studies on these topics have been conducted using microcosms under controlled conditions that do not fully mimic the plethora of factors that affect yeast growth in nature. Developments in fungal transcriptomics, which generate a snapshot of the composition and relative abundance of actively transcribed genes in a given sample [128], may help fill in these research gaps in the future. Nevertheless, given the high complexity of the microbial communities found in some habitats, determining which specific functions are carried out by each particular yeast species will be challenging.

5) Yeast population genomics and phenomics

Recent advances in genome sequencing and high-throughput phenotyping hold promise for investigating how genetic diversity in yeasts is linked to phenotypic variation and habitat adaptation. However, to date, most research in yeast population genomics and phenomics has dealt with model taxa such as *Saccharomyces* spp [129, 130], and *Schizosaccharomyces pombe* [131]. Future research in this field may help to elucidate the natural history of other non-model yeast species and test for genotype-phenotype-habitat associations [132].

CONCLUDING REMARKS

Despite recent progress in the study of yeast ecology, which has been possible because of the concomitant advances in next-generation sequencing-based community analysis and yeast taxonomy, there are many questions in this field that remain to be answered. In particular, most research efforts have focused on cataloging the phylogenetic and physiological diversity of yeasts in natural and human-associated habitats, and on searching for reservoirs of novel yeast taxa. In contrast, little is known regarding the roles of yeasts in such habitats or the relationships that they maintain with other coexisting organisms. Moreover, the role of such interactions with other organisms in yeast evolution remains to be determined. The transition from a mostly descriptive to a more hypothesis-driven ecological study of yeasts may contribute to clarifying these issues.

CONSENT FOR PUBLICATION

Not applicable.

CONFLICT OF INTEREST

The author declares no conflict of interest, financial or otherwise.

ACKNOWLEDGEMENTS

The author acknowledges a 'Ramón y Cajal' contract funded by the Spanish Ministry of Science and Innovation [RYC2018-023847-I]. The funders had no role in the preparation of the manuscript or in the decision to publish. The author thanks Pilar Pérez Sanz and David Jiménez Uceta for their kindness and continuous encouragement.

REFERENCES

[1]   Cantrell SA, Dianese JC, Fell J, Gunde-Cimerman N, Zalar P. Unusual fungal niches. Mycologia 2011; 103(6): 1161-74.
        [http://dx.doi.org/10.3852/11-108] [PMID: 21700639]

[2]   Deák T. Environmental factors influencing yeasts.Biodiversity and Ecophysiology of Yeasts. Berlin, Heidelberg: Springer 2006; pp. 155-74.
        [http://dx.doi.org/10.1007/3-540-30985-3_8]

[3]   Deák T. Ecology.Handbook of Food Spoilage Yeasts. 2nd ed. Boca Raton: CRC Press 2007; pp. 37-58.
        [http://dx.doi.org/10.1201/9781420044942.ch3]

[4]   Lachance MA, Starmer WT. Ecology and yeasts.The Yeasts, a Taxonomic Study. 4th ed. Amsterdam: Elsevier Science 1998; pp. 21-30.
        [http://dx.doi.org/10.1016/B978-044481312-1/50007-1]

[5]   Péter G, Takashima M, Čadež N. Yeast habitats: different but global.Yeasts in Natural Ecosystems: Ecology. Cham: Springer International Publishing AG 2017; pp. 39-71.
        [http://dx.doi.org/10.1007/978-3-319-61575-2_2]

[6]   Goddard MR, Greig D. *Saccharomyces cerevisiae*: a nomadic yeast with no niche? FEMS Yeast Res 2015; 15(3): fov009.
        [http://dx.doi.org/10.1093/femsyr/fov009] [PMID: 25725024]

[7]   Liti G. The fascinating and secret wild life of the budding yeast *S. cerevisiae*. eLife 2015; 4: e05835.
        [http://dx.doi.org/10.7554/eLife.05835] [PMID: 25807086]

[8]   Limtong S, Nasanit R. Phylloplane yeasts in tropical climates.Yeasts in Natural Ecosystems: Diversity. Cham: Springer International Publishing AG 2017; pp. 199-223.
        [http://dx.doi.org/10.1007/978-3-319-62683-3_7]

[9]   Phaff HJ, Starmer WT. Yeasts associated with plants, insects and soil. 1987.

[10]   Nilsson RH, Anslan S, Bahram M, Wurzbacher C, Baldrian P, Tedersoo L. Mycobiome diversity: high-throughput sequencing and identification of fungi. Nat Rev Microbiol 2019; 17(2): 95-109.
        [http://dx.doi.org/10.1038/s41579-018-0116-y] [PMID: 30442909]

[11]   Větrovský T, Morais D, Kohout P, *et al.* GlobalFungi, a global database of fungal occurrences from high-throughput-sequencing metabarcoding studies. Sci Data 2020; 7(1): 228.
        [http://dx.doi.org/10.1038/s41597-020-0567-7] [PMID: 32661237]

[12]   Boynton PJ, Greig D. The ecology and evolution of non-domesticated *Saccharomyces* species. Yeast 2014; 31(12): 449-62.
        [PMID: 25242436]

[13]   Lachance MA, Kaden JE, Phaff HJ, Starmer WT. Phylogenetic structure of the *Sporopachydermia cereana* species complex. Int J Syst Evol Microbiol 2001; 51(Pt 1): 237-47.
        [http://dx.doi.org/10.1099/00207713-51-1-237] [PMID: 11211264]

[14]   Botha A. The importance and ecology of yeasts in soil. Soil Biol Biochem 2011; 43(1): 1-8.
        [http://dx.doi.org/10.1016/j.soilbio.2010.10.001]

[15]   Vadkertiová R, Dudášová H, Balaščáková M. Yeasts in agricultural and managed soils.Yeasts in Natural Ecosystems: Diversity. Cham: Springer International Publishing AG 2017; pp. 117-44.
        [http://dx.doi.org/10.1007/978-3-319-62683-3_4]

[16]   Yurkov A. Yeasts in forest soils.Yeasts in Natural Ecosystems: Diversity. Cham: Springer International Publishing AG 2017; pp. 87-116.
        [http://dx.doi.org/10.1007/978-3-319-62683-3_3]

[17]   Yurkov AM. Yeasts of the soil - obscure but precious. Yeast 2018; 35(5): 369-78.
        [http://dx.doi.org/10.1002/yea.3310] [PMID: 29365211]

[18] Yurkov AM, Kemler M, Begerow D. Assessment of yeast diversity in soils under different management regimes. Fungal Ecol 2012; 5(1): 24-35.
[http://dx.doi.org/10.1016/j.funeco.2011.07.004]

[19] Botha A. Yeasts in soil.Biodiversity and Ecophysiology of Yeasts. Berlin, Heidelberg: Springer 2006; pp. 221-40.
[http://dx.doi.org/10.1007/3-540-30985-3_11]

[20] Grossart HP, Van den Wyngaert S, Kagami M, Wurzbacher C, Cunliffe M, Rojas-Jimenez K. Fungi in aquatic ecosystems. Nat Rev Microbiol 2019; 17(6): 339-54.
[http://dx.doi.org/10.1038/s41579-019-0175-8] [PMID: 30872817]

[21] Fell JW. Yeast in marine environments.In: Jones EBG, Pang KL, Eds. Marine Fungi and Fungal-like Microorganisms Berlin, Boston, de Gruyter. 2012; pp. 91-102.

[22] Hagler AN, Mendonça-Hagler LC, Pagnocca FC. Yeasts in aquatic ecotone habitats.Yeasts in Natural Ecosystems: Diversity. Cham: Springer International Publishing AG 2017; pp. 63-85.
[http://dx.doi.org/10.1007/978-3-319-62683-3_2]

[23] Nagahama T. Yeast biodiversity in freshwater, marine and deep-sea environments.Biodiversity and Ecophysiology of Yeasts. Berlin, Heidelberg: Springer 2006; pp. 241-62.
[http://dx.doi.org/10.1007/3-540-30985-3_12]

[24] Libkind D, Buzzini P, Turchetti B, Rosa CA. Yeasts in continental and seawater.Yeasts in Natural Ecosystems: Diversity. Cham: Springer International Publishing AG 2017; pp. 1-61.
[http://dx.doi.org/10.1007/978-3-319-62683-3_1]

[25] Kutty SN, Philip R. Marine yeasts-a review. Yeast 2008; 25(7): 465-83.
[http://dx.doi.org/10.1002/yea.1599] [PMID: 18615863]

[26] Batista TM, Hilário HO, Moreira RG, *et al.* Draft genome sequence of *Metschnikowia australis* strain UFMG-CM-Y6158, an extremophile marine yeast endemic to Antarctica. Genome Announc 2017; 5(20): e00328-17.
[http://dx.doi.org/10.1128/genomeA.00328-17] [PMID: 28522704]

[27] Furbino LE, Godinho VM, Santiago IF, *et al.* Diversity patterns, ecology and biological activities of fungal communities associated with the endemic macroalgae across the Antarctic peninsula. Microb Ecol 2014; 67(4): 775-87.
[http://dx.doi.org/10.1007/s00248-014-0374-9] [PMID: 24509705]

[28] Brandão LR, Vaz ABM, Espírito Santo LC, *et al.* Diversity and biogeographical patterns of yeast communities in Antarctic, Patagonian and tropical lakes. Fungal Ecol 2017; 28: 33-43.
[http://dx.doi.org/10.1016/j.funeco.2017.04.003]

[29] Bogusławska-Was E, Dabrowski W. The seasonal variability of yeasts and yeast-like organisms in water and bottom sediment of the Szczecin Lagoon. Int J Hyg Environ Health 2001; 203(5-6): 451-8.
[http://dx.doi.org/10.1078/1438-4639-00056] [PMID: 11556149]

[30] Gadanho M, Libkind D, Sampaio JP. Yeast diversity in the extreme acidic environments of the Iberian Pyrite Belt. Microb Ecol 2006; 52(3): 552-63.
[http://dx.doi.org/10.1007/s00248-006-9027-y] [PMID: 17013554]

[31] Russo G, Libkind D, Sampaio JP, van Broock MR. Yeast diversity in the acidic Rio Agrio-Lake Caviahue volcanic environment (Patagonia, Argentina). FEMS Microbiol Ecol 2008; 65(3): 415-24.
[http://dx.doi.org/10.1111/j.1574-6941.2008.00514.x] [PMID: 18537834]

[32] Ekendahl S, O'Neill AH, Thomsson E, Pedersen K. Characterisation of yeasts isolated from deep igneous rock aquifers of the Fennoscandian Shield. Microb Ecol 2003; 46(4): 416-28.
[http://dx.doi.org/10.1007/s00248-003-2008-5] [PMID: 14502418]

[33] Nawaz A, Purahong W, Lehmann R, *et al.* Superimposed pristine limestone aquifers with marked hydrochemical differences exhibit distinct fungal communities. Front Microbiol 2016; 7: 666.

[http://dx.doi.org/10.3389/fmicb.2016.00666] [PMID: 27242696]

[34]   Kandasamy K, Alikunhi NM, Subramanian M. Yeasts in marine and estuarine environments. J Yeast Fungal Res 2012; 3(6): 74-82.

[35]   Nagano Y, Nagahama T, Abe F. Cold-adapted yeasts in deep-sea environments.Cold-Adapted Yeasts. Berlin, Heidelberg: Springer-Verlag 2014; pp. 149-71.
       [http://dx.doi.org/10.1007/978-3-662-45759-7_7]

[36]   Bass D, Howe A, Brown N, *et al.* Yeast forms dominate fungal diversity in the deep oceans. Proc Biol Sci 2007; 274(1629): 3069-77.
       [http://dx.doi.org/10.1098/rspb.2007.1067] [PMID: 17939990]

[37]   Burgaud G, Arzur D, Durand L, Cambon-Bonavita MA, Barbier G. Marine culturable yeasts in deep-sea hydrothermal vents: species richness and association with fauna. FEMS Microbiol Ecol 2010; 73(1): 121-33.
       [http://dx.doi.org/10.1111/j.1574-6941.2010.00881.x] [PMID: 20455940]

[38]   Gadanho M, Sampaio JP. Occurrence and diversity of yeasts in the mid-atlantic ridge hydrothermal fields near the Azores Archipelago. Microb Ecol 2005; 50(3): 408-17.
       [http://dx.doi.org/10.1007/s00248-005-0195-y] [PMID: 16328655]

[39]   Womack AM, Bohannan BJ, Green JL. Biodiversity and biogeography of the atmosphere. Philos Trans R Soc Lond B Biol Sci 2010; 365(1558): 3645-53.
       [http://dx.doi.org/10.1098/rstb.2010.0283] [PMID: 20980313]

[40]   Andreeva IS, Safatov AS, Morozova VV, *et al.* Saprophytic and pathogenic yeasts in atmospheric aerosols of southwestern Siberia. Atmos Oceanic Opt 2020; 33(5): 505-11.
       [http://dx.doi.org/10.1134/S1024856020050024]

[41]   Delort AM, Vaïtilingom M, Amato P, *et al.* A short overview of the microbial population in clouds: potential roles in atmospheric chemistry and nucleation processes. Atmos Res 2010; 98: 249-60.
       [http://dx.doi.org/10.1016/j.atmosres.2010.07.004]

[42]   Ejdys E, Biedunkiewicz A, Dynowska M, Sucharzewska E. Snow in the city as a spore bank of potentially pathogenic fungi. Sci Total Environ 2014; 470-471: 646-50.
       [http://dx.doi.org/10.1016/j.scitotenv.2013.10.045] [PMID: 24176713]

[43]   Segvić Klarić M, Pepeljnjak S. A year-round aeromycological study in Zagreb area, Croatia. Ann Agric Environ Med 2006; 13(1): 55-64.
       [PMID: 16841873]

[44]   Vaïtilingom M, Attard E, Gaiani N, *et al.* Long-term features of cloud microbiology at the puy de Dôme (France). Atmos Environ 2012; 56: 88-100.
       [http://dx.doi.org/10.1016/j.atmosenv.2012.03.072]

[45]   Woo C, An C, Xu S, Yi SM, Yamamoto N. Taxonomic diversity of fungi deposited from the atmosphere. ISME J 2018; 12(8): 2051-60.
       [http://dx.doi.org/10.1038/s41396-018-0160-7] [PMID: 29849168]

[46]   Pulschen AA, de Araujo GG, de Carvalho ACSR, *et al.* Survival of extremophilic yeasts in the stratospheric environment during balloon flights and in laboratory simulations. Appl Environ Microbiol 2018; 84(23): e01942-18.
       [http://dx.doi.org/10.1128/AEM.01942-18] [PMID: 30266724]

[47]   Schmidt SK, Vimercati L, Darcy JL, *et al.* A *Naganishia* in high places: functioning populations or dormant cells from the atmosphere? Mycology 2017; 8(3): 153-63.
       [http://dx.doi.org/10.1080/21501203.2017.1344154] [PMID: 30123637]

[48]   Buzzini P, Margesin R. Cold-adapted yeasts: a lesson from the cold and a challenge for the XXI century.Cold-Adapted Yeasts. Berlin, Heidelberg: Springer-Verlag 2014; pp. 3-22.
       [http://dx.doi.org/10.1007/978-3-662-45759-7_1]

[49]   Buzzini P, Turk M, Perini L, Turchetti B, Gunde-Cimerman N. Yeasts in polar and subpolar habitats.Yeasts in Natural Ecosystems: Diversity. Cham: Springer International Publishing AG 2017; pp. 331-65.
[http://dx.doi.org/10.1007/978-3-319-62683-3_11]

[50]   Sannino C, Tasselli G, Filippucci S, Turchetti B, Buzzini P. Yeasts in nonpolar cold habitats.Yeasts in Natural Ecosystems: Diversity. Cham: Springer International Publishing AG 2017; pp. 367-96.
[http://dx.doi.org/10.1007/978-3-319-62683-3_12]

[51]   Turchetti B, Goretti M, Buzzini P, Margesin R. Cold-adapted yeasts in Alpine and Apennine glaciers.Cold-Adapted Yeasts. Berlin, Heidelberg: Springer-Verlag 2014; pp. 99-122.
[http://dx.doi.org/10.1007/978-3-662-45759-7_5]

[52]   Brandão LR, Libkind D, Vaz AB, *et al.* Yeasts from an oligotrophic lake in Patagonia (Argentina): diversity, distribution and synthesis of photoprotective compounds and extracellular enzymes. FEMS Microbiol Ecol 2011; 76(1): 1-13.
[http://dx.doi.org/10.1111/j.1574-6941.2010.01030.x] [PMID: 21223324]

[53]   Adams RI, Miletto M, Taylor JW, Bruns TD. The diversity and distribution of fungi on residential surfaces. PLoS One 2013; 8(11): e78866.
[http://dx.doi.org/10.1371/journal.pone.0078866] [PMID: 24223861]

[54]   Babič MN, Zupančič J, Gunde-Cimerman N, Zalar P. Yeast in anthropogenic and polluted environments.Yeasts in Natural Ecosystems: Diversity. Cham: Springer International Publishing AG 2017; pp. 145-69.
[http://dx.doi.org/10.1007/978-3-319-62683-3_5]

[55]   Zalar P, Novak M, de Hoog GS, Gunde-Cimerman N. Dishwashers--a man-made ecological niche accommodating human opportunistic fungal pathogens. Fungal Biol 2011; 115(10): 997-1007.
[http://dx.doi.org/10.1016/j.funbio.2011.04.007] [PMID: 21944212]

[56]   Ekowati Y, Ferrero G, Kennedy MD, de Roda Husman AM, Schets FM. Potential transmission pathways of clinically relevant fungi in indoor swimming pool facilities. Int J Hyg Environ Health 2018; 221(8): 1107-15.
[http://dx.doi.org/10.1016/j.ijheh.2018.07.013] [PMID: 30145117]

[57]   Jankowski M, Charemska A, Czajkowski R. Swimming pools and fungi: An epidemiology survey in Polish indoor swimming facilities. Mycoses 2017; 60(11): 736-8.
[http://dx.doi.org/10.1111/myc.12654] [PMID: 28730647]

[58]   Kumar J, Eilertson B, Cadnum JL, *et al.* Environmental contamination with *Candida* species in multiple hospitals including a tertiary care hospital with a *Candida auris* outbreak. Pathog Immun 2019; 4(2): 260-70.
[http://dx.doi.org/10.20411/pai.v4i2.291] [PMID: 31768483]

[59]   Sanna C, Marras L, Desogus A, *et al.* Evaluation of *Rhodotorula* spp. contamination in hospital environments. Environ Monit Assess 2021; 193(3): 152.
[http://dx.doi.org/10.1007/s10661-021-08908-3] [PMID: 33646402]

[60]   Hernández A, Pérez-Nevado F, Ruiz-Moyano S, *et al.* Spoilage yeasts: What are the sources of contamination of foods and beverages? Int J Food Microbiol 2018; 286: 98-110.
[http://dx.doi.org/10.1016/j.ijfoodmicro.2018.07.031] [PMID: 30056262]

[61]   Ocón E, Garijo P, Sanz S, *et al.* Analysis of airborne yeast in one winery over a period of one year. Food Control 2013; 30(2): 585-9.
[http://dx.doi.org/10.1016/j.foodcont.2012.07.051]

[62]   Boucher DH, James S, Keeler KH. The ecology of mutualism. Annu Rev Ecol Syst 1982; 13: 315-47.
[http://dx.doi.org/10.1146/annurev.es.13.110182.001531]

[63]   Bronstein JL. Our current understanding of mutualism. Q Rev Biol 1994; 69(1): 31-51.
[http://dx.doi.org/10.1086/418432]

[64]   Mittelbach M, Vannette RL. Mutualism in yeasts.Yeasts in Natural Ecosystems: Ecology. Cham: Springer International Publishing AG 2017; pp. 155-78.
[http://dx.doi.org/10.1007/978-3-319-61575-2_6]

[65]   Suh SO, McHugh JV, Pollock DD, Blackwell M. The beetle gut: a hyperdiverse source of novel yeasts. Mycol Res 2005; 109(Pt 3): 261-5.
[http://dx.doi.org/10.1017/S0953756205002388] [PMID: 15912941]

[66]   Blackwell M. Yeasts in insects and other invertebrates.Yeasts in Natural Ecosystems: Diversity. Cham: Springer International Publishing AG 2017; pp. 397-433.
[http://dx.doi.org/10.1007/978-3-319-62683-3_13]

[67]   Gonzalez F. Symbiosis between yeasts and insects. Introductory paper at the Faculty of Landscape Architecture, Horticulture and Crop Production Science [serial on the Internet] 2014 . http://urn.kb.se/resolve?urn=urn:nbn:se:slu:epsilon-e-2291

[68]   Blackwell M. Made for each other: ascomycete yeasts and insects 2017.
[http://dx.doi.org/10.1128/9781555819583.ch46]

[69]   Stefanini I. Yeast-insect associations: It takes guts. Yeast 2018; 35(4): 315-30.
[http://dx.doi.org/10.1002/yea.3309] [PMID: 29363168]

[70]   Ganter PF. Yeast and invertebrate associations.Biodiversity and Ecophysiology of Yeasts The Yeast Handbook. Berlin, Heidelberg: Springer 2006; pp. 303-70.
[http://dx.doi.org/10.1007/3-540-30985-3_14]

[71]   Vega FE, Dowd PF. The role of yeasts as insect endosymbionts.Insect-Fungal Associations: Ecology and Evolution. New York: Oxford University Press 2005; pp. 211-43.

[72]   Blackwell M, Suh SO, Nardi JB. Fungi in the hidden environment: the gut of beetles.Fungi in the Environment. New York: Cambridge University Press 2006; pp. 357-70.

[73]   Meriggi N, Di Paola M, Cavalieri D, Stefanini I. *Saccharomyces cerevisiae*-insects association: impacts, biogeography, and extent. Front Microbiol 2020; 11: 1629.
[http://dx.doi.org/10.3389/fmicb.2020.01629] [PMID: 32760380]

[74]   Madden AA, Epps MJ, Fukami T, *et al.* The ecology of insect-yeast relationships and its relevance to human industry. Proc Biol Sci 2018; 285(1875): 20172733.

[75]   Francis F, Jacquemyn H, Delvigne F, Lievens B. From diverse origins to specific targets: role of microorganisms in indirect pest biological control. Insects 2020; 11(8): 533.
[http://dx.doi.org/10.3390/insects11080533] [PMID: 32823898]

[76]   Berasategui A, Shukla S, Salem H, Kaltenpoth M. Potential applications of insect symbionts in biotechnology. Appl Microbiol Biotechnol 2016; 100(4): 1567-77.
[http://dx.doi.org/10.1007/s00253-015-7186-9] [PMID: 26659224]

[77]   Fonseca Á, Inácio J. Phylloplane yeasts.Biodiversity and Ecophysiology of Yeasts. Berlin, Heidelberg: Springer 2006; pp. 263-301.
[http://dx.doi.org/10.1007/3-540-30985-3_13]

[78]   Glushakova AM, Chernov IIu. [Seasonal dynamics of the epiphytic yeast communities]. Mikrobiologiia 2010; 79(6): 832-42.
[PMID: 21774169]

[79]   Kemler M, Witfeld F, Begerow D, Yurkov A. Phylloplane yeasts in temperate climates.Yeasts in Natural Ecosystems: Diversity. Cham: Springer International Publishing AG 2017; pp. 171-97.
[http://dx.doi.org/10.1007/978-3-319-62683-3_6]

[80]   Boynton PJ. The ecology of killer yeasts: Interference competition in natural habitats. Yeast 2019; 36(8): 473-85.
[http://dx.doi.org/10.1002/yea.3398] [PMID: 31050852]

[81]   Klassen R, Schaffrath R, Buzzini P, Ganter PF. Antagonistic interactions and killer yeasts.Yeasts in Natural Ecosystems: Ecology. Cham: Springer International Publishing AG 2017; pp. 229-75.
[http://dx.doi.org/10.1007/978-3-319-61575-2_9]

[82]   Viljoen BC. Yeast ecological interactions. Yeast–yeast, yeast–bacteria, yeast–fungi interactions and yeasts as biocontrol agents.Yeasts in Food and Beverages. Berlin, Heidelberg: Springer-Verlag 2006; pp. 83-110.
[http://dx.doi.org/10.1007/978-3-540-28398-0_4]

[83]   Ganter PF, Morais PB, Rosa CA. Yeasts in cacti and tropical fruit.Yeasts in Natural Ecosystems: Diversity. Cham: Springer International Publishing AG 2017; pp. 225-64.
[http://dx.doi.org/10.1007/978-3-319-62683-3_8]

[84]   Starmer WT, Aberdeen V, Lachance MA. The biogeographic diversity of cactophilic yeasts.Biodiversity and Ecophysiology of Yeasts The Yeast Handbook. Berlin, Heidelberg: Springer 2006; pp. 485-99.
[http://dx.doi.org/10.1007/3-540-30985-3_19]

[85]   Starmer WT, Fogleman JC, Lachance MA. The yeast community of cacti.Microbial Ecology of Leaves. New York: Springer 1991; pp. 158-78.
[http://dx.doi.org/10.1007/978-1-4612-3168-4_8]

[86]   Starmer WT, Lachance M, Phaff HJ, Heed WB. The biogeography of yeasts associated with decaying cactus tissue in North America, the Caribbean, and northern Venezuela. Evol Biol 1990; 24: 253-96.

[87]   Starmer WT, Schmedicke RA, Lachance MA. The origin of the cactus-yeast community. FEMS Yeast Res 2003; 3(4): 441-8.
[http://dx.doi.org/10.1016/S1567-1356(03)00056-4] [PMID: 12748055]

[88]   Anderson TM, Lachance MA, Starmer WT. The relationship of phylogeny to community structure: the cactus yeast community. Am Nat 2004; 164(6): 709-21.
[http://dx.doi.org/10.1086/425372] [PMID: 29641929]

[89]   Fogleman JC, Starmer WT, Heed WB. Larval selectivity for yeast species by *Drosophila mojavensis* in natural substrates. Proc Natl Acad Sci USA 1981; 78(7): 4435-9.
[http://dx.doi.org/10.1073/pnas.78.7.4435] [PMID: 16593060]

[90]   Aleklett K, Hart M, Shade A. The microbial ecology of flowers: an emerging frontier in phyllosphere research. Botany 2014; 92(4): 253-66.
[http://dx.doi.org/10.1139/cjb-2013-0166]

[91]   Klaps J, Lievens B, Álvarez-Pérez S. Towards a better understanding of the role of nectar-inhabiting yeasts in plant-animal interactions. Fungal Biol Biotechnol 2020; 7: 1.
[http://dx.doi.org/10.1186/s40694-019-0091-8] [PMID: 31921433]

[92]   Vannette RL. The floral microbiome: plant, pollinator, and microbial perspectives. Annu Rev Ecol Evol Syst 2020; 51: 363-86.
[http://dx.doi.org/10.1146/annurev-ecolsys-011720-013401]

[93]   Jacquemyn H, Pozo MI, Álvarez-Pérez S, Lievens B, Fukami T. Yeast-nectar interactions: metacommunities and effects on pollinators. Curr Opin Insect Sci 2021; 44: 35-40.
[http://dx.doi.org/10.1016/j.cois.2020.09.014] [PMID: 33065340]

[94]   Pozo M, Lievens B, Jacquemyn H. Impact of microorganisms on nectar chemistry, pollinator attraction and plant fitness.Nectar: Production, Chemical Composition and Benefits to Animals and Plants. New York: Nova Science Publishers, Inc. 2015; pp. 1-40.

[95]   de Vega C, Herrera CM, Johnson SD. Yeasts in floral nectar of some South African plants: quantification and associations with pollinator type. S Afr J Bot 2009; 75: 798-806.
[http://dx.doi.org/10.1016/j.sajb.2009.07.016]

[96]   Herrera CM, de Vega C, Canto A, Pozo MI. Yeasts in floral nectar: a quantitative survey. Ann Bot

(Lond) 2009; 103(9): 1415-23.
[http://dx.doi.org/10.1093/aob/mcp026] [PMID: 19208669]

[97] Brysch-Herzberg M. Ecology of yeasts in plant-bumblebee mutualism in Central Europe. FEMS Microbiol Ecol 2004; 50(2): 87-100.
[http://dx.doi.org/10.1016/j.femsec.2004.06.003] [PMID: 19712367]

[98] Pozo MI, Lachance MA, Herrera CM. Nectar yeasts of two southern Spanish plants: the roles of immigration and physiological traits in community assembly. FEMS Microbiol Ecol 2012; 80(2): 281-93.
[http://dx.doi.org/10.1111/j.1574-6941.2011.01286.x] [PMID: 22224447]

[99] Pozo MI, Bartlewicz J, van Oystaeyen A, *et al.* Surviving in the absence of flowers: do nectar yeasts rely on overwintering bumblebee queens to complete their annual life cycle? FEMS Microbiol Ecol 2018; 94(12): fiy196.
[http://dx.doi.org/10.1093/femsec/fiy196] [PMID: 30285114]

[100] Alvarez-Pérez S, Herrera CM. Composition, richness and nonrandom assembly of culturable bacterial-microfungal communities in floral nectar of Mediterranean plants. FEMS Microbiol Ecol 2013; 83(3): 685-99.
[http://dx.doi.org/10.1111/1574-6941.12027] [PMID: 23057414]

[101] Álvarez-Pérez S, Lievens B, Jacquemyn H, Herrera CM. *Acinetobacter nectaris* sp. nov. and *Acinetobacter boissieri* sp. nov., isolated from floral nectar of wild Mediterranean insect-pollinated plants. Int J Syst Evol Microbiol 2013; 63(Pt 4): 1532-9.
[http://dx.doi.org/10.1099/ijs.0.043489-0] [PMID: 22904213]

[102] Álvarez-Pérez S, Lievens B, Fukami T. Yeast-bacterium interactions: the next frontier in nectar research. Trends Plant Sci 2019; 24(5): 393-401.
[http://dx.doi.org/10.1016/j.tplants.2019.01.012] [PMID: 30792076]

[103] Toju H, Vannette RL, Gauthier MPL, Dhami MK, Fukami T. Priority effects can persist across floral generations in nectar microbial metacommunities. Oikos 2018; 127(3): 345-52.
[http://dx.doi.org/10.1111/oik.04243]

[104] Tucker CM, Fukami T. Environmental variability counteracts priority effects to facilitate species coexistence: evidence from nectar microbes. Proc R Soc B 1778; 281: 20132637.

[105] Chappell CR, Fukami T. Nectar yeasts: a natural microcosm for ecology. Yeast 2018; 35(6): 417-23.
[http://dx.doi.org/10.1002/yea.3311] [PMID: 29476620]

[106] Canto A, Herrera CM, Medrano M, Pérez R, García IM. Pollinator foraging modifies nectar sugar composition in *Helleborus foetidus* (Ranunculaceae):An experimental test. Am J Bot 2008; 95(3): 315-20.
[http://dx.doi.org/10.3732/ajb.95.3.315] [PMID: 21632356]

[107] Vannette RL, Fukami T. Contrasting effects of yeasts and bacteria on floral nectar traits. Ann Bot (Lond) 2018; 121(7): 1343-9.
[http://dx.doi.org/10.1093/aob/mcy032] [PMID: 29562323]

[108] Pozo MI, Mariën T, van Kemenade G, Wäckers F, Jacquemyn H. Effects of pollen and nectar inoculation by yeasts, bacteria or both on bumblebee colony development. Oecologia 2021; 195(3): 689-703.
[http://dx.doi.org/10.1007/s00442-021-04872-4] [PMID: 33582870]

[109] Pozo MI, van Kemenade G, van Oystaeyen A, *et al.* The impact of yeast presence in nectar on bumble bee behavior and fitness. Ecol Monogr 2020; 90(1): e01393.
[http://dx.doi.org/10.1002/ecm.1393]

[110] Hallen-Adams HE, Suhr MJ. Fungi in the healthy human gastrointestinal tract. Virulence 2017; 8(3): 352-8.
[http://dx.doi.org/10.1080/21505594.2016.1247140] [PMID: 27736307]

[111] Huffnagle GB, Noverr MC. The emerging world of the fungal microbiome. Trends Microbiol 2013; 21(7): 334-41.
[http://dx.doi.org/10.1016/j.tim.2013.04.002] [PMID: 23685069]

[112] Inácio J, Daniel HM. Commensalism: the case of the human zymobiome.Yeasts in Natural Ecosystems: Ecology. Cham: Springer International Publishing AG 2017; pp. 211-28.
[http://dx.doi.org/10.1007/978-3-319-61575-2_8]

[113] Schulze J, Sonnenborn U. Yeasts in the gut: from commensals to infectious agents. Dtsch Arztebl Int 2009; 106(51-52): 837-42.
[PMID: 20062581]

[114] Suhr MJ, Hallen-Adams HE. The human gut mycobiome: pitfalls and potentials--a mycologist's perspective. Mycologia 2015; 107(6): 1057-73.
[http://dx.doi.org/10.3852/15-147] [PMID: 26354806]

[115] Urubschurov V, Janczyk P. Biodiversity of yeasts in the gastrointestinal ecosystem with emphasis on its importance for the host.The Dynamical Processes of Biodiversity-Case Studies of Evolution and Spatial Distribution. London: IntechOpen 2011; pp. 277-302.
[http://dx.doi.org/10.5772/24108]

[116] Ianiro G, Bruno G, Lopetuso L, *et al.* Role of yeasts in healthy and impaired gut microbiota: the gut mycome. Curr Pharm Des 2014; 20(28): 4565-9.
[http://dx.doi.org/10.2174/13816128113196660723] [PMID: 24180411]

[117] Hatoum R, Labrie S, Fliss I. Antimicrobial and probiotic properties of yeasts: from fundamental to novel applications. Front Microbiol 2012; 3: 421.
[http://dx.doi.org/10.3389/fmicb.2012.00421] [PMID: 23267352]

[118] Smid EJ, Lacroix C. Microbe-microbe interactions in mixed culture food fermentations. Curr Opin Biotechnol 2013; 24(2): 148-54.
[http://dx.doi.org/10.1016/j.copbio.2012.11.007] [PMID: 23228389]

[119] Sieuwerts S, de Bok FA, Hugenholtz J, van Hylckama Vlieg JE. Unraveling microbial interactions in food fermentations: from classical to genomics approaches. Appl Environ Microbiol 2008; 74(16): 4997-5007.
[http://dx.doi.org/10.1128/AEM.00113-08] [PMID: 18567682]

[120] Roullier-Gall C, David V, Hemmler D, Schmitt-Kopplin P, Alexandre H. Exploring yeast interactions through metabolic profiling. Sci Rep 2020; 10(1): 6073.
[http://dx.doi.org/10.1038/s41598-020-63182-6] [PMID: 32269331]

[121] Hesseltine CW. Mixed-Culture Fermentations.In: National Research Council Applications of Biotechnology in Traditional Fermented Foods Washington, DC, National Academy Press,. 2012; pp. 52-7.

[122] Pettit RK. Mixed fermentation for natural product drug discovery. Appl Microbiol Biotechnol 2009; 83(1): 19-25.
[http://dx.doi.org/10.1007/s00253-009-1916-9] [PMID: 19305992]

[123] Boekhout T. Biodiversity: gut feeling for yeasts. Nature 2005; 434(7032): 449-51.
[http://dx.doi.org/10.1038/434449a] [PMID: 15791239]

[124] Barriga EJC, Libkind D, Briones AI, *et al.* Yeasts biodiversity significance: case studies in natural and human□related environments, ex situ preservation, applications and challenges.Changing Diversity in Changing Environment. Rijeka: IntechOpen 2011; pp. 55-86.
[http://dx.doi.org/10.5772/23906]

[125] Barriuso J, Hogan DA, Keshavarz T, Martínez MJ. Role of quorum sensing and chemical communication in fungal biotechnology and pathogenesis. FEMS Microbiol Rev 2018; 42(5): 627-38.
[http://dx.doi.org/10.1093/femsre/fuy022] [PMID: 29788231]

[126] Jagtap SS, Bedekar AA, Rao CV. Quorum sensing in yeast. In: Dhiman SS, Ed. ACS Symposium Series. 235-50.
[http://dx.doi.org/10.1021/bk-2020-1374.ch013]

[127] Sprague GF Jr, Winans SC. Eukaryotes learn how to count: quorum sensing by yeast. Genes Dev 2006; 20(9): 1045-9.
[http://dx.doi.org/10.1101/gad.1432906] [PMID: 16651650]

[128] Kuske CR, Hesse CN, Challacombe JF, *et al.* Prospects and challenges for fungal metatranscriptomics of complex communities. Fungal Ecol 2015; 14: 133-7.
[http://dx.doi.org/10.1016/j.funeco.2014.12.005]

[129] Gallone B, Steensels J, Prahl T, *et al.* Domestication and divergence of *Saccharomyces cerevisiae* beer yeasts. Cell 2016; 166(6): 1397-1410.e16.
[http://dx.doi.org/10.1016/j.cell.2016.08.020] [PMID: 27610566]

[130] Peter J, De Chiara M, Friedrich A, *et al.* Genome evolution across 1,011 *Saccharomyces cerevisiae* isolates. Nature 2018; 556(7701): 339-44.
[http://dx.doi.org/10.1038/s41586-018-0030-5] [PMID: 29643504]

[131] Jeffares DC, Rallis C, Rieux A, *et al.* The genomic and phenotypic diversity of *Schizosaccharomyces pombe.* Nat Genet 2015; 47(3): 235-41.
[http://dx.doi.org/10.1038/ng.3215] [PMID: 25665008]

[132] Álvarez-Pérez S, Dhami MK, Pozo MI, *et al.* Genetic admixture increases phenotypic diversity in the nectar yeast *Metschnikowia reukaufii.* Fungal Ecol 2021; 49: 101016.
[http://dx.doi.org/10.1016/j.funeco.2020.101016]

CHAPTER 3

Yeast Taxonomy

J. Alfredo Hernández-García[1, 2], Esaú De-la-Vega-Camarillo[2], Lourdes Villa-Tanaca[2] and César Hernández-Rodríguez[2,*]

[1] *Universidad Autónoma de Nuevo León, Facultad de Ciencias Forestales, Departamento de Silvicultura, Carretera Nacional No. 85, Km. 145, Linares, Nuevo León C.P. 67700, México*

[2] *Instituto Politécnico Nacional, Escuela Nacional de Ciencias Biológicas, Departamento de Microbiología, Ciudad de México C.P. 11340, México*

Abstract: The massive parallel sequencing technology, applied to the taxonomy of microorganisms, has been affecting the traditional phenotypic and molecular phylogenies based on the sequence of a single gene or a small handful of genes. The exponential accumulation of new, entire genome sequences of microorganisms in public databases in recent years, especially in the fields of taxonomic and biotechnology, is driving a conceptual revolution in the way of understanding the concepts of species in microorganisms in general and fungi in particular. The problems of drawing species boundaries, reclassification of species, discovering new taxa and clades, recognizing synonyms, and new species for science can now be addressed with genomic approaches. Derived from all this, much more robust high-resolution phylogenies, based on core genomes or broad collections of genes and their deduced proteins, are currently being reconstructed. Although this effort is still far from being a canon in the taxonomy of yeasts, it will gradually turn into a change and challenge that researchers are taking into account due to the great power and reliability of these genomic approaches and bioinformatics tools. Likewise, the complete sequence of the genomes of the strains of microorganisms of industrial or biotechnological interest will allow limiting biopiracy, help protect patents, recognize the appellation of origin, discourage violations of intellectual property rights, and resolve conflicts over the rights of the commercial exploitation of microorganisms. In this chapter, an effort is made to compare conventional taxonomy techniques with the latest work involving genomic sciences as a key tool in yeast taxonomy.

Keywords: Bioinfomatics, Orthologues, Phylogenomics, Yeast, Yeast Taxonomy.

* **Corresponding author César Hernández-Rodríguez:** Instituto Politécnico Nacional, Escuela Nacional de Ciencias Biológicas, Departamento de Microbiología, Ciudad de México C.P. 11340, México; Tel: +52 5557296000; ext: 62554; E-mail: chdez38@hotmail.com

Sérgio Luiz Alves Júnior, Helen Treichel, Thiago Olitta Basso and Boris Ugarte Stambuk (Eds.)
All rights reserved-© 2022 Bentham Science Publishers

INTRODUCTION

Phenotypic Taxonomy of Yeast

Formerly, the techniques used in yeast taxonomy were mainly based on phenotypic traits and physiological characteristics [1]. The identification and description of new species depended on the comparison of morphological features as well as on biochemical and physiological profiles with previously described species [2]. Fermentation and/or assimilation of several sugars, organic acids, alcohols, sugar alcohols, starch, and different nitrogen sources, as well as growth differences, were used for both identification and formal description of new species. Gradually, new biochemical tests, such as polysaccharide composition assays of the cell wall and capsule, mycocin susceptibility, and electrophoretic comparisons of enzymes, are being used to determine subtle differences among closely related strains [3]. However, these approaches require large numbers of tests, chemical standards and substrates, substantial equipment, sensitive techniques, and type strains to identify an isolate or describe a new species; such studies were generally only performed by specialists in yeast taxonomy. Consequently, only a few laboratories in the world had the capacity to fully identify or describe new yeast species. The time-consuming phenotypic characterization of pure cultures and the formal description of the new species became a bottleneck for non-specialist scientists, who resigned themselves to reporting their isolates and strains only at the genus level. In retrospect, we can see that many new yeasts of industrial or ecological interest that were reported before the mid-90s were only characterized to the genus level due to difficulties in analyzing all evidence for proper identification and because of the low number of species formally described.

However, as more strains of each species were isolated and phenotypically characterized, it became evident that a great phenotypic intraspecific diversity was universal to many yeast species, which complicated their adequate identification, species limits, and species concepts [4, 5].

MOLECULAR TAXONOMY OF YEAST

At the global level, the importance of the identification of yeasts with the use of molecular methods is remarkable. Genetic characterization of ascomycete yeasts began with the determination of DNA base composition, traditionally expressed as CG content, which could provide information on the dissimilarity of two organisms but not on their similarity since two organisms that are not phylogenetically related often have a similar CG content [6]. Among yeasts, a ratio between 28 and 50% GC was calculated, whereas, in basidiomycete yeasts, it ranged between 50 and 70% GC. The wide margins of these intervals were insufficient and useless as elements for the identification of species [7].

Later, genetic relatedness using nuclear DNA-DNA reassociation or hybridization of DNA-DNA techniques impacted the taxonomy of fungi in general and that of yeasts in particular because it revealed a clearer picture of similarity and allowed to recognize synonymy of many species, providing a quantitative value as a percentage of the differences [8]. Although DNA-DNA hybridization is a laborious method, the technique had a marked impact on yeast recognition, even if it did not solve the issue of the genetic differentiation of closely related species [9]. An arbitrary minimum cutoff value of 70% of the DNA association of two species was considered sufficient to recognize two strains as belonging to the same species [10]. The DNA-DNA reassociation values were subsequently related to the percentage of similarity of the bacterial 16S rRNA gene. Thus, a DNA-DNA reassociation value of 70% was equivalent to a 97% percent similarity of the 16S rRNA gene [11]. In recent years, an alternative measurement has been adopted: the mean nucleotide identity (ANI), which is calculated by computational comparison of two sequences of complete genomes. An ANI percentage of 95–96% generally corresponds to the threshold of 70% in DNA-DNA hybridization [12, 13].

However, it was not until the advent of sequencing gene and non-coding DNA fragments that molecular biology methods became important due to their ease of performance, economy and universality. Gene sequencing was a relatively easy, quick and powerful method of identifying until species level many microbial groups by phylogenetic reconstruction and nucleotide similitude supported by bioinformatics software tools. As seen in Fig. (**1C**), the yeast ribosomal regions offer several possibilities for sequencing, but domain 2 of large subunit 26S ribosomal RNA (26S r RNA or 26S/28S LSU) was initially used as a molecular marker because it apparently contained sufficiently variable information to distinguish between closely related species [14]. However, the sequencing of both D1 and D2 domains (D1/D2) of 26S rRNA yeast genes (~600 bp) quickly became a popular tool for Ascomycota yeast identification, and the sequence databases suddenly became enriched.

Fig. (1). Phylogeny of ITS1 region of yeasts of biotechnological importance. **A)** A phylogeny constructed from partial sequences of the ITS1 region is shown, using HKY as a nucleotide substitution model, maximum-likelihood as a phylogenetic reconstruction method and 1000 bootstraps. **B)** A fragment of the ITS sequences is illustrated where the conservation and variability of the sequences between strains are observed. **C)** Model of the ordering of regions commonly used for the identification of yeasts and primers commonly used for these regions.

However, when the limitations and lack of depth of the D1/D2 fragments to distinguish between related species, reconstruct phylogenies and recognize new yeast became evident, the internal transcribed spacer (ITS) was introduced as a universal barcode. As can be observed in Fig. (**1B**), the ITS region is frequently shorter than D1/D2 fragment but much more variable, fix mutations more frequently and evolve faster than the 26S rRNA region. Due to these characteristics, the ITS sequence can distinguish between closely related species and even find differences below the species level.

In general terms, isolates with <1% nucleotide differences in D1/D2 or ITS region in aligned sequences can be considered to belong to the same species. But isolates with more than 1% of dissimilitude can be considered as different or probable new species. In this way, the impact of molecular identification and growing D1/D2 and ITS databases permitted the recognizing of a large number of new species principally among environmental isolates [15]. The ribosomal sequences of all known yeasts provided a broad public database for rapid identification, the development of a more suitable taxonomy, and a molecular phylogeny capable of including thousands of strains or species. Both ribosomal regions are used as barcodes for the identification of cultured yeasts or metagenomic DNA from environmental samples [16]. Also, this information turned in an obligate reference to the description of new species and was valuable to propose also a universal phylogenetic taxonomy of yeast [17].

Despite the great advance represented by the use of ribosomal sequences and their gene databases for yeast identification and taxonomy, some limitations began to be recognized. For example, the resolution power of identification based on a single ribosomal gene frequently did not reach the species level with information available in databases. Many nucleotide sequences only allowed certainties at the genus and even family level, although phenotypic differences and/or sequences of other genes were always available to complement the identifications, recognize new species, or solve the limitations of the identification based only on ribosomal sequences.

As seen in Fig. (**1A**), sequencing ITS regions is a good starting point for identifying and constructing phylogenies of fungal species, especially for the identification of completely unknown fungal species [18]. However, ITS primers may produce amplification biases during the Polymerase Chain Reaction (PCR), especially in environmental or metagenomic samples containing mixed templates [19]. Recently, some doubts have arisen about with respect to the use of ITS as the only marker for the delimitation of species in the context of the immense diversity of fungi species that has begun to be detected in the environment [20].

Generally, the ITS region (ITS1, 5S tRNA and ITS) lengths are similar only within a genus. Starting from the family level, the length of the ITS varies very significantly. Thus, for example, among the yeasts of the Saccharomycotina subphylum, the ITS varies between 250 and 980 nucleotides [21]. These huge differences do not affect their use for the identification of most of the yeast at genus and species levels via barcode or similarity. However, they greatly complicate multiple alignments and the derived phylogenies that include members of the taxa of several families or superior taxa. Another problem is that the genotyping of individual strains is also complex using only D1/D2 and ITS regions. Due to the high degree of conservation of ribosomal sequences, characterization of the strains within a species or genotyping of strains is not possible with these molecular markers. Likewise, occasionally, some strains or isolates with a degree of similarity greater than 99% exhibit different phenotypic traits and low percentages of DNA-DNA hybridization [22, 23].

In an attempt to include a greater amount of molecular information to improve genotyping, phylogeny, and diversity studies, the sequences of a set of protein-coding genes were added to the ribosomal information. Over time, various markers have been developed for use in the phylogenetic reconstruction of filamentous and yeast fungi, including several single-copy genes: RNA polymerase II, mitochondrial cytochrome C oxidase, b-tubulin, elongation factor 1 gamma, and actin genes [24 - 27]. However, these genes does not evolve similarly and often provides incongruent phylogenies when they are used separately, and consequently, multigenic analyses are now the preferred choice for the establishment of robust phylogenies [28]. The use of large subunit (LSU) rRNA, SSU rRNA, translation elongation factor 1-alpha (EF-1a), and RNA polymerase II subunit 1 (RPB1) and subunit 2 (RPB2) genes has been proposed to taxonomically classify ascomycetous yeasts [29].

Sequences of additional protein-coding genes were then proposed and adopted for multilocus sequence typing (MLST) analysis. Among yeasts, MLST has been broadly used to study intraspecific diversity, population genetics, and epidemiology of opportunistic pathogens of the *Candida* genus [30 - 32]. Also, MLST of *Saccharomyces cerevisiae* strains from human and natural fermentations have facilitated the recognition of strains associated with the alcoholic fermentation of grape and sake wine from groups that reflect independent "domestication" phenomena [33].

A highly frequent taxonomic problem among yeasts is that many genera were described only with phenotypic criteria and had two separate names for their anamorphic (asexual) and teleomorphic (sexual) states. The arrival of molecular methods allowed to recognize that many genera previously described were polyphyletic anamorphic genera. Likewise, the two morphological states of a species began to be recognized, and their names unified as the same biological entity under the idea of "One fungus = One name" of The Amsterdam Declaration of Fungal Nomenclature [34]. The molecular phylogenies are the guide to update and reassign polyphyletic genera to more proper natural groups, which is a taxonomic task that continues to build up a new classification system [35]. In this way, *Candida, Pichia, Cryptococcus, Rhodotorula, Bullera, Sporobolomyces, Spathaspora,* and *Clavispora,* among others polyphyletic genera, have been relocated to previously described and new genera or species [36 - 39].

In summary, the rapid detection and accurate identification of yeasts are now possible through the use of a variety of molecular methods. The application of these methods will bring a greater degree of clarity to studies in yeast ecology, which previously was not possible when yeasts were identified based on the phenotype. An integral approach to collect more robust data, such as detailed physiological characterization, mating experiments, MLS analyses, and whole-genome comparisons, will be helpful for a reliable taxonomy [40].

GENOMIC AND PHYLOGENOMIC APPROACHES

In recent years, numerous genomes of microorganisms, including fungi, have been deposited in public databases. Sequencing an average yeast genome is now available to everyone, including small mycological laboratories. The genome-sequencing projects that have been uploaded present different degrees of advancement and annotation; therefore, they have to be refined if they are to be used to carry out identifications and reconstruct phylogenies. The information of coding and non-coding sequences of genomes available at this time is spectacular and is accumulating exponentially. Although some recent efforts have attempted to clarify the general principles for the use of genomics in the classification of fungi [41], the applications for the convenient use and analysis of complete genomes are still scarce and are hampered by the lack of clear and simple objectives on the methods of comparison and the criteria for the interpretation of the results [42 - 44]. Undoubtedly, beyond these methodological issues, the genomic approach to biology will have an important impact on the taxonomy and evolution study of yeasts and fungi in general [45]. The ability to compare species at the genomic level has significantly enhanced our understanding of fungal taxonomy, evolution, physiology, and cell biology and allowed us to trace the origin of genes within the fungal kingdom.

The *Saccharomyces cerevisiae* genome was the first fungal genome sequenced [46], a collaborative work that enabled twenty years of innovation and discovery. After nearly a decade, *Candida albicans* soon followed [47], as well as a handful of relatives in both clades and a smattering of taxonomically diverse species [48]. *S. cerevisiae* became a powerful biologic model due to proving ground for new genomic technologies, such as deleting and barcoding genes for functional profiling [49].

The firsts comparative analysis pioneered the now commonplace use of genome sequencing to address specific evolutionary and functional hypotheses, such as using conserved genomic regions to identify functional DNA sequence elements [50]. These genomes began to catalyze research in many other fields, resulting in important advances to our understanding of the evolution of genome content and organismal traits, phylogenetics, and cis-regulatory element prediction [51 - 53]. The evolution of genomes of 26 yeast species allowed the discovery of an important set of evolutionary events such as the duplication of genes, horizontal gene acquisition, introgression, interspecific hybridization, delayed karyogamy, loss of heterozygosity, and de novo gene formation that can be recognized in the yeast genome [54]. With the popularization of next-generation sequencing, the genomes of 80 additional species were published in the last 10 years, while 20 more are available but await formal publications. This implies a great ignorance even of the diversity of yeasts estimated at the genomic level.

In this context, Fig. (**2**) shows a phylogenomic tree containing a total of 75 Saccharomycotina yeast genomes. The genome sequences were downloaded from the NCBI database (https://www.ncbi.nlm.nih.gov/genome/browse#!/overview/), and core proteins were obtained with the M1CR0B1AL1Z3R web server (https://microbializer.tau.ac.il/, accessed on 1 May 2021), using the following cut-offs: 0.01 maximum e-value, 80.0% identity minimal percent, and 100.0% minimum percentage for the core. Core-proteome phylogenetic construction was conducted with a maximum-likelihood algorithm using RAxML [55] based on the inferred core-proteome alignment. Phylogenetic tree visualization was done using FigTree v1.4.4 [56].

Although the members of the Saccharomycotina subphylum are characterized by representing a polyphyletic taxonomic group, the formation of the clades was highly conserved (Fig. **2**). The results obtained through the phylogenomic analysis of the orthologous genes extracted from the complete genomes showed the recovery of the clades previously reported using five molecular markers [57, 58]. We could not recover clade 7 because the genome of *Alloascoidea africana* has not yet been revealed, the only member within this clade. This suggests that for taxonomic identification, at least for yeasts belonging to the subphylum Saccharomycotina, it is sufficient to perform a phylogeny with the five concatenated genes mentioned above.

MAIN TAXONOMIC GROUPS WITH POTENTIAL FOR BIOTECHNOLOGICAL APPLICATION

Because of their metabolic diversity, yeasts colonize a remarkable range of habitats and occupy a wide variety of ecological niches. For example, *Ogataea polymorpha* and *Komagataella pastoris* consume methanol as carbon source, *Scheffersomyces stipitis* and *Spathaspora passalidarum* ferment xylose, *Yarrowia lipolytica* and *Lipomyces starkeyi* produce lipidic acids, *C. albicans,* and many other non-*C. albicans* are important human pathogens, and *Eremothecium gossypii* are cotton pathogens, among many other important species [59]. Most of these metabolic traits are exploited in various biotechnological, food, and beverage industries. For example, *S. cerevisiae* is the workhorse of the multibillion-dollar brewing, wine-making, baking, and biofuel industries, but other commonly used species of economic importance include *Kazachstania exigua* (sourdough), *Cyberlindnera jadinii* (food additives), *K. pastoris* (heterologous protein production), and *E. gossypii* (riboflavin).

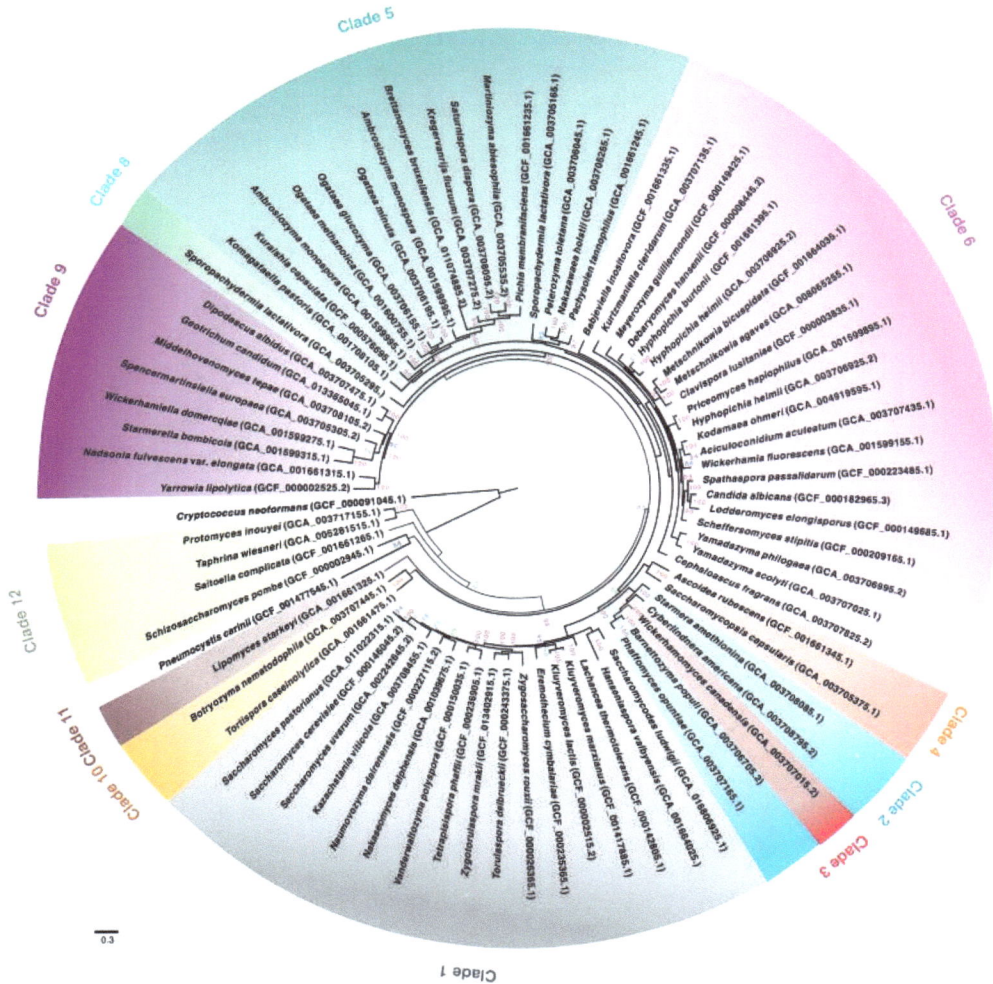

Fig. (2). Phylogenetic relationships among type species of ascomycete yeast of industrial importance and reference taxa determined from maximum likelihood analysis of the complete genome. The basidiomycete *Filobasidiella* (*Cryptococcus*) *neoformans* was the designated outgroup species in the analysis. Bootstrap values (1000 replicates) >50% are given at branch nodes. Genomic data were obtained from the NCBI database based on analyses previously reported by Kurtzman & Robnett [58].

White or industrial yeast biotechnology covers a wide range of research areas such as agriculture, food, medicine and pharmacology, the environment, and climate change. The collection and analysis of data from the fungal genome are necessary to know the biology and behavior of genes of both primary and secondary metabolites [60]. For example, molecular prediction and dynamics between products that are involved in antibiosis, an increase of productivity in crops, interaction networks, and microbe communication, gene regulation of markers, and genome editing can help to improve the bioproduction of molecules of commercial interest; this information can be known and studied from the data obtained through the sequencing techniques of complete genomes which allow us to reconstruct metabolic pathways and their interaction as shown in Fig. (**3**).

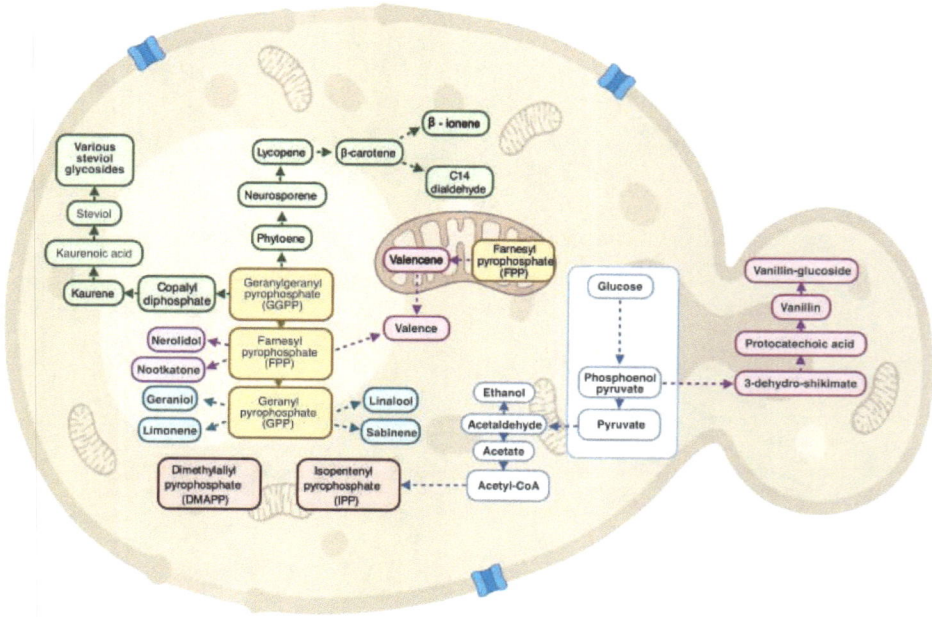

Fig. (3). Simplified central metabolism of *Saccharomyces cerevisiae* from genomic data. The figure illustrates the main bio-processes of biotechnological interest, such as the production of glycosides of food interest, ethanol, among other compounds, information which is of paramount importance for optimizing industrial and technological processes and which can be obtained by genome analysis, data which can be supplemented by laboratory experimentation.

The current yeast taxonomy classification is more and more concordant with Saccharomycotina phylogenomic taxonomies and evolutionary divergences [61]. Finally, there is a need to increase the resolution of yeast phylogeny, and continued efforts are required to sample genomes that encompass the greatest diversity of yeasts of biotechnological interest.

CONCLUSION

Actual access to complete genomes has a relevant impact on taxonomy, phylogeny, evolution, identification, and reclassification of fungal species. Also, the analysis of the genetic heritage of a species allows access to valuable information that will improve the processes of biotechnological importance based on filamentous and yeast.

CONSENT FOR PUBLICATION

Not applicable.

CONFLICT OF INTEREST

The authors declare no conflict of interest, financial or otherwise.

ACKNOWLEDGEMENT

Declare none.

REFERENCES

[1]   Kurtzman CP, Fell JW, Boekhout T. Methods for isolation, phenotypic characterization and maintenance of yeasts. In: CP Kurtzman, JW Fell, T Boekhout, Eds. The yeast: A taxonomic study. 5th. Elsevier 2011; pp. 87-110.
        [http://dx.doi.org/10.1016/B978-0-444-52149-1.00007-0]

[2]   Kurtzman CP, Fell JW, Boekhout T. Gene sequence analyses and other DNA-based methods for yeast species recognition. In: Kurtzman CP, Fell JW, Boekhout T, Eds. The yeast: A taxonomic study. 5th. Elsevier 2010; pp. 137-44.

[3]   Kurtzman CP. Discussion of teleomorphic and anamorphic ascomycetous yeasts and yeast-like taxa. In: CP Kurtzman, JW Fell, T Boekhout, Eds. The yeast: A taxonomic study. 5th. Elsevier 2011; pp. 292-307.
        [http://dx.doi.org/10.1016/B978-0-444-52149-1.00013-6]

[4]   Kurtzman CP, Robnett CJ, Basehoar E, Ward TJ. Four new species of *Metschnikowia* and the transfer of seven *Candida* species to *Metschnikowia* and *Clavispora* as new combinations. Antonie van Leeuwenhoek 2018; 111(11): 2017-35.
        [http://dx.doi.org/10.1007/s10482-018-1095-8] [PMID: 29754318]

[5]   Rambaut A. FigTree v1. 3.1. Institute of Evolutionary Biology, University of Edinburgh, Edinburgg. http://tree. bio. ed ac. uk/software/figtree/.

[6]   Barbu E, Lee KY, Wahl R. Contenu en bases puriques et pyrimidiques des acides désoxyribonucléiques des bactéries. Ann Inst Pasteur (Paris) 1956; 91(2): 212-24.
        [PMID: 13363015]

[7]   Kurtzman CP, Fell JW. Molecular relatedness between the basidiomycetous yeasts *Sporidiobolus ruinenii* and *Sporobolomyces coprophilus.* Mycologia 1991; 83: 107-10.
        [http://dx.doi.org/10.1080/00275514.1991.12025984]

[8]   Kurtzman CP. DNA-DNA hybridization approaches to species identification in small genome organisms. Methods Enzymol. In: Ayala FJ, Ed. Molecular Evolution. 1993; 224: pp. 335-48.
        [http://dx.doi.org/10.1016/0076-6879(93)24025-P] [PMID: 8264397]

[9]   Dobzhansky T. Organismic and molecular aspects of species formation.Sunderland: Sinauer Associates 1976; pp. 95-105.

[10]   Kurtzman C, Fell JW, Boekhout T, Eds. The yeasts: A taxonomic study. 5th ed., London, England: Elsevier Science 2011.

[11]   Sepúlveda VE, Márquez R, Turissini DA, Goldman WE, Matute DR. Genome sequences reveal cryptic speciation in the human pathogen *Histoplasma capsulatum.* MBio 2017; 8(6): e01339-17.
        [http://dx.doi.org/10.1128/mBio.01339-17] [PMID: 29208741]

[12]   Goris J, Konstantinidis KT, Klappenbach JA, Coenye T, Vandamme P, Tiedje JM. DNA-DNA

hybridization values and their relationship to whole-genome sequence similarities. Int J Syst Evol Microbiol 2007; 57(Pt 1): 81-91.
[http://dx.doi.org/10.1099/ijs.0.64483-0] [PMID: 17220447]

[13]   Kim M, Oh HS, Park SC, Chun J. Towards a taxonomic coherence between average nucleotide identity and 16S rRNA gene sequence similarity for species demarcation of prokaryotes. Int J Syst Evol Microbiol 2014; 64(Pt 2): 346-51.
[http://dx.doi.org/10.1099/ijs.0.059774-0] [PMID: 24505072]

[14]   Pärtel K, Baral HO, Tamm H, Põldmaa K. Evidence for the polyphyly of *Encoelia* and Encoelioideae with reconsideration of respective families in Leotiomycetes. Fungal Divers 2017; 82: 183-219.
[http://dx.doi.org/10.1007/s13225-016-0370-0]

[15]   Kurtzman CP, Fell JW. The Yeasts, A Taxonomic Study. 4th ed., Amsterdam: Elsevier 1998.

[16]   Kurtzman CP. *Candida shehatae*-genetic diversity and phylogenetic relationships with other xylose-fermenting yeasts. Antonie van Leeuwenhoek 1990; 57(4): 215-22.
[http://dx.doi.org/10.1007/BF00400153] [PMID: 2353807]

[17]   Kurtzman CP, Robnett CJ. Identification of clinically important ascomycetous yeasts based on nucleotide divergence in the 5′ end of the large-subunit (26S) ribosomal DNA gene. J Clin Microbiol 1997; 35(5): 1216-23.
[http://dx.doi.org/10.1128/jcm.35.5.1216-1223.1997] [PMID: 9114410]

[18]   Salichos L, Rokas A. Inferring ancient divergences requires genes with strong phylogenetic signals. Nature 2013; 497(7449): 327-31.
[http://dx.doi.org/10.1038/nature12130] [PMID: 23657258]

[19]   Bellemain E, Carlsen T, Brochmann C, Coissac E, Taberlet P, Kauserud H. ITS as an environmental DNA barcode for fungi: an *in silico* approach reveals potential PCR biases. BMC Microbiol 2010; 10: 189.
[http://dx.doi.org/10.1186/1471-2180-10-189] [PMID: 20618939]

[20]   Tavanti A, Gow NA, Senesi S, Maiden MCJ, Odds FC. Optimization and validation of multilocus sequence typing for *Candida albicans*. J Clin Microbiol 2003; 41(8): 3765-76.
[http://dx.doi.org/10.1128/JCM.41.8.3765-3776.2003] [PMID: 12904388]

[21]   Kurtzman CP. Molecular taxonomy of the fungi. In: Samson RA, JI Pitt, Eds. Advances in *Penicillum* and *Aspergillus* systematics. Orlando, FL: Academic Press, Inc 1985.
[http://dx.doi.org/10.1016/B978-0-12-088640-1.50008-2]

[22]   Kurtzman CP. Classification of fungi through nucleic acid relatedness. Boston, MA: Springer 1986; pp. 233-54.
[http://dx.doi.org/10.1007/978-1-4757-1856-0_21]

[23]   Samson RA, Houbraken J, Thrane U, Frisvad JC, Andersen B. In Food and Indoor Fungi. Westerdijk Fungal Biodiversity Institute Westerdijk Laboratory Manual Series 2019; p. 2.

[24]   Kwon-Chung KJ, Varma A. Do major species concepts support one, two or more species within *Cryptococcus neoformans*? FEMS Yeast Res 2006; 6(4): 574-87.
[http://dx.doi.org/10.1111/j.1567-1364.2006.00088.x] [PMID: 16696653]

[25]   Belloch C, Querol A, García MD, Barrio E. Phylogeny of the genus *Kluyveromyces* inferred from the mitochondrial cytochrome-c oxidase II gene. Int J Syst Evol Microbiol 2000; 50(Pt 1): 405-16.
[http://dx.doi.org/10.1099/00207713-50-1-405] [PMID: 10826829]

[26]   Daniel HM, Sorrell TC, Meyer W. Partial sequence analysis of the actin gene and its potential for studying the phylogeny of *Candida* species and their teleomorphs. Int J Syst Evol Microbiol 2001; 51(Pt 4): 1593-606.
[http://dx.doi.org/10.1099/00207713-51-4-1593] [PMID: 11491363]

[27]   Daniel HM, Meyer W. Evaluation of ribosomal RNA and actin gene sequences for the identification of ascomycetous yeasts. Int J Food Microbiol 2003; 86(1-2): 61-78.

[http://dx.doi.org/10.1016/S0168-1605(03)00248-4] [PMID: 12892922]

[28]   Kurtzman CP, Robnett CJ. Identification and phylogeny of ascomycetous yeasts from analysis of nuclear large subunit (26S) ribosomal DNA partial sequences. Antonie van Leeuwenhoek 1998; 73(4): 331-71.
[http://dx.doi.org/10.1023/A:1001761008817] [PMID: 9850420]

[29]   Kurtzman CP, Robnett CJ. Phylogenetic relationships among yeasts of the 'Saccharomyces complex' determined from multigene sequence analyses. FEMS Yeast Res 2003; 3(4): 417-32.
[http://dx.doi.org/10.1016/S1567-1356(03)00012-6] [PMID: 12748053]

[30]   Dodgson AR, Pujol C, Denning DW, Soll DR, Fox AJ. Multilocus sequence typing of *Candida glabrata* reveals geographically enriched clades. J Clin Microbiol 2003; 41(12): 5709-17.
[http://dx.doi.org/10.1128/JCM.41.12.5709-5717.2003] [PMID: 14662965]

[31]   Takashima M, Suh-Oui S, Feng-Yan B, Sugita T. Nakase's last tweet: what is the current direction of microbial taxonomy research? FEMS Yeast Res 2019; 19(8): foz066.
[http://dx.doi.org/10.1093/femsyr/foz066]

[32]   Jacobsen MD, Davidson AD, Li SY, Shaw DJ, Gow NA, Odds FC. Molecular phylogenetic analysis of *Candida tropicalis* isolates by multi-locus sequence typing. Fungal Genet Biol 2008; 45(6): 1040-2.
[http://dx.doi.org/10.1016/j.fgb.2008.03.011] [PMID: 18440253]

[33]   Fay JC, Benavides JA. Evidence for domesticated and wild populations of *Saccharomyces cerevisiae*. PLoS Genet 2005; 1(1): 66-71.
[http://dx.doi.org/10.1371/journal.pgen.0010005] [PMID: 16103919]

[34]   Hawksworth DL, Crous PW, Redhead SA, *et al.* The Amsterdam declaration on fungal nomenclature. IMA Fungus 2011; 2(1): 105-12.
[http://dx.doi.org/10.5598/imafungus.2011.02.01.14] [PMID: 22679594]

[35]   Kurtzman CP, Mateo RQ, Kolecka A, Theelen B, Robert V, Boekhout T. Advances in yeast systematics and phylogeny and their use as predictors of biotechnologically important metabolic pathways. FEMS Yeast Res 2015; 15(6): fov050.
[http://dx.doi.org/10.1093/femsyr/fov050]

[36]   Guzmán B, Lachance MA, Herrera CM. Phylogenetic analysis of the angiosperm-floricolous insect-yeast association: have yeast and angiosperm lineages co-diversified? Mol Phylogenet Evol 2013; 68(2): 161-75.
[http://dx.doi.org/10.1016/j.ympev.2013.04.003] [PMID: 23583418]

[37]   Peterson SW, Kurtzman CP. Ribosomal RNA sequence divergence among sibling species of yeasts. Syst Appl Microbiol 1991; 14(2): 124-9.
[http://dx.doi.org/10.1016/S0723-2020(11)80289-4]

[38]   Stamatakis A. RAxML version 8: a tool for phylogenetic analysis and post-analysis of large phylogenies. Bioinformatics 2014; 30(9): 1312-3.
[http://dx.doi.org/10.1093/bioinformatics/btu033] [PMID: 24451623]

[39]   Liu YJ, Whelen S, Hall BD. Phylogenetic relationships among ascomycetes: evidence from an RNA polymerse II subunit. Mol Biol Evol 1999; 16(12): 1799-808.
[http://dx.doi.org/10.1093/oxfordjournals.molbev.a026092] [PMID: 10605121]

[40]   Nguyen HDT, Sultana T, Kesanakurti P, Hambleton S. Genome sequencing and comparison of five *Tilletia* species to identify candidate genes for the detection of regulated species infecting wheat. IMA Fungus 2019; 10: 11.
[http://dx.doi.org/10.1186/s43008-019-0011-9] [PMID: 32355611]

[41]   MacIsaac KD, Wang T, Gordon DB, Gifford DK, Stormo GD, Fraenkel E. An improved map of conserved regulatory sites for *Saccharomyces cerevisiae*. BMC Bioinformatics 2006; 7(1): 113.
[http://dx.doi.org/10.1186/1471-2105-7-113] [PMID: 16522208]

[42]   Gostinčar C, Ohm RA, Kogej T, *et al.* Genome sequencing of four *Aureobasidium pullulans* varieties:

biotechnological potential, stress tolerance, and description of new species. BMC Genomics 2014; 15(1): 549.
[http://dx.doi.org/10.1186/1471-2164-15-549] [PMID: 24984952]

[43]   Matute DR, Sepúlveda VE. Fungal species boundaries in the genomics era. Fungal Genet Biol 2019; 131: 103249.
[http://dx.doi.org/10.1016/j.fgb.2019.103249] [PMID: 31279976]

[44]   Schildkraut CL, Marmur J, Doty P. The formation of hybrid DNA molecules and their use in studies of DNA homologies. J Mol Biol 1961; 3(5): 595-617.
[http://dx.doi.org/10.1016/S0022-2836(61)80024-7] [PMID: 14498380]

[45]   Casaregola S, Weiss S, Morel G. New perspectives in hemiascomycetous yeast taxonomy. C R Biol 2011; 334(8-9): 590-8.
[http://dx.doi.org/10.1016/j.crvi.2011.05.006] [PMID: 21819939]

[46]   Goffeau A, Barrell BG, Bussey H, *et al.* Life with 6000 genes. Science 1996; 274(5287): 546-567, 563-567.
[http://dx.doi.org/10.1126/science.274.5287.546] [PMID: 8849441]

[47]   Jones T, Federspiel NA, Chibana H, *et al.* The diploid genome sequence of *Candida albicans.* Proc Natl Acad Sci USA 2004; 101(19): 7329-34.
[http://dx.doi.org/10.1073/pnas.0401648101] [PMID: 15123810]

[48]   Cliften P, Sudarsanam P, Desikan A, *et al.* Finding functional features in *Saccharomyces* genomes by phylogenetic footprinting. Science 2003; 301(5629): 71-6.
[http://dx.doi.org/10.1126/science.1084337] [PMID: 12775844]

[49]   Giaever G, Chu AM, Ni L, Connelly C, Riles L, Véronneau S, *et al.* Functional profiling of the *Saccharomyces cerevisiae* genome. nature. 2002; 418(6896): 387-91.

[50]   Kellis M, Patterson N, Endrizzi M, Birren B, Lander ES. Sequencing and comparison of yeast species to identify genes and regulatory elements. Nature 2003; 423(6937): 241-54.
[http://dx.doi.org/10.1038/nature01644] [PMID: 12748633]

[51]   Hittinger CT, Rokas A, Carroll SB. Parallel inactivation of multiple GAL pathway genes and ecological diversification in yeasts. Proc Natl Acad Sci USA 2004; 101(39): 14144-9.
[http://dx.doi.org/10.1073/pnas.0404319101] [PMID: 15381776]

[52]   Lv SL, Chai CY, Wang Y, Yan ZL, Hui FL. Five new additions to the genus *Spathaspora* (Saccharomycetales, Debaryomycetaceae) from southwest China. MycoKeys 2020; 75: 31-49.
[http://dx.doi.org/10.3897/mycokeys.75.57192] [PMID: 33223920]

[53]   Rivera FN, González E, Gómez Z, *et al.* Gut-associated yeast in bark beetles of the genus *Dendroctonus* Erichson (Coleoptera: Curculionidae: Scolytinae). 2009. Biol J Linn Soc Lond 2009; 98(2): 325-42.
[http://dx.doi.org/10.1111/j.1095-8312.2009.01289.x]

[54]   Dujon B. Yeast evolutionary genomics. Nat Rev Genet 2010; 11(7): 512-24.
[http://dx.doi.org/10.1038/nrg2811] [PMID: 20559329]

[55]   Stackebrandt E, Goebel BM. Taxonomic note: a place for DNA-DNA reassociation and 16S rRNA sequence analysis in the present species definition in bacteriology. Int J Syst Evol Microbiol 1994; 44(4): 846-9.
[http://dx.doi.org/10.1099/00207713-44-4-846]

[56]   Price CW, Fuson GB, Phaff HJ. Genome comparison in yeast systematics: delimitation of species within the genera *Schwanniomyces, Saccharomyces, Debaryomyces*, and *Pichia.* Microbiol Rev 1978; 42(1): 161-93.
[http://dx.doi.org/10.1128/mr.42.1.161-193.1978] [PMID: 379571]

[57]   Kurtzman CP, Robnett CJ. Relationships among genera of the Saccharomycotina (Ascomycota) from multigene phylogenetic analysis of type species. FEMS Yeast Res 2013; 13(1): 23-33.

[http://dx.doi.org/10.1111/1567-1364.12006] [PMID: 22978764]

[58] Kurtzman CP, Robnett CJ, Yarrow D. Two new anamorphic yeasts: *Candida germanica* and *Candida neerlandica.* Antonie van Leeuwenhoek 2001; 80(1): 77-83.
[http://dx.doi.org/10.1023/A:1012218122038] [PMID: 11761369]

[59] Kurtzman CP. Prediction of biological relatedness among yeasts from comparisons of nuclear DNA complementarity. Stud Mycol 1987; 30: 459-68.

[60] Vrålstad T. ITS, OTUs and beyond—fungal hyperdiversity calls for supplementary solutions. Mol Ecol 2011; 20(14): 2873-5.
[http://dx.doi.org/10.1111/j.1365-294X.2011.05149.x] [PMID: 21861300]

[61] Wayne LG, Brenner DJ, Colwell RR, Grimont PAD, Kandler O, Krichevsky MI, *et al.* Report of the ad hoc committee on reconciliation of approaches to bacterial systematics. Int J Syst Evol Microbiol 1987; 37(4): 463-4.
[http://dx.doi.org/10.1099/00207713-37-4-463]

CHAPTER 4

Saccharomyces: The 5 Ws and One H

Thiago Olitta Basso[1], Thalita Peixoto Basso[2], Sérgio Luiz Alves Júnior[3], Boris U. Stambuk[4,*] and **Luiz Carlos Basso[5,*]**

[1] *Department of Chemical Engineering, University of São Paulo, São Paulo/SP, Brazil*

[2] *Department of Genetics, Escola Superior de Agricultura Luiz de Queiroz, University of São Paulo, Piracicaba/SP, Brazil*

[3] *Laboratory of Biochemistry and Genetics, Federal University of Fronteira Sul, Chapecó/SC, Brazil*

[4] *Department of Biochemistry, Federal University of Santa Catarina, Florianópolis/SC, Brazil*

[5] *Department of Biological Sciences, Escola Superior de Agricultura Luiz de Queiroz, University of São Paulo, Piracicaba/SP, Brazil*

Abstract: The monophyletic *Saccharomyces sensu strictu* genus is composed of 8 species and several interspecies hybrids. Strains of this genus have been used in various processes that form a significant part of human culture and history. These include brewing, baking, production of wine and several other fermented beverages, and more recently, the production of biofuels, drugs, and chemicals. They can be found in the most diverse environments on almost all continents worldwide. A prominent example is the species *S. cerevisiae*, which has a remarkable history with humankind. In the present chapter, we illustrate the habitats of the *Saccharomyces* species and their long-lasting domestication process, as well as the hybridization that occurs between various species of this genus and their underlying industrial applications. We then finalize the text with an emblematic case study of its application in industrial sugarcane-based ethanol production, as performed in Brazil.

Keywords: Diversity, Domestication, Ethanol, Fermented beverages, Fermentation processes, Habitat, Hybridization, Stress, Sugarcane, *Saccharomyces*, yeast.

INTRODUCTION

The monophyletic *Saccharomyces sensu strictu* genus is actually composed of 8 species: *S. arboricola, S. cerevisiae, S. eubayanus, S. jurei, S. kudriavzevii, S. mi-*

---

* **Corresponding authors Boris U. Stambuk and Luiz Carlos Basso:** Department of Biochemistry, Federal University of Santa Catarina, Florianópolis/SC, Brazil and Department of Biological Sciences, Escola Superior de Agricultura Luiz de Queiroz, University of São Paulo, Piracicaba/SP, Brazil; Tel: +55 48 996159566; +55 19 996961150; E-mails: boris.stambuk@ufsc.br, lucbasso@usp.br

Sérgio Luiz Alves Júnior, Helen Treichel, Thiago Olitta Basso and Boris Ugarte Stambuk (Eds.)
All rights reserved-© 2022 Bentham Science Publishers

katae, S. paradoxus, S. uvarum, and several interspecies hybrids, including *S. bayanus* and *S. pastorianus* [1]. *Saccharomyce*s is considered a ubiquitous genus as it can be found in the most diverse natural environments. For example, non-domesticated *Saccharomyces* species have already been isolated on all continents on Earth, except for one (Antarctica). *Saccharomyces* species have been identified both in freshwater and seawater as well as in soil, fruits, and the gastrointestinal tracts of various animals [1 - 5]. The domestication of *Saccharom yces* began even before the domestication of animals by humankind, spanning for thousands of years, during the course of winemaking, brewing, baking, and more recently, the production of fuels and chemicals [6].

Natural or artificial hybridization between strains or species is a very common phenomenon that occurs in almost all sexually reproducing groups of organisms. Moreover, hybrids normally provide a selective advantage in a given environment [7]. Their chimeric genomes usually exhibit unique phenotypic traits that are not necessarily intermediate between those present in the progenitors. Therefore, *Saccharomyces* hybrids have found important industrial applications across industries, including the ones mentioned above.

Humans have consistently exploited one particular species, *S. cerevisiae*, for a myriad of bioprocesses [8, 9]. Specifically, in bioethanol production, this yeast species performs pivotal roles in Brazil, enabling the country to produce renewable and green fuel for transportation, contributing to sustain its renewable energy matrix, which constitutes an interesting case study for yeast industrial biotechnology. In this aspect, Brazil has the most economical and sustainable ethanol fermentation process in the world, with a very favorable energy balance.

SACCHAROMYCES HABITATS: WHERE?

Although *Saccharomyces* yeasts are widely recognized for their biotechnological applications through which they have co-existed with humans for approximately 10,000 years, they can be found in the most diverse natural environments. Due to thousands of years of domestication of the species of this genus (please, refer to the next section), which began even before the discovery of microorganisms [10], the distribution of *Saccharomyces* around the planet is a two-way street with its domesticators: on the one hand, archaeological records and analysis of genetic sequencing point out that similar fermentative processes were conducted by the polyphyletic strains of *Saccharomyces* in different places on the planet, demonstrating that fermentations started by naturally occurring yeasts [11 - 16], while on the other, there are indications that human beings may have transported these yeasts to regions where they were not previously found. Given the close millennial relationship between *Saccharomyces* and humans, the presence of some species in certain places on Earth has been attributed to the migratory movements of humanity [13, 15], as recently reported for the presence of Holarctic strains of *S. uvarum* in Patagonia [17]. As a matter of fact, this region of South America harbors two genetically differentiated populations of *S. uvarum*, with one of them being closely related to the North-hemisphere strains. It is believed that this Holarctic-derived population is a result of the introduction of apple trees by European immigrants in the 16th century [17].

To the best of our knowledge, non-domesticated *Saccharomyces* species have already been isolated on all continents on Earth, with the exception of Antarctica [16, 18, 19]. The City of Fairbanks, in central Alaska, US, accounts for the northernmost point (65°59' N), where a wild strain of *Saccharomyces* (of the species *S. ellipsoideus*) was found [20]. In the southern hemisphere, the Martial Glacier in Ushuaia (Tierra del Fuego, Argentina) appears as the place of isolation of a wild member of *Saccharomyces* (of the species *S. uvarum*) of greater south latitude (54°77' S) [21]. Finally, Auckland, New Zealand is the most easterly point (174°46' E) and the Island of Hawaii, US (155°50' W), the most westerly point, where non-domesticated representatives of this genus (of species *S. cerevisiae* and *S. paradoxus*, respectively) have been reported [22, 23]. However, at the opposite ends of longitude, both places are Pacific islands relatively close to each other (~7200 km). In addition to their wide distribution on the planet, interestingly, a strain of the main species of this genus —*S. cerevisiae*— even surpassed the limits of our biosphere and survived on a 40-day space flight in the Russian space station Mir, despite having presented mutation frequencies up to three times higher than those observed in their parental counterpart strain that stayed on the ground [24].

Therefore, these yeasts are also versatile concerning different temperature conditions. Despite being considered mesophilic microorganisms, in the extreme South and North regions where these yeasts have been found, the average annual temperature is between –1 °C and 2 °C, reaching up to –20 °C in winter (data from CLIMATE-DATA – www.climate-data.org). One of the warmest places where wild *Saccharomyces* strains have been found may be the Amazonian Forest biome. In this environment, Barbosa and co-workers [25] found these yeasts when they carried out a survey of wild *Saccharomyces* populations in the Brazilian state of Roraima, which is cut by the Equator. In this region of Brazil, the average minimum and maximum temperatures range between 22 °C and 34 °C (data from the Weather Forecast and Climate Studies Center of Brazil – www.cptec.inpe.br). These data demonstrate the adaptive diversity within a single genus and the consequent facilitation of its domestication processes.

There are reports of the *Saccharomyces* species both in freshwater [2] and seawater [3]. In the latter case, its physiological characteristics (in particular, higher tolerance to salt) have attracted the attention of the bioprocess industry [26]. For example, in the production of biofuels, even though industrial yeasts are adapted to highly concentrated musts, a large volume of freshwater is required in fermentation tanks. In the production of first-generation ethanol, depending on the crops employed, between 1.4 L and 9.8 L of water are required for each liter of fuel produced [27], and in the second generation, the average consumption is approximately 5.4 $L_{water}/L_{ethanol}$ [28]. Therefore, even considering the strong environmental appeal of biofuel production (due to their potential as substitutes for fossil fuels), this renewable process ends up being responsible for a significant water footprint. Fortunately, halophilic yeasts facilitate, as an alternative, the use of seawater instead of freshwater, and thus guarantee higher water security, without the loss of fermentative efficiency, as shown by the results of Zaky and co-workers [29 - 31].

The soil is also a very representative habitat for *Saccharomyces* [5, 32, 33]. Species of this genus have already been found in the soils of tropical [34 - 36], temperate [37 - 39], and even arid and desert climates, as in the "Zona Alta del Río Mendoza" in the center-west of Argentina [40, 41]. In terms of soil depth, *Saccharomyces* yeasts have already been isolated in the top layers (the upper 10 cm) of different soil types [20, 34, 37 - 39, 42] as well as the deeper layers (30–50 cm) of peat soils in Thailand swamp forests [35]. Despite this diversity, some authors argue that *Saccharomyces* species should be considered transients in the soil, as they spread from the substrates that are already deposited there, such as the bark and exudates from trees as well as the remains of leaves and fruits [32, 43]. This transient association with the soil is particularly noticeable in the regions covered by oak trees (*Quercus* spp.) in temperate forests. In fact, the presence of *Saccharomyces* in oak samples (mostly flux and bark) is widely reported in the literature [25, 32, 44 - 46], with these trees being considered the primary natural ecological niche of *Saccharomyces* [47] and the main source of wild strains of the genus [48]. In any case, when Sniegowski and co-workers [49] surveyed for *S. cerevisiae* and *S. paradoxus* in the soil samples from the specimens of *Quercus* spp. and also in the bark and exudate samples from these trees, they found twice as much representativeness in the soil itself than in the plant samples.

Fruits were possibly one of the first microenvironments responsible for the domestication of yeasts of the genus *Saccharomyces*. The presence of *Saccharomyces* on the fruit surface is a widely documented fact [1, 50 - 52], and the production of fruit-fermented drinks dates back to the Neolithic revolution [13, 53] — this and other domestication processes are described below (please, refer to the next section). The literature indicates the presence of *Saccharomyces* species on different varieties of grapes [19, 50, 54], fig [19], lemon [55], watermelon, orange, and pawpaw [56] as well as cashew and úmbu, a native Caatinga fruit from Brazil [54]. Furthermore, *Saccharomyces* yeasts have also been found in fruits of the following genera of native New Zealand trees: *Pseudopanax*, *Coprosma*, *Rhopalostylis*, *Leptospermum*, *Kunzea*, *Griselinea*, *Prumnopitys*, and *Schefflera* [51].

It is worth noting that different species and strains of the genus in question have already been found within the same vineyard, with the abundance and richness of species varying significantly within a few meters [50]. In fermentation vats, especially considering territorial wines, the use of indigenous yeasts as starter cultures is important [57, 58]; therefore, the same must ends up with yeasts that were previously found on fruits of different shrubs. This fact demonstrates how the development of agriculture and bioprocesses brought different yeasts together, acting as driving forces for the adaptation, domestication, and hybridization of this species (biological events to be discussed in this chapter). Particularly for winemaking, studies show that horizontal gene transfer (facilitated by the proximity among different strains) plays an essential role in microbial selection by conferring adaptive advantages for yeasts [59 - 61].

In addition to being found on the fruit peels, *Saccharomyces* yeasts can be found in the mesocarps [54], in the nectar of flowers of plant species, such as bertam palm (*Eugeissona tristis*) [62], southern-beeches (*Nothofaqus* sp.) [63], brown mustard (*Brassica juncea*), and field mustard (*B. campestres*) [64], and on the surfaces of leaves of trees, such as spruce (*Picea abies*), pine (*Pinus silvestris*), oak (*Quercus robur*), maple (*Acer campestre*), hornbeam (*Carpinus betulus*), linden (*Tilia cordata*), acacia (*Robinia pseudacacia*), and ash (*Fraxinus excelsior*) trees [65]. Since these substrates are food for many animals, they can carry microbial cells to their intestines, which is particularly expressive in the bee microbiota, where *Saccharomyces* can account for up to 99% of the intestinal microbial cells of this insect [66]. In fact, the intestines of herbivorous and nectarivorous insects are usually widely colonized by yeasts, which contribute to the digestion of food [67] and the induction of mechanisms of adaptation and immune-enhancement against intestinal pathogens [68]. The genus *Saccharomyces* is widely associated with the orders Diptera (especially in Drosophilids) and Hymenoptera [69]. It is worth noting that the volatile organic compounds (VOCs) produced by yeasts (from the different substrates that serve as food for insects) often attract these invertebrates to the place where yeasts are found, allowing microbial cells to access the insects' guts. This close association between yeasts and insects makes these animals an excellent means of transport for *Saccharomyces* cells, especially since these microorganisms can also be found on the external surface of these invertebrates. Considering the presence of these yeasts on the legs and bodies of insects, the *Saccharomyces*-Insecta association can also be extended to the orders Lepidoptera and Coleoptera [70].

As yeasts are swallowed along with food, diet can be considered the main reason for the presence of *Saccharomyces* species in the gastrointestinal tracts of animals. However, concerning the appearance of these microorganisms in the animal gut, it is worth noting a significant difference here: (i) there are microbes acquired through the animals' food and remain associated with them in a transitory way (the so-called allochthonous microorganisms) and (ii) there are pioneer microorganisms acquired by the animal during its birth or immediately afterwards that remain associated with it perennially, as part of its "core" microbiome. In the latter case, microorganisms are said to be autochthonous [71]. Undoubtedly, the genus *Saccharomyces* is found in the intestines of representatives of different phyla in the animal kingdom; however, regarding livestock and humans, it is still unclear whether these yeasts would be part of the allochthonous or autochthonous microbiota. In any case, one species in particular, *S. boulardii*, has already proved to be effective in preventive and therapeutic treatments of many diseases of the human digestive tract [4]. Indeed, this species has a series of phenotypic and physiological characteristics that guarantee its success as a probiotic, such as optimal growth temperature of 37 °C [72], resistance to the gastric environment, and viability at low pH [73]. Interestingly, however, this species was discovered by Henri Boulard in the 1920s after having observed that some indigenous people in the Indo-China region consumed fruit peels as antidiarrheal medication [4]. In contrast, through pyrosequencing of breast milk samples, Boix-Amorós and co-workers [74] demonstrated that the genus *Saccharomyces* was the third most prevalent among fungi, accounting for 12% of all sequence reads in these milk samples.

Saccharomyces species are not only associated with animals as commensals or mutualists, but also as opportunistic pathogens [75], causing infections in humans in places, such as the throat, lungs, intestine, vagina, and bones [19, 76]. Recently, two species of this genus, *S. cerevisiae* and *S. boulardii*, became part of the list of emerging pathogens, being recognized as the cause of severe infections in patients with cancer [77]. Literature data suggest that the use of fluconazole and itraconazole for the treatment of candidiasis and other fungal diseases may trigger the selection of new pathogenic yeasts [77, 78]. In *Saccharomyces* infections, the presence of one or more underlying conditions, such as tumours, AIDS, and diabetes, may definitely favour the pathogenicity of the yeasts, but the diagnosis of comorbidities is not *sine qua non* for infections by the species of this genus. In fact, the literature reports cases of invasive infections in immunocompetent bakers, either in the lungs [78], humours [76], or even the eye-sockets [79].

Despite the wide variety of environments reported as *Saccharomyces* habitats, in a recent study Alsammar and co-workers [38] through a targeted-metagenomics approach prioritizing the mycobiome of soils surrounding different trees at various altitudes in the Italian Alps showed that it was possible to verify higher abundance and richness of species and strains in soil samples than previous studies based on culturing methods or pyrosequencing targeting of non-specific *Saccharomyces* species [44, 80]. Alsammar and co-workers approach [38] consisted of carrying out a selectivity step involving the isolation of the internal transcribed spacer DNA specific to this genus. With this methodology, they could identify *S. eubayanus* and *S. mikatae* in their natural habitat for the first time in Europe. The work of these authors suggests that *Saccharomyces* yeasts can be found in even more diverse environments.

Thus, it is likely that the genus *Saccharomyces* is among the most widely distributed on the planet, including both the natural and industrial environments. From the data collected here, it is possible to infer the amplitude of the physiological diversity among species of this genus, which most likely helps to explain the success of these yeasts in several biotechnological applications.

HYSTORY AND DOMESTICATION OF *SACCHAROMYCES*: WHEN AND WHO?

One good indication of how long yeasts have existed within our society is the fact that there are numerous cognates in many languages for the English word, "yeast". There are cognates in French, Italian, Spanish, Dutch, Greek, Danish, Swedish, and even in Medieval Icelandic [6]. In fact, the existence of these cognates is a strong evidence of the involvement of this species with mankind. The term "yeast," and its respective cognates, denotes the occurrence of lifting, frothing, foaming, or bubbling that generally occurs when this species is acting on a particular material or medium. It is believed that wine making was probably the first experience with yeast, once the fermentation process does not necessarily require an external inoculum step, *i.e.*, it can proceed as a natural fermentation process. However, it is no surprise that the concept of yeast, as the microbial agent that carried out the fermentation during wine making and many other applications, was not perceived until circa of 7000 years later with the seminal work of Pasteur [6].

Other alcoholic beverages are believed to have developed quite early, similar to wine. As a prominent example, beer is prepared from malted barley and other cereals and the fermentation step requires the addition of living yeast cells. The procedure of malting leads to the breakdown of starch into fermentable sugars, and this was performed in Egypt and Babylon a long time ago [6]. However, according to Katz and Maytag [81], inoculation in the first fermentations occurred by the addition of fruits, such as grapes and raisins, or even by the addition of previously fermenting wine. Another alternative explanation is that insects, carrying yeasts in their body, landed on the malted grain and inoculated it [6]. A third important experience of human society with this fermenting organism was bread making, that dates back at least 6000 years.

More recently, *cachaça* fermentation was introduced by Portuguese 500 years ago in the colony that today comprises the country Brazil. From there, its production was improved, and it is now considered a typical Brazilian beverage, that is the one of the most popular distilled beverages in our society. Artisanal cachaça producers employed spontaneous fermentation to convert sugarcane juice into a fermentation product that was later distilled. The fermentation step was quite unique as compared to the traditional fermented and distilled beverages worldwide, this was a relatively fast fermentation process, occurring in 18–30 h, and is operated under typical tropical temperatures, where the temperature may reach as high as 40 °C in the fermentation step [82].

One could say that the domestication of *S. cerevisia*e by humankind began before the domestication of horses. In fact, it occurred more than 10,000 years ago, during the course of winemaking, brewing, and baking. We can also state that *S. cerevisiae* has been domesticated both in diverse industrial activities, such as baking, brewing, winemaking, fuel production, pharmaceuticals, *etc.*, as well as in the laboratory environment. In the laboratory, Louis Pasteur investigated yeasts to discover their essential roles in alcoholic fermentation in 1857 [83]. Industrial activities, such as brewing, was the starting point to captivate researchers to study yeast genetics, by crossing different strains to improve important technological traits [84]. However, the idea that *S. cerevisiae* is a domesticated species specialized for the fermentation of alcoholic beverages and other products, and that *S. cerevisiae* found in the wild simply represents the migrants from these processes does not necessarily hold true [85]. Although there are clearly strains of *S. cerevisiae* specialized to produce alcoholic beverages and other products, these have been derived from natural populations unassociated with their production. Yet, *S. cerevisiae* is regarded as an ideal model for the study of the mechanisms of microbe domestication, because this species possesses clear signatures of artificial selection in its genome [15, 23, 85].

In a very interesting study performed by Liti and co-workers [23], neighbor-joining phylogenetic trees were generated based on pairwise SNP differences in the alignment of various *S. cerevisiae* strains, including lab, baking, wine, food spoilage, sake, probiotic, plant, and even pathogenic yeast isolates. They found five lineages that exhibited the same phylogenetic relationship, which were considered 'clean' non-mosaic lineages, which were classified as strains from Malaysia, West Africa, sake and related fermentations (labelled 'Sake'), North America, and a large cluster of mixed sources containing many European and wine strains ('Wine/European'). In addition, phenotypic profiling of these strains (heat, cold, metal and drug stresses) produced results coherent with this general representation of the *S. cerevisiae* population structure. The authors have also found a good correlation between the phenotypic and genotypic similarities. Although some strains are linked to specific geographic origins, many closely related strains are from widely separated locations, that could be a result of human traffic in yeast strains followed by recombination events [23].

As pointed out by Barbosa and co-workers [82], there is an important and still open question regarding the various putative domestication events, whether they all evolved from the artificial selection of local wild *S. cerevisiae* lineages or there was the utilization of a globally dispersed domesticated strain. In their study, they found that cachaça yeasts have derived from wine strains that have undergone a further round of domestication, which was defined as a secondary domestication event. As a result, they believe that cachaça strains combine features of both wine yeasts, such as the presence of genes relevant for wine fermentation as well as beneficial gene(s) inactivation(s), such as the ones present in beer strains, like the resistance to inhibitory compounds that are normally present in molasses [86].

According to Gallone and co-workers [10], yeasts used for beer production experienced a much stronger domestication process as compared to the other applications of yeasts. The combination of time (beer is produced at large-scale for more than 1000 years) and the particularities of the beer production process (since yeast is recycled and it is produced continuously during the entire year) are responsible for this observation. Regarding examples of brewing phenotypes acquired by domestication, the efficient utilization of maltotriose is an emblematic one. This sugar is relevant in the wort environment [87], but not that important in natural yeast habitats. Another interesting trait acquired by yeast cells in beer production is the inability to convert ferulic acid (from barley) to 4-vinyl guaiacol (4VG), an unpleasant aroma-active compound, due to defective genes [88, 89].

Very recently, by combining genomes of Brazilian fuel ethanol strains with various other yeast strains, whole-genome phylogeny using an alignment-free approach indicated that bioethanol yeasts [90 - 92] are monophyletic and closely related to cachaça and wine strains. The authors hypothesized that Brazilian industrial ethanol strains may have been domesticated from a pool of yeasts that were pre-adapted to sugarcane cachaça fermentations [93]. Interestingly, all genomes of bioethanol yeasts have amplified gene clusters for vitamins B1 and B6 biosynthesis, as previously revealed [92]. Moreover, they have also shown the ubiquitous presence of SAM-dependent methyl transferases and quinone oxidoreductases, that are probably linked to nutritional and inhibitory stresses normally found during sugarcane-based industrial scale fermentations [93].

Interestingly, yeast domestication via the use of a relatively small number of starter strains that are routinely used to inoculate industrial fermentations, suggests that these commercial yeasts have been removed from the normal evolutionary processes [6, 94]. This inherently process-driven selective pressure on small, domesticated populations resulted in a reduced genetic diversity. In the context of winemaking, genomic regions relevant for wine production were inherited via horizontal gene transfer (HGT) from other fungi [59, 94]. On the other hand, in beer production, the genomic innovations that led for the selection of new biochemical activities for alcoholic beverages may have appeared rapidly by the hybridization between *S. cerevisiae* and other *Saccharomyces* species [21]. Therefore, it is likely that the wine and brewery processes involved very long and powerful artificial evolution experiments [94].

Another fortunate example of strain domestication is the hybridization between *S. cerevisiae* and *S. eubayanus*, which led to the interspecific hybrid *S. pastorianus*, used worldwide in the brewing industry to produce lager beer [95], that represents 80-90% of the brewing industry market share. These strains are divided in two main distinct lineages, named Saaz (Type 1) and Frohberg (Type 2) [10]. The dominance of *S. pastorianus* in the lager brewing industry suggests an interesting selective advantage of this hybrid over their corresponding parental species. While the *S. eubayanus* genome may confer enhanced cold-tolerance, the *S. cerevisiae* subgenome provides efficient fermentation, such as the use of maltotriose. *Saccharomyces* spp. can also be domesticated in the human body via the ingestion of food and beverages. These colonizers appear to be able to persist in the human gut environment, as is the case of *S. boulardii*, which is marketed as a human probiotic [96].

SACCHAROMYCES HYBRIDS: WHY AND HOW?

Humans have exploited the budding yeast, *S. cerevisiae*, for over ten thousand years for brewing, baking, production of wine and several other fermented beverages, and more recently for the production of biofuels, drugs, and other chemicals [8, 9]. *S. cerevisiae*'s biotechnological usefulness resides in its unique biological characteristics, *i.e.*, its fermentation capacity and dominance in the fermenters, and its resilience to the adverse conditions of high osmolarity, high ethanol, high temperatures, and low pH normally found in industrial processes [9, 88, 92, 97 - 99]. However, it is quite surprising that the first pure yeast brewing strain (*Unterhefe* N°. 1), isolated by Emil Chr. Hansen in 1883 at the Carlsberg

Brewery [100, 101], was not a *S. cerevisiae* strain, but *Saccharomyces pastorianus* (*carlsbergensis*), a hybrid yeast strain member of the *Saccharomyces sensu strictu* genus used for production of Lager beers.

In the last century, yeast taxonomists have used physiological and morphologic characters to study these microorganisms, where even minor phenotypic differences were considered adequate for species delimitations, promoting a continuous rise in the number of accepted *Saccharomyces* species [102]. Nowadays most of them are regarded as conspecific (synonyms) with *S. cerevisiae* [103], and the monophyletic *Saccharomyces sensu strictu* genus actually is composed of just 8 species: *S. arboricola, S. cerevisiae, S. eubayanus, S. jurei, S. kudriavzevii, S. mikatae, S. paradoxus, S. uvarum,* and several interspecies hybrids, including *S. bayanus,* and *S. pastorianus* [1]. All the *Saccharomyces* yeast species (excluding the hybrids) have 16 chromosomes (haploid strains) with similar gene content and synteny, a genome size of 11.3 to 12.2 Mb, and between 5,413 (*S. arboricola*) to 6,834 (*S. mikatae*) predicted protein-coding genes [104]. The genomes can have up to 4% sequence divergence within the same species, but when comparing the genomes from different species we find from 7% (Se x Su) to almost 20% (Sc x Su) sequence divergence [104].

The *Saccharomyces* species exhibit similar fermentation and basic physiological parameters, although they can have different levels of tolerance to some stresses [105]. Indeed, both the genes and promoters of the glycolytic and fermentative genes can be swapped between *S. kudriavzevii* or *S. eubayanus* into *S. cerevisiae*, and efficiently complement the native pathway [106, 107]. A significant difference between *S. cerevisiae* the other *Saccharomyces* yeast species is related to thermotolerance: all species are cryophilic (perform better at temperatures below 20 °C, and have maximum growth temperatures ranging from 33–38 °C), while *S. cerevisiae* performs bad at temperatures below 20°C, but can easily tolerate over 40 °C [108 - 110]. Consequently, in competitions between *S. cerevisiae* and the other *Saccharomyces* yeasts the former will in general outcompete the other yeasts at temperatures above 25–30 °C, while the cryophilic yeasts will dominate fermentations at temperatures below 20 °C [108 - 110]. Indeed, tolerance to high temperatures is a trait evolved during domestication of *S. cerevisiae* strains to industrial fermentations [111]. Importantly, the genomic comparison of *S. cerevisiae* with the cryophilic yeasts shows that glycolytic enzymes were among the proteins exhibiting higher than expected divergence, and thus the glycolytic fluxes and glycolytic enzymatic activities parallels the growth temperature profiles of the different *Saccharomyces* species [108].

Natural or artificial hybridization between strains or species is a very common phenomenon that occurs in almost all sexually reproducing group of organisms, including bacteria, plants, animals, and yeasts. *Saccharomyces* yeasts exhibit asexual and sexual reproductive cycles, whereas the most common mode of vegetative growth is asexual reproduction by budding (through mitosis) of both haploid and diploid cells (Fig. **1A**). However, in a stressful environment (*e.g.* starvation), the diploid cells will sporulate (meiosis) forming an ascus with four ascospores, 2 of the **a**-mating type, and 2 of the **α**-mating type. Haploid cells of different mating types will mate, restoring the normal diploid cell. But, if the haploid cells are of two different *Saccharomyces* species, then a diploid hybrid will be form (Fig. **1A**). Indeed, all *Saccharomyces sensu strictu* species have identical predicted **a** pheromones and nearly identical predicted **α** pheromones [112], and thus there is no pre-zygotic barrier between these species, but the hybrids show post-zygotic isolation and hybrids cannot mate, sporulate, or have very poor (<1%) spore viability [113, 114]. Fig. (**1B**) shows other possibilities for hybrid construction through rare-mating, and Fig. (**1C**) shows protoplast fusion that can take place in the gut of insects [115, 116], and can be used in the laboratory to create new hybrid strains. Although they are totally or partially sterile, these hybrids can be maintained by asexual reproduction. As we will discuss further, hybrid *Saccharomyces* yeasts are usually found in industrial settings, but since the *Saccharomyces* species are found in a variety of natural environments, including tree bark, soil, fruits, and insects guts, usually sharing the same space [1, 38]. There is recent evidence of the existence of *S. cerevisiae* x *S. paradoxus* hybrids isolated from the wild, in Brazilian forests, Mexican agave, and the French Guiana [25, 117, 118].

Hybrid genomes can provide a selective advantage in a given environment, a phenomenon also known as hybrid vigor [7]. Hybrids have chimeric genomes and may exhibit unique phenotypic traits that are not necessarily intermediate between those present in the progenitors (also known as transgressive phenotypes). Furthermore, the hybrid genomes are usually unstable and prone to postzygotic changes either during vegetative propagation of the sterile alloploid cells, or during sporulation of the allopolyploid cells upon the breakdown of their sterility. Processes, such as the loss of heterozygosity (LOH), whole genome duplication (WGD), appearance of aneuploidies, gene loss, genome re-arrangements, and translocations, have been reported during the evolution of hybrid strains [114, 119 - 121].

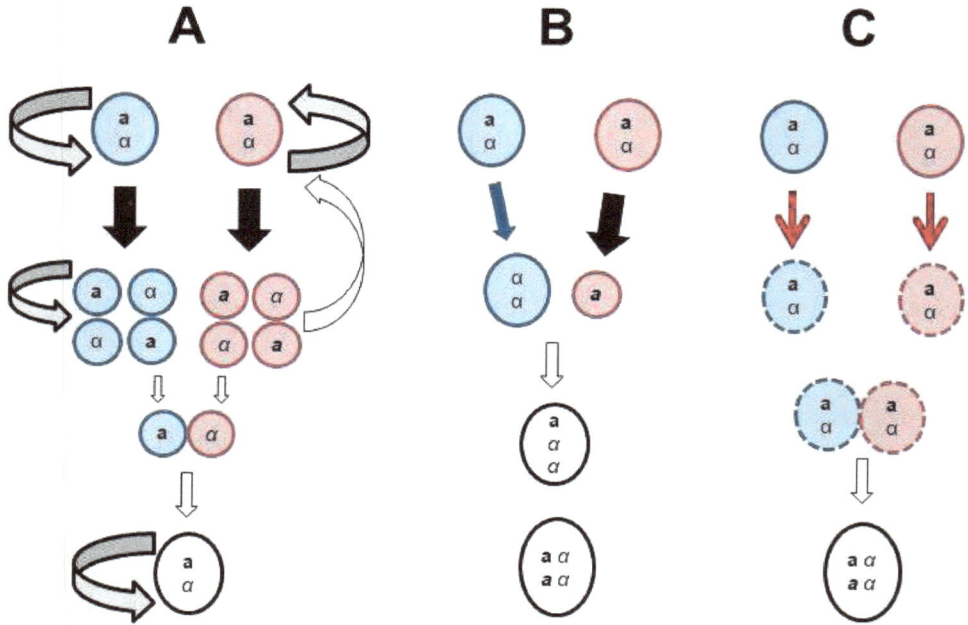

Fig. (1). Formation of *Saccharomyces* hybrids. (**A**) Diploid and haploid yeast cells can reproduce asexually by mitotic budding (grey arrows), but under certain circumstances diploid cells will sporulate (black arrows), producing an ascus with four spores with defined mating types (**a** or α). When two spores of different mating type encounter, they conjugate and form a diploid cell again (white arrows), but if the spores are from different species (cells with different colors), they will produce a hybrid diploid and sterile cell (white cells). (**B**) Rare-mating. Diploid cells can become competent to conjugate by the conversion of the *MAT* locus (loss of heterozygosity, blue arrow), and if it encounters a cell of opposite mating type, it will conjugate and produce a triploid hybrid (if mating with a haploid spore) or a tetraploid hybrid (if mating with a converted diploid of opposite mating type, not shown). (**C**) Protoplast fusion. Diploid cells while passing through insect guts will have its cell wall digested (red arrows), allowing the direct fusion of cells from different species producing a tetraploid hybrid.

Table **1** shows some examples of known *Saccharomyces* yeast hybrids and their industrial applications. In most cases hybrids have a better fermentation performance, especially with sugars that are important for the brewing (*e.g.* maltotriose fermentation [122 - 124]) or wine (*e.g.* fructose utilization [125, 126]) industries, including those that perform fermentations at cold temperatures or possess high-sugar content [127 - 130].

Table 1. *Saccharomyces* hybrids and their industrial applications.

Hybrid Yeast	Process, and Characteristics of Hybrids	References
S. pastorianus[a] / *calbergensis* (Saaz) (1n *S. cerevisiae* x 2n *S. eubayanus*)	Lager beers, cold fermentation temperature (7-10°C), non-maltotriose utilization, lower alcohol and ester production	[1, 100, 131–133]
S. pastorianus (Frohberg) (2n *S. cerevisiae* x 2n *S. eubayanus*)	Lager beers, cold fermentation temperature, maltotriose utilization	
S. bayanus (*S. cerevisiae* x *S. uvarum* x *S. eubayanus*)	Beer, Chardonnay wines and ice-cider, high sugar fermentation, cool climate vineyards	[134, 135]
S. cerevisiae x *S. arboricola*	Beer, maltotriose fermentation, increased production of aroma compounds	[136]
S. cerevisiae x *S. eubayanus*	Beer, cider and wine, improved fermentation power at low temperatures, no off-flavours are produced, maltotriose fermentation, increased production of aroma compounds	[136–141]
S. cerevisiae x *S. kudriavzevii*	Beer, cider and wine, higher fermentation rate, higher ethanol concentration, less residual sugar, increased flavor and aroma diversity	[127, 142–145]
S. cerevisiae x *S. mikatae*	Beer and wine, maltotriose fermentation, increased production of aroma compounds, glycerol, and lower production of acetic acid	[127, 136, 145, 146]
S. cerevisiae x *S. paradoxus*	Olives and olive products, decreased production of organic acids, better growth under environmental stresses, dominance during competition	[145, 147, 148]
S. cerevisiae x *S. uvarum*	Beer, cider and wine, maltotriose fermentation, increased production of aroma compounds, broader temperature range; intermediate levels of stress sensitivity and higher production of ethanol and glycerol, higher growth and fermentation rate	[127, 136, 149–152]

[a]For *S. pastorianus*, the best characterized hybrids, two different groups exist with different ploidies: Saaz are triploids, and Frohberg are tetraploids. There are several discussions on how theses hybrid genomes were generated. For the other hybrids in the table the ploidy of the strains is not described/known.

SACCHAROMYCES IN SUGARCANE-BASED BIOETHANOL PRODUCTION: WHAT AND HOW?

Ethanol in Brazil

Ethanol is largely used in Brazil as a biofuel for transportation, contributing to the world's highest renewable energy fraction in its energy matrix. Brazil is also a sugar cane producer (700 Mi ton/season) as this crop is used as the feedstock for sugar and ethanol production, and together with USA both countries are responsible for ca. 84% of the world ethanol production. This crop was introduced in Brazil by Portuguese colonizers in the beginning of 15th century to supply sugar for Europe. The spent cane juice was accidently and naturally fermented, and was soon shown to be useful as feed for animals and appreciated as a beverage by slaves. Later on, the juice and the residual waste from sugar crystallization were fermented and distilled, resulting in the "cachaça," a Brazilian spirit with increasing economic and cultural importance. Industrial ethanol and even fuel ethanol was later produced since the first decade of the last century, sometimes stimulated by economical events (as the Great Depression in 1929), but always limited by technological advances occurring in Europe regarding ethanol distillation and dehydration [153 - 155].

Due to the Petroleum Crises in 1973, the Brazilian Government launched an Ethanol Program to increase the ethanol production to be used as transportation fuel and to relieve the country's economy due to the high prices of oil. Since then, a great effort was devoted by Research Institutes, Universities, Distilleries itself and private Institutions, aiming at increasing sugar cane production, increasing ethanol yield, costs reduction, value addition to byproducts, energy savings, electricity co-generation and overall industrial improvements. As a result of these improvements, Brazil has the most economical and sustainable fermentation process with an energy balance (units of energy in the produced ethanol related to the total energy used for its production) of 8:1, very high in comparison with that of corn ethanol in USA (less than 2:1). Currently bioethanol is produced by ca. 400 distilleries, mostly located at Sugar Mills (using cane juice and molasses – a byproduct of sugar refining), but some others only use sugar cane juice as substrate [153, 156].

The Fermentation Process

Sugar cane proved to be a suitable crop for bioenergy production due to its efficiency, sustainability, and eco-friendly nature. Its high biomass productivity, as a C-4 photosynthetic species, amounts to 80–120 ton/ha/year with an ethanol production of 8,000 liters/ha, much higher when compared to 3,000 liters/ha from maize. Endophytic nitrogen-fixing bacteria recently found in this species supply at least 60% of the plant's nitrogen requirement when growing in low fertility soil. Economic and environmental benefits of sugar cane result from the fact that nitrogen fertilizers are expensive and consume fossil energy for its production. Drought tolerance and the perennial nature of this crop add others agricultural advantages [153].

Sugar cane stalks are harvested, shopped, and pressed for juice extraction leaving a fiber residue (*cane bagasse*). This feedstock has on a wet weigh base, 13–14% of fiber and 12–17% of readily fermented sugars (90% as sucrose and 10% as mixture of equal amounts of glucose and fructose). The collected juice is clarified and concentrated by evaporation for sucrose crystallization. Sucrose crystals are removed by centrifugation, leaving a sucrose saturated viscous phase called *cane molasses* with 45–60% of sucrose and 5–20% of glucose and fructose, in mass fraction. Both molasses and juice are used, in varying proportions, as substrate for fermentation. A mixture of both sources is a good option, since molasses has high salt concentration and some yeast inhibitors while juice normally has mineral deficiencies and lower buffer capacity. The mineral composition of sugarcane substrates has a pronounced effect on fermentation efficiency, varying widely in terms of nitrogen, phosphorus, potassium, magnesium, sulfur, and calcium levels due to differences in the molasses proportion in the must preparation, cane variety and maturity, soil, climate, and cane juice processing [153, 154].

In general, the fermentation proceeds according to a so-called "Melle-Boinot" process, as a fed-batch mode (75% of the distilleries) or as a continuous version, both utilizing yeast cell recycling. In this process, yeast cells are collected at the end of the batch by centrifugation and totally re-used in a next fermentation cycle. Up to 90–95% of yeast cells are recycled, resulting in high cell densities inside the fermenter of almost 10–14% (wet weigh basis/v). Cell reuse reduces yeast propagation, and less sugar is deviated for biomass formation, although a 5–10% biomass increase is noted in each fermentation cycle. Although restricted, this cell growth is of paramount importance to replace yeast loss mainly in the centrifugation step. A high biomass content in the fermenter promotes a very short fermentation time (6–10 h), when compared to 40–50 h in corn fermentation process. The temperature is normally kept at 32–35 °C, but high fermentation rates can generate heat in such amount that cooling system could not be efficient enough and temperature may reach 40°C, especially during summer [154].

In the fermentation step, the fed-batch mode (Fig. **2**) starts by adding the must (prepared with molasses and cane juice at any proportion with pH within 5.0–5.5), which contains 18–25% (w/v) total reducing sugars (TRS) to a yeast suspension (with 30% of yeast cell, on wet basis). This yeast suspension comprises 25–30% of the total volume of fermentation that is performed in tanks of 300–3,000 m³. A large inoculum is normally prepared by mixing 2–12 ton of bakers' yeast to 10 to 300 kg of selected *Saccharomyces cerevisiae* strains in form of active dry yeast. This inoculum is prepared only for the first cycle in the beginning of the season, knowing that baker's yeast will be replaced by the selected strain in a couple of weeks of cell recycling. This is because baker's yeast is cheaper and contributes for fermentation during the short time required for selected strains being implanted. The feeding of the must is carried out for 4–6 h and the batch is terminated within 8–10 h, resulting in ethanol titers of 8–12% *(v/v)*. After that, yeast cells are centrifuged and the resulting concentrated yeast cell suspension with 60–70% cells (wet weigh basis/v) is collected in a separate vat. This yeast slurry is diluted with water (1:1) and treated with sulfuric acid (pH 1.8-2.5, for 1 h) to reduce bacterial contamination. Then, the treated slurry is used as a starter for the next fermentation batch. With that, yeast cells are reused at least twice a day over the course of the production season (200–250 days). After the centrifugation step, the resulting yeast-free "beer" or "wine" is conveyed to distillation for ethanol recovery. The resulting liquid stream is called stillage or "vinasse", generating 10–15 L of vinasse for each L of ethanol. This stream is delivered into the cane plantations as water and nutrient supplies, in a process known as fertirrigation [153, 156, 157].

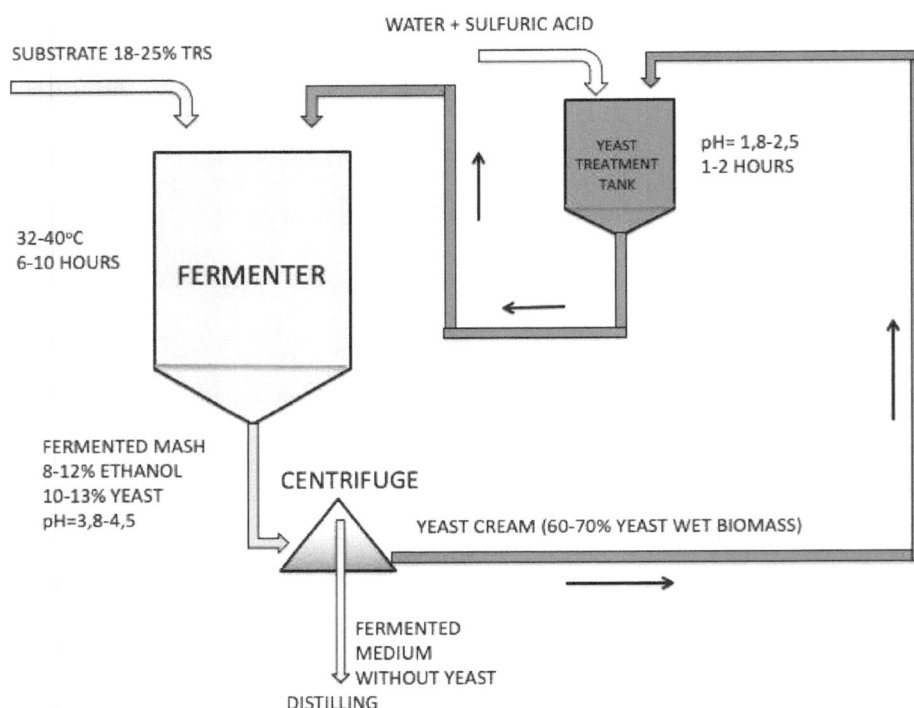

Fig. (2). Ethanol production process using under fed-batch with cell recycle (Adapted from Basso and co-workers [160]).

The Various Stresses in the Industrial Scenario

Industrial yeast strains encounter several stressing conditions during ethanol production, including high ethanol titers, high osmotic pressure, acidic conditions, high temperatures, and others, that are intensified by the yeast cell recycle, a characteristic of the Brazilian process (Fig. 3).

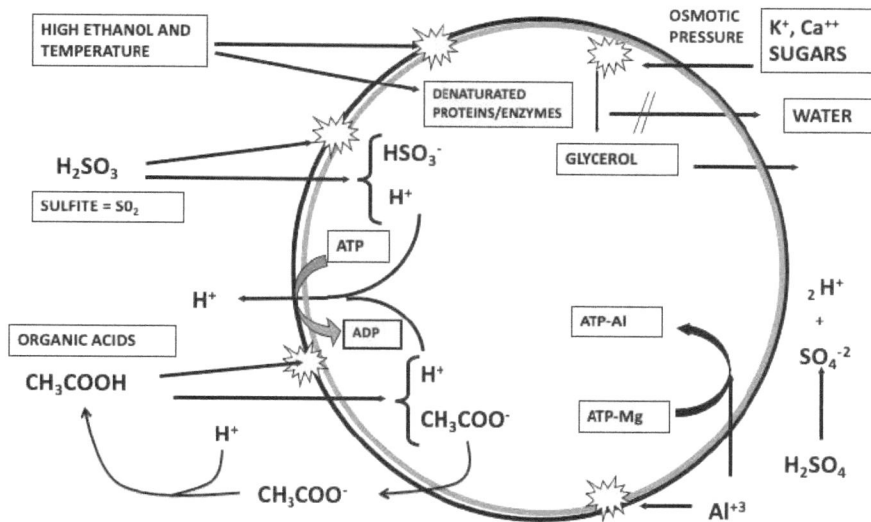

Fig. (3). Schematic representation of stress conditions faced by yeast cells during sugarcane-based ethanol fermentation (Adapted from Basso and co-workers [160]).

Ethanol Stress

Due to the high ethanol titers obtained at the end of each fermentation batch (8-12%, v/v), it is considered a major stress factor that affects yeast cells. Although the inhibitory role of ethanol on *S. cerevisiae* is not fully understood, the cytoplasmatic membrane is known to be the main target [158, 159]. Membrane fluidity is extremely altered in the presence of ethanol, thus membrane permeability to some compounds is significantly enhanced. As these compounds permeate the cells, there is a dissipation of the electro-chemical gradient, which in turn affects the maintenance of the proton motive force. It also affects the yeast membrane composition, growth inhibition, and enzymatic inactivation, which all lead to decreased cell viability and impaired fermentation performance [86].

Yeast cell recycling, a technique routinely used during sugarcane ethanol fermentation, contributes to a cumulative ethanol stress in fermenting industrial *Saccharomyces* strains [90, 154, 161 - 163]. Therefore, it is imperative that yeast cells keep high viabilities at the end of each batch to withstand the next one. For this reason, in industry, ethanol titers rarely exceed 11–12% *(v/v)* due to the limited tolerance of the strains during recycling condition. On the other hand, if the yeast physiological condition at the end of fermentation, namely its viability, is not a matter of concern (without cell reuse), ethanol titers of 17-23% *(v/v)* are easily achieved, as in corn and others cereals fermentations [164, 165].

Although ethanol is considered an important stress factor for *Saccharomyces*, high ethanol titers are highly desirable in the industrial context. High titers lead to reduced water consumption, as well as savings in energy costs during the distillation step [156]. High-gravity fermentations will also favor the energy balance and improve the sustainability of the industrial process. In most of the distilleries in Brazil, the final ethanol content is kept in the range of 8–9% *(v/v)* due to the limited tolerance of the strain [86, 90]. In addition to sole ethanol stress, the perturbations can be intensified by high temperature and acidic conditions faced by fermenting yeast cells [166].

Continuous efforts have been made both in academia and industry to search for ethanol tolerant *Saccharomyces* yeast strains, specifically for the case of cell recycling [86, 90, 156]. A direct approach for this purpose is quite challenging, since the genetic basis for ethanol tolerance is polygenic and very complex [159, 167]. The majority of the genes related to ethanol tolerance are linked to energy and biosynthetic metabolism in yeasts. Therefore, the search for yeasts with high ethanol tolerance is a difficult task, and various strategies, such as genome shuffling, mutagenesis, and adaptive evolution, might be useful to select more tolerant strains [168, 169].

Acidic Stress

Although it is well known that yeasts generally tolerate low pH conditions, the sulfuric acid treatment of the "yeast cream" to reduce contamination leads to detrimental effects in yeast cells. This is observed by leakages of K, P, Mg and organic nitrogenous compounds in parallel with drops in yeast cellular trehalose content and cell viability [170]. Usually, yeasts that are resistant to the stress conditions normally present higher trehalose content [86, 90].

As the plasmatic membrane is permeable to undissociated forms of weak organic acids present in the substrate or produced by contaminating bacteria (as acetic or lactic acids, with pKa around 4.7) they enter the cell and reduce the intracellular pH value. Industrial fermentation starts with the addition of substrate (pH = 5.0–5.5) over the treated yeast suspension (pH = 2.6–3.0). As a fed-batch process the initial pH (around 2.6–3.0) starts to increase during the feeding phase (4–6 h) reaching pH 3.7–4.0 at the end of fermentation. During the whole fermentation process pH values are always below the pKa of most of the organic acids found in the medium, rendering them in its protonated (and permeable) forms. Residual levels of sulphite (SO_2), an inorganic weak acid used for juice clarification, can also exert toxic effect on yeast when in levels above 200 mg/L, but when at 100 mg/L, it can show desirable effects due to reduction in bacterial contamination [171].

Osmotic Stress

It is intuitive to imagine that yeasts would be exposed to very high levels of sugars during industrial fermentation since high ethanol concentrations are needed. Nevertheless, the fed-batch nature of the industrial process minimizes the osmotic stress of sugars, despite the high sugar content of the substrate, with about 18–25% *(m/v)* total reducing sugar. In other words, during the feeding phase, sugars are added and consumed simultaneously, reducing the exposure of yeast cells to high sugar concentrations as depicted in Fig. (**4**).

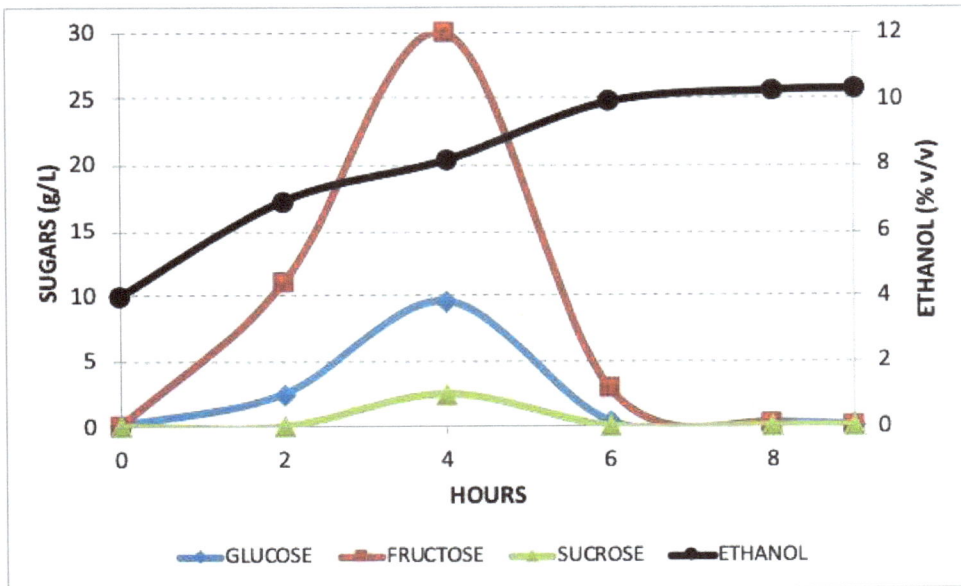

Fig. (4). Sugars and ethanol concentration during a typical sugarcane-based fermentation batch performed at lab-scale mimicking the industrial process, including a feed phase of 4 hours with molasses broth containing 210 sucrose, 18 glucose and 18 fructose (g/L) (Adapted from Basso and co-workers [160]).

The sugar concentrations are governed by the feeding time and the rates at which sucrose is hydrolyzed and the resulting products (glucose plus fructose) are metabolized. From Fig. (**4**), it is noticed that sucrose hydrolysis is faster than glucose and fructose absorptions by yeast and that glucose is preferably metabolized compared with fructose. As a result, there is a peak of glucose and fructose at the end of the feeding phase (4 h) but with higher levels of fructose. In this condition, final ethanol titer higher than 10% *(v/v)* was obtained while the maximum sugar concentrations experienced by yeast cell at the end of feeding phase was 2.7% fructose, 1% glucose and 0.3% sucrose, and fermentation was finished after 8 h. When changing the feeding time to 6 h a great decrease in sugar was observed, mainly in fructose and fermentation was finished in 8 h with the same ethanol titer (data not shown). Glycerol formation was also decreased, suggesting that the feeding time of 6 h alleviated the osmotic stress toward yeast, without compromising ethanol yield and productivity (unpublished results).

The osmotic stress caused by high salt concentrations in molasses media is a matter of concern, and the levels of potassium, calcium, and magnesium exceed the yeast requirements. Average levels of potassium induce yeast stress responses as higher glycerol formation, lower cellular trehalose accumulation and reduced ethanol yield [154]. Increased glycerol formation occurs during osmotic stress, bacterial contamination (with heterofermentative *Lactobacilli* strains), and in other conditions, suggesting that this byproduct could be a helpful parameter indicating general stressing conditions for yeast fermentation. Yeast strains with higher invertase activity cause higher glycerol production, probably due to osmotic stress by the rapidly sucrose hydrolysis, and when operating with such yeasts a low feeding rate is recommended.

Biotic Stress

In industrial fuel ethanol production process, aseptic conditions are difficult to achieve due to its nature and presence of large volumes. Therefore, bacterial contamination is inevitably found in sugarcane fermentations. Not only sugars are diverted from ethanol formation but bacterial metabolites decrease yeast performance, resulting in low ethanol yield, intense yeast cell flocculation, and uncontrolled foam formation [172 - 175]. Yeast flocculation, which can be caused by bacterial contamination, impairs the centrifuge efficiency. Foam formation increases costs due to antifoam consumption [176]. The antibiotics used to decrease contamination results in increased costs and their residual levels make dried yeast improper for commercialization [177] and may result in alterations in the soil microbial dynamics [178].

Most of the bacterial contaminants in the fermentation tanks of ethanol production are lactic acid bacteria (LAB) [179 - 183]. It was also showed that heterofermentative *Lactobacilli* were more deleterious for yeast fermentation than homofermentative strains, leading to a higher formation of glycerol, lower yeast cell viability and reduced ethanol yield. It is suggested that the higher fructose content than glucose observed during the fed-batch fermentation process (Figs. **2** and **4**) may favor the heterofermentative strain over homofermentative one, since the former consumes fructose faster than glucose [184].

Table **2** illustrates the typical detrimental effects of *Lactobacillus* contamination on several technological parameters on yeast fermentation, when mimicking the industrial fed-batch process at the lab-scale. From these data it is possible to deduce that approximately 3.7% of metabolized sugars is diverted to undesirable products formed by the bacterial metabolism. As such, since fermentation byproducts are more precisely estimated than the formation of ethanol in industries, it can be suggested that the overall effect on ethanol yield is underestimated (2.4% is the difference of ethanol yield observed in Table **2**). These observations encourage the use of produced bacterial metabolites titers to access the impact of contamination on ethanol yield and could explain the great variability of this parameter in the literature when based on ethanol content only.

Table 2. The effect of *Lactobacillus* contamination (10^8 cells/mL) on the fermentation stoichiometry, evaluated by sugar deviation (as % of sugars diverted to metabolic pathways leading to products or biomass) for each fermentation product simulating the industrial fed batch process with molasses medium, *S cerevisiae* PE-2, at 32°C. Each value is the average of 5 fermentation cycles. Data obtained from Vital [199].

Fermentation Product	Without Contamination	With Contamination
Ethanol (% of maximal yield)	92.40	90.00
Biomass	1.51	1.13
Glycerol	2.23	2.47
Acetic acid	0.14	1.11
Succinic acid	0.17	0.24
Lactic acid	0.22	3.21
Mannitol	0.52	0.83

Quorum Sensing & Stress

The high cell densities with continuous recycling used by the Brazilian sugarcane fuel ethanol industries have several advantages in terms of productivity of the bioprocess, but it can also trigger cell density dependent cell-to-cell communication via extracellular signaling molecules, a phenomenon called quorum sensing [185 - 187]. Indeed, quorum sensing plays an important role in the development of stress tolerance in high cell density yeast cultures through the production of quorum sensing molecules (QSM), such as phenylethanol, tryptophol, tyrosol, farnesol, and various other alcohols, by-products of amino acid metabolism [188 - 190]. With the increase in population density, QSM produced by the cells accumulate in the extracellular environment, and once their concentration reaches a given threshold, many genes are regulated through specific signaling pathways, promoting fast adaptation to diverse conditions and resulting, for example, in increased cell fitness and survival [191, 192]. In *S. cerevisiae* QSM are involved in the transition between single cells to flocculation, filamentous growth and biofilm formation, where the aromatic alcohols phenylethanol and tryptophol promote filamentous growth by induction of *FLO11* via the Ras-cAMPK protein kinase A pathway, and tryptophol induces flocculation by induction of *FLO1* [193 - 197]. Indeed, the analysis of hundreds of *S. cerevisiae* strains isolated from industrial sugarcane fuel ethanol producing plants in Brazil revealed that less than 20% of the strains did not present such undesirable features, including premature flocculation, biofilm formation, filamentation or excessive foam production [90, 198].

QSM may also have some other effects in *S. cerevisiae* fermentation processes [200], since several reports have shown that high-density cultures of *S. cerevisiae* promote the early death of non-*Saccharomyces* yeasts present in the medium, including *K. thermotolerans, T. delbrueckii, H. guilliermondii, H. uvarum* and other yeasts species [201 - 204], and recent data indicate that the known QSM tyrosol has several effects in the fermentation performance of several yeasts [205]. Another QSM, farnesol, was shown to inhibit *S. cerevisiae* growth due to a signalling-mediated cell cycle arrest [206]. Finally, recent data also show that a bacterial QSM, *N*-(3-oxododecanoyl)-L-homoserine lactone, produced by the Gram-negative and opportunistic human pathogen *Pseudomonas aeruginosa*, inhibits *S. cerevisiae* growth and promotes a clear stress response mediated by the *MSN2* and *MSN4* stress-responsive transcription factors [207].

Heavy Metal Stress

Aluminum is found in sugarcane substrates, and its presence is responsible for decreasing the fermentation performance. In view of the acidic nature of sugarcane-based fermentations, aluminum is mainly present as the trivalent toxic form (Al^{3+}). It is alleged to deleteriously affect yeast cell viability, intracellular trehalose levels, fermentation rate and ethanol yield [86]. Interestingly, aluminum detrimental effects can be reduced by magnesium ions and a molasses-based medium. In addition, industrial yeast strains vary regarding aluminum tolerance. For example, *S. cerevisiae* CAT-1 is more robust than PE-2 and baker's yeasts, and the negative effects are more pronounced in juice than in molasses [153, 161].

Although low levels of heavy metals are found in sugarcane plants, the cell recycling process and the ability of yeasts to bioaccumulate such compounds have led to detrimental effects on yeast fermentation. Vinasse, when used as a complement, was found to reduce the toxic effects caused by both aluminum [208] and cadmium [209], probably due to the presence of chelating compounds present in this substrate.

Selection of Yeast Strains for Industrial Fermentation

For a long time, Brazilian distilleries operated its mills with baker's yeast, while other *S. cerevisiae* strains (such as IZ-1904 and TA) were believed to be superior for industrial practice. The main criterion for this evaluation was ethanol yield presented by a given strain, and in fact IZ-1904 strain showed higher ethanol yield in laboratory conditions, making IZ-1904 a widely used strain in Brazilian distilleries. Many ethanol producers exalted the strain, while others depreciated it. Most of the evaluation assays (performed in laboratory conditions) did not use a cell recycling protocol and important physiological traits of the strains were dismissed regarding industrial fermentation. In 1987 these strains were, for the first time, evaluated in conditions of cell recycling, mimicking the industrial process as far as possible in the laboratory. The results unraveled physiological traits suggesting that IZ-1904 would not be tolerant enough to cope with industrial conditions, mainly due to the viability loss and biomass decrease, making it improper for a cell recycling process (Table 3). Moreover, its lower tolerance to fermentation stress correlated with lower trehalose and glycogen cellular levels, and to temporary higher ethanol yield solely at the first cycles. In other words, ethanol was produced at the expense of cellular carbohydrate reserves and yeast biomass during cell re-use typically employed in industrial processes. This observation pointed out that ethanol yield, *per se,* is not an adequate parameter for yeast evaluation when a cell recycling process is considered. Noteworthy, only after 6 years, using the karyotyping technic, these conclusions were confirmed in industrial practice [210].

Table 3. Physiological and technological parameters of yeast strains evaluated in fed-batch conditions simulating the industrial process with cell recycle, in two different stressing conditions. Values are the average of 6 fermentation cycles. Data obtained from Basso and co-workers [210].

Physiological & Technological Parameters	7-8% *(v/v)* Final Ethanol, 30°C [210]		9% *(v/v)* Final Ethanol, 33°C [90]	
	Baker's Yeast	IZ-1904	Baker's Yeast	PE-2
Ethanol yield (%)[1]	88.8	90.0	89.5	93.2
Glycerol yield (%)[1]	4.6	5.1	5.7	3.7
Biomass gain (%)[2]	41	-10	35	49
Cell viability (%)[3]	80	28	48	97
Glycogen (% cell DCW)[3]	16	7	9	16
Trehalose (% cell DCW[3])	7	0	4	9

[1] Fraction of consumed sugar converted into ethanol or glycerol [2] Biomass gain from the first to the last cycle [3] Values from yeast biomass at the end of the last fermentation cycle

Thus, the karyotyping technique revealed that not only the IZ-1904 strain, but also any other starter strain (such as baker's yeast and TA) was not able to withstand the fermentation process with cell recycling. The only strain that could be detected at the end of the season was an indigenous *S. cerevisiae* strain isolated from one of the distilleries involved in the study [210]. This was the starting point to a systematic tracking of the population dynamics of *S. cerevisiae* strains in various distilleries, with monthly collected samples for 12 years. During this period, strains with prevalence (representing a high proportion within a given yeast population) and persistency (able to survive all long the production season) were selected and evaluated in laboratory for desirable fermentation traits relevant for industrial performance. Whenever a promising strain was selected, it was immediately introduced in distilleries and tracked by karyotyping. Many of these selected strains displayed prevalence and persistency in distilleries, and some of them were outstanding, being found at the end of the process for various production seasons [90]. Table **3** illustrates the performance of one of these selected strains (PE-2), which is still used in many distilleries. It is noteworthy to mention that PE-2 strain, as compared to baker's yeast, produces less glycerol while presenting a more vigorous growth during fermentation with cell recycle. In addition, higher cellular reserve carbohydrates are correlated to higher viabilities. The remarkable performance of these selected strains allowed them to be successfully implemented in industrial processes, reassuring that the industrial process itself is a tremendous source of yeast biodiversity. Due to their unique abilities, the selected strains, PE-2 and CAT-1, are widely used in various others fermentation processes, including cachaça, tequila, rum, and cereal alcohol production. They have been also genetically engineered for different purposes and are currently used in research for genetic dissection of stress tolerance [86, 92, 155].

CONCLUDING REMARKS

Saccharomyces yeasts are used in several industrial applications, although these microorganisms can also be found in nature associated with the soil, bark, trees, flowers, and even in our own gut microbiome. Humans have isolated and used these yeasts for millennia; however, only recently we have begun to understand the characteristics of these organisms, especially in relation to their superior fermentation performance, thereby highlighting the various improvements that can be made. Industrial bioethanol production from sugarcane presents not only higher efficiency but also has economic and sustainable advantages. *Saccharomyces* strains with desirable tolerance profiles evolve from this industrial process, and its recycling nature makes it an important source of obtaining new variants, even for other industrial applications. Bioethanol production in   Brazil is still a growing sector, and there is enough room for further industrial process improvements.

CONSENT FOR PUBLICATION

Not applicable.

CONFLICT OF INTEREST

The authors declare no conflict of interest, financial or otherwise.

ACKNOWLEDGEMENTS

The authors would like to acknowledge the following Brazilian funding agencies: Conselho Nacional de Desenvolvimento Científico e Tecnológico – CNPq (grant numbers 454215/2014-2, 305258/2018-4, 429029/2018-7, and 308389/2019-0), Coordenação de Aperfeiçoamento de Pessoal de Nível Superior – CAPES (grant number 359/14), Financiadora de Estudos e Projetos – FINEP (grant number 01.09.0566.00/1421-08), Fundação de Amparo à Pesquisa do Estado de São Paulo – FAPESP (grant numbers 2018/17172-2 and 2019/08845-6), and Fundação de Amparo à Pesquisa do Estado de Santa Catarina – FAPESC (grant numbers 17293/2009-6 and 749/2016 T.O. 2016TR2188).

REFERENCES

[1]   Alsammar H, Delneri D. 2020; An update on the diversity, ecology and biogeography of the Saccharomyces genus. FEMS Yeast Res 20: foaa013.
        [http://dx.doi.org/10.1093/femsyr/foaa013]

[2]   Medeiros AO, Missagia BS, Brandão LR, Callisto M, Barbosa FAR, Rosa CA. Water quality and diversity of yeasts from tropical lakes and rivers from the Rio Doce basin in Southeastern Brazil. Braz J Microbiol 2012; 43(4): 1582-94.
        [http://dx.doi.org/10.1590/S1517-83822012000400043] [PMID: 24031990]

[3]   Zaky AS, Tucker GA, Daw ZY, Du C. Marine yeast isolation and industrial application. FEMS Yeast Res 2014; 14(6): 813-25.
        [http://dx.doi.org/10.1111/1567-1364.12158] [PMID: 24738708]

[4]   Garcia-Mazcorro JF, Ishaq SL, Rodriguez-Herrera MV, Garcia-Hernandez CA, Kawas JR, Nagaraja TG. Review: Are there indigenous *Saccharomyces* in the digestive tract of livestock animal species? Implications for health, nutrition and productivity traits. Animal 2020; 14(1): 22-30.
        [http://dx.doi.org/10.1017/S1751731119001599] [PMID: 31303186]

[5]   Arroyo-López FN, Salvadó Z, Tronchoni J, Guillamón JM, Barrio E, Querol A. Susceptibility and resistance to ethanol in Saccharomyces strains isolated from wild and fermentative environments. Yeast 2010; 27(12): 1005-15.
        [http://dx.doi.org/10.1002/yea.1809] [PMID: 20824889]

[6]   Mortimer RK. Evolution and variation of the yeast (Saccharomyces) genome. Genome Res 2000; 10(4): 403-9.
        [http://dx.doi.org/10.1101/gr.10.4.403] [PMID: 10779481]

[7]   Lopandic K. Saccharomyces interspecies hybrids as model organisms for studying yeast adaptation to

stressful environments. Yeast 2018; 35(1): 21-38.
[http://dx.doi.org/10.1002/yea.3294] [PMID: 29131388]

[8]   Parapouli M, Vasileiadis A, Afendra AS, Hatziloukas E. *Saccharomyces cerevisiae* and its industrial applications. AIMS Microbiol 2020; 6(1): 1-31.
[http://dx.doi.org/10.3934/microbiol.2020001] [PMID: 32226912]

[9]   Stambuk B. Yeasts: The leading figures on Bioethanol production.Ethanol as a Green Alternative Fuel: Insight and Perspectives. 1st ed. Hauppauge, NY, USA: Nova Science Publishers Inc 2019; pp. 57-91.

[10]   Gallone B, Mertens S, Gordon JL, Maere S, Verstrepen KJ, Steensels J. Origins, evolution, domestication and diversity of Saccharomyces beer yeasts. Curr Opin Biotechnol 2018; 49: 148-55.
[http://dx.doi.org/10.1016/j.copbio.2017.08.005] [PMID: 28869826]

[11]   Aouizerat T, Gutman I, Paz Y, *et al.* Isolation and characterization of live yeast cells from ancient vessels as a tool in bio-archaeology. MBio 2019; 10(2): e00388-19.
[http://dx.doi.org/10.1128/mBio.00388-19] [PMID: 31040238]

[12]   Brüssow H. Bioarchaeology: a profitable dialogue between microbiology and archaeology. Microb Biotechnol 2020; 13(2): 406-9.
[http://dx.doi.org/10.1111/1751-7915.13527] [PMID: 32053292]

[13]   Sicard D, Legras JL. Bread, beer and wine: yeast domestication in the Saccharomyces sensu stricto complex. C R Biol 2011; 334(3): 229-36.
[http://dx.doi.org/10.1016/j.crvi.2010.12.016] [PMID: 21377618]

[14]   Bigey F, Segond D, Friedrich A, *et al.* Evidence for Two Main Domestication Trajectories in *Saccharomyces cerevisiae* Linked to Distinct Bread-Making Processes. Curr Biol 2021; 31(4): 722-732.e5.
[http://dx.doi.org/10.1016/j.cub.2020.11.016] [PMID: 33301710]

[15]   Legras JL, Merdinoglu D, Cornuet JM, Karst F. Bread, beer and wine: *Saccharomyces cerevisiae* diversity reflects human history. Mol Ecol 2007; 16(10): 2091-102.
[http://dx.doi.org/10.1111/j.1365-294X.2007.03266.x] [PMID: 17498234]

[16]   Han DY, Han PJ, Rumbold K, *et al.* Adaptive Gene Content and Allele Distribution Variations in the Wild and Domesticated Populations of *Saccharomyces cerevisiae.* Front Microbiol 2021; 12: 631250.
[http://dx.doi.org/10.3389/fmicb.2021.631250] [PMID: 33679656]

[17]   Gonzalez Flores M, Rodríguez ME, Peris D, Querol A, Barrio E, Lopes CA. Human-associated migration of Holarctic Saccharomyces uvarum strains to Patagonia. Fungal Ecol 2020; 48: 100990.
[http://dx.doi.org/10.1016/j.funeco.2020.100990]

[18]   Boynton PJ, Greig D. The ecology and evolution of non-domesticated Saccharomyces species. Yeast 2014; 31(12): 449-62.
[PMID: 25242436]

[19]   Tilakaratna V, Bensasson D. Habitat predicts levels of genetic admixture in *Saccharomyces cerevisiae.* G3 Genes, Genomes. G3 (Bethesda) 2017; 7(9): 2919-29.
[http://dx.doi.org/10.1534/g3.117.041806] [PMID: 28696926]

[20]   Capriotti A. Yeasts from U.S.A. soils. Arch Mikrobiol 1967; 57(4): 406-13.
[http://dx.doi.org/10.1007/BF00416939] [PMID: 5607401]

[21]   Almeida P, Gonçalves C, Teixeira S, *et al.* A Gondwanan imprint on global diversity and domestication of wine and cider yeast Saccharomyces uvarum. Nat Commun 2014; 5: 4044.
[http://dx.doi.org/10.1038/ncomms5044] [PMID: 24887054]

[22]   Zhang H, Skelton A, Gardner RC, Goddard MR. Saccharomyces paradoxus and *Saccharomyces cerevisiae* reside on oak trees in New Zealand: evidence for migration from Europe and interspecies hybrids. FEMS Yeast Res 2010; 10(7): 941-7.
[http://dx.doi.org/10.1111/j.1567-1364.2010.00681.x] [PMID: 20868381]

[23]   Liti G, Carter DM, Moses AM, *et al.* Population genomics of domestic and wild yeasts. Nature 2009; 458(7236): 337-41.
[http://dx.doi.org/10.1038/nature07743] [PMID: 19212322]

[24]   Fukuda T, Fukuda K, Takahashi A, *et al.* Analysis of deletion mutations of the rpsL gene in the yeast *Saccharomyces cerevisiae* detected after long-term flight on the Russian space station Mir. Mutat Res 2000; 470(2): 125-32.
[http://dx.doi.org/10.1016/S1383-5742(00)00054-5] [PMID: 11027966]

[25]   Barbosa R, Almeida P, Safar SVB, *et al.* Evidence of natural hybridization in Brazilian wild lineages of *Saccharomyces cerevisiae*. Genome Biol Evol 2016; 8(2): 317-29.
[http://dx.doi.org/10.1093/gbe/evv263] [PMID: 26782936]

[26]   Zaky AS, Greetham D, Louis EJ, Tucker GA, Du C. A new isolation and evaluation method for marine-derived yeast spp. with potential applications in industrial biotechnology. J Microbiol Biotechnol 2016; 26(11): 1891-907.
[http://dx.doi.org/10.4014/jmb.1605.05074] [PMID: 27435537]

[27]   Gerbens-Leenes W, Hoekstra AY, van der Meer TH. The water footprint of bioenergy. Proc Natl Acad Sci USA 2009; 106(25): 10219-23.
[http://dx.doi.org/10.1073/pnas.0812619106] [PMID: 19497862]

[28]   Wu M, Mintz M, Wang M, Arora S. Water consumption in the production of ethanol and petroleum gasoline. Environ Manage 2009; 44(5): 981-97.
[http://dx.doi.org/10.1007/s00267-009-9370-0] [PMID: 19774326]

[29]   Zaky AS, French CE, Tucker GA, Du C. Improving the productivity of bioethanol production using marine yeast and seawater-based media. Biomass Bioenergy 2020; 139: 105615.
[http://dx.doi.org/10.1016/j.biombioe.2020.105615]

[30]   Greetham D, Zaky AS, Du C. Exploring the tolerance of marine yeast to inhibitory compounds for improving bioethanol production. Sustain Energy Fuels 2019; 3: 1545-53.
[http://dx.doi.org/10.1039/C9SE00029A]

[31]   Zaky AS, Greetham D, Tucker GA, Du C. The establishment of a marine focused biorefinery for bioethanol production using seawater and a novel marine yeast strain. Sci Rep 2018; 8(1): 12127.
[http://dx.doi.org/10.1038/s41598-018-30660-x] [PMID: 30108287]

[32]   Sampaio JP, Gonçalves P. Biogeography and ecology of the genus Saccharomyces.Yeasts Nat Ecosyst Ecol. Springer International Publishing 2017; pp. 131-53.
[http://dx.doi.org/10.1007/978-3-319-61575-2_5]

[33]   Sylvester K, Wang QM, James B, Mendez R, Hulfachor AB, Hittinger CT. Temperature and host preferences drive the diversification of Saccharomyces and other yeasts: a survey and the discovery of eight new yeast species. FEMS Yeast Res 2015; 15(3): fov002.
[http://dx.doi.org/10.1093/femsyr/fov002] [PMID: 25743785]

[34]   Simões MF, Antunes A, Ottoni CA, *et al.* Soil and Rhizosphere Associated Fungi in Gray Mangroves (Avicennia marina) from the Red Sea--A Metagenomic Approach. Genomics Proteomics Bioinformatics 2015; 13(5): 310-20.
[http://dx.doi.org/10.1016/j.gpb.2015.07.002] [PMID: 26549842]

[35]   Nasanit R, Imklin N, Limtong S. Assessment of yeasts in tropical peat swamp forests in Thailand. Mycol Prog 2020; 19: 1559-73.
[http://dx.doi.org/10.1007/s11557-020-01646-9]

[36]   Aljohani R, Samarasinghe H, Ashu T, Xu J. Diversity and relationships among strains of culturable yeasts in agricultural soils in Cameroon. Sci Rep 2018; 8(1): 15687.
[http://dx.doi.org/10.1038/s41598-018-34122-2] [PMID: 30356081]

[37]   Boynton PJ, Kowallik V, Landermann D, Stukenbrock EH. Quantifying the efficiency and biases of forest Saccharomyces sampling strategies. Yeast 2019; 36(11): 657-68.

[http://dx.doi.org/10.1002/yea.3435] [PMID: 31348543]

[38]   Alsammar HF, Naseeb S, Brancia LB, Gilman RT, Wang P, Delneri D. Targeted metagenomics approach to capture the biodiversity of Saccharomyces genus in wild environments. Environ Microbiol Rep 2019; 11(2): 206-14.
[http://dx.doi.org/10.1111/1758-2229.12724] [PMID: 30507071]

[39]   Brysch-Herzberg M, Seidel M. Distribution patterns of Saccharomyces species in cultural landscapes of Germany. FEMS Yeast Res 2017; 17(5): fox033.
[http://dx.doi.org/10.1093/femsyr/fox033] [PMID: 28520895]

[40]   González ML, Sturm ME, Lerena MC, *et al*. Persistence and reservoirs of *Saccharomyces cerevisiae* biodiversity in different vineyard niches. Food Microbiol 2020; 86: 103328.
[http://dx.doi.org/10.1016/j.fm.2019.103328] [PMID: 31703883]

[41]   Catania CD, del Monte SA, Uliarte EM, del Monte RF, Tonietto J. El clima vitícola de las regiones productoras de uvas para vinos de Argentina.Caracter Climática Regiões Vitivinícolas Ibero-americanas. 1st ed. Bento Gonçalves, Brasil: Embrapa Uva e Vinho 2007; p. 64.

[42]   Nally MC, Pesce VM, Maturano YP, *et al*. Biocontrol of Botrytis cinerea in table grapes by non-pathogenic indigenous *Saccharomyces cerevisiae* yeasts isolated from viticultural environments in Argentina. Postharvest Biol Technol 2012; 64: 40-8.
[http://dx.doi.org/10.1016/j.postharvbio.2011.09.009]

[43]   Yurkov AM. Yeasts of the soil - obscure but precious. Yeast 2018; 35(5): 369-78.
[http://dx.doi.org/10.1002/yea.3310] [PMID: 29365211]

[44]   Kowallik V, Miller E, Greig D. The interaction of Saccharomyces paradoxus with its natural competitors on oak bark. Mol Ecol 2015; 24(7): 1596-610.
[http://dx.doi.org/10.1111/mec.13120] [PMID: 25706044]

[45]   Naseeb S, James SA, Alsammar H, *et al*. Saccharomyces jurei sp. nov., isolation and genetic identification of a novel yeast species from Quercus robur. Int J Syst Evol Microbiol 2017; 67(6): 2046-52.
[http://dx.doi.org/10.1099/ijsem.0.002013] [PMID: 28639933]

[46]   Iurkov AM. First isolation of the yeast Saccharomyces paradoxus in Western Siberia. Mikrobiologiia 2005; 74(4): 533-6.
[PMID: 16211858]

[47]   Sampaio JP, Gonçalves P. Natural populations of Saccharomyces kudriavzevii in Portugal are associated with oak bark and are sympatric with *S. cerevisiae* and *S. paradoxus*. Appl Environ Microbiol 2008; 74(7): 2144-52.
[http://dx.doi.org/10.1128/AEM.02396-07] [PMID: 18281431]

[48]   Kowallik V, Greig D. A systematic forest survey showing an association of *Saccharomyces paradoxus* with oak leaf litter. Environ Microbiol Rep 2016; 8(5): 833-41.
[http://dx.doi.org/10.1111/1758-2229.12446] [PMID: 27481438]

[49]   Sniegowski PD, Dombrowski PG, Fingerman E. *Saccharomyces cerevisiae* and *Saccharomyces paradoxus* coexist in a natural woodland site in North America and display different levels of reproductive isolation from European conspecifics. FEMS Yeast Res 2002; 1(4): 299-306.
[PMID: 12702333]

[50]   Vaudano E, Quinterno G, Costantini A, Pulcini L, Pessione E, Garcia-Moruno E. Yeast distribution in Grignolino grapes growing in a new vineyard in Piedmont and the technological characterization of indigenous Saccharomyces spp. strains. Int J Food Microbiol 2019; 289: 154-61.
[http://dx.doi.org/10.1016/j.ijfoodmicro.2018.09.016] [PMID: 30245288]

[51]   Gayevskiy V, Goddard MR. *Saccharomyces eubayanus* and *Saccharomyces arboricola* reside in North Island native New Zealand forests. Environ Microbiol 2016; 18(4): 1137-47.
[http://dx.doi.org/10.1111/1462-2920.13107] [PMID: 26522264]

[52] Taylor MW, Tsai P, Anfang N, Ross HA, Goddard MR. Pyrosequencing reveals regional differences in fruit-associated fungal communities. Environ Microbiol 2014; 16(9): 2848-58.
[http://dx.doi.org/10.1111/1462-2920.12456] [PMID: 24650123]

[53] McGovern PE, Zhang J, Tang J, *et al.* Fermented beverages of pre- and proto-historic China. Proc Natl Acad Sci USA 2004; 101(51): 17593-8.
[http://dx.doi.org/10.1073/pnas.0407921102] [PMID: 15590771]

[54] Gava CAT, de Castro APC, Pereira CA, Fernandes-Júnior PI. Isolation of fruit colonizer yeasts and screening against mango decay caused by multiple pathogens. Biol Control 2018; 117: 137-46.
[http://dx.doi.org/10.1016/j.biocontrol.2017.11.005]

[55] Dabassa Koricha A, Han DY, Bacha K, Bai FY. Occurrence and molecular identification of wild yeasts from Jimma Zone, South West Ethiopia. Microorganisms 2019; 7(12): 633.
[http://dx.doi.org/10.3390/microorganisms7120633] [PMID: 31801247]

[56] Mailafia S, Okoh GR, Olabode HOK, Osanupin R. Isolation and identification of fungi associated with spoilt fruits vended in Gwagwalada market, Abuja, Nigeria. Vet World 2017; 10(4): 393-7.
[http://dx.doi.org/10.14202/vetworld.2017.393-397] [PMID: 28507410]

[57] Tristezza M, Fantastico L, Vetrano C, *et al.* Molecular and technological characterization of *Saccharomyces cerevisiae* strains isolated from natural fermentation of susumaniello grape must in Apulia, Southern Italy. Int J Microbiol 2014; 2014: 897428.
[http://dx.doi.org/10.1155/2014/897428] [PMID: 24672552]

[58] Šuranská H, Vránová D, Omelková J. Isolation, identification and characterization of regional indigenous *Saccharomyces cerevisiae* strains. Braz J Microbiol 2016; 47(1): 181-90.
[http://dx.doi.org/10.1016/j.bjm.2015.11.010] [PMID: 26887243]

[59] Almeida P, Barbosa R, Zalar P, *et al.* A population genomics insight into the Mediterranean origins of wine yeast domestication. Mol Ecol 2015; 24(21): 5412-27.
[http://dx.doi.org/10.1111/mec.13341] [PMID: 26248006]

[60] Marsit S, Sanchez I, Galeote V, Dequin S. Horizontally acquired oligopeptide transporters favour adaptation of *Saccharomyces cerevisiae* wine yeast to oenological environment. Environ Microbiol 2016; 18(4): 1148-61.
[http://dx.doi.org/10.1111/1462-2920.13117] [PMID: 26549518]

[61] Marsit S, Mena A, Bigey F, *et al.* Evolutionary advantage conferred by an eukaryote-to-eukaryote gene transfer event in wine yeasts. Mol Biol Evol 2015; 32(7): 1695-707.
[http://dx.doi.org/10.1093/molbev/msv057] [PMID: 25750179]

[62] Bozdag GO, Greig D. The genetics of a putative social trait in natural populations of yeast. Mol Ecol 2014; 23(20): 5061-71.
[http://dx.doi.org/10.1111/mec.12904] [PMID: 25169714]

[63] Naumov GI, Serpova EV, Naumova ES. A genetically isolated population of *Saccharomyces cerevisiae* in Malaysia. Mikrobiologiia 2006; 75(2): 245-9.
[PMID: 16758873]

[64] Sandhu DK, Waraich MK. Yeasts associated with pollinating bees and flower nectar. Microb Ecol 1985; 11(1): 51-8.
[http://dx.doi.org/10.1007/BF02015108] [PMID: 24221239]

[65] Sláviková E, Vadkertiová R, Vránová D. Yeasts colonizing the leaf surfaces. J Basic Microbiol 2007; 47(4): 344-50.
[http://dx.doi.org/10.1002/jobm.200710310] [PMID: 17645279]

[66] Yun JH, Jung MJ, Kim PS, Bae JW. Social status shapes the bacterial and fungal gut communities of the honey bee. Sci Rep 2018; 8(1): 2019.
[http://dx.doi.org/10.1038/s41598-018-19860-7] [PMID: 29386588]

[67]   Alves SL, Müller C, Bonatto C, *et al.* Bioprospection of Enzymes and Microorganisms in Insects to Improve Second-Generation Ethanol Production. Ind Biotechnol (New Rochelle NY) 2019; 15: 336-49.
        [http://dx.doi.org/10.1089/ind.2019.0019]

[68]   Meriggi N, Di Paola M, Vitali F, *et al. Saccharomyces cerevisiae* Induces Immune Enhancing and Shapes Gut Microbiota in Social Wasps. Front Microbiol 2019; 10: 2320.
        [http://dx.doi.org/10.3389/fmicb.2019.02320] [PMID: 31681197]

[69]   Stefanini I. Yeast-insect associations: It takes guts. Yeast 2018; 35(4): 315-30.
        [http://dx.doi.org/10.1002/yea.3309] [PMID: 29363168]

[70]   Meriggi N, Di Paola M, Cavalieri D, Stefanini I. *Saccharomyces cerevisiae* - Insects Association: Impacts, Biogeography, and Extent. Front Microbiol 2020; 11: 1629.
        [http://dx.doi.org/10.3389/fmicb.2020.01629] [PMID: 32760380]

[71]   Yeoman CJ, Ishaq SL, Bichi E, Olivo SK, Lowe J, Aldridge BM. Biogeographical differences in the influence of maternal microbial sources on the early successional development of the bovine neonatal gastrointestinal tract. Sci Rep 2018; 8(1): 3197.
        [http://dx.doi.org/10.1038/s41598-018-21440-8] [PMID: 29453364]

[72]   Fietto JLR, Araújo RS, Valadão FN, *et al.* Molecular and physiological comparisons between *Saccharomyces cerevisiae* and Saccharomyces boulardii. Can J Microbiol 2004; 50(8): 615-21.
        [http://dx.doi.org/10.1139/w04-050] [PMID: 15467787]

[73]   Pais P, Almeida V, Yılmaz M, Teixeira MC. Saccharomyces boulardii: What makes it tick as successful probiotic? J Fungi (Basel) 2020; 6(2): 1-15.
        [http://dx.doi.org/10.3390/jof6020078] [PMID: 32512834]

[74]   Boix-Amorós A, Martinez-Costa C, Querol A, Collado MC, Mira A. Multiple approaches detect the presence of fungi in human breastmilk samples from healthy mothers. Sci Rep 2017; 7(1): 13016.
        [http://dx.doi.org/10.1038/s41598-017-13270-x] [PMID: 29026146]

[75]   Siedlarz P, Sroka M, Dyląg M, Nawrot U, Gonchar M, Kus-Liśkiewicz M. Preliminary physiological characteristics of thermotolerant *Saccharomyces cerevisiae* clinical isolates identified by molecular biology techniques. Lett Appl Microbiol 2016; 62(3): 277-82.
        [http://dx.doi.org/10.1111/lam.12542] [PMID: 26693946]

[76]   Seng P, Cerlier A, Cassagne C, Coulange M, Legré R, Stein A. *Saccharomyces cerevisiae* osteomyelitis in an immunocompetent baker. IDCases 2016; 5: 1-3.
        [http://dx.doi.org/10.1016/j.idcr.2016.05.002] [PMID: 27347482]

[77]   Zembower TR. Epidemiology of infections in cancer patients. Cancer Treat Res 2014; 161: 43-89.
        [http://dx.doi.org/10.1007/978-3-319-04220-6_2] [PMID: 24706221]

[78]   Enache-Angoulvant A, Hennequin C. Invasive Saccharomyces infection: a comprehensive review. Clin Infect Dis 2005; 41(11): 1559-68.
        [http://dx.doi.org/10.1086/497832] [PMID: 16267727]

[79]   González Nafría N, Redondo Robles L, Lara Lezama LB. *Saccharomyces cerevisiae* infection in an immunocompetent host. Med Clin (Barc) 2019; 152(3): 122-3.
        [PMID: 29980287]

[80]   Mašínová T, Bahnmann BD, Větrovský T, Tomšovský M, Merunková K, Baldrian P. Drivers of yeast community composition in the litter and soil of a temperate forest. FEMS Microbiol Ecol 2017; 93(2): 223.
        [http://dx.doi.org/10.1093/femsec/fiw223] [PMID: 27789535]

[81]   Katz S, Maytag F. Brewing an ancient beer. Archaeology 1991; 44: 22-3.

[82]   Barbosa R, Pontes A, Santos RO, *et al.* Multiple rounds of artificial selection promote microbe secondary domestication-the case of cachaça yeasts. Genome Biol Evol 2018; 10(8): 1939-55.

[http://dx.doi.org/10.1093/gbe/evy132] [PMID: 29982460]

[83]   Liti G. The fascinating and secret wild life of the budding yeast S. cerevisiae. eLife 2015; 4: 1-9.
        [http://dx.doi.org/10.7554/eLife.05835] [PMID: 25807086]

[84]   Barnett JA. A history of research on yeasts 10: foundations of yeast genetics. Yeast 2007; 24(10): 799-
        845.
        [http://dx.doi.org/10.1002/yea.1513] [PMID: 17638318]

[85]   Fay JC, Benavides JA. Evidence for domesticated and wild populations of *Saccharomyces cerevisiae*.
        PLoS Genet 2005; 1(1): 66-71.
        [http://dx.doi.org/10.1371/journal.pgen.0010005] [PMID: 16103919]

[86]   Walker GM, Basso TO. Mitigating stress in industrial yeasts. Fungal Biol 2020; 124(5): 387-97.
        [http://dx.doi.org/10.1016/j.funbio.2019.10.010] [PMID: 32389301]

[87]   Alves SL, Herberts RA, Hollatz C, Miletti LC, Stambuk BU. Maltose and maltotriose active transport
        and fermentation by *Saccharomyces cerevisiae*. J Am Soc Brew Chem 2007; 65: 99-104.
        [http://dx.doi.org/10.1094/ASBCJ-2007-0411-01]

[88]   Gonçalves M, Pontes A, Almeida P, *et al.* Distinct domestication trajectories in top-fermenting beer
        yeasts and wine yeasts. Curr Biol 2016; 26(20): 2750-61.
        [http://dx.doi.org/10.1016/j.cub.2016.08.040] [PMID: 27720622]

[89]   Gallone B, Steensels J, Prahl T, *et al.* Domestication and Divergence of *Saccharomyces cerevisiae*
        Beer Yeasts. Cell 2016; 166(6): 1397-1410.e16.
        [http://dx.doi.org/10.1016/j.cell.2016.08.020] [PMID: 27610566]

[90]   Basso LC, de Amorim HV, de Oliveira AJ, Lopes ML. Yeast selection for fuel ethanol production in
        Brazil. FEMS Yeast Res 2008; 8(7): 1155-63.
        [http://dx.doi.org/10.1111/j.1567-1364.2008.00428.x] [PMID: 18752628]

[91]   Argueso JL, Carazzolle MF, Mieczkowski PA, *et al.* Genome structure of a *Saccharomyces cerevisiae*
        strain widely used in bioethanol production. Genome Res 2009; 19(12): 2258-70.
        [http://dx.doi.org/10.1101/gr.091777.109] [PMID: 19812109]

[92]   Stambuk BU, Dunn B, Alves SL Jr, Duval EH, Sherlock G. Industrial fuel ethanol yeasts contain
        adaptive copy number changes in genes involved in vitamin B1 and B6 biosynthesis. Genome Res
        2009; 19(12): 2271-8.
        [http://dx.doi.org/10.1101/gr.094276.109] [PMID: 19897511]

[93]   Jacobus AP, Stephens TG, Youssef P, *et al.* Comparative genomics supports that brazilian bioethanol
        Saccharomyces cerevisiae comprise a unified group of domesticated strains related to cachaça spirit
        yeasts. Front Microbiol 2021; 12: 644089.
        [http://dx.doi.org/10.3389/fmicb.2021.644089] [PMID: 33936002]

[94]   Eberlein C, Leducq JB, Landry CR. The genomics of wild yeast populations sheds light on the
        domestication of man's best (micro) friend. Mol Ecol 2015; 24(21): 5309-11.
        [http://dx.doi.org/10.1111/mec.13380] [PMID: 26509691]

[95]   Libkind D, Hittinger CT, Valério E, *et al.* Microbe domestication and the identification of the wild
        genetic stock of lager-brewing yeast. Proc Natl Acad Sci USA 2011; 108(35): 14539-44.
        [http://dx.doi.org/10.1073/pnas.1105430108] [PMID: 21873232]

[96]   Rizzetto L, De Filippo C, Cavalieri D. Richness and diversity of mammalian fungal communities
        shape innate and adaptive immunity in health and disease. Eur J Immunol 2014; 44(11): 3166-81.
        [http://dx.doi.org/10.1002/eji.201344403] [PMID: 25257052]

[97]   Steenwyk J, Rokas A. Extensive copy number variation in fermentation-related genes among
        Saccharomyces cerevisiae wine strains. G3 Genes, Genomes. G3 (Bethesda) 2017; 7(5): 1475-85.
        [http://dx.doi.org/10.1534/g3.117.040105] [PMID: 28292787]

[98]   Legras JL, Galeote V, Bigey F, *et al.* Adaptation of S. Cerevisiae to fermented food environments

reveals remarkable genome plasticity and the footprints of domestication. Mol Biol Evol 2018; 35(7): 1712-27.
[http://dx.doi.org/10.1093/molbev/msy066] [PMID: 29746697]

[99]   Goddard MR. Quantifying the complexities of *Saccharomyces cerevisiae*'s ecosystem engineering via fermentation. Ecology 2008; 89(8): 2077-82.
[http://dx.doi.org/10.1890/07-2060.1] [PMID: 18724717]

[100]   Walther A, Hesselbart A, Wendland J. Genome sequence of *Saccharomyces carlsbergensis*, the world's first pure culture lager yeast. G3 (Bethesda) 2014; 4(5): 783-93.
[http://dx.doi.org/10.1534/g3.113.010090] [PMID: 24578374]

[101]   Hansen E. Recherches sur la physiologie et la morphologie des ferments alcooliques. V. Methodes pour obtenir des cultures pures de Saccharomyces et de mikroorganismes analogues. C R Trav Lab Carlsberg 1883; 2: 92-105.

[102]   Van der Walt J. Saccharomyces (Meyen) emend. Reess. In: Lodder J, Ed. Yeasts, A Taxon Study. Amsterdam 1970; pp. 555-718.

[103]   Pontes A, Hutzler M, Brito PH, Sampaio JP. Revisiting the taxonomic synonyms and populations of *Saccharomyces cerevisiae*—phylogeny, phenotypes, ecology and domestication. Microorganisms 2020; 8(6): 1-23.
[http://dx.doi.org/10.3390/microorganisms8060903] [PMID: 32549402]

[104]   Dujon BA, Louis EJ. Genome diversity and evolution in the budding yeasts (Saccharomycotina). Genetics 2017; 206(2): 717-50.
[http://dx.doi.org/10.1534/genetics.116.199216] [PMID: 28592505]

[105]   Zemančíková J, Kodedová M, Papoušková K, Sychrová H. Four Saccharomyces species differ in their tolerance to various stresses though they have similar basic physiological parameters. Folia Microbiol (Praha) 2018; 63(2): 217-27.
[http://dx.doi.org/10.1007/s12223-017-0559-y] [PMID: 29052811]

[106]   Kuijpers NGA, Solis-Escalante D, Luttik MAH, *et al.* Pathway swapping: Toward modular engineering of essential cellular processes. Proc Natl Acad Sci USA 2016; 113(52): 15060-5.
[http://dx.doi.org/10.1073/pnas.1606701113] [PMID: 27956602]

[107]   Boonekamp FJ, Dashko S, van den Broek M, Gehrmann T, Daran J-M, Daran-Lapujade P. The genetic makeup and expression of the glycolytic and fermentative pathways are highly conserved within the *saccharomyces* genus. Front Genet 2018; 9: 504.
[http://dx.doi.org/10.3389/fgene.2018.00504] [PMID: 30505317]

[108]   Gonçalves P, Valério E, Correia C, de Almeida JMGCF, Sampaio JP. Evidence for divergent evolution of growth temperature preference in sympatric Saccharomyces species. PLoS One 2011; 6(6): e20739.
[http://dx.doi.org/10.1371/journal.pone.0020739] [PMID: 21674061]

[109]   Salvadó Z, Arroyo-López FN, Guillamón JM, Salazar G, Querol A, Barrio E. Temperature adaptation markedly determines evolution within the genus Saccharomyces. Appl Environ Microbiol 2011; 77(7): 2292-302.
[http://dx.doi.org/10.1128/AEM.01861-10] [PMID: 21317255]

[110]   Weiss CV, Roop JI, Hackley RK, *et al.* Genetic dissection of interspecific differences in yeast thermotolerance. Nat Genet 2018; 50(11): 1501-4.
[http://dx.doi.org/10.1038/s41588-018-0243-4] [PMID: 30297967]

[111]   Duan SF, Han PJ, Wang QM, *et al.* The origin and adaptive evolution of domesticated populations of yeast from Far East Asia. Nat Commun 2018; 9(1): 2690.
[http://dx.doi.org/10.1038/s41467-018-05106-7] [PMID: 30002370]

[112]   Scannell DR, Zill OA, Rokas A, *et al.* The awesome power of yeast evolutionary genetics: New genome sequences and strain resources for the Saccharomyces sensu stricto genus. G3 Genes,

Genomes. G3 (Bethesda) 2011; 1(1): 11-25.
[http://dx.doi.org/10.1534/g3.111.000273] [PMID: 22384314]

[113]   Hittinger CT. Saccharomyces diversity and evolution: a budding model genus. Trends Genet 2013; 29(5): 309-17.
[http://dx.doi.org/10.1016/j.tig.2013.01.002] [PMID: 23395329]

[114]   Gabaldón T. 2020.>Hybridization and the origin of new yeast lineages. FEMS Yeast Res 20:foaa040
[http://dx.doi.org/10.1093/femsyr/foaa040]

[115]   Reuter M, Bell G, Greig D. Increased outbreeding in yeast in response to dispersal by an insect vector. Curr Biol 2007; 17(3): R81-3.
[http://dx.doi.org/10.1016/j.cub.2006.11.059] [PMID: 17276903]

[116]   Di Paola M, Meriggi N, Cavalieri D. Applications of wild isolates of *saccharomyces* yeast for industrial fermentation: the gut of social insects as niche for yeast hybrids' production. Front Microbiol 2020; 11: 578425.
[http://dx.doi.org/10.3389/fmicb.2020.578425] [PMID: 33193200]

[117]   Peter J, De Chiara M, Friedrich A, *et al.* Genome evolution across 1,011 *Saccharomyces cerevisiae* isolates. Nature 2018; 556(7701): 339-44.
[http://dx.doi.org/10.1038/s41586-018-0030-5] [PMID: 29643504]

[118]   D'Angiolo M, De Chiara M, Yue JX, *et al.* A yeast living ancestor reveals the origin of genomic introgressions. Nature 2020; 587(7834): 420-5.
[http://dx.doi.org/10.1038/s41586-020-2889-1] [PMID: 33177709]

[119]   Dunn B, Paulish T, Stanbery A, *et al.* Recurrent rearrangement during adaptive evolution in an interspecific yeast hybrid suggests a model for rapid introgression. PLoS Genet 2013; 9(3): e1003366.
[http://dx.doi.org/10.1371/journal.pgen.1003366] [PMID: 23555283]

[120]   Lairón-Peris M, Pérez-Través L, Muñiz-Calvo S, *et al.* Differential contribution of the parental genomes to a *s. cerevisiae* × *s. uvarum* hybrid, inferred by phenomic, genomic, and transcriptomic analyses, at different industrial stress conditions. Front Bioeng Biotechnol 2020; 8: 129.
[http://dx.doi.org/10.3389/fbioe.2020.00129] [PMID: 32195231]

[121]   Morard M, Benavent-Gil Y, Ortiz-Tovar G, *et al.* Genome structure reveals the diversity of mating mechanisms in *Saccharomyces cerevisiae* x *Saccharomyces kudriavzevii* hybrids, and the genomic instability that promotes phenotypic diversity. Microb Genom 2020; 6(3): 6.
[http://dx.doi.org/10.1099/mgen.0.000333] [PMID: 32065577]

[122]   Stambuk BU, Alves SL Jr, Hollatz C, Zastrow CR. Improvement of maltotriose fermentation by *Saccharomyces cerevisiae*. Lett Appl Microbiol 2006; 43(4): 370-6.
[http://dx.doi.org/10.1111/j.1472-765X.2006.01982.x] [PMID: 16965366]

[123]   Zastrow CR, Mattos MA, Hollatz C, Stambuk BU. Maltotriose metabolism by *Saccharomyces cerevisiae*. Biotechnol Lett 2000; 22: 455-9.
[http://dx.doi.org/10.1023/A:1005691031880]

[124]   Cousseau FEM, Alves SL Jr, Trichez D, Stambuk BU. Characterization of maltotriose transporters from the *Saccharomyces eubayanus* subgenome of the hybrid Saccharomyces pastorianus lager brewing yeast strain Weihenstephan 34/70. Lett Appl Microbiol 2013; 56(1): 21-9.
[http://dx.doi.org/10.1111/lam.12011] [PMID: 23061413]

[125]   Tronchoni J, Gamero A, Arroyo-López FN, Barrio E, Querol A. Differences in the glucose and fructose consumption profiles in diverse Saccharomyces wine species and their hybrids during grape juice fermentation. Int J Food Microbiol 2009; 134(3): 237-43.
[http://dx.doi.org/10.1016/j.ijfoodmicro.2009.07.004] [PMID: 19632733]

[126]   Zuchowska M, Jaenicke E, König H, Claus H. Allelic variants of hexose transporter Hxt3p and hexokinases Hxk1p/Hxk2p in strains of *Saccharomyces cerevisiae* and interspecies hybrids. Yeast 2015; 32(11): 657-69.

[http://dx.doi.org/10.1002/yea.3087] [PMID: 26202678]

[127] Lopandic K, Pfliegler WP, Tiefenbrunner W, Gangl H, Sipiczki M, Sterflinger K. Genotypic and phenotypic evolution of yeast interspecies hybrids during high-sugar fermentation. Appl Microbiol Biotechnol 2016; 100(14): 6331-43.
[http://dx.doi.org/10.1007/s00253-016-7481-0] [PMID: 27075738]

[128] Sipiczki M. Interspecies hybridisation and genome chimerisation in Saccharomyces: Combining of gene pools of species and its biotechnological perspectives. Front Microbiol 2018; 9: 3071.
[http://dx.doi.org/10.3389/fmicb.2018.03071] [PMID: 30619156]

[129] Sipiczki M. Yeast two- and three-species hybrids and high-sugar fermentation. Microb Biotechnol 2019; 12(6): 1101-8.
[http://dx.doi.org/10.1111/1751-7915.13390] [PMID: 30838806]

[130] Origone AC, González Flores M, Rodríguez ME, Querol A, Lopes CA. Inheritance of winemaking stress factors tolerance in *Saccharomyces uvarum/S. eubayanus* × *S. cerevisiae* artificial hybrids. Int J Food Microbiol 2020; 320: 108500.
[http://dx.doi.org/10.1016/j.ijfoodmicro.2019.108500] [PMID: 32007764]

[131] Dunn B, Sherlock G. Reconstruction of the genome origins and evolution of the hybrid lager yeast Saccharomyces pastorianus. Genome Res 2008; 18(10): 1610-23.
[http://dx.doi.org/10.1101/gr.076075.108] [PMID: 18787083]

[132] Duval EH, Alves SL Jr, Dunn B, Sherlock G, Stambuk BU. Microarray karyotyping of maltose-fermenting Saccharomyces yeasts with differing maltotriose utilization profiles reveals copy number variation in genes involved in maltose and maltotriose utilization. J Appl Microbiol 2010; 109(1): 248-59.
[http://dx.doi.org/10.1111/j.1365-2672.2009.04656.x] [PMID: 20070441]

[133] Gibson BR, Storgårds E, Krogerus K, Vidgren V. Comparative physiology and fermentation performance of Saaz and Frohberg lager yeast strains and the parental species *Saccharomyces eubayanus*. Yeast 2013; 30(7): 255-66.
[http://dx.doi.org/10.1002/yea.2960] [PMID: 23695993]

[134] Nguyen HV, Boekhout T. Characterization of Saccharomyces uvarum (Beijerinck, 1898) and related hybrids: assessment of molecular markers that predict the parent and hybrid genomes and a proposal to name yeast hybrids. FEMS Yeast Res 2017; 17(2): fox014.
[http://dx.doi.org/10.1093/femsyr/fox014] [PMID: 28334169]

[135] Kelly J, Yang F, Dowling L, *et al.* Characterization of *Saccharomyces bayanus* CN1 for fermenting partially dehydrated grapes grown in cool climate winemaking regions. Fermentation (Basel) 2018; 4: 77.
[http://dx.doi.org/10.3390/fermentation4030077]

[136] Nikulin J, Krogerus K, Gibson B. Alternative Saccharomyces interspecies hybrid combinations and their potential for low-temperature wort fermentation. Yeast 2018; 35(1): 113-27.
[http://dx.doi.org/10.1002/yea.3246] [PMID: 28755430]

[137] Hebly M, Brickwedde A, Bolat I, *et al.* S. cerevisiae × S. eubayanus interspecific hybrid, the best of both worlds and beyond. FEMS Yeast Res 2015; 15(3): fov005.
[http://dx.doi.org/10.1093/femsyr/fov005] [PMID: 25743788]

[138] Krogerus K, Magalhães F, Vidgren V, Gibson B. New lager yeast strains generated by interspecific hybridization. J Ind Microbiol Biotechnol 2015; 42(5): 769-78.
[http://dx.doi.org/10.1007/s10295-015-1597-6] [PMID: 25682107]

[139] Diderich JA, Weening SM, van den Broek M, Pronk JT, Daran JG. Selection of Pof- *Saccharomyces eubayanus* variants for the construction of *S. cerevisiae* × *S. eubayanus* hybrids with reduced 4-Vinyl guaiacol formation. Front Microbiol 2018; 9: 1640.
[http://dx.doi.org/10.3389/fmicb.2018.01640] [PMID: 30100898]

[140] Magalhães F, Krogerus K, Vidgren V, Sandell M, Gibson B. Improved cider fermentation performance and quality with newly generated *Saccharomyces cerevisiae* × *Saccharomyces eubayanus* hybrids. J Ind Microbiol Biotechnol 2017; 44(8): 1203-13.
[http://dx.doi.org/10.1007/s10295-017-1947-7] [PMID: 28451838]

[141] Magalhães F, Krogerus K, Castillo S, Ortiz-Julien A, Dequin S, Gibson B. Exploring the potential of *Saccharomyces eubayanus* as a parent for new interspecies hybrid strains in winemaking. FEMS Yeast Res 2017; 17(5): fox049.
[http://dx.doi.org/10.1093/femsyr/fox049] [PMID: 28810703]

[142] González SS, Gallo L, Climent MA, Barrio E, Querol A. Enological characterization of natural hybrids from *Saccharomyces cerevisiae and S. kudriavzevii.* Int J Food Microbiol 2007; 116(1): 11-8.
[http://dx.doi.org/10.1016/j.ijfoodmicro.2006.10.047] [PMID: 17346840]

[143] González SS, Barrio E, Querol A. Molecular characterization of new natural hybrids of *Saccharomyces cerevisiae and S. kudriavzevii* in brewing. Appl Environ Microbiol 2008; 74(8): 2314-20.
[http://dx.doi.org/10.1128/AEM.01867-07] [PMID: 18296532]

[144] Gangl H, Batusic M, Tscheik G, Tiefenbrunner W, Hack C, Lopandic K. Exceptional fermentation characteristics of natural hybrids from *Saccharomyces cerevisiae* and S. kudriavzevii. N Biotechnol 2009; 25(4): 244-51.
[http://dx.doi.org/10.1016/j.nbt.2008.10.001] [PMID: 19026772]

[145] Bellon JR, Eglinton JM, Siebert TE, *et al.* Newly generated interspecific wine yeast hybrids introduce flavour and aroma diversity to wines. Appl Microbiol Biotechnol 2011; 91(3): 603-12.
[http://dx.doi.org/10.1007/s00253-011-3294-3] [PMID: 21538112]

[146] Bellon JR, Schmid F, Capone DL, Dunn BL, Chambers PJ. Introducing a new breed of wine yeast: interspecific hybridisation between a commercial *Saccharomyces cerevisiae* wine yeast and Saccharomyces mikatae. PLoS One 2013; 8(4): e62053.
[http://dx.doi.org/10.1371/journal.pone.0062053] [PMID: 23614011]

[147] Stelkens RB, Brockhurst MA, Hurst GDD, Miller EL, Greig D. The effect of hybrid transgression on environmental tolerance in experimental yeast crosses. J Evol Biol 2014; 27(11): 2507-19.
[http://dx.doi.org/10.1111/jeb.12494] [PMID: 25262771]

[148] Bernardes JP, Stelkens RB, Greig D. Heterosis in hybrids within and between yeast species. J Evol Biol 2017; 30(3): 538-48.
[http://dx.doi.org/10.1111/jeb.13023] [PMID: 27933674]

[149] Pfliegler WP, Atanasova L, Karanyicz E, *et al.* Generation of new genotypic and phenotypic features in artificial and natural yeast hybrids. Food Technol Biotechnol 2014; 52: 46-57.

[150] Bellon JR, Yang F, Day MP, Inglis DL, Chambers PJ. Designing and creating Saccharomyces interspecific hybrids for improved, industry relevant, phenotypes. Appl Microbiol Biotechnol 2015; 99(20): 8597-609.
[http://dx.doi.org/10.1007/s00253-015-6737-4] [PMID: 26099331]

[151] da Silva T, Albertin W, Dillmann C, *et al.* Hybridization within Saccharomyces genus results in homoeostasis and phenotypic novelty in winemaking conditions. PLoS One 2015; 10(5): e0123834.
[http://dx.doi.org/10.1371/journal.pone.0123834] [PMID: 25946464]

[152] Origone AC, Rodríguez ME, Oteiza JM, Querol A, Lopes CA. *Saccharomyces cerevisiae* × *Saccharomyces uvarum* hybrids generated under different conditions share similar winemaking features. Yeast 2018; 35(1): 157-71.
[http://dx.doi.org/10.1002/yea.3295] [PMID: 29131448]

[153] Basso L, Basso T, Rocha S. Ethanol Production in Brazil: The Industrial Process and Its Impact on Yeast Fermentation. In: Biofuel production-recent developments and prospects Intechopen, Croatia, . 2011; pp. 85-100.

[154] Della-Bianca BE, Basso TO, Stambuk BU, Basso LC, Gombert AK. What do we know about the yeast strains from the Brazilian fuel ethanol industry? Appl Microbiol Biotechnol 2013; 97(3): 979-91.
[http://dx.doi.org/10.1007/s00253-012-4631-x] [PMID: 23271669]

[155] Jacobus AP, Gross J, Evans JH, Ceccato-Antonini SR, Gombert AK. *Saccharomyces cerevisiae* strains used industrially for bioethanol production. Essays Biochem 2021; 65(2): 147-61.
[http://dx.doi.org/10.1042/EBC20200160] [PMID: 34156078]

[156] Lopes ML, Paulillo SC de L, Godoy A, *et al.* Ethanol production in Brazil: a bridge between science and industry. Braz J Microbiol 2016; 47 (Suppl. 1): 64-76.
[http://dx.doi.org/10.1016/j.bjm.2016.10.003] [PMID: 27818090]

[157] Amorim HV, Lopes ML, de Castro Oliveira JV, Buckeridge MS, Goldman GH. Scientific challenges of bioethanol production in Brazil. Appl Microbiol Biotechnol 2011; 91(5): 1267-75.
[http://dx.doi.org/10.1007/s00253-011-3437-6] [PMID: 21735264]

[158] Thomas DS, Hossack JA, Rose AH. Plasma-membrane lipid composition and ethanol tolerance in *Saccharomyces cerevisiae*. Arch Microbiol 1978; 117(3): 239-45.
[http://dx.doi.org/10.1007/BF00738541] [PMID: 358937]

[159] Alexandre H, Ansanay-Galeote V, Dequin S, Blondin B. Global gene expression during short-term ethanol stress in *Saccharomyces cerevisiae*. FEBS Lett 2001; 498(1): 98-103.
[http://dx.doi.org/10.1016/S0014-5793(01)02503-0] [PMID: 11389906]

[160] Basso LC, Basso TP, Basso TO. 2020.Aspectos fisiológicos e bioquímicos da fermentação etanólica nas destilarias brasileiras. 2020, 2nd Edition.
[http://dx.doi.org/10.5151/9788521218982-08]

[161] Della-Bianca BE, Gombert AK. Stress tolerance and growth physiology of yeast strains from the Brazilian fuel ethanol industry. Antonie van Leeuwenhoek 2013; 104(6): 1083-95.
[http://dx.doi.org/10.1007/s10482-013-0030-2] [PMID: 24062068]

[162] Della-Bianca BE, de Hulster E, Pronk JT, van Maris AJA, Gombert AK. Physiology of the fuel ethanol strain *Saccharomyces cerevisiae* PE-2 at low pH indicates a context-dependent performance relevant for industrial applications. FEMS Yeast Res 2014; 14(8): 1196-205.
[http://dx.doi.org/10.1111/1567-1364.12217] [PMID: 25263709]

[163] Gombert AK, van Maris AJ. Improving conversion yield of fermentable sugars into fuel ethanol in 1st generation yeast-based production processes. Curr Opin Biotechnol 2015; 33: 81-6.
[http://dx.doi.org/10.1016/j.copbio.2014.12.012] [PMID: 25576737]

[164] Jones AM, Thomas KC, Ingledew WM. Ethanolic fermentation of blackstrap molasses and sugarcane juice using very high gravity technology. J Agric Food Chem 1994; 42: 1242-6.
[http://dx.doi.org/10.1021/jf00041a037]

[165] Bayrock DP, Michael Ingledew W. Application of multistage continuous fermentation for production of fuel alcohol by very-high-gravity fermentation technology. J Ind Microbiol Biotechnol 2001; 27(2): 87-93.
[http://dx.doi.org/10.1038/sj.jim.7000167] [PMID: 11641766]

[166] Dorta C, Oliva-Neto P, de-Abreu-Neto MS, Nicolau-Junior N, Nagashima AI. Synergism among lactic acid, sulfite, pH and ethanol in alcoholic fermentation of *Saccharomyces cerevisiae* (PE-2 and M-26). World J Microbiol Biotechnol 2006; 22: 177-82.
[http://dx.doi.org/10.1007/s11274-005-9016-1]

[167] Hu XH, Wang MH, Tan T, *et al.* Genetic dissection of ethanol tolerance in the budding yeast *Saccharomyces cerevisiae*. Genetics 2007; 175(3): 1479-87.
[http://dx.doi.org/10.1534/genetics.106.065292] [PMID: 17194785]

[168] Stephanopoulos G. Metabolic engineering by genome shuffling. Two reports on whole-genome shuffling demonstrate the application of combinatorial methods for phenotypic improvement in bacteria. Nat Biotechnol 2002; 20(7): 666-8.

[http://dx.doi.org/10.1038/nbt0702-666] [PMID: 12089547]

[169] Giudici P, Solieri L, Pulvirenti AM, Cassanelli S. Strategies and perspectives for genetic improvement of wine yeasts. Appl Microbiol Biotechnol 2005; 66(6): 622-8.
[http://dx.doi.org/10.1007/s00253-004-1784-2] [PMID: 15578179]

[170] Ferreira LV, Amorim HV, Basso LC. Fermentação de trealose e glicogênio endógenos em *Saccharomyces cerevisiae*. Food Sci Technol (Campinas) 1999; 19: 29-32.
[http://dx.doi.org/10.1590/S0101-20611999000100008]

[171] Basso LC, Alves DMG, Amorim HV. The antibacterial action of succinic acid produced by yeast during fermentation. Rev Microbiol 1997; 28: 77-82.

[172] Narendranath NV, Hynes SH, Thomas KC, Ingledew WM. Effects of lactobacilli on yeast-catalyzed ethanol fermentations. Appl Environ Microbiol 1997; 63(11): 4158-63.
[http://dx.doi.org/10.1128/aem.63.11.4158-4163.1997] [PMID: 9361399]

[173] Bayrock DP, Ingledew WM. Inhibition of yeast by lactic acid bacteria in continuous culture: nutrient depletion and/or acid toxicity? J Ind Microbiol Biotechnol 2004; 31(8): 362-8.
[http://dx.doi.org/10.1007/s10295-004-0156-3] [PMID: 15257443]

[174] Eggleston G, Basso LC, de Amorim HV, Paulillo SCL, Basso TO. Mannitol as a sensitive indicator of sugarcane deterioration and bacterial contamination in fuel alcohol production. Zuckerindustrie 2007; 132: 33-9.

[175] Basso TO, Lino FSO. Clash of Kingdoms: How do bacterial contaminants thrive in and interact with yeasts during ethanol production?Fuel Ethanol Prod from Sugarcane. London, UK: IntechOpen 2018.

[176] Nielsen JC, Senne de Oliveira Lino F, Rasmussen TG, Thykær J, Workman CT, Basso TO. Industrial antifoam agents impair ethanol fermentation and induce stress responses in yeast cells. Appl Microbiol Biotechnol 2017; 101(22): 8237-48.
[http://dx.doi.org/10.1007/s00253-017-8548-2] [PMID: 28993899]

[177] Ceccato-Antonini SR. Conventional and nonconventional strategies for controlling bacterial contamination in fuel ethanol fermentations. World J Microbiol Biotechnol 2018; 34(6): 80.
[http://dx.doi.org/10.1007/s11274-018-2463-2] [PMID: 29802468]

[178] Braga LPP, Alves RF, Dellias MTF, Navarrete AA, Basso TO, Tsai SM. Vinasse fertirrigation alters soil resistome dynamics: an analysis based on metagenomic profiles. BioData Min 2017; 10: 17.
[http://dx.doi.org/10.1186/s13040-017-0138-4] [PMID: 28546829]

[179] Kandler O. Carbohydrate metabolism in lactic acid bacteria. Antonie van Leeuwenhoek 1983; 49(3): 209-24.
[http://dx.doi.org/10.1007/BF00399499] [PMID: 6354079]

[180] Skinner KA, Leathers TD. Bacterial contaminants of fuel ethanol production. J Ind Microbiol Biotechnol 2004; 31(9): 401-8.
[http://dx.doi.org/10.1007/s10295-004-0159-0] [PMID: 15338420]

[181] Lucena BTL, dos Santos BM, Moreira JL, *et al.* Diversity of lactic acid bacteria of the bioethanol process. BMC Microbiol 2010; 10: 298.
[http://dx.doi.org/10.1186/1471-2180-10-298] [PMID: 21092306]

[182] Gallo CR. 1989; Determinação da microbiota bacteriana de mosto e de dornas na fermentação alcoolica. PhD Thesis, Faculdade de Engenharia de Alimentos, Universidade Estadual de Campinas, 388.

[183] Cabrini KT, Gallo CR. Identificação de leveduras no processo de fermentação alcoólica em usina do Estado de São Paulo, Brasil. Sci Agric 1999; 56: 207-16.
[http://dx.doi.org/10.1590/S0103-90161999000100028]

[184] Basso TO, Gomes FS, Lopes ML, de Amorim HV, Eggleston G, Basso LC. Homo- and heterofermentative lactobacilli differently affect sugarcane-based fuel ethanol fermentation. Antonie

van Leeuwenhoek 2014; 105(1): 169-77.
[http://dx.doi.org/10.1007/s10482-013-0063-6] [PMID: 24198118]

[185] Sprague GF Jr, Winans SC. Eukaryotes learn how to count: quorum sensing by yeast. Genes Dev 2006; 20(9): 1045-9.
[http://dx.doi.org/10.1101/gad.1432906] [PMID: 16651650]

[186] Albuquerque P, Casadevall A. Quorum sensing in fungi--a review. Med Mycol 2012; 50(4): 337-45.
[http://dx.doi.org/10.3109/13693786.2011.652201] [PMID: 22268493]

[187] Brexó RP, Sant'Ana AS. Microbial interactions during sugar cane must fermentation for bioethanol production: does quorum sensing play a role? Crit Rev Biotechnol 2018; 38(2): 231-44.
[http://dx.doi.org/10.1080/07388551.2017.1332570] [PMID: 28574287]

[188] Wongsuk T, Pumeesat P, Luplertlop N. Fungal quorum sensing molecules: Role in fungal morphogenesis and pathogenicity. J Basic Microbiol 2016; 56(5): 440-7.
[http://dx.doi.org/10.1002/jobm.201500759] [PMID: 26972663]

[189] Padder SA, Prasad R, Shah AH. Quorum sensing: A less known mode of communication among fungi. Microbiol Res 2018; 210: 51-8.
[http://dx.doi.org/10.1016/j.micres.2018.03.007] [PMID: 29625658]

[190] Avbelj M, Zupan J, Kranjc L, Raspor P. Quorum-Sensing Kinetics in *Saccharomyces cerevisiae*: A Symphony of ARO Genes and Aromatic Alcohols. J Agric Food Chem 2015; 63(38): 8544-50.
[http://dx.doi.org/10.1021/acs.jafc.5b03400] [PMID: 26367540]

[191] Hlavácek O, Kucerová H, Harant K, Palková Z, Váchová L. Putative role for ABC multidrug exporters in yeast quorum sensing. FEBS Lett 2009; 583(7): 1107-13.
[http://dx.doi.org/10.1016/j.febslet.2009.02.030] [PMID: 19250938]

[192] Wuster A, Babu MM. Transcriptional control of the quorum sensing response in yeast. Mol Biosyst 2010; 6(1): 134-41.
[http://dx.doi.org/10.1039/B913579K] [PMID: 20024075]

[193] Lorenz MC, Cutler NS, Heitman J. Characterization of alcohol-induced filamentous growth in *Saccharomyces cerevisiae*. Mol Biol Cell 2000; 11(1): 183-99.
[http://dx.doi.org/10.1091/mbc.11.1.183] [PMID: 10637301]

[194] Dickinson JR. 'Fusel' alcohols induce hyphal-like extensions and pseudohyphal formation in yeasts. Microbiology 1996; 142(Pt 6): 1391-7.
[http://dx.doi.org/10.1099/13500872-142-6-1391] [PMID: 8704979]

[195] Chen H, Fink GR. Feedback control of morphogenesis in fungi by aromatic alcohols. Genes Dev 2006; 20(9): 1150-61.
[http://dx.doi.org/10.1101/gad.1411806] [PMID: 16618799]

[196] Smukalla S, Caldara M, Pochet N, *et al*. *FLO1* is a variable green beard gene that drives biofilm-like cooperation in budding yeast. Cell 2008; 135(4): 726-37.
[http://dx.doi.org/10.1016/j.cell.2008.09.037] [PMID: 19013280]

[197] Chauhan NM, Mohan Karuppayil S. Dual identities for various alcohols in two different yeasts. Mycology 2020; 12(1): 25-38.
[http://dx.doi.org/10.1080/21501203.2020.1837976] [PMID: 33628606]

[198] de Figueiredo CM, Hock DH, Trichez D, *et al*. High foam phenotypic diversity and variability in flocculant gene observed for various yeast cell surfaces present as industrial contaminants. Fermentation (Basel) 2021; 7: 127.
[http://dx.doi.org/10.3390/fermentation7030127]

[199] Vital S. Estudo das relações tróficas entre leveduras e bactérias láticas na fermentação alcoólica. MSc Thesis, University of São Paulo, Brazil 2013.
[http://dx.doi.org/10.11606/D.11.2013.tde-16122013-150236]

[200] Avbelj M, Zupan J, Raspor P. Quorum-sensing in yeast and its potential in wine making. Appl Microbiol Biotechnol 2016; 100(18): 7841-52.
[http://dx.doi.org/10.1007/s00253-016-7758-3] [PMID: 27507587]

[201] Nissen P, Arneborg N. Characterization of early deaths of non- Saccharomyces yeasts in mixed cultures with *Saccharomyces cerevisiae*. Arch Microbiol 2003; 180(4): 257-63.
[http://dx.doi.org/10.1007/s00203-003-0585-9] [PMID: 12898132]

[202] Nissen P, Nielsen D, Arneborg N. Viable *Saccharomyces cerevisiae* cells at high concentrations cause early growth arrest of non-Saccharomyces yeasts in mixed cultures by a cell-cell contact-mediated mechanism. Yeast 2003; 20(4): 331-41.
[http://dx.doi.org/10.1002/yea.965] [PMID: 12627399]

[203] Pérez-Nevado F, Albergaria H, Hogg T, Girio F. Cellular death of two non-Saccharomyces wine-related yeasts during mixed fermentations with *Saccharomyces cerevisiae*. Int J Food Microbiol 2006; 108(3): 336-45.
[PMID: 16564103]

[204] Wang C, Mas A, Esteve-Zarzoso B. The interaction between *Saccharomyces cerevisiae* and Non-Saccharomyces yeast during alcoholic fermentation is species and strain specific. Front Microbiol 2016; 7: 502.
[http://dx.doi.org/10.3389/fmicb.2016.00502] [PMID: 27148191]

[205] Nath BJ, Mishra AK, Sarma HK. Assessment of quorum sensing effects of tyrosol on fermentative performance by chief ethnic fermentative yeasts from northeast India. J Appl Microbiol 2021; 131(2): 728-42.
[http://dx.doi.org/10.1111/jam.14908] [PMID: 33103297]

[206] Machida K, Tanaka T, Yano Y, Otani S, Taniguchi M. Farnesol-induced growth inhibition in *Saccharomyces cerevisiae* by a cell cycle mechanism. Microbiology 1999; 145(Pt 2): 293-9.
[http://dx.doi.org/10.1099/13500872-145-2-293] [PMID: 10075411]

[207] Delago A, Gregor R, Dubinsky L, *et al.* A bacterial quorum sensing molecule elicits a general stress response in *Saccharomyces cerevisiae*. Front Microbiol 2021; 12: 632658.
[http://dx.doi.org/10.3389/fmicb.2021.632658] [PMID: 34603220]

[208] de Souza Oliveira RP, Rivas Torres B, Zilli M, de Araújo Viana Marques D, Basso LC, Converti A. Use of sugar cane vinasse to mitigate aluminum toxicity to *Saccharomyces cerevisiae*. Arch Environ Contam Toxicol 2009; 57(3): 488-94.
[http://dx.doi.org/10.1007/s00244-009-9287-x] [PMID: 19184166]

[209] Mariano-da-Silva S, Basso LC. Efeitos do cádmio sobre o crescimento das leveduras *Saccharomyces cerevisiae* PE-2 e *Saccharomyces cerevisiae* IZ-1904, e a capacidade da vinhaça em atenuar a toxicidade. Food Sci Technol (Campinas) 2004; 24: 16-22.
[http://dx.doi.org/10.1590/S0101-20612004000100004]

[210] Basso L, Oliveira A, Orelli V, Campos A, Gallo C, Amorim H. Dominância das leveduras contaminantes sobre as linhagens industriais avaliada pela técnica da cariotipagem. An Congr Nac STAB 1993; 5: 246-50.

CHAPTER 5

Candida

Atrayee Chattopadhyay[1] and **Mrinal K. Maiti**[1,*]

[1] *Department of Biotechnology, Indian Institute of Technology Kharagpur, Kharagpur-721302, India*

Abstract: The genus *Candida* represents a huge repertoire of fungal species with diverse functions. These unicellular yeasts possess great clinical significance because of the high level of pathogenicity exhibited by some members of the genus. Moreover, several species of *Candida* are highly valued industrially as microbial platforms to produce commercial products. Therefore, there is a persistent need to describe the genus as a whole, considering its immense applications in both medical and biotechnological grounds. The genus is being continuously explored, with new species regularly emerging as pathogenic. However, since most of the in-depth research has been focused on a few species of these yeasts, therefore, only the pathogenic species have been described in this chapter, reviewing the underlying characteristics that label their pathogenicity, which include their incidence of occurrence among the population, spread of infection, factors affecting the host immune system, and disease control. Further, the major studies identifying the biotechnological potential of the species have been discussed.

Keywords: Adhesin, Biodegradation, Biofilm, Bioremediation, Biosurfactants, Candidiasis, CTG clade, Drug resistance, Enzymes, Hyphae, Non-albicans *Candida*, Parasexual, Pathogenic, Phagocytes, Pseudohyphae, Single-cell oil, Single-cell protein, Transcription factor, Virulence, Yeast.

INTRODUCTION

The genus *Candida* contains a mixed population of unicellular yeasts, which are difficult to group by their morphological or physiological characteristics. The species belonging to this genus do not share similar types of distinctive features. The term '*Candida*' loosely refers to imperfect fungi as the members do not exhibit a clearly defined sexual cycle. However, the name originates from the Latin word 'candidus' meaning white, as pigmentation is usually absent in these yeasts [1]. In 1923, the Danish microbiologist Christine Berkhout first classified

* **Corresponding author Mrinal K. Maiti:** Department of Biotechnology, Indian Institute of Technology Kharagpur, Kharagpur-721302, India; Tel: 91-3222-283796; E-mail: maitimk@bt.iitkgp.ac.in

Sérgio Luiz Alves Júnior, Helen Treichel, Thiago Olitta Basso and Boris Ugarte Stambuk (Eds.)
All rights reserved-© 2022 Bentham Science Publishers

nine yeast species, originally assigned to the *Monilla* genus, to this new taxon based on certain morphological, biochemical, and physiological characteristics [2]. Since then, the genus was reassigned several times, and finally, with the aid of molecular biology tools, a definite classification has been possible. Different species of these yeasts are widely distributed in natural ecosystems and are usually found in human microflora as saprophytes. However, some of them can become opportunistic pathogens given a congenial host environment.

(I). *CANDIDA*: A POLYPHYLETIC GENUS

The genus currently comprises around 200 species of ascomycetous yeasts (kingdom: Fungi, division: Ascomycota, subdivision: Saccharomycotina) grouped under Hemiascomycetes class and Saccharomycetaceae family [1]. The cell shape and size vary from species to species and also on the environmental conditions, but usually range from ovoid to elongate, and (1-8) μm x (1-6) μm [3, 4]. Cells may or may not form mycelium; pseudohyphae and occasionally true hyphae are observed [1, 5]. The usual mode of reproduction is budding, but the sexual cycle is observed in some cases [6]. Typically, *Candida* species are aerobic, glucose-fermenting yeasts, unable to assimilate nitrate or inositol, and do not synthesize carotenoids [1, 4]. A small number of *Candida* species are efficient pentose-utilizers that have been exploited for the effective use of hemicellulosic wastes. Numerous studies also focused on these yeasts metabolizing plant byproducts and subsequent bioconversion to important biotechnological compounds such as antibiotics, vitamins, complex alkanes, and biofuel [7]. Several species like *C. tropicalis*, *C. maltosa*, *C. famata*, *C. guilliermondii,* and *C. krusei* have been extensively employed for the industrial production of value-added metabolites [8]. Despite such immense biotechnological potential, the genus has gained its importance because of the pathogenesis and clinical significance of many species, the foremost being *C. albicans*, closely followed by *C. tropicalis, C. parapsilosis,* and *C. glabrata*.

Despite the numerous attempts at reclassification, a common evolutionary origin has not been established among the members of this genus; hence the genus *Candida* appears to be a polyphyletic group posing a substantial problem in characterizing the species. A breakthrough came in 1989 when Kawaguchi *et al.* described a different codon usage for CTG (or CUG) codon in *C. cylindraceae* (currently known as *C. rugosa*), which was found to code for serine instead of the usual leucine [9]. Further investigation by Sugita and Nakase identified 67 such species exhibiting alternative codon usage [10]. Currently, phylogenetic clustering based on whole-genome sequencing has divided the Saccharomycotina yeasts into two major groups- 1) **CTG clade** with species exhibiting non-classical translation of CUG into serine (Fig. **1**), and 2) non-CTG clade including **whole-genome duplication (WGD) clade** comprising yeasts which have undergone genome duplication, including *Saccharomyces* species and the pathogenic *C. glabrata.* While many of the *Candida* species belong to the CTG clade, including *C. albicans, C. dubliniensis, C. tropicalis, C. parapsilosis, C. maltosa, C. famata, C. guilliermondii, C. lusitaniae, C. oleophila,* and *C. rugosa,* which are mostly pathogenic, the non-*Candida* yeast *Pichia stipitis* also falls in this group [11 - 14]. Such reassignment of codon has been correlated with the length of the isoprene chain present in coenzyme Q9 (Co-Q9), the predominant ubiquinone found in these species [10].

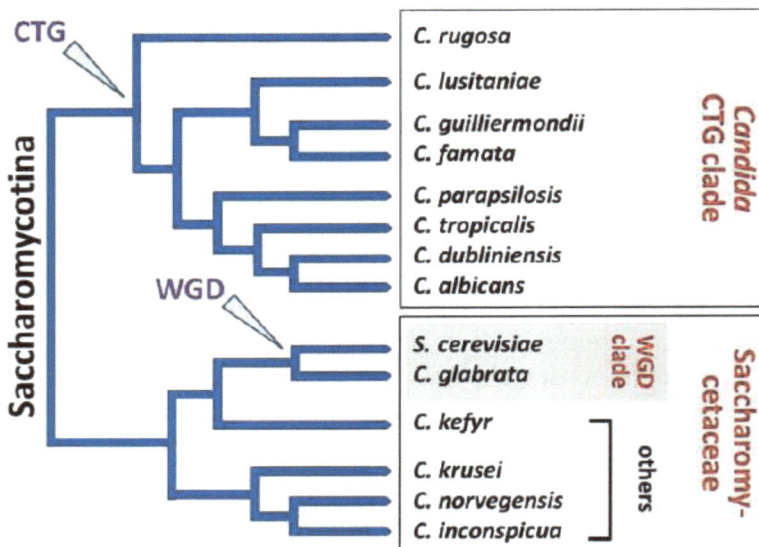

Fig. (1). Phylogenetic distribution of *Candida* species of yeasts. Most of the important pathogens of the genus fall into the CTG clade translating the CUG codon into serine, except for the highly frequent *C. glabrata,* which falls in the WGD clade and is more closely related to *S. cerevisiae.* The figure is taken from Papon *et al.*, 2013 [15].

Normally mistranslation of codons is seen in organisms as an adaptive response to various stresses. For example, the mistranslation of UGA stop codon to selenocysteine in response to oxidative stress in neurons has been reported [16]. Likewise, the mistranslation of CUG in *C. albicans* is an example of stress-induced response, although 3% of codons are still translated as leucine in the yeast following standard codon usage pattern [17]. Misincorporation of serine initiates a cascade of morphological changes in the organism, altering its pathogenicity and susceptibility to the host immune system and morphogenic transition. Miranda *et al.* reported that CUG mistranslation dramatically changes the cell wall structure of *C. albicans*, reducing the exposure of cell surface β-glucans, as a result of which cells become less susceptible to recognition by host macrophages thereby altering host-pathogen interaction [18]. In addition, the mistranslated cell surface protein adhesin exhibits increased adherence to a broad range of host substrates, aiding in the colonization of the yeast cells [18]. Although most of the pathogenic *Candida* species lie within the CTG clade, there are exceptions of *C. glabrata* and *C. krusei* [19]. *C. bracarensis* and *C. nivariensis* are two new additions to this group, which are also pathogenic, but the isolation frequency is pretty low [20, 21]. *C. krusei* is moderately pathogenic and more well-known for its fermentation ability which has been exploited for bioethanol production [22]. Other *Candida* species not belonging to the CTG clade that are albeit mildly pathogenic are *C. kefyr* (*Kluyveromyces marxianus*), *C. norvegensis* (*Pichia norvegensis*), and *C. inconspicua* (*Pichia cactophila*), the latter two being closely related phylogenetically [23]. *C. lipolytica* (syn. *Yarrowia lipolytica*) is rarely found among infected cases and is responsible for only about 0.1% of *Candida* infections [23]. This yeast is distantly related to both the CTG and *Saccharomycetaceae* clades.

(II). CANDIDIASIS IN HUMAN AND ANIMALS

The term 'Candidiasis' collectively refers to chronic fungal infections in humans caused by *Candida* species of yeasts, and is currently considered the fourth leading cause contributing to 8–10% of all nosocomial infections occurring worldwide [24]. Usually, these yeasts reside in human microflora as commensals and are frequently found in skin, genitourinary and gastrointestinal tracts [25]. However, an appropriate host environment such as an impaired immune system can trigger their proliferation resulting in infection ranging from local mucosal to severe systemic disease. Depending on the site of infection, candidiasis can be divided into three groups: 1) cutaneous, infections that are restricted to the skin, 2) mucosal, when the fungi infect oropharynx, oesophagous, and vagina, and lastly, 3) systemic, spreading across the internal organs and to the bloodstream. The latter is called candidaemia, for which the mortality rate ranges from 15–35%, and is the most common type of invasive candidiasis [24]. Although *C. albicans* remains the most isolated pathogen in candidiasis, most of the candidaemia cases worldwide are caused by five predominant species — *C. albicans*, *C. glabrata*, *C. parapsilosis*, *C. tropicalis,* and *C. krusei* —, while the less predominant species associated with other candidiasis infections include *C. guilliermondii*, *C. lusitaniae*, *C. kefyr*, *C. famata*, *C. inconspicua*, *C. rugosa*, *C. dubliniensis*, and *C. norvegensis* [8, 23]. These non-albicans *Candida* (NAC) species have shown a gradual emergence in the last two decades, accounting for 35–65% of global candidiasis cases [26]. *C. auris*, another NAC species, was first reported in Japan in 2009 as a cause of nosocomial infection with a high mortality rate. Since then, *C. auris* infection has been reported in Asia, UK, the US, and many other countries [27 - 29]. Two more species — *C. bracarensis* and *C. nivariensis* — have been identified later as pathogenic both being closely related to *C. glabrata* [20, 21, 30].

Cutaneous Candidiasis

It is usually a chronic infection occurring in the skin and nails of predisposed patients, commonly isolated from diabetic and obese individuals [31].

Mucosal Candidiasis

Oropharyngeal candidiasis (OPC) is the most common oral infection found in HIV patients, usually at the onset of AIDS. Although *C. albicans* is the most common cause of these infections, NAC species are increasingly being found to be associated as well, and they mostly include *C. glabrata, C. parapsilosis, C. krusei, C. tropicalis, C. dubliniensis,* and *C. guilliermondii* [32]. Infection of the vaginal surface is called vulvovaginal candidiasis (VVC), and the infection affects about 75% of women globally. *C. albicans* is the most frequent colonizer in the vagina, but there are increasing cases of *C. glabrata* infection for VVC, followed by other NAC species such as *C. tropicalis*, *C. krusei,* and *C. parapsilosis* [33, 34].

Systemic Candidiasis

Under these conditions, the yeast cells spread across the body, invading the vital internal organs from the bloodstream such as the liver, spleen, kidneys, heart, brain. Thus, three major predisposing conditions expose humans to invasive infections- 1) repetitive use of broad-spectrum antibiotics, which removes the natural gut microbiota that releases protective factors against *Candida* spp. [35], 2) invasive surgical procedures that might breach the cutaneous or gastrointestinal barriers, 3) suppression of the immune system by chemotherapy or corticosteroid therapy [36, 37].

Epidemiology

Among the most frequently isolated NAC species- *C. parapsilosis*, *C. glabrata*, *C. tropicalis*, the severity of infection depends on several demographic factors such as age group and location, as well as certain predisposing conditions such as immunodeficiency, neutropenia, and diabetes. While *C. glabrata* is most frequently isolated from elderly patients, *C. parapsilosis* predominates among neonates and is also isolated from patients who have undergone invasive surgical procedures [38 - 41]. The infected individuals may develop endocarditis (inflammation of endocardium of the heart), endophthalmitis (inflammation of intraocular fluids of the eye), peritonitis (inflammation of the inner abdominal lining), fungemia (fungal infection in the blood), and septic arthritis (infection in the joint space) [38]. *C. tropicalis* infections, on the other hand, are most severe in neutropenic patients and those with cancer, particularly leukemia [19, 42, 43]. *C. rugosa* accounts for 1.1% of candidiasis cases and is usually isolated from the oral cavity of patients with diabetes [44]. The less common pathogen *C. dubliniensis* is reported in the oral cavities of HIV positive individuals, and more recently, from other parts of both HIV positive and negative patients [45 - 48]. However, the pathogen is not known to cause invasive disease which is attributed to its inability to form hyphae and invade underlying tissues [49].

Geographically, *C. glabrata* is the most commonly occurring NAC species in North America and Europe in contrast to Latin American countries, India, and Pakistan, where *C. parapsilosis* and *C. tropicalis* are more common, followed by *C. guilliermondi* and *C. rugosa* [23, 39, 50, 51]. The frequency of recovery of *C. parapsilosis* from clinical isolates is 3-fold higher in North America than in Europe, whereas *C. inconspicua* and *C. novigensis* are frequently recovered in Europe [23]. These NAC species also exhibit differential resistance to antifungal drugs; *C. glabrata* shows multidrug resistance to a large range of antibiotics, making it a growing concern among medical researchers, whereas resistance to azole compounds is largely found in *C. krusei*, *C. inconspicua*, *C. rugosa*, and *C. norvegensis*. As opposed to the trend, *C. parapsilosis* and *C. guilliermondi* show increasing resistance towards echinocandins [8, 52]. The underlying mechanisms for such varied susceptibility are very poorly understood due to inadequately developed genetic and molecular resources. Two main problems in this regard are- the absence of sexual reproduction in most of the pathogenic *Candida* species and alternate codon usage limiting the scope of using molecular tools designed in model organisms like *E. coli* or *S. cerevisiae*. Albeit the pathogenesis and drug resistance mechanisms of *C. albicans* have been intensively studied since the end of the 20th century, revealing several molecular targets associated with antifungal resistance in the yeast. Moreover, with the dramatic upsurge of NAC species among candidiasis patients globally with their reduced susceptibility to antifungals, a great deal of medical interest is focused on deciphering the molecular mechanisms associated with virulence and drug resistance in these yeasts.

Antifungal Drug Resistance

Since 1990, fluconazole has been considered a standard antifungal drug for the treatment of candidiasis. Azole compounds are effective against both *C. tropicalis* and *C. parapsilosis*, although the susceptibility of *C. tropicalis* towards fluconazole is relatively lower than that of *C. albicans* [53]. *C. lusitaniae* is also susceptible to azoles but shows higher resistance towards amphotericin B, which is concerning as this yeast accounts for 1-2% of candidaemia worldwide [54]. However, *C. glabrata* is infamously resistant to this empirical drug which is reflected in the elevated minimum inhibitory concentration (MIC) during its treatment. MIC distribution of fluconazole for each of the seven invasive *Candida* species is presented by Pfaller [55]. With time, the propensity of resistance by *C. glabrata* has spread to both azole and echinocandin group of antifungal agents, giving rise to multidrug-resistant (MDR) strains which are of serious medical concern. The antifungal azole compounds affect the ergosterol synthesis (a major constituent of the fungal cell membrane) by interfering with the activity of the target enzyme lanosterol 14-α-sterol-demethylase. Resistance to azole drugs is associated with enhanced expression of *MDR* and *CDR* genes that encode efflux pumps in *C. albicans*, *C. glabrata,* and *C. dubliniensis*, or point mutation in the *ERG11* gene encoding the target enzyme in these yeasts. The latter reduces the affinity for binding of azoles to the target [55]. Echinocandins inhibit the enzyme 1,3-β-D-glucan synthase, responsible for the β-glucan formation in fungal cell walls. *Candida* species with reduced susceptibility to echinocandins, such as *C. albicans*, *C. glabrata*, *C. krusei*, *C. tropicalis,* and *C. dubliniensis* have all shown point mutations in the *FKS1* gene encoding the major catalytic subunit of the enzyme, thereby altering enzyme kinetics, reducing its sensitivity towards inhibition [56]. Mutations in the *FKS2* gene and a certain *MSH2* gene encoding a DNA mismatch repair protein have also been detected in clinical isolates of MDR *C. glabrata* [57, 58]. The MD transporter genes *CDR* are regulated by the transcription factor PDR1 in the yeast, and the gene encoding PDR1 showed a *gain-of-function* mutation in the drug-resistant isolates [59, 60].

Immune Response

The host immune system is activated through a myriad of signaling due to *Candida* spp. infections. Mucosal candidiasis activates the T-cell-dependent response; invasive candidiasis triggers a lymphocyte-independent response involving myeloid phagocytes, i.e., neutrophils, monocytes, macrophages, and dendritic cells [61 - 63]. Pappas *et al.* have excellently reviewed the various effector-mediated signaling mechanisms activated by myeloid phagocytic cells in the *Candida* spp.-infected tissues [61]. Myeloid phagocytes express soluble and membrane-bound receptors called pattern recognition receptors (PRRs) that are activated by the pathogen-associated molecular patterns (PAMPs) of the invading pathogen. Additionally, inflammation complexes NLRP3 (NOD-like receptor family pyrin domain containing 3) and NLRP10 (NOD-like receptor family pyrin domain containing 10) are also activated that in turn produce pro-inflammatory cytokines IL-1β, IFN-γ, and IL-17 [64 - 66]. These cytokines further promote phagocytosis, activate T_H cell responses and ROS generation. The latter induces an oxidative process of killing *Candida* spp. cells by recruiting the NADPH oxidase complex at the phagosomal membrane and activating myeloperoxidase for ROS generation [67]. The dendritic cell-associated dectin 1-calcineurin pathway activation also promotes the oxidative killing of the fungal cells [61]. Neutrophils are particularly important at the early stage of infection, usually after 24–48 h, and exhibit the most potent phagocytic activities [63]. They release extracellular antimicrobial elastase and calprotectin that damage the large filamentous forms of *Candida* spp. or undergo degranulation themselves by the action of jagunal homolog 1 (JAGN1) and the chemokine receptor CXCR1. Dendritic cells produce cytokine IL-23 which triggers natural killer cells to release granulocyte-macrophage colony-stimulating factor (GM-CSF), which in turn activates neutrophils in the *Candida*-infected cells. Another cytokine, IL-15, produced by inflammatory monocytes similarly activates neutrophils [61].

(III). *CANDIDA* IN NATURE

Among the 200 species currently identified in the genus, *C. albicans* is the most dominant pathogen accounting for half of all disseminated candidiasis. Overall 95% of all candidiasis cases are caused by *C. albicans*, *C. tropicalis*, *C. parapsilosis*, and *C. glabrata* [68 - 70]. Other less relevant species include *C. lusitaniae*, *C. guilliermondii*, *C. krusei*, and *C. dubliniensis* [71]. All of these species have different mechanisms for virulence and pathogenesis, which need to be understood individually.

Pathogenesis

By far, *C. albicans* is the most studied example of *Candida* pathogen and considered as a model for all host-pathogen interaction studies. The yeast is well known for its dimorphic character, i.e., unicellular yeast form and filamentous mold form, and virulence of *C. albicans* is associated with its morphogenic transition between the two forms as well as its phenotypic switching between white and opaque cells [72 - 75]. Such transitions are highly regulated, and each type possesses a different level of virulence. For example, the white cells show more virulence in systemic infections, while opaque cells are in cutaneous ones, and also the latter is more efficient in evading host cell immune response [76, 77]. Generally, the filamentous morphologies aid the organisms colonization, adherence to the host cell surface, rapid dissemination to host tissues, biofilm formation; thereby the hyphal or pseudohyphal forms are more associated with virulence. In *C. albicans,* the hyphal formation promotes virulence in the following ways: 1) Hyphae mechanically invade the epithelial tissue; 2)

Branching of hyphae further damages the endothelial cells; 3) Hyphal growth can lyse macrophages and neutrophils from inside following phagocytosis [78, 79].

The transition from yeast form to hyphal form in *C. albicans* is highly regulated at the transcriptional level, and a direct correlation has been drawn between the gradual transition of single cells-pseudohyphae-hyphae formation with the expression of the transcriptional regulator UME6. The yeast form does not express UME6, while the pseudohyphal form expresses a low level of UME6, and the highest level of expression is detected only in true hyphal morphology [80]. On the contrary, the transcription factor gene *NRG1* represses the filament-specific genes (*UME6*, *HWP1*, *ECE1*, etc.), and its expression is negatively correlated with the transition. Deletion of the gene has resulted in constitutive filamentous growth in *C. albicans* [81]. Another TF gene, *EFG1*, influences genes responsible for hyphal growth like *HGC1* and controls adhesion and secretion of protease, lipase, and other hydrolytic enzymes [82 - 85]. The latter enzymes are responsible for host tissue destruction seen during an invasion. Several proteases and lipases are identified in *Candida* pathogens, the most important being aspartic proteinase (Sap), which are reportedly involved in host tissue adherence and damage [86]. They have been found in *C. albicans*, *C. parapsilosis*, *C. tropicalis,* and *C. guillermondii* [87].

Moreover, secretion of extracellular lipases has been correlated with virulence in many pathogenic species. They are thought to help the pathogen adhere to the host tissues, interfere with host immune response, and facilitate colonization by destroying competing microflora [88, 89]. Besides, the function of lipase in biofilm formation has been demonstrated in *C. parapsilosis* [89].

Having understood the roles of hyphae in pathogenesis, the formation of the intermediate pseudohyphae puts forth a pertinent question of its evolutionary significance. Why transition to a less virulent morphology, when attaining a more virulent hyphal form is possible? One plausible explanation is that the pseudohyphae allow the yeast to invade tissues with sub-optimal nutrient resources since the highly branched pseudohyphae have better nutrient scavenging and disseminating properties [90]. Among the NAC species, while many possess the ability to form pseudohyphae, only *C. dubliniensis* and *C. tropicalis* can form hyphae like *C. albicans* [91]. Consistent with this observation, the above three *Candida* species appear to be more virulent with increased adherence ability to host cells and secretion of protease than other less-pathogenic species [90, 91]. *C. dubliniensis* is genetically closest to *C. albicans* and exhibits yeast-hyphal switching. Expression of *UME6* during hyphal growth is also observed in this yeast [92]. The diploid *C. tropicalis* exhibits white-opaque switching and is controlled by the TF Wor1 as observed in *C. albicans*, but EFG1 is absent in the yeast. The role of such transition in the virulence of *C. tropicalis* is not yet established [93, 94]. This phenomenon is, however, absent in a relatively distant species, *C. parapsilosis*, although it secretes lipase during the invasion of the host, a trait shared with *C. albicans*. The yeast also shows biofilm formation as an important sign of its virulence like *C. albicans*, and the process is mainly controlled by the TF Bcr1 [95, 96]. Instead of white-opaque switching, in this case, the colonies undergo other forms of switching, which are regulated by EFG1 [97, 98]. The TF UME6 regulates pseudohyphal formation in *C. tropicalis*, *C. parapsilosis,* and *C. guilliermondii,* suggesting its importance in regulating the morphology of CTG clade yeasts [99]. The less common causes of infection, *C. lusitaniae,* and *C. guilliermondii* exist both in yeast and pseudohyphal forms in the infected tissues. The latter is of high biotechnological interest; it produces riboflavin and xylitol [100]. The haploid genome of *C. guilliermondii* has facilitated the development of genetic engineering tools in the yeast [8].

A major deviation from the CTG clade yeasts is *C. glabrata*, which despite being a non-member, is highly virulent and of great concern because of its rapidly increasing drug resistance. This yeast is considered the most pathogenic NAC species belonging to the WGD clade, followed by the less pathogenic *C. bracarensis* and *C. nivariensis,* both of which have been grouped in the same clade much later [20, 21]. *C. glabrata* has a haploid genome whose sequence was published in 2003 as the first published genome of a *Candida* species together with *C. famata* [101]. Contrary to the yeasts described above, *C. glabrata* is a pathogen in unicellular yeast form and shows no morphogenic switching. Despite that, a high level of virulence is imparted to the yeast by a large number of secretory adhesin proteins, which are important for biofilm formation [102]. The TF PDR1 described before is responsible for both adherence and virulence in this yeast [59, 60].

Sexuality

As discussed earlier, *Candida* is a diverse genus of yeasts, not only with respect to pathogenicity and morphology, but also the exhibition of different ploidy and modes of sexuality (Table **1**). Generally, the haploid species are relatively less pathogenic such as *C. lusitaniae, C. guillermondii, C. famata,etc.*, and the highly pathogenic species like *C. albicans*, *C. tropicalis*, *C. parapsilosis* are diploid, although there are exceptions of pathogenic *C. glabrata*, *C. krusei*, which are also haploid. The sexual cycle is described for most of the prominent species but is mostly characterized for *C. albicans*. Evidence of sexual cycle in the yeast is suggested by the presence of mating-type-like (*MTL*) locus, and mating between two diploid strains containing *MTL***a** and *MTL*α alleles gives rise to tetraploid cells. The frequency of this mating phenomenon is very low in the yeast, and is found to be governed by the characteristic white-to-opaque switching observed in its case, and that the opaque cells can undergo mating more efficiently. Such regulation of the sexual cycle might be important for protecting the mating cells from the host immune response as opaque cells are more efficient in evading immune cells.

In many cases, although white cells do not participate in mating directly, they promote mating in the opaque cells by secreting pheromones and mediating cell-to-cell fusion. However, the mating cycle of the yeast is not completed by the conventional meiosis as observed in fungal sexual cycles. Instead, the tetraploid cells undergo a parasexual mechanism of chromosome loss giving rise to diploid and aneuploid cells [103]. Several studies suggest that this chromosomal instability is an adaptive response to various factors such as stress, antifungal drugs, *etc* [19, 104], and plays a key role in generating the white-opaque transition required for mating as well as many other phenotypes of the yeast.

Table 1. Characteristics of *Candida* species.

Name	Clade	Pathogenicity[a]	Ploidy	Morphology	Sexuality
C. albicans	CTG	High	Diploid	Yeast, pseudohyphae, hyphae	P
C. glabrata	WGD	High	Haploid	Yeast, pseudohyphae	Mating not observed
C. tropicalis	CTG	High	Diploid	Yeast, pseudohyphae, hyphae	P
C. parapsilosis	CTG	High	Diploid	Yeast, pseudohyphae	Mating not observed
C. krusei	Others	Moderate	Haploid	Yeast, pseudohyphae	Het
C. dubliniensis	CTG	Low	Diploid	Yeast, pseudohyphae, hyphae	P
C. guilliermondii (syn. *Meyerozyma guilliermondii*)	CTG	Low	Haploid	Yeast, pseudohyphae	Het
C. lusitaniae (syn. *Clavispora lusitaniae*)	CTG	Low	Haploid	Yeast, pseudohyphae	Het
C. orthopsilosis	CTG	Low	Diploid	Yeast, pseudohyphae	Mating not observed
C. kefyr (syn. *Kluyveromyces marxianus*)	Others	Low	Haploid	Yeast, pseudohyphae	Ho
C. famata (syn. *Debaryomyces hansenii*)	CTG	Low	Haploid	Yeast, pseudohyphae	Ho

(Table 1) cont.....

Name	Clade	Pathogenicity[a]	Ploidy	Morphology	Sexuality
C. rugosa	CTG	Low	Haploid	Yeast, pseudohyphae	Mating not observed
C. inconspicua	WGD	Low	Diploid	Yeast, pseudohyphae	Sexual
C. norvegensis	WGD	Low	Diploid	Yeast, pseudohyphae	Sexual

[a]Pathogenicity refers to the frequency of isolation; High: >5%; Moderate: 1-5%; Low: <1%; P, parasexual; Het, Heterothallic; Ho, Homothallic

An interesting aspect of sexual reproduction in the yeast is manifested by same-sex mating observed in some cases. This type of homothallic mating has been observed in opaque cells and is thought to be mediated by the fact that *C. albicans* **a** cells secrete both **a** and **α** type pheromones, and the latter is normally degraded by a protease encoded by the *BAR1* gene. However, the absence of this protease leads to the accumulation of the **α** pheromone, which triggers autocrine signaling and eventually self-mating of **a** cells. The products of such self-fertilization are **a-a** tetraploid cells which can also undergo a parasexual cycle of chromosome loss and return to being normal diploid cells [105]. Both types of mating in the yeast are illustrated in Fig. (**2**).

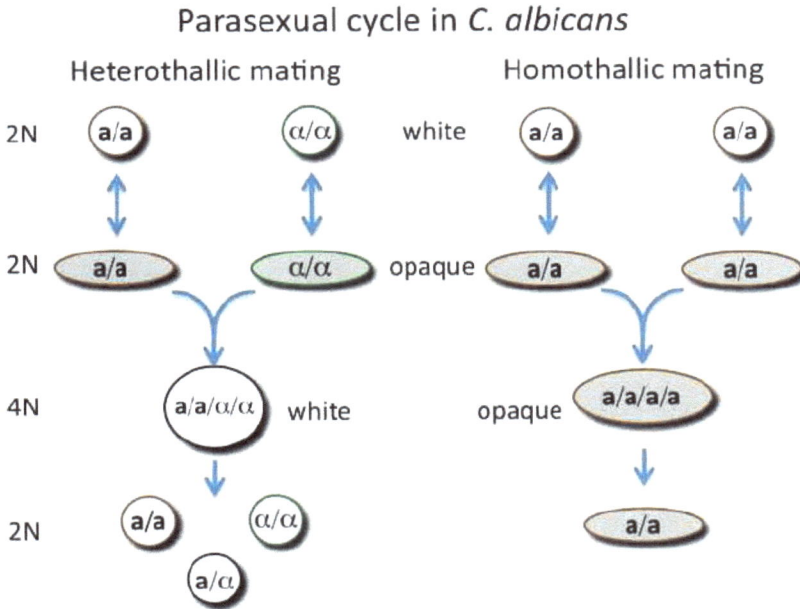

Fig. (**2**). **Parasexual mating cycles observed in *C. albicans*.** Usually, opaque diploid **a** and **α** cells fuse in heterothallic mating to form white tetraploid **a/α** cells. However, certain conditions may also trigger homothallic mating between two identical types of opaque cells to form opaque **a/a** tetraploid cells or **α/α** tetraploid cells (not shown in the figure). The figure is taken from Bennet, 2010 [106].

The *MTL* locus is conserved in most *Candida* species, and the parasexual cycle has also been characterized in the diploid *C. tropicalis* but not in *C. parapsilosis* [93, 107]. Interestingly, a complete sexual cycle is observed in haploid species like *C. lusitaniae* and *C. guilliermondii* that ends with conventional meiosis but is usually absent in species like *C. tropicalis* and *C. parapsilosis* even though they contain many genes involved in mating and meiosis [11]. The sexual cycle is also not described in *C. glabrata* despite the presence of all the genes responsible for mating and meiosis, and even though mating-type switching is also observed in the yeast [19, 101, 108].

Biofilm

A critical feature of most pathogenic *Candida* species that attributes to their pathogenicity is their ability to form biofilms. These are surface-associated microbial cells that are usually covered by an extracellular matrix composed of carbohydrates and proteins. Due to structural complexity and metabolic heterogeneity, cells constituting the biofilms exhibit much higher resistance to antifungal drugs than individually. From a medical perspective, they are highly important as about half of all nosocomial infections are caused by biofilms [109]. An important characteristic of the *C. albicans* biofilm is its ability to attach to both biotic (host mucosal lining) and abiotic (medical devices) surfaces, contributing significantly to its adverse pathogenicity. The *C. albicans* biofilm is well-characterized among the *Candida* species, although biofilm formation has also been detected in *C. tropicalis*, *C. parapsilosis*, and *C. glabrata* infections [110]. The cell wall of *C. albicans* contains numerous proteins giving rise to a complex architecture, which is suited for various physiological roles, including biofilm formation. The glycosylated proteins belonging to the adhesin family are primarily responsible for the adhesion of the yeast to biotic and abiotic surfaces, particularly, Als1, Als3, and Als5 display adhesive properties, along with some other cell-surface proteins such as Sap6 and Rbt1. A recent study suggested these proteins self-aggregate and form amyloid structures that promote biofilm formation in the yeast [111]. *C. albicans* and *C. parapsilosis* biofilms have shown increasing resistance to fluconazole, flucytosine, amphotericin B, capsofungin, nystatin, and such common antifungal drugs [112, 113]. This is explained at the molecular level by the upregulation of various gene products in the biofilm responsible for drug resistance and adherence. For example, the multidrug resistance transporter genes *MDR1*, *CDR1*, *CDR2*, or the ergosterol biosynthetic *ERG* gene, and the adherence gene *ALS1* have shown to be highly upregulated in the *C. albicans* biofilm. Additionally, several cell wall biogenesis genes may contribute to the extracellular matrix formation. Accordingly, expression of β-1,3 glucan synthesis genes *FKS1*, *BGL2*, and *XOG1* are reportedly induced in the biofilm of *C. albicans* [114].

Generally, biofilm formation requires the presence of both the unicellular yeast form and hyphal form of cells, although true hyphae are not found in *C. tropicalis, C. parapsilosis,* and *C. glabrata* biofilms [110]. Initially, the yeast cells attach to the surface and grow horizontally by budding, forming the basal layer. Subsequently, the formation of hyphal cells is induced, which constitute the upper layer. Finally, the cells secrete ECM, which covers the biofilms and aids in cohesion [113]. The formation of biofilm is regulated at the transcriptional level by numerous proteins, of which the most critical are Bcr1, Ace2, and Efg1 that are identified in *C. albicans* [95]. Bcr1 and Ace2 are zinc finger TFs, whose orthologs in *C. parapsilosis* and *C. glabrata* are also found to be functionally conserved [96, 115]. Efg1, as discussed earlier, is a global regulator influencing genes that control adhesion and hyphal growth and is also required for biofilm formation in these yeasts [116]. Other TFs such as Cph1, Efh1, Rap1, Ino4, and Tec1 have also been reported to regulate biofilm formation in *C. albicans* [117 - 119]. Further, it is observed in *C. albicans* that the genes up-regulated in white cells due to the release of mating pheromone primarily encode cell surface proteins, which contribute to biofilm formation [120]. Several findings suggest that different biofilm-forming pathogens communicate among themselves by secreting small signaling molecules. One such molecule is indole-3-acetic acid that is shown to be induced in the biofilm of *C. tropicalis* in a recent study and is also documented to promote biofilm formation in the yeast [121]. The recently emerged pathogen *C. auris* also forms a robust biofilm associated with its virulence, even though it grows predominantly in the yeast form. Expression of the NRG1 protein, previously described for its association with yeast-hyphal transition in *C. albicans*, has been found to act as a cell surface protein during biofilm growth in *C. auris*. However, it was not found to play any role in the morphological transition of the yeast [122].

(IV). *CANDIDA* IN BIOPROCESSES

Although the *Candida* species of yeasts are mostly known for their pathogenicity, their natural biodiversity is also exploited for their various industrial applications; hence several species are widely used in bioprocesses to produce commercial compounds in food, medicine, cosmetics, and many other applications. Several species of the genus, for instance, have been identified to be efficient pentose fermenters, and this property has been widely exploited for the cost-effective production of metabolites of industrial significance by growing these yeasts on lignocellulosic substrates. The methylotrophic xylose-fermenting *C. boidinii* strain was selected for ethanol production from hemicellulosic hydrolysates showing potential use in biorefineries [123]. The high xylose content of sugarcane bagasse was efficiently converted naturally by *C. guilliermondii* to significant amounts of xylitol, which is widely popular for artificial sweeteners and other applications in chemical and pharmaceutical industries [124]. One specific strain, SY005, of the well-known industrial yeast *C. tropicalis* has been extensively investigated for its efficacy in xylose utilization by exploring the xylose transport activities of the cells [125, 126]. *C. tropicalis* has been genetically engineered several times to produce xylitol, DCA, and other high-value metabolites [127 - 129].

SCP Production

Due to the increasing rate of protein deficiency worldwide, single-cell protein (SCP) production from microbes has been long considered a promising alternative acting as a high-value source of nutrition and extensively used in food and feed supplements. Usually, SCP is referred to as the high protein-containing microbial biomass obtained by cultivating non-pathogenic microbial species, including bacteria, fungi, or algae. They also contain other nutritionally important biomolecules such as carbohydrates, fats, vitamins, *etc.* The benefit of cultivating yeast species for SCP production is that it contains all the amino acids essential to humans. The composition of these amino acids can also be altered according to need, either by modulating the cultivation conditions or by genetic manipulations. *C. utilis* is commonly used for SCP production because of its high content of intracellular proteins, with the highest prevalence of lysine, an essential amino acid that is usually not found in plant biomass. Similarly, another *Candida* yeast *C. langeronii* shows high intracellular protein content and lysine content (7.8%) greater than that found in soybeans [130]. Apart from amino acids, yeast biomass also contains several biologically active microelements, which play crucial roles in maintaining human health. Selenium and magnesium are two such examples. Selenium present in the body as selenoprotein helps in the functioning of various enzymes like glutathione peroxidase, glutathione reductase, *etc.*, and also serves as an important antioxidant. Magnesium, on the other hand, acts as a cofactor for several important enzymes and proteins in the body, a key component of DNA molecules and ribosomes, and is involved in lipid metabolism, ATP synthesis, and protein translation. *Candida* species of yeast have an inherent ability to bind to various elements from their substrate, thereby enriching the biomass with vital elements. Hence, these yeast biomasses are used as feedstock supplements or additives in animal feeds. *C. utilis* was reported to bind and accumulate magnesium and selenium in high amounts when cultivated in magnesium sulfate solution or on selenium as substrate [131, 132].

Moreover, the cell walls of these yeasts are composed of various components that have important therapeutic applications. For example, glucomannan is an important component of the *C. utilis* cell wall and has proven antioxidant properties. It can capture free radicals, chelate iron ions, interact with immune receptors, and modulate signaling pathways. They also inhibit DNA damage and exhibit photo-protective properties, shielding the cells from harmful effects of UV-B radiation [1]. Besides, anti-mutagenic, anti-toxic, and anti-genotoxic properties of cell wall polysaccharides of *C. utilis* have been described in several other studies [133, 134]. The yeast cell wall also contains β-(1,3)-glucan polymer, which is not water-soluble. However, chemically solubilized β-glucan compounds, as well as glucomannans and mannoproteins from *C. utilis*, have shown immunostimulatory activities. These compounds act as PAMPs that activate immune cells' receptors, thereby stimulating them to trigger a response against pathogens.

Furthermore, the anticancer effects of β-glucan and mannans of *Candida* yeasts were determined by Tizard *et al.* on malignant tumors in mice [135]. The exopolysaccharides (EPSs) are usually highly viscous, and due to their pseudoplastic behavior, they are considered suitable biopolymers in pharmaceutical and cosmetic industries [1]. Besides *C. utilis*, EPS composition has been characterized in *C. famata*, *C. guilliermondii*, and *C. boidinii* [136, 137].

SCO Production

Among oleaginous microorganisms capable of producing single-cell oil (SCO), the unicellular morphology of yeast has proven to be beneficial for extraction. These SCOs find huge applications in the food and healthcare industries. Moreover, the saturated fatty acid composition of most of the oleaginous yeast species makes them potential candidates for biofuel production. *Candida* species of yeasts exhibit the ability to utilize different carbon sources, hence are considered to aid in cheap SCO production from agro-industrial wastes. Several of these species have been studied for SCO production by growing on cheap substrates like molasses, glycerol, and industrial wastewater [138 - 140].

Furthermore, the high stearic acid content of *C. tropicalis* reported in one study makes it promising for cocoa-butter substitutes [141]. The exceptional biodegradability of *C. vishwanathii* and *C. tropicalis* has been investigated on hydrophobic and phenolic wastes, respectively, for SCO production [142, 143]. However, the lack of molecular biology tools has limited the scope of engineering these species for lipid production. However, in a recent study, *C. tropicalis* has been engineered genetically to increase lipid production by 60% [144].

Sugar Alcohols

Because of the alarming incidences of diabetes globally, sugar consumption is condemned for many people; thus, sugar alcohols pose a safer alternative to regular sugars such as sucrose. *Candida* yeasts synthesize these alcohols as metabolic byproducts, and many studies have aimed to exploit them as a potential source for large-scale production of these alcohols. Moreover, xylitol and erythritol have also exhibited beneficial effects on dental health, mainly attributed to the inability of oral bacteria to ferment these alcohols. Xylitol is already used in the market to sweeten chewing gums and candies [145, 146].

Xylitol

This 5-C sugar alcohol is largely recommended in the diet of diabetic people because of its very low glycemic index. *Candida* species of yeasts produce this alcohol as a byproduct of the xylose metabolic pathway. The first step of the pathway catalyzed by xyloreductase is NADPH-dependent, while the second step catalyzed by xylitol dehydrogenase is NAD^+-dependent. Due to this difference, an imbalance of cofactor arises leading to the accumulation of the intermediate product xylitol. Among yeasts of the *Candida* genus, *C. tropicalis* has been particularly used to naturally produce xylitol in significant amounts from cheap raw materials such as sugarcane bagasse or corncob hydrolysates [147 - 149], while many genetic engineering strategies have also been implemented to increase the yield [150 - 153]. Recently, fermentation of xylose from rice straw and wheat bran hydrolysates by *C. boidinii* was optimized, and the conditions favoring the maximum xylitol yield were evaluated [154].

Erythritol

It is a 4-C sugar alcohol produced from erythrose, a product of the pentose phosphate (PP) pathway in yeast, by the action of the erythrose reductase enzyme. *Candida* (syn. *Yarrowia*) *lipolytica* has been a promising host for converting cheap carbon sources like molasses and glycerol to erythritol. A high titer (224 g/L) was obtained by cultivating this yeast in repeated batch culture [155]. Mirończuk *et al.* overexpressed the PP pathway genes encoding transketolase (*TKL1*), transaldolase (*TAL1*), glucose-6-phosphate dehydrogenase (*ZWF1*), and 6-phosphogluconate dehydrogenase (*GND1*) to enhance the titer substantially, and a maximum increase of 67% was reported [155]. A newly isolated yeast, *C. sorbosivorans,* has been found to be a potent erythritol producer and exerts a significant inhibitory effect on the biofilm formation of *Streptococcus mutans* [156].

Citric Acid

It is an important organic compound commonly used in food, pharmaceutical, chemical, and other industries. Because of its ability to bind to metal ions and inactivate enzymes such as oxidase, it is used as an antioxidant in processed fruits and vegetables, additive for fruit juice and confectionaries, stabilizer in wineries and fat industry, and also as an anti-clotting agent for the preservation of blood [157]. Usually, the production of this acid is enhanced by modulating the C/N ratio of the medium being utilized. The principal mechanism lies in the fact that under low nitrogen conditions, intracellular AMP is rapidly converted into IMP and NH_4^+ ions to compensate for the deficit of N in the yeast cells. Low AMP concentration, in turn, reduces the activity of the TCA cycle enzyme isocitrate dehydrogenase, which converts isocitrate to α-ketoglutarate. As a result, isocitrate, and eventually citrate accumulates in the mitochondria, is subsequently released into the cytosol and taken up in other biochemical pathways. Such enhanced citrate production by altering the C/N ratio was described in *C. zeylanoides*, *C. lipolytica*, *C. tropicalis*, and *C. oleophila* [158].

Dicarboxylic Acids

These are usually derived from long-chain fatty acids (FA) and find widespread industrial applications in manufacturing polymers and cosmetics. Usually, dicarboxylic acids (DCAs) are produced by alkane oxidizing microbes where alkanes are first converted to α-monocarboxylic acids (MCA) by α-oxidation, followed by oxidation of MCA by ω-oxidation. *C. tropicalis* is well-known for its robust ω-oxidation pathway and has long been considered a suitable platform for the industrial production of long-chain DCA. This yeast has been engineered several times to enhance DCA production. Since DCA is metabolized by the β-oxidation pathway, impairing the pathway genes has been a major strategy to improve the DCA yield in the organism [128]. Later, in an alternative approach, Cao *et al*. manipulated the acetyl-CoA transportation in *C. tropicalis* by deleting a copy of the carnitine acetyltransferase (CAT) gene, interfering with β-oxidation and demonstrating an improved yield of α,ω-tridecanedioic acid (DCA13), an important intermediate used in industries [159]. Recently, the synthesis of long-chain DCAs has been improved by engineering fatty acid transporter protein Fat1p and peroxisomal transporter Pxa1 of the yeast [160]. Other alkane-utilizing yeasts of the genus have also been reported to produce DCA by ω-oxidation pathway, *viz.C. sorbophila* and *C. lipolytica*, although the yield is not as high as found in *C. tropicalis* [161, 162]. However, due to the ease of genetic engineering, the latter has been engineered to improve DCA production in a multigene editing approach where β-oxidation of the organism was compromised and ω-oxidation was up-regulated [163].

Ethanol

Ethanol is usually produced by conventional yeasts like *S. cerevisiae* by fermenting glucose. However, many *Candida* species possess the ability to ferment a wide variety of other sugars, such as xylose, arabinose, lactose, *etc.*, to produce ethanol, and this property is exploited in a significant number of studies to optimize ethanol production. *C. tropicalis*, for example, is well-known for utilizing a diverse range of substrates and the yeast has been cultivated together with an ethanol-fermenting bacteria, *Zymomonas mobilis*, on various fruits and vegetable wastes to produce a high quantity of ethanol [164]. *C. tropicalis* has also been immobilized to produce ethanol with unsaccharified corn starch [165]. *C. shehatae* and *C. boidinii* are also being studied for their potent ethanol-producing ability from lignocellulose hydrolysates [123, 166]. Besides, *C. krusei* is well-known for its fermenting ability and bioethanol production [22].

Enzymes

Biocatalysts of different classes, such as protease, amylase, lipase, *etc.*, are required during the processing of raw materials to develop industrial products. In many cases, like in textile, paper, food, and detergent industries, the requirement of enzymes is huge, whereas small quantities of different enzymes are needed in the pharmaceutical industry. Several natural isolates and genetically engineered microorganisms are employed to produce different classes of enzymes for various applications.

Lipase

These enzymes find increasing applications in modifying the structure of lipid and lipid-derived compounds used in food, cosmetics, and oleochemical industries to manufacture surfactants, polymers, *etc.* Extracellular lipase activity has been detected in numerous studies on *Candida* sp. *viz.C. rugosa, C. antarctica, C. vishwanathii,etc* [167]. One important compound synthesized by microbial lipases is monoacylglycerol, which is used as a food emulsifier as well as in cosmetics [168]. In the pharmaceutical industry, *Candida* lipases are often used for selective hydrolysis to resolve racemic mixtures of acids or alcohols. *C. antarctica* lipase CAL-B is commercially produced as Novozyme 435, Chirazyme L2-C2 for such purposes [169, 170]. CAL-B has been used for the resolution of the racemic mixture of 2-pentanol. *S*-(+)-2-pentanol is an important chiral intermediate in the synthesis of drugs used to treat Alzheimer's [167, 171, 172]. Similarly, *C. rugosa* lipase is responsible for the resolution of profens, a group of anti-inflammatory drugs whose activity largely depends on their optical purity, and usually, the S-isomers are more effective than the R-isomers [173].

Invertase

These enzymes are particularly important in the food industry to produce an inverted syrup containing an equimolar mixture of glucose and fructose formed by the hydrolysis of sucrose. Since the monomeric mixture is sweeter and more hygroscopic than the disaccharide itself, it is widely used in the food industry for the manufacture of candies, artificial honey, and also in cosmetics, pharmaceuticals, and paper industries [174, 175]. Invertase production by *C. utilis* has been reported [176].

Biosurfactants

These are compounds known to lower surface tension and find widespread industrial applications. In the food industry, they are used for the production of mayonnaise and sauces, while their ability to form emulsions is exploited in cosmetics and pharmaceutical industries. In addition, their amphiphilic properties are used for detergent and surface cleaning uses. *Candida* yeasts are well-known for biosurfactant production from hydrophobic substrates. For example, *C. glabrata* produced biosurfactants using plant fats as substrates [177]. Several of the *Candida* spp. produces mannosylerythritol lipids that are widely used in the pharmaceutical industry [178, 179]. *C. tropicalis* and *C. bombicola* were reported to produce sophorolipids having biosurfactant properties [180, 181]. *C. lipolytica* reportedly produces liposan, a lipopolysaccharide commonly known for emulsifying vegetable oils [182]. Besides, biosurfactants obtained from several species of the *Candida* genus exhibits bioremediation properties. For example, those produced by *C. lipolytica* and *C. sphaerica* have been successfully employed for the removal of heavy metals from soil and aqueous effluent [183 - 185].

Food Applications

Various *Candida* yeasts are known to enhance the flavor of fermented food products. Many species, for instance, participate in the fermentation of milk during kefyr formation, influencing the organoleptic properties of kefyr. These yeasts include *C. famata*, *C. guilliermondii*, *C. kefyr*, and *C. inconspicua*. In addition, *C. parapsilosis* is often isolated from different fermented products of soy that are traditionally used in many states of India [186, 187]. Besides, the various lipolytic enzymes secreted from *Candida* yeasts enhance the organoleptic properties of meat and can be used for sausage maturation and similar other uses in meat processing industries [188].

Biocontrol

Certain *Candida* yeasts belonging to the CTG clade, such as *C. guilliermondii* and *C. oleophila,* have been used as biocontrol agents to prevent fungal spoilage of food harvested from plants. Although the molecular mechanism underlying such activity has not been understood so far, the secretion of chitinolytic enzymes has been considered a common characteristic among these biocontrol yeasts [189, 190]. Furthermore, the involvement of extracellular glucanase has also been reported in the biocontrol activity of *C. oleophila* [189]. Thus, exploring a suitable consortium of microorganisms as biocontrol agents has a promising future to minimize the use of synthetic chemicals that are hazardous to the environment and human health.

Biodegradation

The exceptional biodegradability of *Candida* species has been studied extensively for waste utilization, bioconversion of cheap substrates, and combat heavy metal contamination in the environment. *C. tropicalis* is known to degrade aromatic hydrocarbons such as phenolic compounds, pyrenes, *etc.* Several phenol-degrading strains of *C. tropicalis* have been recently identified, which effectively reduce the toxicity load in industrial wastewater [142, 191]. Pyrene is a major representative of polycyclic aromatic hydrocarbons that are well-known environmental pollutants. This yeast has been reported to degrade pyrene, and the efficiency of the process has been enhanced in a recently developed sonication-assisted technique leading to 70% degradation of the compound [192]. Many other species have also been studied for pyrene degradation with comparable performances, such as *C. vishwanathii* (60.77%) and *Candida* sp. S1 (75%) [193, 194].

CONCLUDING REMARKS

Apart from the clinical importance of the genus, the potential of many *Candida* species to develop biofilm and chelate heavy metals, produce industrially useful enzymes and biosurfactants, as well as efficient biodegrading properties, have all led to this genus being regarded as an industrial repository. Therefore, a great deal of scope remains to screen more isolates and species of this yeast genus to come up with useful applications.

CONSENT FOR PUBLICATION

Not applicable.

CONFLICT OF INTEREST

The authors declare no conflict of interest, financial or otherwise.

ACKNOWLEDGEMENTS

Authors acknowledge the Indian Institute of Technology Kharagpur and Ministry of Human Resource and Development, Govt. of India for providing facilities and infrastructure supporting this work.

REFERENCES

[1] Kieliszek M, Kot AM, Bzducha-Wróbel A. Błażejak S, Gientka I, Kurcz A. Biotechnological use of *Candida* yeasts in the food industry: A review. Fungal Biol Rev 2017; 31: 185-98.
[http://dx.doi.org/10.1016/j.fbr.2017.06.001]

[2] Barnett JA. A history of research on yeasts 8: taxonomy. Yeast 2004; 21(14): 1141-93.
[http://dx.doi.org/10.1002/yea.1154] [PMID: 15515119]

[3] Kurtzman CP, Fell JW. The yeasts - a taxonomic study. 4th ed. Amsterdam, The Netherlands: . 1998.

[4] Singleton P, Sainsbury D. Dictionary of Microbiology and Molecular Biology (2nd Edition). Dict Microbiol Mol Biol. 1987; 16, Wiley: pp. 653-4.
[http://dx.doi.org/10.1042/bst0160653a]

[5] Lachance M-A, Boekhout T, Scorzetti G, Fell JW, Kurtzman CP. Candida Berkhout (1923) The Yeasts. Elsevier 2011; pp. 987-1278.
[http://dx.doi.org/10.1016/B978-0-444-52149-1.00090-2]

[6] Yarrow D. Four new combinations in yeasts. Antonie van Leeuwenhoek 1972; 38(3): 357-60.
[http://dx.doi.org/10.1007/BF02328105] [PMID: 4538626]

[7] Akinterinwa O, Khankal R, Cirino PC. Metabolic engineering for bioproduction of sugar alcohols. Curr Opin Biotechnol 2008; 19(5): 461-7.
[http://dx.doi.org/10.1016/j.copbio.2008.08.002] [PMID: 18760354]

[8] Papon N, Courdavault V, Clastre M, Simkin AJ, Crèche J, Giglioli-Guivarc'h N. Deus ex *Candida* genetics: overcoming the hurdles for the development of a molecular toolbox in the CTG clade. Microbiology 2012; 158(Pt 3): 585-600.
[http://dx.doi.org/10.1099/mic.0.055244-0] [PMID: 22282522]

[9] Kawaguchi Y, Honda H, Taniguchi-Morimura J, Iwasaki S. The codon CUG is read as serine in an asporogenic yeast *Candida cylindracea*. Nature 1989; 341(6238): 164-6.
[http://dx.doi.org/10.1038/341164a0] [PMID: 2506450]

[10] Sugita T, Nakase T. Non-universal usage of the leucine CUG codon and the molecular phylogeny of the genus *Candida*. Syst Appl Microbiol 1999; 22(1): 79-86.
[http://dx.doi.org/10.1016/S0723-2020(99)80030-7] [PMID: 10188281]

[11] Butler G, Rasmussen MD, Lin MF, *et al.* Evolution of pathogenicity and sexual reproduction in eight *Candida* genomes. Nature 2009; 459(7247): 657-62.
[http://dx.doi.org/10.1038/nature08064] [PMID: 19465905]

[12] Fitzpatrick DA, Logue ME, Stajich JE, Butler G. A fungal phylogeny based on 42 complete genomes derived from supertree and combined gene analysis. BMC Evol Biol 2006; 6: 99.
[http://dx.doi.org/10.1186/1471-2148-6-99] [PMID: 17121679]

[13] Fitzpatrick DA, Logue ME, Butler G. Evidence of recent interkingdom horizontal gene transfer between bacteria and *Candida parapsilosis*. BMC Evol Biol 2008; 8: 181.
[http://dx.doi.org/10.1186/1471-2148-8-181] [PMID: 18577206]

[14] Fitzpatrick DA, O'Gaora P, Byrne KP, Butler G. Analysis of gene evolution and metabolic pathways using the *Candida* Gene Order Browser. BMC Genomics 2010; 11: 290.
[http://dx.doi.org/10.1186/1471-2164-11-290] [PMID: 20459735]

[15] Papon N, Courdavault V, Clastre M, Bennett RJ. Emerging and emerged pathogenic *Candida* species: beyond the *Candida albicans* paradigm. PLoS Pathog 2013; 9(9): e1003550.
[http://dx.doi.org/10.1371/journal.ppat.1003550] [PMID: 24086128]

[16] Morley SJ, Willett M. Kinky binding and SECsy insertions. Mol Cell 2009; 35(4): 396-8.
[http://dx.doi.org/10.1016/j.molcel.2009.08.001] [PMID: 19716783]

[17] Suzuki T, Ueda T, Watanabe K. The 'polysemous' codon--a codon with multiple amino acid assignment caused by dual specificity of tRNA identity. EMBO J 1997; 16(5): 1122-34.
[http://dx.doi.org/10.1093/emboj/16.5.1122] [PMID: 9118950]

[18] Miranda I, Silva-Dias A, Rocha R, *et al*. *C. albicans* CUG mistranslation is a mechanism to create cell surface variation. MBio 2013; 4(4): 1-9.
[http://dx.doi.org/10.1128/mBio.00285-13] [PMID: 23800396]

[19] Turner SA, Butler G. The *Candida* pathogenic species complex. Cold Spring Harb Perspect Med 2014; 4(9): a019778.
[http://dx.doi.org/10.1101/cshperspect.a019778] [PMID: 25183855]

[20] Alcoba-Flórez J, Méndez-Alvarez S, Cano J, Guarro J, Pérez-Roth E, del Pilar Arévalo M. Phenotypic and molecular characterization of *Candida nivariensis* sp. nov., a possible new opportunistic fungus. J Clin Microbiol 2005; 43(8): 4107-11.
[http://dx.doi.org/10.1128/JCM.43.8.4107-4111.2005] [PMID: 16081957]

[21] Correia A, Sampaio P, James S, Pais C. Candida bracarensis sp. nov., a novel anamorphic yeast species phenotypically similar to *Candida glabrata*. Int J Syst Evol Microbiol 2006; 56(Pt 1): 313-7.
[http://dx.doi.org/10.1099/ijs.0.64076-0] [PMID: 16403904]

[22] Dandi ND, Dandi BN, Chaudhari AB. Bioprospecting of thermo- and osmo-tolerant fungi from mango pulp-peel compost for bioethanol production. Antonie van Leeuwenhoek 2013; 103(4): 723-36.
[http://dx.doi.org/10.1007/s10482-012-9854-4] [PMID: 23180376]

[23] Pfaller MA, Diekema DJ, Gibbs DL, *et al*. Results from the ARTEMIS DISK Global Antifungal Surveillance Study, 1997 to 2007: a 10.5-year analysis of susceptibilities of *Candida* Species to fluconazole and voriconazole as determined by CLSI standardized disk diffusion. J Clin Microbiol 2010; 48(4): 1366-77.
[http://dx.doi.org/10.1128/JCM.02117-09] [PMID: 20164282]

[24] Pfaller MA, Diekema DJ. Epidemiology of invasive candidiasis: a persistent public health problem. Clin Microbiol Rev 2007; 20(1): 133-63.
[http://dx.doi.org/10.1128/CMR.00029-06] [PMID: 17223626]

[25] Kabir MA, Ahmad Z. *Candida* infections and their prevention. ISRN Prev Med 2012; 2013: 763628.
[http://dx.doi.org/10.5402/2013/763628] [PMID: 24977092]

[26] Krcmery V, Barnes AJ. Non-albicans *Candida* spp. causing fungaemia: pathogenicity and antifungal resistance. J Hosp Infect 2002; 50(4): 243-60.
[http://dx.doi.org/10.1053/jhin.2001.1151] [PMID: 12014897]

[27] Lockhart SR, Etienne KA, Vallabhaneni S, *et al*. Simultaneous emergence of multidrug-resistant *Candida* auris on 3 continents confirmed by whole-genome sequencing and epidemiological analyses. Clin Infect Dis 2017; 64(2): 134-40.
[http://dx.doi.org/10.1093/cid/ciw691] [PMID: 27988485]

[28] Chowdhary A, Sharma C, Meis JF. *Candida Auris*: A rapidly emerging cause of hospital-acquired multidrug-resistant fungal infections globally. PLoS Pathog 2017; 13(5): e1006290.
[http://dx.doi.org/10.1371/journal.ppat.1006290] [PMID: 28542486]

[29] Clancy CJ, Nguyen MH. Emergence of *Candida Auris*: An international call to arms. Clin Infect Dis 2017; 64(2): 141-3.
[http://dx.doi.org/10.1093/cid/ciw696] [PMID: 27989986]

[30] Lockhart SR, Messer SA, Gherna M, *et al*. Identification of *Candida nivariensis* and Candida bracarensis in a large global collection of *Candida glabrata* isolates: comparison to the literature. J Clin Microbiol 2009; 47(4): 1216-7.
[http://dx.doi.org/10.1128/JCM.02315-08] [PMID: 19193845]

[31] Dabas PS. An approach to etiology, diagnosis and management of different types of candidiasis. J Yeast Fungal Res 2013; 4: 63-74.

[http://dx.doi.org/10.5897/JYFR2013.0113]

[32]   Martins M, Henriques M, Ribeiro AP, Fernandes R, Gonçalves V, Seabra Á, *et al.* Oral colonization by *Candida* in patients attending a dental clinic in Braga, Portugal Rev Iberoam Micol 2010; 27: 119-24.
[http://dx.doi.org/10.1016/j.riam.2010.03.007] [PMID: 20403455]

[33]   Ray D, Goswami R, Banerjee U, *et al.* Prevalence of *Candida glabrata* and its response to boric acid vaginal suppositories in comparison with oral fluconazole in patients with diabetes and vulvovaginal candidiasis. Diabetes Care 2007; 30(2): 312-7.
[http://dx.doi.org/10.2337/dc06-1469] [PMID: 17259500]

[34]   Beigi RH, Meyn LA, Moore DM, Krohn MA, Hillier SL. Vaginal yeast colonization in nonpregnant women: a longitudinal study. Obstet Gynecol 2004; 104(5 Pt 1): 926-30.
[http://dx.doi.org/10.1097/01.AOG.0000140687.51048.73] [PMID: 15516380]

[35]   Fan D, Coughlin LA, Neubauer MM, *et al.* Activation of HIF-1α and LL-37 by commensal bacteria inhibits *Candida albicans* colonization. Nat Med 2015; 21(7): 808-14.
[http://dx.doi.org/10.1038/nm.3871] [PMID: 26053625]

[36]   McCarty TP, Pappas PG. Invasive Candidiasis. Infect Dis Clin North Am 2016; 30(1): 103-24.
[http://dx.doi.org/10.1016/j.idc.2015.10.013] [PMID: 26739610]

[37]   Kullberg BJ, Arendrup MC. Invasive Candidiasis. N Engl J Med 2015; 373(15): 1445-56.
[http://dx.doi.org/10.1056/NEJMra1315399] [PMID: 26444731]

[38]   Cantón E, Pemán J, Quindós G, *et al.* Prospective multicenter study of the epidemiology, molecular identification, and antifungal susceptibility of *Candida parapsilosis*, Candida orthopsilosis, and Candida metapsilosis isolated from patients with candidemia. Antimicrob Agents Chemother 2011; 55(12): 5590-6.
[http://dx.doi.org/10.1128/AAC.00466-11] [PMID: 21930869]

[39]   Bohner F, Gacser A, Toth R. Epidemiological Attributes of Candida Species in Tropical Regions. Curr Trop Med Rep 2021; 8: 59-68.
[http://dx.doi.org/10.1007/s40475-021-00226-5]

[40]   Lerma A, Cantero E, Soriano M, Orden B, Muñez E, Ramos-Martinez A. Clinical presentation of candidaemia in elderly patients: experience in a single institution. Rev Esp Quimioter 2017; 30(3): 207-12.
[PMID: 28361527]

[41]   Pammi M, Holland L, Butler G, Gacser A, Bliss JM. *Candida parapsilosis* is a significant neonatal pathogen: a systematic review and meta-analysis. Pediatr Infect Dis J 2013; 32(5): e206-16.
[http://dx.doi.org/10.1097/INF.0b013e3182863a1c] [PMID: 23340551]

[42]   Colombo AL, Guimarães T, Silva LRBF, *et al.* Prospective observational study of candidemia in São Paulo, Brazil: incidence rate, epidemiology, and predictors of mortality. Infect Control Hosp Epidemiol 2007; 28(5): 570-6.
[http://dx.doi.org/10.1086/513615] [PMID: 17464917]

[43]   Silva S, Negri M, Henriques M, Oliveira R, Williams DW, Azeredo J. *C. glabrata, Candida parapsilosis* and *C. tropicalis*: biology, epidemiology, pathogenicity and antifungal resistance. FEMS Microbiol Rev 2012; 36(2): 288-305.
[http://dx.doi.org/10.1111/j.1574-6976.2011.00278.x] [PMID: 21569057]

[44]   Pires-Gonçalves RH, Miranda ET, Baeza LC, Matsumoto MT, Zaia JE, Mendes-Giannini MJS. Genetic relatedness of commensal strains of *Candida albicans* carried in the oral cavity of patients' dental prosthesis users in Brazil. Mycopathologia 2007; 164(6): 255-63.
[http://dx.doi.org/10.1007/s11046-007-9052-5] [PMID: 17906942]

[45]   Tintelnot K, Haase G, Seibold M, *et al.* Evaluation of phenotypic markers for selection and identification of *Candida dubliniensis*. J Clin Microbiol 2000; 38(4): 1599-608.

[http://dx.doi.org/10.1128/JCM.38.4.1599-1608.2000] [PMID: 10747150]

[46]   Lasker BA, Elie CM, Lott TJ, *et al.* Molecular epidemiology of *Candida albicans* strains isolated from the oropharynx of HIV-positive patients at successive clinic visits. Med Mycol 2001; 39(4): 341-52.
[http://dx.doi.org/10.1080/mmy.39.4.341.352] [PMID: 11556764]

[47]   Loreto ÉS, Scheid LA, Nogueira CW, Zeni G, Santurio JM, Alves SH. *Candida dubliniensis*: epidemiology and phenotypic methods for identification. Mycopathologia 2010; 169(6): 431-43.
[http://dx.doi.org/10.1007/s11046-010-9286-5] [PMID: 20490751]

[48]   Khan Z, Ahmad S, Joseph L, Chandy R. *Candida dubliniensis*: an appraisal of its clinical significance as a bloodstream pathogen. PLoS One 2012; 7(3): e32952.
[http://dx.doi.org/10.1371/journal.pone.0032952] [PMID: 22396802]

[49]   Jackson AP, Gamble JA, Yeomans T, *et al.* Comparative genomics of the fungal pathogens *Candida dubliniensis* and *Candida albicans*. Genome Res 2009; 19(12): 2231-44.
[http://dx.doi.org/10.1101/gr.097501.109] [PMID: 19745113]

[50]   Guinea J. Global trends in the distribution of Candida species causing candidemia. Clin Microbiol Infect 2014; 20 (Suppl. 6): 5-10.
[http://dx.doi.org/10.1111/1469-0691.12539] [PMID: 24506442]

[51]   Nucci M, Queiroz-Telles F, Alvarado-Matute T, *et al.* Epidemiology of candidemia in Latin America: a laboratory-based survey. PLoS One 2013; 8(3): e59373.
[http://dx.doi.org/10.1371/journal.pone.0059373] [PMID: 23527176]

[52]   Walker LA, Gow NAR, Munro CA. Elevated chitin content reduces the susceptibility of Candida species to caspofungin. Antimicrob Agents Chemother 2013; 57(1): 146-54.
[http://dx.doi.org/10.1128/AAC.01486-12] [PMID: 23089748]

[53]   Sardi JCO, Scorzoni L, Bernardi T, Fusco-Almeida AM, Mendes Giannini MJS. Candida species: current epidemiology, pathogenicity, biofilm formation, natural antifungal products and new therapeutic options. J Med Microbiol 2013; 62(Pt 1): 10-24.
[http://dx.doi.org/10.1099/jmm.0.045054-0] [PMID: 23180477]

[54]   Cruciani M, Serpelloni G. Management of Candida infections in the adult intensive care unit. Expert Opin Pharmacother 2008; 9(2): 175-91.
[http://dx.doi.org/10.1517/14656566.9.2.175] [PMID: 18201143]

[55]   Pfaller MA. Antifungal drug resistance: mechanisms, epidemiology, and consequences for treatment. Am J Med 2012; 125(1) (Suppl.): S3-S13.
[http://dx.doi.org/10.1016/j.amjmed.2011.11.001] [PMID: 22196207]

[56]   Perlin DS. Resistance to echinocandin-class antifungal drugs. Drug Resist Updat 2007; 10(3): 121-30.
[http://dx.doi.org/10.1016/j.drup.2007.04.002] [PMID: 17569573]

[57]   Healey KR, Zhao Y, Perez WB, *et al.* Prevalent mutator genotype identified in fungal pathogen *Candida glabrata* promotes multi-drug resistance. Nat Commun 2016; 7: 11128.
[http://dx.doi.org/10.1038/ncomms11128] [PMID: 27020939]

[58]   Katiyar S, Pfaller M, Edlind T. *Candida albicans* and *Candida glabrata* clinical isolates exhibiting reduced echinocandin susceptibility. Antimicrob Agents Chemother 2006; 50(8): 2892-4.
[http://dx.doi.org/10.1128/AAC.00349-06] [PMID: 16870797]

[59]   Vermitsky JP, Earhart KD, Smith WL, Homayouni R, Edlind TD, Rogers PD. Pdr1 regulates multidrug resistance in *Candida glabrata*: gene disruption and genome-wide expression studies. Mol Microbiol 2006; 61(3): 704-22.
[http://dx.doi.org/10.1111/j.1365-2958.2006.05235.x] [PMID: 16803598]

[60]   Tsai HF, Krol AA, Sarti KE, Bennett JE. *Candida glabrata* PDR1, a transcriptional regulator of a pleiotropic drug resistance network, mediates azole resistance in clinical isolates and petite mutants. Antimicrob Agents Chemother 2006; 50(4): 1384-92.
[http://dx.doi.org/10.1128/AAC.50.4.1384-1392.2006] [PMID: 16569856]

[61]   Pappas PG, Lionakis MS, Arendrup MC, Ostrosky-Zeichner L, Kullberg BJ. Invasive candidiasis. Nat Rev Dis Primers 2018; 4: 18026.
[http://dx.doi.org/10.1038/nrdp.2018.26] [PMID: 29749387]

[62]   Quintin J, Saeed S, Martens JHA, *et al. Candida albicans* infection affords protection against reinfection *via* functional reprogramming of monocytes. Cell Host Microbe 2012; 12(2): 223-32.
[http://dx.doi.org/10.1016/j.chom.2012.06.006] [PMID: 22901542]

[63]   Romani L, Mencacci A, Cenci E, Del Sero G, Bistoni F, Puccetti P. An immunoregulatory role for neutrophils in CD4+ T helper subset selection in mice with candidiasis. J Immunol 1997; 158(5): 2356-62.
[PMID: 9036985]

[64]   Gringhuis SI, Kaptein TM, Wevers BA, *et al.* Dectin-1 is an extracellular pathogen sensor for the induction and processing of IL-1β *via* a noncanonical caspase-8 inflammasome. Nat Immunol 2012; 13(3): 246-54.
[http://dx.doi.org/10.1038/ni.2222] [PMID: 22267217]

[65]   Gross O, Poeck H, Bscheider M, *et al.* Syk kinase signalling couples to the Nlrp3 inflammasome for anti-fungal host defence. Nature 2009; 459(7245): 433-6.
[http://dx.doi.org/10.1038/nature07965] [PMID: 19339971]

[66]   Joly S, Eisenbarth SC, Olivier AK, *et al.* Cutting edge: Nlrp10 is essential for protective antifungal adaptive immunity against *Candida albicans.* J Immunol 2012; 189(10): 4713-7.
[http://dx.doi.org/10.4049/jimmunol.1201715] [PMID: 23071280]

[67]   Reeves EP, Lu H, Jacobs HL, *et al.* Killing activity of neutrophils is mediated through activation of proteases by K+ flux. Nature 2002; 416(6878): 291-7.
[http://dx.doi.org/10.1038/416291a] [PMID: 11907569]

[68]   Wisplinghoff H, Ebbers J, Geurtz L, *et al.* Nosocomial bloodstream infections due to Candida spp. in the USA: species distribution, clinical features and antifungal susceptibilities. Int J Antimicrob Agents 2014; 43(1): 78-81.
[http://dx.doi.org/10.1016/j.ijantimicag.2013.09.005] [PMID: 24182454]

[69]   Diekema D, Arbefeville S, Boyken L, Kroeger J, Pfaller M. The changing epidemiology of healthcare-associated candidemia over three decades. Diagn Microbiol Infect Dis 2012; 73(1): 45-8.
[http://dx.doi.org/10.1016/j.diagmicrobio.2012.02.001] [PMID: 22578938]

[70]   Pfaller M, Neofytos D, Diekema D, *et al.* Epidemiology and outcomes of candidemia in 3648 patients: data from the Prospective Antifungal Therapy (PATH Alliance®) registry, 2004-2008. Diagn Microbiol Infect Dis 2012; 74(4): 323-31.
[http://dx.doi.org/10.1016/j.diagmicrobio.2012.10.003] [PMID: 23102556]

[71]   Priest SJ, Lorenz MC. Characterization of virulence-related phenotypes in Candida species of the CUG clade. Eukaryot Cell 2015; 14(9): 931-40.
[http://dx.doi.org/10.1128/EC.00062-15] [PMID: 26150417]

[72]   Liu H. Co-regulation of pathogenesis with dimorphism and phenotypic switching in *Candida albicans*, a commensal and a pathogen. Int J Med Microbiol 2002; 292(5-6): 299-311.
[http://dx.doi.org/10.1078/1438-4221-00215] [PMID: 12452278]

[73]   Huang G. Regulation of phenotypic transitions in the fungal pathogen *Candida albicans*. Virulence 2012; 3(3): 251-61.
[http://dx.doi.org/10.4161/viru.20010] [PMID: 22546903]

[74]   Kvaal C, Lachke SA, Srikantha T, Daniels K, McCoy J, Soll DR. Misexpression of the opaque-phase-specific gene *PEP1 (SAP1)* in the white phase of *Candida albicans* confers increased virulence in a mouse model of cutaneous infection. Infect Immun 1999; 67(12): 6652-62.
[http://dx.doi.org/10.1128/IAI.67.12.6652-6662.1999] [PMID: 10569787]

[75]   Kvaal CA, Srikantha T, Soll DR. Misexpression of the white-phase-specific gene WH11 in the opaque

phase of *Candida albicans* affects switching and virulence. Infect Immun 1997; 65(11): 4468-75.
[http://dx.doi.org/10.1128/iai.65.11.4468-4475.1997] [PMID: 9353021]

[76] Geiger J, Wessels D, Lockhart SR, Soll DR. Release of a potent polymorphonuclear leukocyte chemoattractant is regulated by white-opaque switching in *Candida albicans*. Infect Immun 2004; 72(2): 667-77.
[http://dx.doi.org/10.1128/IAI.72.2.667-677.2004] [PMID: 14742507]

[77] Sasse C, Hasenberg M, Weyler M, Gunzer M, Morschhäuser J. White-opaque switching of *Candida albicans* allows immune evasion in an environment-dependent fashion. Eukaryot Cell 2013; 12(1): 50-8.
[http://dx.doi.org/10.1128/EC.00266-12] [PMID: 23125350]

[78] Lo HJ, Köhler JR, DiDomenico B, Loebenberg D, Cacciapuoti A, Fink GR. Nonfilamentous *C. albicans* mutants are avirulent. Cell 1997; 90(5): 939-49.
[http://dx.doi.org/10.1016/S0092-8674(00)80358-X] [PMID: 9298905]

[79] Korting HC, Hube B, Oberbauer S, *et al*. Reduced expression of the hyphal-independent *Candida albicans* proteinase genes *SAP1* and *SAP3* in the *efg1* mutant is associated with attenuated virulence during infection of oral epithelium. J Med Microbiol 2003; 52(Pt 8): 623-32.
[http://dx.doi.org/10.1099/jmm.0.05125-0] [PMID: 12867554]

[80] Carlisle PL, Banerjee M, Lazzell A, Monteagudo C, López-Ribot JL, Kadosh D. Expression levels of a filament-specific transcriptional regulator are sufficient to determine *Candida albicans* morphology and virulence. Proc Natl Acad Sci USA 2009; 106(2): 599-604.
[http://dx.doi.org/10.1073/pnas.0804061106] [PMID: 19116272]

[81] Murad AMA, Leng P, Straffon M, *et al*. *NRG1* represses yeast-hypha morphogenesis and hypha-specific gene expression in *Candida albicans*. EMBO J 2001; 20(17): 4742-52.
[http://dx.doi.org/10.1093/emboj/20.17.4742] [PMID: 11532938]

[82] Liu H. Transcriptional control of dimorphism in *Candida albicans*. Curr Opin Microbiol 2001; 4(6): 728-35.
[http://dx.doi.org/10.1016/S1369-5274(01)00275-2] [PMID: 11731326]

[83] Zheng X, Wang Y, Wang Y. Hgc1, a novel hypha-specific G1 cyclin-related protein regulates *Candida albicans* hyphal morphogenesis. EMBO J 2004; 23(8): 1845-56.
[http://dx.doi.org/10.1038/sj.emboj.7600195] [PMID: 15071502]

[84] Biswas S, Van Dijck P, Datta A. Environmental sensing and signal transduction pathways regulating morphopathogenic determinants of *Candida albicans*. Microbiol Mol Biol Rev 2007; 71(2): 348-76.
[http://dx.doi.org/10.1128/MMBR.00009-06] [PMID: 17554048]

[85] Ernst JF. Transcription factors in *Candida albicans* - environmental control of morphogenesis. Microbiology 2000; 146(Pt 8): 1763-74.
[http://dx.doi.org/10.1099/00221287-146-8-1763] [PMID: 10931884]

[86] Naglik JR, Challacombe SJ, Hube B. *Candida albicans* secreted aspartyl proteinases in virulence and pathogenesis. Microbiol Mol Biol Rev 2003; 67(3): 400-28.
[http://dx.doi.org/10.1128/MMBR.67.3.400-428.2003] [PMID: 12966142]

[87] Zaugg C, Borg-Von Zepelin M, Reichard U, Sanglard D, Monod M. Secreted aspartic proteinase family of *Candida tropicalis*. Infect Immun 2001; 69(1): 405-12.
[http://dx.doi.org/10.1128/IAI.69.1.405-412.2001] [PMID: 11119531]

[88] Stehr F, Felk A, Gácser A, *et al*. Expression analysis of the *Candida albicans* lipase gene family during experimental infections and in patient samples. FEMS Yeast Res 2004; 4(4-5): 401-8.
[http://dx.doi.org/10.1016/S1567-1356(03)00205-8] [PMID: 14734020]

[89] Gácser A, Trofa D, Schäfer W, Nosanchuk JD. Targeted gene deletion in *Candida parapsilosis* demonstrates the role of secreted lipase in virulence. J Clin Invest 2007; 117(10): 3049-58.
[http://dx.doi.org/10.1172/JCI32294] [PMID: 17853941]

[90]   Thompson DS, Carlisle PL, Kadosh D. Coevolution of morphology and virulence in Candida species. Eukaryot Cell 2011; 10(9): 1173-82.
[http://dx.doi.org/10.1128/EC.05085-11] [PMID: 21764907]

[91]   Moran GP, Sullivan DJ, Coleman DC. Emergence of non-Candida albicans Candida species as pathogens Candida and candidiasis. Washington, DC: ASM Press 2002; pp. 37-54.

[92]   O'Connor L, Caplice N, Coleman DC, Sullivan DJ, Moran GP. Differential filamentation of *Candida albicans* and *Candida dubliniensis* Is governed by nutrient regulation of UME6 expression. Eukaryot Cell 2010; 9(9): 1383-97.
[http://dx.doi.org/10.1128/EC.00042-10] [PMID: 20639413]

[93]   Porman AM, Alby K, Hirakawa MP, Bennett RJ. Discovery of a phenotypic switch regulating sexual mating in the opportunistic fungal pathogen *Candida tropicalis*. Proc Natl Acad Sci USA 2011; 108(52): 21158-63.
[http://dx.doi.org/10.1073/pnas.1112076109] [PMID: 22158989]

[94]   Porman AM, Hirakawa MP, Jones SK, Wang N, Bennett RJ. MTL-independent phenotypic switching in *C. tropicalis* and a dual role for Wor1 in regulating switching and filamentation. PLoS Genet 2013; 9(3): e1003369.
[http://dx.doi.org/10.1371/journal.pgen.1003369] [PMID: 23555286]

[95]   Nobile CJ, Andes DR, Nett JE, Smith FJ, Yue F, Phan QT, *et al.* Critical role of Bcr1-dependent adhesins in *C. albicans* biofilm formation *in vitro* and *in vivo*. PLoS Pathog 2006.
[http://dx.doi.org/10.1371/journal.ppat.0020063]

[96]   Ding C, Butler G. Development of a gene knockout system in *Candida parapsilosis* reveals a conserved role for BCR1 in biofilm formation. Eukaryot Cell 2007; 6(8): 1310-9.
[http://dx.doi.org/10.1128/EC.00136-07] [PMID: 17586721]

[97]   Kim SK, Bissati KE, Mamoun CB. Amino acids mediate colony and cell differentiation in the fungal pathogen *Candida parapsilosis*. Microbiology 2006; 152(Pt 10): 2885-94.
[http://dx.doi.org/10.1099/mic.0.29180-0] [PMID: 17005970]

[98]   Connolly LA, Riccombeni A, Grózer Z, *et al.* The APSES transcription factor Efg1 is a global regulator that controls morphogenesis and biofilm formation in *Candida parapsilosis*. Mol Microbiol 2013; 90(1): 36-53.
[http://dx.doi.org/10.1111/mmi.12345] [PMID: 23895281]

[99]   Lackey E, Vipulanandan G, Childers DS, Kadosh D. Comparative evolution of morphological regulatory functions in *Candida* species. Eukaryot Cell 2013; 12(10): 1356-68.
[http://dx.doi.org/10.1128/EC.00164-13] [PMID: 23913541]

[100]   Papon N, Savini V, Lanoue A, *et al.* Candida guilliermondii: biotechnological applications, perspectives for biological control, emerging clinical importance and recent advances in genetics. Curr Genet 2013; 59(3): 73-90.
[http://dx.doi.org/10.1007/s00294-013-0391-0] [PMID: 23616192]

[101]   Wong S, Fares MA, Zimmermann W, Butler G, Wolfe KH. Evidence from comparative genomics for a complete sexual cycle in the 'asexual' pathogenic yeast *Candida glabrata*. Genome Biol 2003; 4(2): R10.
[http://dx.doi.org/10.1186/gb-2003-4-2-r10] [PMID: 12620120]

[102]   Kaur R, Domergue R, Zupancic ML, Cormack BP. A yeast by any other name: *Candida glabrata* and its interaction with the host. Curr Opin Microbiol 2005; 8(4): 378-84.
[http://dx.doi.org/10.1016/j.mib.2005.06.012] [PMID: 15996895]

[103]   Bennett RJ, Uhl MA, Miller MG, Johnson AD. Identification and characterization of a *Candida albicans* mating pheromone. Mol Cell Biol 2003; 23(22): 8189-201.
[http://dx.doi.org/10.1128/MCB.23.22.8189-8201.2003] [PMID: 14585977]

[104]   Berman J, Hadany L. Does stress induce (para)sex? Implications for *C. albicans* evolution. Trends

Genet 2012; 28(5): 197-203.
[http://dx.doi.org/10.1016/j.tig.2012.01.004] [PMID: 22364928]

[105] Alby K, Schaefer D, Bennett RJ. Homothallic and heterothallic mating in the opportunistic pathogen *Candida albicans*. Nature 2009; 460(7257): 890-3.
[http://dx.doi.org/10.1038/nature08252] [PMID: 19675652]

[106] Bennett RJ. Coming of age--sexual reproduction in Candida species. PLoS Pathog 2010; 6(12): e1001155.
[http://dx.doi.org/10.1371/journal.ppat.1001155] [PMID: 21203475]

[107] Sai S, Holland LM, McGee CF, Lynch DB, Butler G. Evolution of mating within the *Candida parapsilosis* species group. Eukaryot Cell 2011; 10(4): 578-87.
[http://dx.doi.org/10.1128/EC.00276-10] [PMID: 21335529]

[108] Brockert PJ, Lachke SA, Srikantha T, Pujol C, Galask R, Soll DR. Phenotypic switching and mating type switching of *Candida glabrata* at sites of colonization. Infect Immun 2003; 71(12): 7109-18.
[http://dx.doi.org/10.1128/IAI.71.12.7109-7118.2003] [PMID: 14638801]

[109] Kojic EM, Darouiche RO. Candida infections of medical devices. Clin Microbiol Rev 2004; 17(2): 255-67.
[http://dx.doi.org/10.1128/CMR.17.2.255-267.2004] [PMID: 15084500]

[110] Silva S, Negri M, Henriques M, Oliveira R, Williams DW, Azeredo J. Adherence and biofilm formation of non-*Candida albicans* Candida species. Trends Microbiol 2011; 19(5): 241-7.
[http://dx.doi.org/10.1016/j.tim.2011.02.003] [PMID: 21411325]

[111] Mourer T, El Ghalid M, d'Enfert C, Bachellier-Bassi S. Involvement of amyloid proteins in the formation of biofilms in the pathogenic yeast *Candida albicans*. Res Microbiol 2021; 172(3): 103813.
[http://dx.doi.org/10.1016/j.resmic.2021.103813] [PMID: 33515679]

[112] Fanning S, Mitchell AP. Fungal biofilms. PLoS Pathog 2012; 8(4): e1002585.
[http://dx.doi.org/10.1371/journal.ppat.1002585] [PMID: 22496639]

[113] Kabir MA, Hussain MA, Ahmad Z. *Candida albicans*: A Model Organism for Studying Fungal Pathogens. ISRN Microbiol 2012; 2012: 538694.
[http://dx.doi.org/10.5402/2012/538694] [PMID: 23762753]

[114] Nett JE, Lepak AJ, Marchillo K, Andes DR. Time course global gene expression analysis of an in vivo Candida biofilm. J Infect Dis 2009; 200(2): 307-13.
[http://dx.doi.org/10.1086/599838] [PMID: 19527170]

[115] Mulhern SM, Logue ME, Butler G. *Candida albicans* transcription factor Ace2 regulates metabolism and is required for filamentation in hypoxic conditions. Eukaryot Cell 2006; 5(12): 2001-13.
[http://dx.doi.org/10.1128/EC.00155-06] [PMID: 16998073]

[116] Finkel JS, Mitchell AP. Genetic control of *Candida albicans* biofilm development. Nat Rev Microbiol 2011; 9(2): 109-18.
[http://dx.doi.org/10.1038/nrmicro2475] [PMID: 21189476]

[117] Nobile CJ, Nett JE, Hernday AD, *et al.* Biofilm matrix regulation by *Candida albicans* Zap1. PLoS Biol 2009; 7(6): e1000133.
[http://dx.doi.org/10.1371/journal.pbio.1000133] [PMID: 19529758]

[118] Norice CT, Smith FJ Jr, Solis N, Filler SG, Mitchell AP. Requirement for *Candida albicans* Sun41 in biofilm formation and virulence. Eukaryot Cell 2007; 6(11): 2046-55.
[http://dx.doi.org/10.1128/EC.00314-07] [PMID: 17873081]

[119] Blankenship JR, Mitchell AP. How to build a biofilm: a fungal perspective. Curr Opin Microbiol 2006; 9(6): 588-94.
[http://dx.doi.org/10.1016/j.mib.2006.10.003] [PMID: 17055772]

[120] Sahni N, Yi S, Daniels KJ, Srikantha T, Pujol C, Soll DR. Genes selectively up-regulated by

pheromone in white cells are involved in biofilm formation in *Candida albicans*. PLoS Pathog 2009; 5(10): e1000601.
[http://dx.doi.org/10.1371/journal.ppat.1000601] [PMID: 19798425]

[121] Miyagi M, Wilson R, Saigusa D, *et al.* Indole-3-acetic acid synthesized through the indole-3-pyruvate pathway promotes *Candida tropicalis* biofilm formation. PLoS One 2020; 15(12): e0244246.
[http://dx.doi.org/10.1371/journal.pone.0244246] [PMID: 33332404]

[122] Paudyal A, Vediyappan G. Cell Surface Expression of Nrg1 Protein in Candida Auris 2021; 1-14.
[http://dx.doi.org/10.3390/jof7040262]

[123] Gonçalves DB, Batista AF, Rodrigues MQRB, Nogueira KMV, Santos VL. Ethanol production from macaúba (Acrocomia aculeata) presscake hemicellulosic hydrolysate by Candida boidinii UFMG14. Bioresour Technol 2013; 146: 261-6.
[http://dx.doi.org/10.1016/j.biortech.2013.07.075] [PMID: 23941709]

[124] Vaz de Arruda P, dos Santos JC, de Cássia Lacerda Brambilla Rodrigues R, da Silva DDV, Yamakawa CK, de Moraes Rocha GJ, *et al.* Scale up of xylitol production from sugarcane bagasse hemicellulosic hydrolysate by Candida guilliermondii FTI 20037. J Ind Eng Chem 2017; 47: 297-302.
[http://dx.doi.org/10.1016/j.jiec.2016.11.046]

[125] Chattopadhyay A, Maiti MK. Efficient xylose utilization leads to highest lipid productivity in *C. tropicalis* SY005 among six yeast strains grown in mixed sugar medium. Appl Microbiol Biotechnol 2020; 104(7): 3133-44.
[http://dx.doi.org/10.1007/s00253-020-10443-z] [PMID: 32076780]

[126] Chattopadhyay A, Singh R, Das AK, Maiti MK. Characterization of two sugar transporters responsible for efficient xylose uptake in an oleaginous yeast *C. tropicalis* SY005. Arch Biochem Biophys 2020; 695: 108645.
[http://dx.doi.org/10.1016/j.abb.2020.108645] [PMID: 33122161]

[127] Ko BS, Kim J, Kim JH. Production of xylitol from D-xylose by a xylitol dehydrogenase gene-disrupted mutant of *C. tropicalis*. Appl Environ Microbiol 2006; 72(6): 4207-13.
[http://dx.doi.org/10.1128/AEM.02699-05] [PMID: 16751533]

[128] Picataggio S, Rohrer T, Deanda K, *et al.* Metabolic engineering of *C. tropicalis* for the production of long-chain dicarboxylic acids. Biotechnology (N Y) 1992; 10(8): 894-8.
[PMID: 1368984]

[129] Zhang L, Zhang H, Liu Y, *et al.* A CRISPR-Cas9 system for multiple genome editing and pathway assembly in *C. tropicalis*. Biotechnol Bioeng 2020; 117(2): 531-42.
[http://dx.doi.org/10.1002/bit.27207] [PMID: 31654413]

[130] Nigam JN. Cultivation of Candida langeronii in sugar cane bagasse hemicellulosic hydrolyzate for the production of single cell protein. World J Microbiol Biotechnol 2000; 16: 367-72.
[http://dx.doi.org/10.1023/A:1008922806215]

[131] Bzducha-Wróbel A, Błazejak S, Tkacz K. Cell wall structure of selected yeast species as a factor of magnesium binding ability. Eur Food Res Technol 2012; 235: 355-66.
[http://dx.doi.org/10.1007/s00217-012-1761-4]

[132] Kieliszek M, Błażejak S, Płaczek M. Spectrophotometric evaluation of selenium binding by *Saccharomyces cerevisiae* ATCC MYA-2200 and Candida utilis ATCC 9950 yeast. J Trace Elem Med Biol 2016; 35: 90-6.
[http://dx.doi.org/10.1016/j.jtemb.2016.01.014] [PMID: 27049131]

[133] Miadoková E, Svidová S, Vlcková V, *et al.* Diverse biomodulatory effects of glucomannan from Candida utilis. Toxicol In Vitro 2006; 20(5): 649-57.
[http://dx.doi.org/10.1016/j.tiv.2005.12.001] [PMID: 16413741]

[134] Vlcková V, Dúhová V, Svidová S, *et al.* Antigenotoxic potential of glucomannan on four model test systems. Cell Biol Toxicol 2004; 20(6): 325-32.

[http://dx.doi.org/10.1007/s10565-004-0089-7] [PMID: 15868477]

[135] Tizard IR, Carpenter RH, McAnalley BH, Kemp MC. The biological activities of mannans and related complex carbohydrates. Mol Biother 1989; 1(6): 290-6.
[PMID: 2692629]

[136] Gientka I, Bzducha-Wróbel A, Stasiak-Różańska L, Bednarska AA, Błażejak S. The exopolysaccharides biosynthesis by Candida yeast depends on carbon sources. Electron J Biotechnol 2016; 22: 31-7.
[http://dx.doi.org/10.1016/j.ejbt.2016.02.008]

[137] Petersen GR, Schubert WW, Richards GF, Nelson GA. Yeasts producing exopolysaccharides with drag-reducing activity. Enzyme Microb Technol 1990; 12: 255-9.
[http://dx.doi.org/10.1016/0141-0229(90)90096-9]

[138] Karatay SE, Dönmez G. Improving the lipid accumulation properties of the yeast cells for biodiesel production using molasses. Bioresour Technol 2010; 101(20): 7988-90.
[http://dx.doi.org/10.1016/j.biortech.2010.05.054] [PMID: 20542422]

[139] Raimondi S, Rossi M, Leonardi A, Bianchi MM, Rinaldi T, Amaretti A. Getting lipids from glycerol: new perspectives on biotechnological exploitation of *C. freyschussii*. Microb Cell Fact 2014; 13: 83.
[http://dx.doi.org/10.1186/1475-2859-13-83] [PMID: 24906383]

[140] Thangavelu K, Sundararaju P, Srinivasan N, Muniraj I, Uthandi S. Simultaneous lipid production for biodiesel feedstock and decontamination of sago processing wastewater using *C. tropicalis* ASY2. Biotechnol Biofuels 2020; 13: 35.
[http://dx.doi.org/10.1186/s13068-020-01676-1] [PMID: 32158499]

[141] Dey P, Maiti MK. Molecular characterization of a novel isolate of *C. tropicalis* for enhanced lipid production. J Appl Microbiol 2013; 114(5): 1357-68.
[http://dx.doi.org/10.1111/jam.12133] [PMID: 23311514]

[142] Dias B, Lopes M, Ramôa R, Pereira AS, Belo I. *Candida tropicalis* as a Promising Oleaginous Yeast for Olive Mill Wastewater Bioconversion. Energies 2021; 14: 640.
[http://dx.doi.org/10.3390/en14030640]

[143] Ayadi I, Kamoun O, Trigui-Lahiani H, *et al.* Single cell oil production from a newly isolated Candida viswanathii Y-E4 and agro-industrial by-products valorization. J Ind Microbiol Biotechnol 2016; 43(7): 901-14.
[http://dx.doi.org/10.1007/s10295-016-1772-4] [PMID: 27114386]

[144] Chattopadhyay A, Gupta A, Maiti MK. Engineering an oleaginous yeast *C. tropicalis* SY005 for enhanced lipid production. Appl Microbiol Biotechnol 2020; 104(19): 8399-411.
[http://dx.doi.org/10.1007/s00253-020-10830-6] [PMID: 32820371]

[145] Ly KA, Milgrom P, Rothen M. Xylitol, sweeteners, and dental caries. Pediatr Dent 2006; 28(2): 154-63.
[PMID: 16708791]

[146] Burt BA. The use of sorbitol- and xylitol-sweetened chewing gum in caries control. J Am Dent Assoc 2006; 137(2): 190-6.
[http://dx.doi.org/10.14219/jada.archive.2006.0144] [PMID: 16521385]

[147] Rao RS, Jyothi ChP, Prakasham RS, Sarma PN, Rao LV. Xylitol production from corn fiber and sugarcane bagasse hydrolysates by *C. tropicalis*. Bioresour Technol 2006; 97(15): 1974-8.
[http://dx.doi.org/10.1016/j.biortech.2005.08.015] [PMID: 16242318]

[148] Kumar V, Krishania M, Preet Sandhu P, Ahluwalia V, Gnansounou E, Sangwan RS. Efficient detoxification of corn cob hydrolysate with ion-exchange resins for enhanced xylitol production by *C. tropicalis* MTCC 6192. Bioresour Technol 2018; 251: 416-9.
[http://dx.doi.org/10.1016/j.biortech.2017.11.039] [PMID: 29276111]

[149] Cheng KK, Zhang JA, Ling HZ, *et al.* Optimization of pH and acetic acid concentration for

bioconversion of hemicellulose from corncobs to xylitol by *C. tropicalis.* Biochem Eng J 2009; 43: 203-7.
[http://dx.doi.org/10.1016/j.bej.2008.09.012]

[150] Jeon WY, Yoon BH, Ko BS, Shim WY, Kim JH. Xylitol production is increased by expression of codon-optimized *Neurospora crassa* xylose reductase gene in *C. tropicalis.* Bioprocess Biosyst Eng 2012; 35(1-2): 191-8.
[http://dx.doi.org/10.1007/s00449-011-0618-8] [PMID: 21922311]

[151] Jeon WY, Shim WY, Lee SH, Choi JH, Kim JH. Effect of heterologous xylose transporter expression in *C. tropicalis* on xylitol production rate. Bioprocess Biosyst Eng 2013; 36(6): 809-17.
[http://dx.doi.org/10.1007/s00449-013-0907-5] [PMID: 23411871]

[152] Ahmad I, Shim WY, Kim JH. Enhancement of xylitol production in glycerol kinase disrupted *C. tropicalis* by co-expression of three genes involved in glycerol metabolic pathway. Bioprocess Biosyst Eng 2013; 36(9): 1279-84.
[http://dx.doi.org/10.1007/s00449-012-0872-4] [PMID: 23232964]

[153] Lee JK, Koo BS, Kim SY. Cloning and characterization of the *xyl1* gene, encoding an NADH-preferring xylose reductase from *C. parapsilosis*, and its functional expression in *C. tropicalis.* Appl Environ Microbiol 2003; 69(10): 6179-88.
[http://dx.doi.org/10.1128/AEM.69.10.6179-6188.2003] [PMID: 14532079]

[154] Bedő S, Fehér A, Khunnonkwao P, Jantama K, Fehér C. Optimized bioconversion of xylose derived from pre-treated crop residues into xylitol by using candida boidinii. Agronomy (Basel) 2021; 11: 79.
[http://dx.doi.org/10.3390/agronomy11010079]

[155] Mirończuk AM, Rakicka M, Biegalska A, Rymowicz W, Dobrowolski A. A two-stage fermentation process of erythritol production by yeast Y. lipolytica from molasses and glycerol. Bioresour Technol 2015; 198: 445-55.
[http://dx.doi.org/10.1016/j.biortech.2015.09.008] [PMID: 26409857]

[156] Saran S, Mukherjee S, Dalal J, Saxena RK. High production of erythritol from Candida sorbosivorans SSE-24 and its inhibitory effect on biofilm formation of Streptococcus mutans. Bioresour Technol 2015; 198: 31-8.
[http://dx.doi.org/10.1016/j.biortech.2015.08.146] [PMID: 26363499]

[157] Soccol CR, Vandenberghe LPS, Rodrigues C, Pandey A. New perspectives for citric acid production and application. Food Technol Biotechnol 2006; 44: 141-9.

[158] Kim KH, Lee HY, Lee CY. Pretreatment of sugarcane molasses and citric acid production by candida zeylanoides. Korean J Microbiol Biotechnol 2015; 43: 164-8.
[http://dx.doi.org/10.4014/mbl.1503.03006]

[159] Cao Z, Gao H, Liu M, Jiao P. Engineering the acetyl-CoA transportation system of *C. tropicalis* enhances the production of dicarboxylic acid. Biotechnol J 2006; 1(1): 68-74.
[http://dx.doi.org/10.1002/biot.200500008] [PMID: 16892226]

[160] Zhang L, Xiu X, Wang Z, *et al.* Increasing Long-Chain Dicarboxylic Acid Production in *Candida tropicalis* by Engineering Fatty Transporters. Mol Biotechnol 2021; 63(6): 544-55.
[http://dx.doi.org/10.1007/s12033-021-00319-6] [PMID: 33786739]

[161] Fickers P, Benetti PH, Waché Y, *et al.* Hydrophobic substrate utilisation by the yeast *Yarrowia lipolytica*, and its potential applications. FEMS Yeast Res 2005; 5(6-7): 527-43.
[http://dx.doi.org/10.1016/j.femsyr.2004.09.004] [PMID: 15780653]

[162] Lee H, Sugiharto YEC, Lee S, *et al.* Characterization of the newly isolated ω-oxidizing yeast Candida sorbophila DS02 and its potential applications in long-chain dicarboxylic acid production. Appl Microbiol Biotechnol 2017; 101(16): 6333-42.
[http://dx.doi.org/10.1007/s00253-017-8321-6] [PMID: 28589225]

[163] Abghari A, Madzak C, Chen S. Combinatorial engineering of Yarrowia lipolytica as a promising cell

biorefinery platform for the *de novo* production of multi-purpose long chain dicarboxylic acids. Fermentation (Basel) 2017; 3.
[http://dx.doi.org/10.3390/fermentation3030040]

[164] Patle S, Lal B. Ethanol production from hydrolysed agricultural wastes using mixed culture of Zymomonas mobilis and *Candida tropicalis*. Biotechnol Lett 2007; 29(12): 1839-43.
[http://dx.doi.org/10.1007/s10529-007-9493-4] [PMID: 17657407]

[165] Jamai L, Ettayebi K, El Yamani J, Ettayebi M. Production of ethanol from starch by free and immobilized *Candida tropicalis* in the presence of α-amylase. Bioresour Technol 2007; 98(14): 2765-70.
[http://dx.doi.org/10.1016/j.biortech.2006.09.057] [PMID: 17127052]

[166] Yuvadetkun P, Boonmee M. Ethanol production capability of *Candida shehatae* in mixed sugars and rice straw hydrolysate. Sains Malays 2016; 45: 581-7.

[167] Żymańczyk-Duda E, Brzezińska-Rodak M, Klimek-Ochab M, Duda M, Zerka A. Yeast as a Versatile Tool in Biotechnology Yeast - Ind Appl. InTech 2017; Vol. 32: pp. 137-44.
[http://dx.doi.org/10.5772/intechopen.70130]

[168] Kaewthong W, Sirisansaneeyakul S, Prasertsan P. H-Kittikun A. Continuous production of monoacylglycerols by glycerolysis of palm olein with immobilized lipase. Process Biochem 2005; 40: 1525-30.
[http://dx.doi.org/10.1016/j.procbio.2003.12.002]

[169] Weber N, Weitkamp P, Mukherjee KD. Steryl and stanyl esters of fatty acids by solvent-free esterification and transesterification in vacuo using lipases from Rhizomucor miehei, Candida antarctica, and Carica papaya. J Agric Food Chem 2001; 49(11): 5210-6.
[http://dx.doi.org/10.1021/jf0107407] [PMID: 11714305]

[170] Anderson EM, Larsson KM, Kirk O. One biocatalyst - many applications: The use of Candida antarctica B-lipase in organic synthesis. Biocatal Biotransform 1998; 16: 181-204.
[http://dx.doi.org/10.3109/10242429809003198]

[171] Patel RN, Banerjee A, Nanduri V, Goswami A, Comezoglu FT. Enzymatic resolution of racemic secondary alcohols by lipase B from Candida antarctica. JAOCS. J Am Oil Chem Soc 2000; 77: 1015-9.
[http://dx.doi.org/10.1007/s11746-000-0161-y]

[172] Hamilton. Method of using neurotrophic sulfonamide compounds. US005721256A 1998.

[173] Hutt AJ, Caldwell J. The importance of stereochemistry in the clinical pharmacokinetics of the 2-arylpropionic acid non-steroidal anti-inflammatory drugs. Clin Pharmacokinet 1984; 9(4): 371-3.
[http://dx.doi.org/10.2165/00003088-198409040-00007] [PMID: 6467769]

[174] Kotwal SM, Shankar V. Immobilized invertase. Biotechnol Adv 2009; 27(4): 311-22.
[http://dx.doi.org/10.1016/j.biotechadv.2009.01.009] [PMID: 19472508]

[175] Kulshrestha S, Tyagi P, Sindhi V, Yadavilli KS. Invertase and its applications – A brief review. J Pharm Res 2013; 7: 792-7.
[http://dx.doi.org/10.1016/j.jopr.2013.07.014]

[176] Chiura H, Iizuka M, Yamamoto T. A glucomannan as an extracellular product of candida utilis ii. structure of a glucomannan: characterization of oligosaccharides obtained by partial hydrolysis. Agric Biol Chem 1982; 46: 1733-42.
[http://dx.doi.org/10.1271/bbb1961.46.1733]

[177] de Gusmão CAB, Rufino RD, Sarubbo LA. Laboratory production and characterization of a new biosurfactant from *Candida glabrata* UCP1002 cultivated in vegetable fat waste applied to the removal of hydrophobic contaminant. World J Microbiol Biotechnol 2010; 26: 1683-92.
[http://dx.doi.org/10.1007/s11274-010-0346-2]

[178] Kim KS, Yun HS. Production of soluble β-glucan from the cell wall of *Saccharomyces cerevisiae*.

Enzyme Microb Technol 2006; 39: 496-500.
[http://dx.doi.org/10.1016/j.enzmictec.2005.12.020]

[179] Arutchelvi JI, Bhaduri S, Uppara PV, Doble M. Mannosylerythritol lipids: a review. J Ind Microbiol Biotechnol 2008; 35(12): 1559-70.
[http://dx.doi.org/10.1007/s10295-008-0460-4] [PMID: 18716809]

[180] Solaiman DKY, Ashby RD, Nuñez A, Foglia TA. Production of sophorolipids by *C. bombicola* grown on soy molasses as substrate. Biotechnol Lett 2004; 26(15): 1241-5.
[http://dx.doi.org/10.1023/B:BILE.0000036605.80577.30] [PMID: 15289681]

[181] Develter DWG, Lauryssen LML. Properties and industrial applications of sophorolipids. Eur J Lipid Sci Technol 2010; 112: 628-38.
[http://dx.doi.org/10.1002/ejlt.200900153]

[182] Alizadeh-Sani M, Hamishehkar H, Khezerlou A, *et al.* Bioemulsifiers derived from microorganisms: applications in the drug and food industry. Adv Pharm Bull 2018; 8(2): 191-9.
[http://dx.doi.org/10.15171/apb.2018.023] [PMID: 30023320]

[183] Luna JM, Rufino RD, Sarubbo LA. Biosurfactant from *Candida sphaerica* UCP0995 exhibiting heavy metal remediation properties. Process Saf Environ Prot 2016; 102: 558-66.
[http://dx.doi.org/10.1016/j.psep.2016.05.010]

[184] Menezes CTB, Barros EC, Rufino RD, Luna JM, Sarubbo LA. Replacing synthetic with microbial surfactants as collectors in the treatment of aqueous effluent produced by acid mine drainage, using the dissolved air flotation technique. Appl Biochem Biotechnol 2011; 163(4): 540-6.
[http://dx.doi.org/10.1007/s12010-010-9060-7] [PMID: 20714828]

[185] Rufino RD, Rodrigues GIB, Campos-Takaki GM, Sarubbo LA, Ferreira SRM. Application of a yeast biosurfactant in the removal of heavy metals and hydrophobic contaminant in a soil used as slurry barrier. Appl Environ Soil Sci 2011; 2011: 1-7.
[http://dx.doi.org/10.1155/2011/939648]

[186] Sarkar PK, Tamang JP, Cook PE, Owens JD. Kinema - a traditional soybean fermented food: proximate composition and microflora. Food Microbiol 1994; 11: 47-55.
[http://dx.doi.org/10.1006/fmic.1994.1007]

[187] Sohliya I, Joshi SR, Bhagobaty RK, Kumar R. Tungrymbai - A traditional fermented soybean food of the ethnic tribes of Meghalaya. Indian J Tradit Knowl 2009; 8: 559-61.

[188] Zalacain I, Zapelena MJ, Astiasaran I, Bello J. Addition of lipase from *Candida cylindracea* to a traditional formulation of a dry fermented sausage. Meat Sci 1996; 42(2): 155-63.
[http://dx.doi.org/10.1016/0309-1740(95)00033-X] [PMID: 22060681]

[189] Bar-Shimon M, Yehuda H, Cohen L, *et al.* Characterization of extracellular lytic enzymes produced by the yeast biocontrol agent *Candida oleophila*. Curr Genet 2004; 45(3): 140-8.
[http://dx.doi.org/10.1007/s00294-003-0471-7] [PMID: 14716497]

[190] Zhang D, Spadaro D, Garibaldi A, Gullino ML. Potential biocontrol activity of a strain of *Pichia guilliermondii* against grey mold of apples and its possible modes of action. Biol Control 2011; 57: 193-201.
[http://dx.doi.org/10.1016/j.biocontrol.2011.02.011]

[191] Gong Y, Ding P, Xu MJ, Zhang CM, Xing K, Qin S. Biodegradation of phenol by a halotolerant versatile yeast *Candida tropicalis* SDP-1 in wastewater and soil under high salinity conditions. J Environ Manage 2021; 289: 112525.
[http://dx.doi.org/10.1016/j.jenvman.2021.112525] [PMID: 33836438]

[192] Kashyap N, Moholkar VS. Intensification of pyrene degradation by native *Candida tropicalis* MTCC 184 with sonication: Kinetic and mechanistic investigation. Chem Eng Process 2021; 164: 108415.
[http://dx.doi.org/10.1016/j.cep.2021.108415]

[193] Hesham AEL, Alamri SA, Khan S, Mahmoud ME, Mahmoud HM. Isolation and molecular genetic

characterization of a yeast strain able to degrade petroleum polycyclic aromatic hydrocarbons. Afr J Biotechnol 2009; 8: 2218-23.
[http://dx.doi.org/10.4314/ajb.v8i10.60559]

[194] Hadibarata T, Khudhair AB, Kristanti RA, Kamyab H. Biodegradation of pyrene by *Candida* sp. S1 under high salinity conditions. Bioprocess Biosyst Eng 2017; 40(9): 1411-8.
[http://dx.doi.org/10.1007/s00449-017-1798-7] [PMID: 28612166]

CHAPTER 6

Pichia: From Supporting Actors to the Leading Roles

Rosicler Colet[1], Guilherme Hassemer[1], Sérgio Luiz Alves Júnior[2], Natalia Paroul[1], Jamile Zeni[1], Geciane Toniazzo Backes[1], Eunice Valduga[1] and **Rogerio Luis Cansian[1,*]**

[1] *Universidade Regional Integrada do Alto Uruguai e das Missões, Campus Erechim, Avenida Sete de Setembro, Erechim/RS, Brazil*

[2] *Laboratory of Biochemistry and Genetics, Federal University of Fronteira Sul, Chapecó/SC, Brazil*

Abstract: *Pichia pastoris* are heterotrophic yeasts able to use many carbon sources such as glucose, glycerol, and methanol; they are unable, however, to metabolize lactose. Their methylotrophic properties, high yield, efficient post-translational modifications, and secretion of recombinant proteins, alongside a lack of hyperglycosylation, a post-translational process similar to that of mammals, and low maintenance costs for large-scale applications, make this yeast a promising alternative to produce recombinant proteins. The main recombinant products obtained from *P. pastoris* include vaccines and other biopharmaceuticals, enzymes, proteins, and pigments. *Pichia* spp. are also used in ethanol production and many other foods such as fermentation of coffee, cocoa, and olives, as well as alcoholic beverages. The use of *Pichia* yeasts in wastewater treatment and in fungal control of stored grains and fruit has also been reported. This chapter will discuss the environmental diversity of many species of *Pichia*, especially *P. pastoris*. Furthermore, the main uses of *Pichia* spp. in many bioprocesses will also be explored.

Keywords: Alcoholic beverages, Alcoholic beverages, Bioprocesses, Biocontrol systems, Carotenoids, Cocoa fermentation, Ethanol, Enzymes, Environmental diversity, Fermentation, Hyaluronic acid, Isobutanol, Pharmaceuticals, *Pichia pastoris*, Recombinant proteins, Ricinoleic acid, Vaccines, Wastewater treatment, Xylitol, Yeast.

---

* **Corresponding author Rogerio Luis Cansian:** Universidade Regional Integrada do Alto Uruguai e das Missões, Campus Erechim, Avenida Sete de Setembro, Erechim/RS, Brazil; Tel: +55 54 999763183; E-mail: cansian@uricer.edu.br

Sérgio Luiz Alves Júnior, Helen Treichel, Thiago Olitta Basso and Boris Ugarte Stambuk (Eds.)
All rights reserved-© 2022 Bentham Science Publishers

INTRODUCTION

The methylotrophic yeast *Pichia pastoris,* also known as *Komagataella pastoris,* has been commercialized by the Phillips Petroleum Company as a source of single-cell protein (SCP) destined for animal feed. It grows using methanol as a carbon source, causing overexpression of alcohol oxidase enzyme (AOX1) [1]. However, the increase in oil prices around 1970 negatively affected the use of *P. pastoris* as SCP [2]. Later on, Phillips Petroleum contacted Salk Institute Biotechnology/Industrial Associates, Inc. (SIBIA) seeking to develop a *Pichia* strain that could be used as a host cell for recombinant protein production [3, 4]. Based on the success this strain has shown as host, many different companies and research groups refined the initial protein expression system seeking to improve the recombinant protein expression rate. Its potential applications now include synthetic biology and whole-cell biotransformation.

The first record of protein production through biological systems for human use was a protein-based smallpox vaccine developed by Edward Jenner in 1796. From 1990 onwards, the biotechnology industry has been using microbial fermentation techniques to obtain products to be used in many different areas, such as the production of cleaning agents, fabrics, medicines, plastics, and even nutrition supplements. With the advent of recombinant DNA, it is now possible to use cultures of yeast, mold, bacteria, mammal cells, and even bugs in recombinant protein production (RPP) [5].

Escherichia coli is one of the most commonly used microorganisms in recombinant protein research, mostly due to its quick duplication time, high cell density, fully mapped genome, and low cost. However, the use of *E. coli* also has disadvantages, such as the lack of post-translational processing (glycosylation), reduced yield of recombinant products, presence of inactive proteins, and potential production of cytotoxic compounds [6, 7]. Many proteins are not able to be expressed in *E. coli* strains, as they require exact levels of post-translational maturity and, as such, must be produced by methylotrophic yeasts [8].

In this regard, yeasts such as *Pichia pastoris*, *Saccharomyces cerevisiae*, *Hansenula polymorpha,* and *Kluyveromyces lactis* are the most prominent [9, 10]. These yeasts tend to be applied in the production of heterologous proteins, mostly due to their high yield, strain stability, rapid growth, high cell density, and post-translational processing similar to that of mammals [11]; however, their glycosylation pattern remains different from that of human cells [12, 13]. The use of non-conventional yeasts has become a promising alternative, merging microbial advantages and eukaryotic protein processing while displaying several advantages over *S. cerevisiae* regarding pathway requirements, product profiles, and overall cell physiology [6, 14].

Regarding the protein expression system using recombinant DNA techniques, *P. pastoris* displays improved productivity rates, more efficient post-translational modifications, better secretion of recombinant proteins, lack of hyper-glycosylation, reduced costs for large-scale production and maintenance, as well the ability to grow under high cell density conditions (up to 130 g/L) when compared to *S. cerevisiae* [4, 15, 16]. *P. pastoris* also secretes low amounts of endogenous proteins, which helps with the purification process, and adapts well to genetic manipulation, which allows the use of advanced genetic modification tools (*i.e.,* CRISPR/Cas9). These factors, in tandem with *P. pastoris*' low production costs [17, 18], make it a versatile option for biotechnological expression systems.

Pichia pastoris has been successfully used in the production of many recombinant heterologous proteins [4, 19 - 21] and multiple enzymes, such as α-amylase [19], β-mannanase [22], and β-glucosidase [23] to be used in chemical, pharmaceutical, and food industries. Furthermore, the increase in knowledge of *P. pastoris*' properties, the availability of genome data, and the development of new tools for cloning multiple genes have expanded its applications in industrial processes and in the production of important chemical compounds *via* metabolic engineering. These compounds and processes include xanthophylls [24], carotenoids [25], hyaluronic acid [26], ricinoleic acid [27], ethanol [28, 29], isobutanol [30], xylitol [31 - 33], cocoa fermentation [34], vaccine production [35 - 42], biocontrol systems [43 - 45], and removal of dyes and colorings from wastewaters [31 - 33].

PICHIA: A DIVERSITY OF ENVIRONMENTS

More than 100 species of *Pichia* have been discovered, most of them found in rotting plants or symbiotically with insects; some, however, can be found in the necrotic tissues of some cacti (*Pichia cactophila*) [46], in cured cheeses (*P. membranifaciens*) [47, 48], raw milk and fresh cheese (*P. anômala*) [49], and even some citruses [50]. Many of these *Pichia* species might even be present in many foods, beverages, and products with high sugar content, as undesirable contaminants. Some of these species are able to utilize organic acids present in many foods, causing spoilage, usually as a thin film on the surface of pickles [51], beer, and wines. Among all yeasts, *Pichia, Candida, Saccharomyces,* and *Rhodotorula* are the genera most commonly associated with spoilage of juices and wines, as these products contain microflora naturally resistant to the product's acidity. The main effects of the spoilage brought by these microorganisms include

CO_2 production, the presence of unpleasant flavors and smells, as well as changes in the product's color and texture [50, 52, 53].

Pichia pastoris is a methylotrophic yeast, usually found in American Chestnut trees (*Castanea dentata*) and in California Black Oak (*Quercus kelloggii*) [54]. Some *Pichia* yeasts are also found in dirt, fresh water, insects, exudates of plants and fruits, as well as contaminants in a variety of foods and beverages. Nevertheless, some of these species might actually offer positive effects during the initial stages of wine and cheese production, while others have been described as pathogenic to humans [55, 56].

Almeida and co-workers [34] have found *Pichia fermentans, P. kudriavzevii, P. manshurica, S. cerevisiae,* and *Zygosaccharomyces bailii* in spontaneously fermented cocoa samples in Brazil; Daniel and co-workers [57] found *P. kudriavzevii, P. caribbica, P. manshurica* e *P. mexicana* in fermented cocoa beans in Ghana. Araújo [58] isolated yeasts from ice cream-bean (*Inga edulis*), guavira (*Campomanesia adamantium*), and sugar-apple (*Annona squamosa*), with *Candida citri, Wickerhamomyces (Pichia) ciferrii, W. (Pichia) kudriavzevii, Meyerozyma (Pichia) caribbica,* and *Saccharomyces* sp. found among them.

Jutakanoke and co-workers [59] isolated 72 strains of sugar-cane juice from sugar processing plants in Thailand and identified *Pichia kudriavzevii* as a potential bioethanol producer. Lima and co-workers [60] evaluated ethanol production using cassava roots and yeast strains isolated from sugarcane juice samples of an alcohol producing plant in northeast Brazil; they found that *Pichia* spp. represented 21.7% of the total 69 isolated strains tested.

Nevertheless, some *Pichia* strains, such as *P. ohmeri* are classified as opportunistic pathogens to immunocompromised humans, usually causing fungemia [61]. The prolonged use of catheters may also increase one's susceptibility to infection by *P. ohmeri*; some studies have also reported that the use of mechanical ventilators, dialysis, and tracheostomies [62], as well as the removal of catheters from major blood vessels [63], might also increase the chances of infection.

THE USE OF *PICHIA* YEASTS IN BIOPROCESSES

Methylotrophic yeasts use methanol as their sole carbon source. All methylotrophic yeasts belong to four genera: *Pichia, Hansenula, Candida,* and *Torulopsis* [64]. These yeasts are used in many biotechnological processes, especially in recombinant protein production, in which initial reactions happen in the peroxisome, a membrane-bound organelle, when using methanol as carbon source [65, 66].

The genus *Pichia*, when compared to *S. cerevisiae*, displays certain advantages regarding protein glycosylation and total protein yield, because, in many cases, *Pichia* yeasts do not hyperglycosylate proteins. As such, the chain length of the oligosaccharides added to the protein after the transduction process is much smaller than those found when using *S. cerevisiae* [67].

Pichia, formerly known as *Hansenula* or *Hyphopichia*, is a yeast that belongs to the *Saccharomycetaceae* family, its cells can be spherical, elliptical or oblong acuminate in shape. During sexual reproduction, *Pichia* cells form round, hat-shaped, or hemispherical ascospores. When reproduction occurs asexually, it is done *via* multilateral budding alongside the formation of a pseudomycelium [68].

Yeasts belonging to the *Pichia* family are able to thrive in foods with high sugar content, such as concentrated juices, but they are unable to consume lactose. *Pichia guilliermondii* is an osmotolerant yeast able to consume glucose, sucrose, raffinose, and trehalose; it also has the ability to assimilate carbonic compounds derived from galactose, sucrose, maltose, cellobiose, trehalose, raffinose, and xylose, among others, however, it is unable to assimilate nitrates [59] and does not grow in media lacking vitamins [67, 69, 70]. *P. guilliermondii* is also used as an antagonistic microorganism in the biological control of fungal diseases in harvested fruits [71, 72].

Pichia angusta (*Hansenula polymorpha*) is used in the study of peroxisomal functions and molecular biology [70]. It is able to grow quickly using glucose, fructose and sucrose, and does not generate fermentation-related products when grown under aerobic conditions [73].

Pichia anomala is an ascomycota that can act as deteriorating agent in high sugar content products and even silage. It shows ability to grow in low pH, high osmotic pressure, and low oxygen conditions [74]. Despite its status as a deteriorating microorganism, it can be used in wines in order to enhance the beverage's aroma. *Pichia anomala* can also be used as biocontrol agent, due to its potential to inhibit the growth of many fungi in harvested apples [75] and grains during storage [76 - 78]. *P. anomala* also contributes to reduce the amount of greenhouse gases released into the environment due to its ability to assimilate a large range of nitrogen-based compounds, converting nitrate and ammonium into proteins. Furthermore, this species also displays considerable phytase activity (myo-inositol-hexaphosphate phosphohydrolase, E.C. 3.1.3.8. and 3.1.3.26), making it able to digest phytate, the main phosphate storing compound in many plants, improving the nutritional value of animal feeds and reducing the amount of phosphorus present in animal manure [79].

Pichia membranifaciens is a species widely found in nature; it is also considered a deteriorating agent in foods and beverages, especially in vegetables and fruits, as well as fermented beverages [80]. *P. membranifaciens* has also been described as a potential biological agent in the control of post-harvest diseases in many fruits [81 - 84], it was also identified as part of starter cultures in spontaneous fermentation processes for production of distilled liquors in China [85, 86], wines [87, 88], and fermented black [89, 90] and green olives [91].

Pichia pastoris is a species of methylotrophic, facultative anaerobic yeast [92], Many uses in recombinant protein and enzyme production for pharmaceutical and food industries [93 - 95] were found. *P. pastoris* is able to produce high amounts of intra and extracellular proteins, being widely used in molecular biology and biotechnological processes as expression agent [96]. It is able to ferment glucose and assimilates carbon compounds derived from trehalose, rhamnose, mannitol and succinic acid. *P. pastoris*, however, is not able to assimilate nitrates, requires the presence of vitamins, and a temperature close to 37 °C [69].

Recombinant Proteins

P. pastoris' expression system offers high yield of recombinant proteins [35]. Furthermore, it offers some advantages over other prokaryotic and eukaryotic expression systems, such as:

1. Genome: when compared to mammal and insect cells, *P. pastoris* displays a relatively simple and easily manipulated genome [97, 98]; the target gene is integrated to the yeast's genome, ensuring the stability and reproducibility of the expression system [97, 99]; through homologous recombination, linear plasmids may be easily added to the cell [100]; also, the genome sequences and metabolic model of the strains are already known [101].

2. Metabolism: *P. pastoris* displays a strong preference for aerobic conditions, but it does not produce fermentation compounds under these conditions, minimizing the risk of ethanol accumulation in the system [101]; it also has the ability to grow in high cell density environments [20, 97, 98, 100].

3. Post-translational modifications: these are necessary for target proteins that require glycosylation, formation of disulfide bridges, methylation, acylation, phosphorylation, or proteolytic cleavage, as well as for proteins that must be directed to subcellular locations [97, 100].

4. Production: it is able to express large quantities of intra and extracellular proteins; it also displays ease of genetic manipulation, and the ability to reach high cell densities using carbon sources such as glucose and glycerol [4].

5. Alcohol oxidase 1 (AOX1) gene promoter: it is a strong promoter used to transcribe heterologous genes, from which AOX1 derives. This promoter is transcriptionally regulated by methanol. In addition to requiring methanol, the AOX1 system needs lack of glucose to be fully active. According to Pennell and Eldin [99], AOX1 and AOX2 enzymes display 97% identity in the aminoacid chain sequence and enzymatic activity, however, close to 95% of alcohol oxidase in *P. pastoris* is relegated to the AOX1 promoter. This enzyme catalyzes methanol conversion into formaldehyde and hydrogen peroxide. Subsequently, catalase breaks the hydrogen peroxide into oxygen and water. The formaldehyde is released from the peroxisome to the cytoplasm and quickly oxidized into formiate and carbon dioxide, by the actions of dehydrogenase enzymes, and is then used either to produce energy [68].

6. Safety: it is classified as GRAS (Generally Recognized as Safe) by the Food and Drug Administration (FDA) since 2006; and it does not produce endotoxins or pyrogen [102].

7. Purification: it displays the capacity to quickly and effectively segregate target recombinant proteins in the culture's supernatant with a surprisingly low amount of extracellular endogenous proteins, creating high purity products. As such, Pichia not only allows for cultures with high volumetric yield but also simplifies the downstream process, making the recombinant protein production system more economically feasible [103].

Still, some disadvantages do exist, such as the requirement of methanol, which is very flammable and may cause severe health and safety risks if not properly handled; furthermore, methanol may not be ideal for food products [4]; during the culture, high protease expression levels might be achieved, which may degrade the recombinant proteins present in the system [98]; the size of the protein to be expressed may also be a limiting factor, as proteins expressed usually have a molecular weight below 10 kDa [99].

Production of Vaccines and Biopharmaceutical Products

The biopharmaceutical market has its origins around 1970, following the discovery of recombinant DNA techniques. The Food and Drug Administration (FDA) approved the very first biopharmaceutical in 1982, Humulin, a compound analogous to human insulin, developed by Herbert Boyer in 1977 and acquired by Eli Lilly and Company.

Pichia pastoris is the most widely used species of microorganisms in the development of vaccines, for, by using the yeast's gene expression system, it is possible to obtain proteins related to diseases such as hepatitis B, influenza, cancer, atherosclerosis, human papillomavirus (HPV), and diabetes [35 - 42], turning the process cheaper and more productive.

P. pastoris secretes a high number of heterologous proteins to the culture media, with little amounts of endogenous proteins, facilitating the recovery and purification of recombinant proteins [104]. *P. pastoris* is also able to eliminate endotoxins, displays low susceptibility of bacteriophage contamination, and presents no pathogenicity to humans [105].

Pichia's gene expression system is used to clone and express recombinant proteins from different strains, promoters and markers, because it is easily manipulated, displaying strong AOX1 promoter activity, high expression and secretion rates, as well as allowing for insertion of multiple gene copies [20]. Table **1** presents some examples of recombinant proteins produced by *Pichia* with medicinal applications.

Table 1. Examples of heterologous proteins expressed by *Pichia* strains [35, 106].

Protein	*Pichia* Strain	Total Yield
Vaccine against recombinant influenza H1N1	His + *Pichia* strain	50-70 mg / L
Dengue fever vaccines, DENV-3 and VLPs	KM71H	10–20 mg / L
HCV E1E2 vaccine	SMD 1168	30-40 mg / L
HBsAg vaccine	GS115	5–10 g / L
Kunitz protease inhibitor	XIa	1 g/L
Human parathyroid hormone	Methylotrophic *Pichia* strains	0.2-0.5 mg / mL
Tetanus toxin C fragment	HIS4	12 g/L
Human plasmin	GS115	500–520 mg / L
Stage III anticancer agent	Y-2448	ND
Human insulin	SuperMan5	2–5 mg / L
	SuperMan5	5 mg/L
	X-33	1.5 g/L
IgG antibodies	YGLY27435 YGLY27431	100–250 mg / L
Hepatitis B surface antigen	HBsAg	0.3 g/L
α1 recombinant bovine interferon	GS115	0.1–0.5 mg/ mL
Atopic dermatitis	MGL 1304	ND
Tumor necrosis factor	GS115	8 g/L

Hepatitis B is considered an infectious disease in humans, with no therapy or treatment available once infected. New advances in vaccine research are necessary in order to help reduce the spread of hepatitis B [107]. Many studies point out that hepatitis B vaccine, while unable to cure the disease after it has been contracted, has 95% efficacy in preventing its development in a host. The vaccine is also considered an alum adjuvant, being composed of purified hepatitis B surface antigens (HBsAg). These antigens are obtained from plasma or produced in mammal or yeast cells using recombinant protein production techniques. HBsAg particles can be commercially produced in *P. pastoris*, *S. cerevisiae* and *Hansenula polymorpha* [108].

Pichia pastoris stands out as the best host for heterologous expression of functionally active recombinant proteins, as it allows for multiple copies of HBsAg associated gene to be implants, while being closely monitored by the AOX1 promoter [109].

Among biopharmaceuticals, scFv antibody fragments such as Nanobody ALX-0061 (IL6 anti-receptor) and ALX-00171 (anti-RSV) stand out; these antibodies were recently approved for treatment of rheumatoid arthritis and against respiratory syncytial virus (RSV), respectively [110, 111].

Streptokinase is a well-established therapeutic molecule with good cost benefit for the treatment of thromboembolic complications. Adivitiva and co-workers [112] evaluated a tag-free variant of streptokinase with a native N-terminus (N-rSK), developed using the *Pichia pastoris* expression system. A clone with three copies of the gene was able to secrete 1062 mg/L of N-rSK when using complex culture medium. The concomitant feeding of 1g/L/h of hydrolysed soy flour during the protein synthesis phase yielded 2.29 g/L of N-rSK, and when supplementing the medium with urea, both growth and induction phases shown significant improvements in protein stability and resulted in N-rSK concentration of 4.03 g/L. Under optimized conditions, volumetric productivity and specific yield were 52.33 mg/L/h and 33.24 mg/g, respectively.

Pichia's RCT expression platform allows the synthesis of a wide variety of products, both for therapeutic and industrial uses, some of which were approved for human use by the FDA. Table **2** lists some commercial products obtained from *Pichia*, by Industries registered at the Research Corporation Technologies, as well as their uses.

Table 2. Commercial products developed from *Pichia* by Industries registered at the Research Corporation Technologies and their uses [113].

Product	Company	Uses
Kalbitor® (DX-88 ecallantide, a kallikrein inhibiting recombinant protein)	Dyax (Cambridge, MA)	Treatment of hereditary angioedema
Insugen® (recombinant human insulin)	Biocon (India)	Diabetes therapy
Medway (recombinant human serum albumin)	Mitsubishi Tanabe Pharma (Japan)	Blood volume expander
Shanvac ™ (recombinant hepatitis B vaccine)	Shantha/Sanofi (India)	Hepatitis B prevention
Shanferon ™ (recombinant α-2b interferon)	Shantha/Sanofi (India)	Treatment of hepatitis C and cancer
Ocriplasmin (recombinant microplasmin)	ThromboGenics (Belgium)	Vitreomacular adhesion (VMA) treatment
Nanobody® ALX-0061 (recombinant fragment of single-domain antibody of anti-IL6 receptor)	Ablynx (Belgium)	Rheumatoid arthritis treatment
Nanobody® ALX00171 (recombinant fragment of single-domain antibody of anti-RSV)	Ablynx (Belgium)	Treatment of respiratory syncytial virus (RSV)
Heparin-binding EGF-like growth factor (HB-EGF)	Trillium (Canada)	Treatment of interstitial cystitis (IC), also known as bladder pain syndrome (BPS)
Purifine (recombinant phospholipase C)	Verenium/DSM (San Diego, CA/Netherlands)	Degumming of oils with high phosphorus content
Recombinant trypsin	Roche Applied Science (Germany)	Protein digestion
Recombinant collagen	FibroGen (San Francisco, CA)	Medical research reagents / skin filler
AQUAVAC IPN (recombinant capsid protein of necrotic pancreatic infection virus)	Merck/Schering Plough Animal Health (Summit, NJ)	Vaccine for necrotic pancreatic infection in salmon
Recombinant phytase	Phytex, LLC (Sheridan, IN)	Animal feed supplement
Superior Stock (recombinant nitrate reductase)	The Nitrate Elimination Co. (Lake Linden, MI)	Enzyme-based products for water testing and treatment
Recombinant human cystatin C	Scipac (United Kingdom)	Research reagent

Protein and Enzymes

Due to its status as a GRAS microorganism, *Pichia* has been used to produce more than 500 proteins of pharmaceutical interest and more than 1000 recombinant proteins since 2009 [114], including, α-amylase, D-alanine carboxypeptidase, glucoamylase, L-catalase, hexose oxidase, α-mannosidase acid, B-phytochrome, acid phosphatase, green fluorescent protein, bovine β-casein, aprotinin, B-gelatinase, serum albumin, and thrombomodulin [20, 21].

Some of *P. pastoris'* most commonly produced proteins include ocriplasmin (under the trade name Jetrea), kallikrein, and ecallantide (under the trade name Kalbitor) [115].

Zhu and co-workers [116] optimized culture medium for high yield production of human serum albumin (HSA) in *P. pastoris*, obtaining a production rate of 17.47 g/L. By using a three-step purification system, 67% of the proteins were recovered, with 99% purity. Other than HSA, other proteins can be obtained from *P. pastoris*, such as recombinant human interleukin-2-serum albumin (rhIL- 2-HSA) fusion protein [117 - 122].

Industrial enzymes have been widely used in many different sectors, such as production of chemical compounds, pharmaceuticals, fabrics, cosmetics, food and beverages, among others. These are considered more economically viable and environmentally safe when compared to whole cells. Enzyme production is quite important, as they act as high efficiency, high specificity catalysts which display great conversion potential and high activity levels under moderate conditions [123].

Zhang and co-workers [124] evaluated the behavior of a protease on the hydrolysis of zein (ZDP) from *Zea mays* heterologously expressed in *Pichia pastoris*, as well as its behavior and effects on corn starch hydrolysis. Ideal temperature and pH conditions for ZDP were 40 °C and pH of 5.0. Dithiothreitol and β-mercaptoethanol (5 mM) increased ZDP's activity by 102.1% and 60.7%, respectively. As such, recombinant ZDP may be useful in helping to better utilize available corn during different processes.

Other studies using *P. pastoris* were able to obtain an increase in α-antitrypsin [125], recombinant β-lipase [126 - 128], and lacase [129] production. Araújo, Oliveira and Cereda (2015) tested the biotechnological potential of yeasts isolated from ice cream-bean (*Inga edulis*), guavira (*Campomanesia adamantium*), and sugar-apple (*Annona squamosa*), where they were able to obtain 15.83 UI/mL of invertase produced by *Meyerozyma (Pichia) caribbica*. Enzymes pectinase and β-glucosidase were also produced by the same strain in concentrations of 2.8 and 1.84 UI/mL, respectively.

Information regarding the production of industrial enzymes *viaPichia pastoris* as a cell factory in bioreactor fermentations focusing on different enzyme types, area of application and production process details was compiled by Duman-Özdamar and co-workers [123].

Pigments

The number of biotechnological applications of carotenoids such as β-carotene, lycopene, canthaxanthin, lutein or astaxanthin, has been steadily increasing, especially in the fields of dietary supplements for humans and animals, as well as in biomedical areas [130].

Microorganisms are a strong alternative for carotenoid production [131 - 134]. The use of genetically modified yeasts as expression hosts shows many advantages over other microorganisms, especially due to their GRAS classification as well as for displaying efficient isoprenoid metabolism. Some promising examples include *Saccharomyces cerevisiae* [135], *Candida utilis* [136], *Rhodotorula toruloides* [137], *Mucor circinelloides* [138] and *Pichia pastoris* X-33 [139].

The bioengineering of *P. pastoris* resulted in an efficient method to obtain a new source of carotenoids, adding to actual sources. Bhataya *et al.* [25] developed a strain of *P. pastoris* able to produce lycopene by introducing a synthesis pathway consisting of three heterologous genes from lycopene production, with each gene being expressed either under GAP or a mutant promoter. The genes from the synthesis pathway were assembled in a plasmid and high lycopene concentrations were obtained (73.9 mg/L). A similar strategy was used by Araya-Garay *et al.* [140] to create *P. pastoris* strains able to produce both lycopene and β-carotene, while Araya-Garay *et al.* [24] were able to develop an additional extension of the β-carotene pathway, resulting in astaxanthin production.

Non-genetically modified strains of *Pichia fermentans* were studied as well regarding carotenoid production. Otero *et al.* [141] isolated three microorganisms (*Sporiodiobolus pararoseus*, *Rhodotorula mucilaginosa* and *Pichia fermentans*) to be used as potential carotenoid producers. *Pichia fermentans* was able to produce 500 μg/L of total carotenoids in a culture medium composed of agroindustry byproducts (sugarcane molasses). Cipolatti *et al.* [142] also used P. *fermentans,* grown in agroindustry byproducts as well, and were able to produce 201.6 μg/L of total carotenoids, with 30.7% identified as astaxanthin.

Xylitol and Oligosaccharide Production

The most important strains for xylitol synthesis are *Candida guilliermondii* [143] *Candida tropicalis* [144], *Pichia guilliermondii* [145]; *Pichia stipitis* and *Pachysolen tannophilus* [146]. Xylitol has a sweetening power similar to that of sucrose, low caloric value and has several applications in the pharmaceutical, chemical, food and odontological fields [146].

Many lignocellulosic materials such as agroindustry byproducts, sugarcane bagasse, rice straw, wheat straw and eucalyptus chips [147] might be used in biotechnological processes for xylitol production. The process has become possible with the discovery of pentose metabolizing yeasts [148], especially D-xylose, the main sugar found in hemicellulose hydrolysates. The first step of xylose metabolism is its transportation through the cell membrane. Once inside the cell, xylose is then reduced to xylitol *via* the action of the xylose reductase enzyme (EC1.1.1.21) alongside NADPH or NADH coenzymes [149]. Xylitol is then secreted to the outside of the cell or oxidized into xylulose by xylitol dehydrogenase (EC 1.1.1.9) alongside NAD+ [150]. Xylulose is phosphorylated into xylulose 5-phosphate, which can be converted into pyruvate through the transketolase-transaldolase enzymatic complex, which connects the phosphopentose pathway to EMP cycle [151]. Xylose reductase and xylitol dehydrogenase enzymes differ in their specificity from NADPH, NADH, NAD+ and NADP+ coenzymes. This fact was the precursor to the use of lignocellulosic materials to obtain different biotechnological products [152].

The *Pichia stipitis* FPL-YS30 strain is a D-xylulokinase deletion mutant with high xylitol productivity (0.0451 g xylitol/g cell/h). The YS30 mutant uses a kinase bypass to metabolize xylose while producing negligible amounts of ethanol. Xylitol formation is influenced by initial cell and xylose concentrations, temperature, pH and oxygen transfer rate [153].

Zou and co-workers [145] evaluated the xylitol production from xylose by a self-isolated strain of *Pichia guilliermondii* able to assimilate 5-hydroxymethylfurfural, an inhibiting compound present in lignocellulosic material. When using a low volumetric oxygen transfer coefficient (0.075 h^{-1}) they were able to obtain a maximum xylitol yield of 0.61 g/g.

Other products synthesized by *Pichia* include oligosaccharides such as xylo-oligosaccharides (XOS), cello-oligosaccharides (COS), fructo-oligosaccharides (FOS) and manno-oligosaccharides (MOS). These compounds display high prebiotic activity and may be added to many foods such as sweets and baked goods, or as dietary supplements. Xylo-oligosaccharides (XOS) selectively stimulate the growth of probiotic bacteria from the *Lactobacillus* and

Bifidobacterium genera in the intestinal flora [154]. Xylan, the precursor of XOS is the second most abundant polysaccharide in nature, present in the cell wall of plant cells as well as in lignocellulosic residue (LCR) such as sugarcane bagasse [155].

Nascimento [155] has used recombinant xylanase GH10 from *Thermoascus aurantiacus* expressed by *Pichia pastoris* to hydrolyze xylan extracted from sugarcane bagasse. The process' optimization was performed by using 25% xylan and 220 U of xylanase GH10, resulting in a XOS concentration of 32.33 g/L, with most of them being identified as xylobiose, xylotriose and xylotetraose. In another study, Martins and co-workers [156] produced XOS from sugarcane straw employing recombinant thermostable endoxylanase from *Cryptococcus flavescens,* expressed in *Pichia pastoris* GS115, using the pGAPZαA vector synergically associated with α-l-arabinofuranosidase GH 51 (Megazyme®). They were able to find a xylan conversion rate of 72.56%; the main identified compounds were xylobiose and xylotriose, short chain-length polysaccharides favored by probiotic cells.

Tao and co-workers [157] evaluated the production of recombinant endoglucanase by the EG I gene of *Trichoderma reesei* expressed in *Pichia pastoris* in order to produce oligosaccharides from different biomass-related substrates (paper pulp, carboxymethyl cellulose, oat spelt xylan, birchwood xylan, and alkali-extracted corn cob). Birchwood xylan displayed the highest yield of XOS (69.5%) and ReEG I was able to simultaneously produce XOS alongside cello-oligosaccharides (COS). However, when alkali-extracted corn cob residue was used as substrate, the COS yield was found to be much lower than that of XOS, with 2.7% after 72 h.

Trujillo and co-workers [158] evaluated fructo-oligosaccharide production of levansucrase from *Gluconacetobacter diazotrophicus* expressed in *P. pastoris* using sucrose as carbon source. The recombinant LsdA was glycosylated and presented optimal pH and temperature for enzyme activity, similar to those of the native enzyme, however, the thermal stability of the recombinant enzyme was increased. Incubation of recombinant LsdA in sucrose (500 g/L) yielded 43% (w/w) of total sugar as 1-kestose, with a conversion efficiency of about 70%. The recombinant *P. pastoris* cells also converted sucrose into FOS, although only for a 30% efficiency.

Manno-oligosaccharides are functional additives that have gained significant interest, mostly due to their beneficial effects, which include the promotion of beneficial microflora as well as fat absorption reduction [159, 160]. They are produced when β-mannanases (EC 3.2.1.78) cleave the β-1,4-mannosidic bond in mannan [161]. Li and co-workers [162] engineered β-mannanase (mRmMan5A) from *Rhizomucor miehei* expressed in *Pichia pastoris* for manno-oligosaccharide production using palm kernel cake (PKC) pretreated *via* steam explosion. To this end, PCK was pretreated *via* steam explosion at 200 °C for 7.5 min, and then hydrolyzed by mRmMan5A. The mRmMan5A displayed a maximum activity of 79.68 U/mL at pH 4.5 and 65 °C, and exhibited many specific activities towards mannans. The total manno-oligosaccharide yield reached 34.8 g/100 g dry PKC, indicating that 80.6% of total mannan in PKC was hydrolyzed and a total of 261.3 g of manno-oligosaccharides were produced from 1.0 kg of dry PKC.

In a similar study, Li and co-workers [163] produced manno-oligosaccharides from cassia gum using a mutated glycoside hydrolase family 134 β-mannanase gene (mRmMan134A) from *Rhizopus microsporus* var. *rhizopodiformis* F518 expressed *Pichia pastoris*. They were able to obtain high protein expression levels (3680 U/mL), with mRmMan134A showing maximum enzymatic activity at pH 5.5 and 50 °C. The hydrolysis of cassia gum produced 70.6% (w/w) of manno-oligosaccharides, with some identified as mannose, mannobiose, galactose, mannotriose, mannotetraose, 61-α-d-galactosyl-β-d-mannobiose, and mannopentaose. Furthermore, the manno-oligosaccharides were able to promote the growth of three *Bifidobacterium* and six *Lactobacillus* strains.

Michalak and co-workers [164] used potato pulp polysaccharides, rich in β-1,--galactan (> 100 kDa), as substrate for production of potentially probiotic bacteria (*Bifidobacterium longum* and *Lactobacillus acidophilus*) as well as antibacterial agents (*Clostridium perfringens*). The authors sought to produce endo-1,4-β-galactanase from *Emericella nidulans* (also known as *Aspergillus nidulans*), GH family 53, expressed in *P. pastoris*. Enzyme activity was similar to natural 1,4-β-galactanase from *Aspergillus niger*, with a yield of 82% and total production of 1.6 g/L.

Other β-mannanase genes have been successfully expressed in *Pichia pastoris* (Table **3**) to meet its growing demand in many industries such as food, paper, animal feed, cleaning products, biofuels and oil drilling [165].

Table 3. Heterologous cloning and gene expression of endo-mannanase in *Pichia*.

Source Organism	Heterologous Host	Vector	Gene Size (bp)	Amino Acids	Molecular Weight (kDa)	pI	Endo-Mannanase Production
Aspergillus nidulans FGSC A4	*P.pastoris X*33	pPICZαA	-	-	56	-	NR

(Table 3) cont.....

Source Organism	Heterologous Host	Vector	Gene Size (bp)	Amino Acids	Molecular Weight (kDa)	pI	Endo-Mannanase Production
Aspergillus niger LW1	*P. pastoris* GS115	pPICZαA	1152	383 (21) [17]	37.5	4.15	29 U/mL
Chaetomium sp. CQ31	*P. pastoris* GS115	pPIC9K	1251	451	50	4.5	50.030 U/mL
Neosartorya fischeri P1	*P. pastoris* GS115	pPIC9	1187	373	39.5	5.93	101.5 U/mL
Penicillium sp. C6	*P. pastoris* GS115	pPIC9	1155	384 (26)	39	NR	2.5 g/L
Penicillium oxalicum GZ-2	*P. pastoris* GS115	pPICZαA	1380	426	45	5.09	NR
Mytilus edulis	*P. pastoris* with *Saccharomyces cerevisiae* α-factor	pPICZa B	1104	368	40	6.85-8.15	900 mg/mL
Bacillus subtilis WD23	*Pichia pastoris* GS115	pPIC9K	1089	362	41	5.81	5156.742 U/mL
Reticulitermes speratus	*P. pastoris*	pPICZa-A	390	NR	40	NR	0.67 g/L

Source: https://doi.org/10.1016/j.biotechadv.2016.11.001.

Wang and co-workers [166] evaluated the β-mannanase gene of *Aspergillus niger* (AnMan26), heterologously expressed in *Pichia pastoris*, to degrade mannans. Purified AnMan26 displayed a maximum enzymatic activity of 22.100 U/mL, remaining stable between pH levels of 2.5-7.0 and up to 40 °C. Moreover, when applying the enzyme to locust bean gum, a maximum specific activity of 2869.0 U/mg was described, with the enzyme being able to efficiently degrade various mannans to manno-oligosaccharides, hydrolysed mannotetraose and mannopentaose. Furthermore, AnMan26 hydrolysed fenugreek gum to produce partially hydrolysed fenugreek gum (PHFG) with an average molecular weight of 1.8 KDa, with the hydrolysis ratio of galactomannan in fenugreek gum being of 82.2% was described.

Ethanol

Researchers have explored and characterized thermotolerant yeasts for ethanol production such as *Kluyveromyces marxianus*, *Pichia* sp., *Candida* sp. as well as some *Saccharomyces cerevisiae* strains [29, 167 - 169].

There have been reports of ethanol production under high temperatures using *Pichia kudriavzevii* (formerly known as *Issatchenkia orientalis*) [80, 170 - 173], *P. kudriavzevii* MF-121 [174] and *P. kudriavzevii* DMKU 3-ET15 [175]. In addition, some *P. kudriavzevii* (MF-121 strains also displayed significant resistance to stress conditions such as acidity, and ethanol and salt content. *S. cerevisiae*, the yeast most commonly used in ethanol production displays both high fermentation efficiency and high tolerance to ethanol [176, 177].

In comparison, *P. kudriavzevii* exhibits greater thermotolerance than that of *S. cerevisiae* [178, 179]. Chamnipa and co-workers [28] isolated *P. kudriavzevii* RZ8-1 and observed that it was the highest ethanol producing strain (35.09 a 69.85 g/L) when grown at high temperatures (37 to 45 °C), using glucose as carbon source (160 g/L) and even displaying tolerance to ethanol and acetic acid.

According to existing data, there are a large number of xylose metabolizing yeasts, however, only 1% of them are able to convert xylose into ethanol [180], among these, *Pichia membranifaciens* [181] and *Pichia stipitis* [182, 183] stand out.

Ribeiro and co-workers [181], evaluated ideal growth conditions for *Pichia membranifaciens* LJ04 in synthetic medium added with pentose from hemicellulosic hydrolysate. They have found that ideal pH ranges from 4.0-4.5 while the optimal temperature found was of 32 °C, moreover, an increased consumption of xylose was observed, alongside high ethanol production, reduced number of secondary metabolites and great cell viability during the fermentation process.

Isobutanol production by *Pichia pastoris* was also evaluated. By overexpressing three native genes (ILV2, ILV5 and ILV3) together with heterologous expression of *S. cerevisiae*'s alcohol dehydrogenase gene (ADH7) and 2-keto-acid decarboxylase (KIVD) from *Lactobacillus lactis*, the modified strain was able to produce 2.2 g/L of isobutanol in shake-flask cultures [30].

Patra and co-workers [95], overexpressed valine and leucine production pathways in *Pichia pastoris* to increase 2-ketoisocaproate production, obtaining 191.0 mg/L of 3-methyl-1-butanol.

Nosrati-Ghods and co-workers [183], observed production kinetics of ethanol from xylose using *Pichia stipitis*, and the highest ethanol yield and productivity was 0.45 g/g and 0.75 g/L/h, respectively. Similarly, Raina and co-workers [184] used lignocellulosic biomass from *Shorea robusta* to produce bioethanol employing *Pichia stipitis* (NCIM-3498) and were able to obtain 43 g/L and 97% conversion rate efficiency.

Alcoholic Beverages

The continuous growth of the alcoholic beverages market and the consumer's interest has directed efforts into the development of new products. In this regard, non-*Saccharomyces* yeasts have been gaining traction as tools to improve the quality of beers, wines and other beverages.

Burini and co-workers [185] described the application of different non-conventional yeast species for beer production, including *Brettanomyces*, *Torulaspora*, *Lachancea*, *Wickerhamomyces*, *Mrakia* and *Pichia*, as well as some non-*cerevisiae* species of *Saccharomyces*. *Pichia kluyveri* in particular was only able to consume the glucose present in the wort, producing very low amounts of ethanol. This strain, however, was able to produce large amounts of isoamyl acetate as well as other compounds related to different fruit aromas, such as ethyl propionate and butyrate (pineapple), ethyl valerate (apple), ethyl decanoate (apple), and ethyl octanoate (pineapple, apple, tropical, and brandy); this behavior could make of *P. kluyveri* a promising agent to produce fruity aromas in beers. Moreover, it has been reported that *P. kluyveri* is able to convert precursor compounds found in hops into polyfunctional thiols that confer the aroma of tropical fruit, thus improving the beer's flavor [186].

Some non-*Saccharomyces* yeasts have been reported to display improved tolerance to brewing conditions [187, 188], with many genera producing a variety of enzymes (β-glucosidase, β-xylosidase, protease and pectinase) that assist in the formation of aromatic compounds, with *Pichia* yeasts producing many of them [189] and can also be used in low alcohol content wines [190]. Yeasts belonging to the *Pichia* genus can use many substrates, including some organic acids e most sugars, even xylose [191, 192]. It has been confirmed, as reported by Caputo and co-workers [192], that *P. fermentans* plays a positive effect on the flavor of wines, especially by releasing 2-phenylethanol *via* catabolism of L-phenylalanine. Božič and co-workers [193] investigated 95 strains of *Saccharomyces* and non-*Saccharomyces* yeasts for the production of pyranoanthocyanins. All strains were tested regarding hydroxycinnamate decarboxylase (HCDC) activity. *Pichia guilliermondii* ZIM624 and *Wickerhamomyces anomalus* S138 synthesized vinylphenolic pyranoanthocyanins in concentrations of 40.2 and 38.5 mg/L, respectively.

Zhong and co-workers [194] used *Pichia fermentans* JT-1-3 to reduce the acidity of blueberry wine; they reported a reduction in citric, malic, and tartaric acid levels by 43.35, 35.29 and 44.68%, respectively. Sensorial evaluation has shown that the wine produced by *P. fermentans* had improved flavor, lower acidity and bitterness. The main volatile compounds in blueberry wine that confer its aroma were 2-phenylethanol and 9-ethyl hexadecanoate. Zhong and co-workers [188], have observed *P. fermentans'* potential to degrade citric acid, improving the quality of kiwi wine. *Pichia fermentans* also displayed good performance when used to produce a low-ethanol content drink made from oranges and grapes, improving the formation of ethyl acetate and higher alcohols [195]. Méndez-Zamora and co-workers [196], evaluated 21 strains of *Pichia kluyveri* isolated from different fermentative processes in the production of aromatic volatile compounds during alcoholic fermentation; the authors were able to find significant differences between the strains, especially with regard to the variety of produced compounds as well as their concentration.

Frootzen® [197] was the first commercial product obtained from a strain of *Pichia kluyveri*, a yeast present in naturally fermented wine. *P. kluyveri* was chosen due to its ability to synthesize fruity compounds, from flavor precursors naturally found in grape peels, into volatile thiols that create the aroma of tropical fruits such as passion fruit, in white wines, and gooseberry in red wines.

Another alcoholic beverage that uses *Pichia* in its fermentation is sugarcane firewater, also known as "cachaça". Although *Saccharomyces cerevisiae* is the main responsible for causing the fermentation process, many other genera of yeasts may be present, including *Kluyveromyces*, *Pichia*, *Hanseniaspora*, *Debaryomyces* and some *Candida* species [198, 199]. Portugal and co-workers [200] observed four native yeast species (*Meyerozyma guilliermondii*, *Hanseniaspora guilliermondii*, *Pichia fermentans* and *Saccharomyces cerevisiae*) and found that *P. fermentans* in particular, when associated with *S. cerevisiae,* displays limited persistence [201], positively contributing to a floral bouquet [202] and improving the production of volatile aromatic compounds [203]. However, the unbalanced growth of the yeasts may produce high concentrations of acetic acid, acetaldehyde and glycerol, which can negatively impact the product's quality [68, 198].

Coffee Fermentation

The removal of mucilage from coffee beans *via* fermentative process is an interesting strategy to improve the coffee's quality, as it aids in the formation of aromatic compounds. The use of a combination of yeast and lactic bacteria can improve the fermentative process' efficiency and positively influence the chemical makeup of coffee beans [204 - 208]. Da Silva Vale and co-workers [205] evaluated the effect of co-inoculation with *Pichiafermentans* and *Pediococcus acidilactici* in the fermentation of coffee beans; they found increase in sugar consumption as well as increase in lactic acid, ethanol, ethyl acetate, benzeneacetaldehyde, 2-heptanol, and benzyl alcohol production. Furthermore, a positive synergistic interaction between both microbial groups was observed. Green coffee bean fermentation by *P. fermentans*, however, impacted 2-hexanol, nonanal and D-limonene production.

Yeasts are a very important group of native microorganisms found in coffee beans, especially due to the roles they play during the processing of said beans [209]. Many yeast species from various genera, such as *Pichia, Candida, Saccharomyces* and *Torulaspora*, were detected during the entire process [206, 210 - 212] and the addition of these yeasts during coffee fermentation improved the product's sensorial characteristics [213 - 218]. *Pichia fermentans* YC5.2 [215] and *P. kluyveri* [217], in particular, were able to improve the fruity aroma present in roasted coffee beans

Cocoa Fermentation

The fermentation of cocoa seeds is a natural process caused by many microorganisms, including yeasts, as well as acetic and lactic bacteria. Yeasts are responsible for ethanol production from sugars present in cocoa pulp, facilitating oxygen intake by degrading the pulp and releasing enzymes that produce higher esters and alcohols, reported as essential volatile compounds for high quality chocolate [219, 220].

Research with *S. cerevisiae* and *P. kudriavzevii* inoculants has yielded positive results regarding the quality of fermented cocoa beans, such as being able to produce beans with fruity, sweet, and floral aromas, while also reducing the amount of undesired compounds, such as alkanes and acids, lowering the beans' acidity, increasing the amount of phenolic compounds and methylxanthine levels, and reducing the overall duration of the fermentation process [221 - 224].

Santos and co-workers [225], evaluated the influence of starter cultures of *Candida parapsilosis, Torulaspora delbrueckii* and *Pichia kluyveri* in the fermentation of cocoa beans. The strains that displayed more changes in the amount of free amino acids, acidity and reducing sugar content were *Candida parapsilosis*, followed by *Torulaspora delbrueckii* and *Pichia kluyveri*. Fernandez-Maura and co-workers [226], checked the diversity of naturally present yeasts in the spontaneous fermentation of cocoa beans in eastern Cuba; a total of 435 isolates were identified, and *Pichia manshurica, Hanseniaspora opuntiae*, and *Pichia kudriavzevii* were the most common strains. The authors noted that tolerance to higher ethanol levels was more frequently observed in *Pichia* spp. and *S. cerevisiae* strains when compared to *Hanseniaspora* spp.; they also point out that the ability to consume ethanol may confer a selective advantage to *Pichia* spp.

Olive Fermentation

The quality of olives relies heavily on how the fermentation process was conducted, in which many microorganisms play a major role, including *Pichia* sp. Romo-Sánchez and co-workers [227], identified the presence of *Pichia holstii* and *Pichia caribbica* in two varieties of olives. These yeasts produce cellulases and polygalacturonases, enzymes that can increase the amount of antioxidant compounds in olive oil, improving its shelf life and overall productivity, as they aid in breaking down polysaccharides from the olive's cell wall, facilitating the extraction process.

Montaño and co-workers [228] evaluated the behavior of 5 yeast strains (*Nakazawaea molendinolei* NC168.1, *Zygotorulaspora mrakii* NC168.2, *Pichia manshurica* NC168.3, *Candida adriatica* NC168.4 and *Candida boidinii* NC168.5) isolated from olive brine. When these yeasts were grown individually in an olive-based culture medium for 7 days at 25 °C, the number of volatile compounds produced ranged from 22, for *P. manshurica* NC168.3, to 60, for *C. adriatica* NC168.4. Similarly, Lucena-Padrós and co-workers [229] studied the microbial biogeography of the fermentation of green olives in Seville (Spain). A grand total of 951 isolates were obtained from 30 fermenters. Distinct genotypes were also identified, with 41 species belonging to bacteria and 16 to yeasts, out of which one was a *Pichia manshurica* strain.

Biodegradation of Dyes

Synthetic dyes used as ingredients in the production of foods, cosmetics, paper, paints and fabrics represent a major environmental problem due to their high potential to contaminate bodies of water such as lakes and rivers. These compounds contain large quantities of aromatic rings, azo couplings, amines and sulfonic groups, which increases their toxicological potential and reduces their biodegradability [34].

Yeasts have shown good results when used to degrade these dyes, becoming a low-cost, environmentally safe alternative to treat effluents. Studies focusing on the degradation of dyes by *Pichia* sp., *Candida* sp. and *Magnusiomyces* sp. have shown that these microorganisms display potential to not only degrade dyes, but to mineralize them as well [230 - 232].

Delane and co-workers [233], evaluated the discoloration rate of Reactive Black 5 (RB5) by *Pichia kudriavzevii* SD5, with assays demonstrating a reduction in toxicity after 24 h alongside considerable degradative discoloration. Saravanan and co-workers [234] evaluated the bioremediation of two synthetic fabric dyes, Reactive Red 11 (RR11) and Acid Green 1 (AG1), using *P. pastoris*; they observed that *P. pastoris* was able to successfully degrade both dyes, especially AG1.

Nanotechnology is also being explored to reduce the environmental impact of dyes, with noble metals displaying notable catalytic response, especially in wastewater treatment. As such, Eze and Nwabor [235] evaluated the use of residual culture medium for heterologous protein expression by *Pichia pastoris* to obtain silver nanoparticles (PSM-AgNPs); said particles were efficient in eliminating radicals, inhibiting tyrosinase and pathogenic bacteria (*B. cereus* and *E. coli*), as well as catalyzing the degradation of two highly toxic dyes, methyl orange and Congo red, in wastewaters.

Biocontrol

The antagonistic action that some yeast species display over fungi has been studied by some authors, especially to improve the storage of different grains [43 - 45]. Petersson and Schnürer [44] evaluated *Pichia anomala*'s impact on the growth of *Penicillium roqueforti* in rye, wheat, and barley stored at high humidity. They report that a cell concentration of 5 x 10^4 UFC/g of *P. anomala* was able to fully inhibit the growth of *P. roqueforti*. Ramos and co-workers [236] studied the antagonistic potential of *Debaryomyces hansenii* (UFLACF 889 and UFLACF 847) and *Pichia anomala* (UFLACF 710 and UFLACF 951), isolated from coffee beans, in a mixed culture with *Aspergillus ochraceus*, *Aspergillus. parasiticus* and *P. roqueforti*. The authors observed that UFLACF 71 inhibited the spore production of *A. ochraceus* by 60% and that of *P. roqueforti* by 75.6%.

Pichia, in particular *P. guilliermondii*, *P. membranifaciens* and *P. anomala*, have been used as bioagents in the control of postharvest diseases in fruits and vegetables such as citrus [237 - 239], apples and pears [240 - 242], bananas [243], grape tomatoes [241], among others [244, 245]. The biological control mechanism employed is based on ecological interactions, such as competition for nutrients and space, macroparasitism, antibiosis and defense response induction to produce lysins [246 - 249].

Codified enzymes, particularly endolysin, have been used as a novel strategy for efficient biocontrol and treatment of bacterial diseases caused by *Vibrio parahaemolyticus* in seafood [250]. Srinivasan and co-workers [250] pointed out that the cell biocatalyst of *Pichia pastoris* X-33 for endolysin expression displayed strong antivibriose potential, anti-biofilm activity as well as protective activity against *V. parahaemolyticus*, helping to reduce bacterial infection outbreaks in aquaculture systems.

Fungi present in the oral cavity of humans, even in low amounts, play a significant role in the maintenance and balance of the mouth's microbiota through their interaction with other microorganisms. *Pichia* yeasts have been reported to act as antagonists against *Candida*, *Aspergillus* and *Fusarium*, which are known to cause oral infections [251].

REASSIGNMENTS FROM AND TO *PICHIA*

The genus *Pichia* was described by Hansen in 1904 [252, 253]. Since then, many species have been assigned to it, making this genus one of the most biodiverse in the family Pichiaceae, order Saccharomycetales. In contrast, following multigene phylogeny development [254, 255], many species were reassigned, either from *Pichia* to other genera or from those to *Pichia*. As in the examples systematized in Table **4**, most reassignments occurred from the genus *Hansenula* to *Pichia* and from *Pichia* to *Ogataea* and *Wickerhamomyces*. It is also worth noting that, while some yeasts were just reallocated to another genus, others also had their specific epithet changed [256]. Even though most of these scientific names were changed over a decade ago, their basionym *Pichia* is still the most referred to in the literature.

Table 4. Examples of species reassignments from and to the genus Pichia.

First Assignment	Reassignment	Second/Third Reassignment	References
Debaryomyces toletanus	*Pichia toletanus*	*Peterozyma toletanus*	[257]
Hansenula alni	*Pichia alni*	-	[258]
Hansenula canadensis	*Pichia canadensis*	*Wickerhamomyces canadensis*	[258, 259]
Hansenula capsulata	*Pichia capsulata*	-	[258]
Hansenula ciferrii	*Pichia ciferrii*	*Wickerhamomyces ciferrii*	[255, 258–260]
Hansenula euphorbiaphila	*Pichia euphorbiaphila*	-	[258]
Hansenula lynferdii	*Pichia lynferdii*	*Wickerhamomyces lynferdii*	[259, 260]
Hansenula nonfermentans	*Pichia minuta var. nonfermentans*	*Ogataea nonfermentans*	[257, 258]
Hansenula ofunaensis	*Pichia ofunaensis*	*Zygoascus ofunaensis*	[261]
Hansenula polymorpha	*Pichia angusta*	*Ogataea polymorpha/ Ogataea angusta*	[261, 262, 263]
Hansenula silvicola	*Pichia silvicola*	*Wickerhamomyces silvicola*	[258, 259]
Hansenula subpelliculosa	*Pichia subpelliculosa*	*Wickerhamomyces subpelliculosa*	[258, 259]

(Table 4) cont.....

Hansenula sydowiorum	Pichia sydowiorum	Wickerhamomyces sydowiorum	[255]
Hansenula wickerhamii	Pichia finlandica	Ogataea wickerhamii	[256, 258]
Issatchenka orientalis	Pichia kudriavzevii	-	[256]
Pichia acaciae	Millerozyma acaiae	-	[256]
Pichia angophorae	Ambrosiozyma angophorae	-	[257]
Pichia antillensis	Phaffomyces antillensis	-	[264]
Pichia bispora	Wickerhamomyces bisporus	-	[256]
Pichia caribaea	Starmera caribaea	-	[264]
Pichia caribbica	Meyerozyma caribbica	-	[256]
Pichia chambardii	Wickerhamomyces chambardii	-	[255]
Pichia dorogensis	Ogataea dorogensis	-	[265]
Pichia farinosa	Millerozyma farinosa	-	[266, 267]
Pichia guilliermondii	Meyerozyma guilliermondii	-	[256]
Pichia haplophila	Priceomyces haplophilus	-	[256]
Pichia koratensis	Millerozyma koratensis	-	[268]
Pichia media	Priceomyces medius	-	[256]
Pichia methanolica	Ogataea methanolica	-	[257]
Pichia methylivora	Ogataea methylivora	-	[257]
Pichia mexicana	Yamadozyma mexicana	-	[256]
Pichia mucosa	Waltiozyma mucosa	Williopsis mucosa/ Wickerhamomyces mucosus	[259]
Pichia naganishii	Ogataea naganishii	-	[257]
Pichia pastoris	Komagataella pastoris	-	[269]
Pichia philodendra	Ogataea philodendri	-	[256]
Pichia pilisensis	Ogataea pilisensis	-	[257]
Pichia pinus	Ogataea pini	-	[256]
Pichia pseudopastoris	Komagataella pseudopastoris	-	[270]
Pichia ramenticola	Ogataea ramenticola	-	[257]
Pichia siamensis	Ogataea siamensis	-	[271]
Pichia tannicola	Zygoascus tannicolus	-	[261]
Pichia thermomethanolica	Ogataea thermomethanolica	-	[271]
Pichia thermotolerans	Phaffomyces thermotolerans	-	[256]
Pichia trehaloabstinens	Ogataea trehaloabstinens	-	[265]
Pichia trehalophila	Ogataea trehalophila	-	[257]

(Table 4) cont.....

Pichia triangularis	Yamadozyma triangularis	-	[256]
Pichia xylosa	Pterozyma xylosa	-	[257]
Pichia zsoltii	Ogataea zsoltii	-	[265]
Saccharomyces anomalus	Pichia anomala	Wickerhamomyces anomalus	[260]
Yamadazyma segobiensis	Pichia segobiensis	Scheffersomyces segobiensis	[256, 272]
Yamadazyma spartinae	Pichia spartinae	Scheffersomyces spatinae	[256, 272]
Yamadazyma stipitis	Pichia stipitis	Scheffersomyces stipitis	[272]

CONCLUDING REMARKS

P. pastoris is considered an excellent alternative for industrial production of metabolites and recombinant proteins, where the efficient production of specific compounds relies on multilevel optimization strategies using promoters, codons, signal peptides, gene does and culture strategies.

Pichia's versatility was outstanding for its potential to be applied in the production of other high value materials, such as vaccines, pigments, enzymes, sugars such as xylitol and other oligosaccharides, and ethanol. Furthermore, it may also be applied in fermentation processes of alcoholic beverages, cocoa, coffee and olives, as well as employed as a natural biodegrading or biocontrol agent.

The growing interest in *P. pastoris* as a promising host organism for metabolic engineering has been fueled by advances in synthetic biotechnology, with *P. pastoris* remaining to this day a strong tool for research and industrial applications.

CONSENT FOR PUBLICATION

Not applicable.

CONFLICT OF INTEREST

The authors declare no conflict of interest, financial or otherwise.

ACKNOWLEDGEMENT

Declare none.

REFERENCES

[1]   Couderc R, Baratti J. Oxidation of methanol by the yeast *Pichia pastoris*: purification and properties

of alcohol oxidase. Agric Biol Chem 1980; 44: 2279-89.

[2]   Cos O, Ramón R, Montesinos JL, Valero F. Operational strategies, monitoring and control of heterologous protein production in the methylotrophic yeast *Pichia pastoris* under different promoters: a review. Microb Cell Fact 2006; 5: 17.
[http://dx.doi.org/10.1186/1475-2859-5-17] [PMID: 16600031]

[3]   Wegner EH. Biochemical conversions by yeast fermentation at high cell densities. US Patent 4617274. 1986.

[4]   Cereghino JL, Cregg JM. Heterologous protein expression in the methylotrophic yeast *Pichia pastoris*. FEMS Microbiol Rev 2000; 24(1): 45-66.
[http://dx.doi.org/10.1111/j.1574-6976.2000.tb00532.x] [PMID: 10640598]

[5]   Demain AL, Vaishnav P. Production of recombinant proteins by microbes and higher organisms. Biotechnol Adv 2009; 27(3): 297-306.
[http://dx.doi.org/10.1016/j.biotechadv.2009.01.008] [PMID: 19500547]

[6]   Nieto-Taype MA, Garcia-Ortega X, Albiol J, Montesinos-Seguí JL, Valero F. Continuous cultivation as a tool toward the rational bioprocess development with *Pichia pastoris* cell factory. Front Bioeng Biotechnol 2020; 8: 632. [Review].
[http://dx.doi.org/10.3389/fbioe.2020.00632] [PMID: 32671036]

[7]   Chen R. Bacterial expression systems for recombinant protein production: *E. coli* and beyond. Biotechnol Adv 2012; 30(5): 1102-7.
[http://dx.doi.org/10.1016/j.biotechadv.2011.09.013] [PMID: 21968145]

[8]   Monsalve RI, Lu G, King TP. Expressions of recombinant venom allergen, antigen 5 of yellowjacket *(Vespula vulgaris)* and paper wasp *(Polistes annularis)*, in bacteria or yeast. Protein Expr Purif 1999; 16(3): 410-6.
[http://dx.doi.org/10.1006/prep.1999.1082] [PMID: 10425162]

[9]   Çelik E, Çalik P. Production of recombinant proteins by yeast cells. Biotechnol Adv 2012; 30(5): 1108-18.
[http://dx.doi.org/10.1016/j.biotechadv.2011.09.011] [PMID: 21964262]

[10]   Baghban R, Farajnia S, Rajabibazl M, *et al.* Yeast expression systems: overview and recent advances. Mol Biotechnol 2019; 61(5): 365-84.
[http://dx.doi.org/10.1007/s12033-019-00164-8] [PMID: 30805909]

[11]   Damasceno LM, Huang CJ, Batt CA. Protein secretion in *Pichia pastoris* and advances in protein production. Appl Microbiol Biotechnol 2012; 93(1): 31-9.
[http://dx.doi.org/10.1007/s00253-011-3654-z] [PMID: 22057543]

[12]   Jung E, Williams KL. The production of recombinant glycoproteins with special reference to simple eukaryotes including Dictyostelium discoideum. Biotechnol Appl Biochem 1997; 25(1): 3-8.
[http://dx.doi.org/10.1111/j.1470-8744.1997.tb00407.x] [PMID: 9032931]

[13]   Hinnen A, Buxton F, Chaudhuri B, *et al.* Gene expression in recombinant yeast. A Smith (Ed), Gene Expression in Recombinant Micro-organisms, Dekker M, New York. 1994; pp. 121-93.

[14]   Wagner JM, Alper HS. Biologia sintética e genética molecular em leveduras não convencionais: ferramentas atuais e avanços futuros. Fungal Genet Biol 2012; 89: 126-36.
[http://dx.doi.org/10.1016/j.fgb.2015.12.001] [PMID: 26701310]

[15]   Wegner GH. Emerging applications of the methylotrophic yeasts. FEMS Microbiol Rev 1990; 7(3-4): 279-83.
[http://dx.doi.org/10.1111/j.1574-6968.1990.tb04925.x] [PMID: 2094288]

[16]   Gasser B, Mattanovich D. A yeast for all seasons - Is *Pichia pastoris* a suitable chassis organism for future bioproduction? FEMS Microbiol Lett 2018; 365(17).
[http://dx.doi.org/10.1093/femsle/fny181] [PMID: 30052876]

[17]   Cregg JM, Vedvick TS, Raschke WC. Recent advances in the expression of foreign genes in *Pichia pastoris*. Biotechnology (N Y) 1993; 11(8): 905-10.
[PMID: 7763913]

[18]   Weninger A, Fischer JE, Raschmanová H, Kniely C, Vogl T, Glieder A. Expandindo o kit de ferramentas CRISPR / Cas9 para *Pichia pastoris* com integração eficiente de doadores e marcadores de resistência alternativos. J Cell Biochem 2018; 119: 3183-98.
[http://dx.doi.org/10.1002/jcb.26474] [PMID: 29091307]

[19]   Yuan SF, Brooks SM, Nguyen AW, *et al.* Bioproduced proteins on demand (Bio-POD) in hydrogels using *Pichia pastoris*. Bioact Mater 2021; 6(8): 2390-9.
[http://dx.doi.org/10.1016/j.bioactmat.2021.01.019] [PMID: 33553823]

[20]   Macauley-Patrick S, Fazenda ML, McNeil B, Harvey LM. Heterologous protein production using the *Pichia pastoris* expression system. Yeast 2005; 22(4): 249-70.
[http://dx.doi.org/10.1002/yea.1208] [PMID: 15704221]

[21]   Cregg JM. DNA-mediated transformation.Methods in molecular biology: Pichia protocols. 2nd ed. Totowa, NJ: Human Press 2008; pp. 27-42.

[22]   Chen F, Lin D, Wang J, *et al.* Heterologous expression of the *Monilinia fructicola* CYP51 (MfCYP51) gene in *Pichia pastoris* confirms the mode of action of the novel fungicide, SYP-Z048. Front Microbiol 2015; 6: 457.
[http://dx.doi.org/10.3389/fmicb.2015.00457] [PMID: 26042103]

[23]   Liu T, Zou W, Liu L, Chen J. A constraint-based model of *Scheffersomyces stipitis* for improved ethanol production. Biotechnol Biofuels 2012; 5(1): 72.
[http://dx.doi.org/10.1186/1754-6834-5-72] [PMID: 22998943]

[24]   Araya-Garay JM, Ageitos JM, Vallejo JA, Veiga-Crespo P, Sánchez-Pérez A, Villa TG. Construction of a novel *Pichia pastoris* strain for production of xanthophylls. AMB Express 2012; 2(1): 24.
[http://dx.doi.org/10.1186/2191-0855-2-24] [PMID: 22534340]

[25]   Bhataya A, Schmidt-Dannert C, Lee PC. Metabolic engineering of *Pichia pastoris* X-33 for lycopene production. Process Biochem 2009; 44(10): 1095-02.
[http://dx.doi.org/10.1016/j.procbio.2009.05.012]

[26]   Jeong E, Shim WY, Kim JH. Metabolic engineering of *Pichia pastoris* for production of hyaluronic acid with high molecular weight. J Biotechnol 2014; 185: 28-36.
[http://dx.doi.org/10.1016/j.jbiotec.2014.05.018] [PMID: 24892811]

[27]   Meesapyodsuk D, Chen Y, Ng SH, Chen J, Qiu X. Metabolic engineering of *Pichia pastoris* to produce ricinoleic acid, a hydroxy fatty acid of industrial importance. J Lipid Res 2015; 56(11): 2102-9. [S].
[http://dx.doi.org/10.1194/jlr.M060954] [PMID: 26323290]

[28]   Chamnipa N, Thanonkeo S, Klanrit P, Thanonkeo P. The potential of the newly isolated thermotolerant yeast *Pichia kudriavzevii* RZ8-1 for high-temperature ethanol production. Braz J Microbiol 2018; 49(2): 378-91.
[http://dx.doi.org/10.1016/j.bjm.2017.09.002] [PMID: 29154013]

[29]   Nuanpeng S, Thanonkeo S, Yamada M, Thanonkeo P. Ethanol production from sweet sorghum juice at high temperatures using a newly isolated thermotolerant yeast *Saccharomyces cerevisiae* DBKKU Y-53. Energies 2016; 9: 253.
[http://dx.doi.org/10.3390/en9040253]

[30]   Siripong W, Wolf P, Kusumoputri TP, *et al.* Metabolic engineering of *Pichia pastoris* for production of isobutanol and isobutyl acetate. Biotechnol Biofuels 2018; 11: 1.
[http://dx.doi.org/10.1186/s13068-017-1003-x] [PMID: 29321810]

[31]   Branco RF, Santos JC, Silva SS. A novel use for sugarcane bagasse hemicellulosic fraction: Xylitol enzymatic production. Biomass Bioenergy 2011; 35(7): 3241-6.

[http://dx.doi.org/10.1016/j.biombioe.2011.02.014]

[32]   Huang CF, Jiang YF, Guo GL, Hwang WS. Development of a yeast strain for xylitol production without hydrolysate detoxification as part of the integration of co-product generation within the lignocellulosic ethanol process. Bioresour Technol 2011; 102(3): 3322-9.
[http://dx.doi.org/10.1016/j.biortech.2010.10.111] [PMID: 21095119]

[33]   Lourenço M. Seleção de leveduras para bioconversão de D-xilose em xilitol [master's thesis]. Piracicaba: Microbiologia agrícola, Escola superior de agricultura Luiz de Queiroz, 2010; 79.

[34]   Almeida SFO, Silva LRC, Junior GCAC, *et al.* Diversidade de leveduras durante a fermentação do cacau em dois locais da Amazônia brasileira. Acta Amazon 2019; 49(1): 64-70.
[http://dx.doi.org/10.1590/1809-4392201703712]

[35]   Kuruti K, Vittaladevaram V, Urity SV, Palaniappan P, Bhaskar RU. Evolution of *Pichia pastoris* as a model organism for vaccines production in healthcare industry. Gene Rep 2020; 21: 1-6.
[http://dx.doi.org/10.1016/j.genrep.2020.100937]

[36]   Montoliu-Gava L, Esquerda-Canals G, Bronsoms S, Villegas S. Production of an anti-Aβ antibody fragment in *Pichia pastoris* and *in vitro* and *in vivo* validation of its therapeutic effect. PLoS One 2017; 12(8): 1-19.

[37]   Ben Azoun S, Belhaj AE, Göngrich R, Gasser B, Kallel H. Molecular optimization of rabies virus glycoprotein expression in *Pichia pastoris*. Microb Biotechnol 2016; 9(3): 355-68.
[http://dx.doi.org/10.1111/1751-7915.12350] [PMID: 26880068]

[38]   Yu P, Zhu Q, Chen K, Lv X. Improving the secretory production of the heterologous protein in Pichia pastoris by focusing on protein folding. Appl Biochem Biotechnol 2015; 175(1): 535-48.
[http://dx.doi.org/10.1007/s12010-014-1292-5] [PMID: 25326186]

[39]   Subathra M, Santhakumar P, Narasu ML, Beevi SS, Lal SK. Evaluation of antibody response in mice against avian influenza A (H5N1) strain neuraminidase expressed in yeast *Pichia pastoris*. J Biosci 2014; 39(3): 443-51.
[http://dx.doi.org/10.1007/s12038-014-9422-3] [PMID: 24845508]

[40]   Athmaram TN, Saraswat S, Singh AK, *et al.* Influence of copy number on the expression levels of pandemic influenza hemagglutinin recombinant protein in methylotrophic yeast *Pichia pastoris*. Virus Genes 2012; 45(3): 440-51.
[http://dx.doi.org/10.1007/s11262-012-0809-7] [PMID: 22940846]

[41]   Li Z, Hong G, Wu Z, Hu B, Xu J, Li L. Optimization of the expression of hepatitis B virus e gene in *Pichia pastoris* and immunological characterization of the product. J Biotechnol 2008; 138(1-2): 1-8.
[http://dx.doi.org/10.1016/j.jbiotec.2008.07.1989] [PMID: 18721834]

[42]   Gasser B, Saloheimo M, Rinas U, *et al.* Protein folding and conformational stress in microbial cells producing recombinant proteins: a host comparative overview. Microb Cell Fact 2008; 7: 11.
[http://dx.doi.org/10.1186/1475-2859-7-11] [PMID: 18394160]

[43]   Masoud W, Kaltoft CH. The effects of yeasts involved in the fermentation of Coffea arabica in East Africa on growth and ochratoxin A (OTA) production by *Aspergillus ochraceus*. Int J Food Microbiol 2006; 106(2): 229-34.
[http://dx.doi.org/10.1016/j.ijfoodmicro.2005.06.015] [PMID: 16213049]

[44]   Petersson S, Schnürer J. *Pichia anomala* as a biocontrol agent of *Penicillium roqueforti* in high moisture wheat, rye, barley, and oats stored under airtight conditions. Can J Microbiol 1998; 44(5): 471-6.
[http://dx.doi.org/10.1139/w98-018]

[45]   Petersson S, Hansen MW, Axberg K, Hult K, Schnürer J. Ochratoxin A accumulation in cultures of *Penicillium verrucosum* with the antagonistic yeast *Pichia anomala* and *Saccharomyces cerevisiae*. Mycol Res 1998; 102(8): 1003-6.
[http://dx.doi.org/10.1017/S0953756297006047]

[46] Ganter PF, Morais PB, Rosa CA. Yeast in cacti and tropical fruit. In: Buzzini P, Lachance MA, Yurkov A (eds) Yeasts in Natural Ecosystems: Diversity Springer, Champ,. 2017; pp. 225-64. [http://dx.doi.org/10.1007/978-3-319-62683-3_8]

[47] Irlinger F, Layec S, Hélinck S, Dugat-Bony E. Cheese rind microbial communities: diversity, composition and origin. FEMS Microbiol Lett 2015; 362(2): 1-11. [http://dx.doi.org/10.1093/femsle/fnu015] [PMID: 25670699]

[48] Abrantes MR, Campêlo MCS, Freire DAC, Assis APP, Lima PO, Silva JBA. Influence of sanitary inspection on the physicochemical quality of coalho cheese. Acta Vet Bras 2020; 14: 36-40. [http://dx.doi.org/10.21708/avb.2020.14.2.8393]

[49] Passoth V, Olstorpe M, Schnürer J. Past, present and future research directions with *Pichia anomala*. Antonie van Leeuwenhoek 2011; 99(1): 121-5. [http://dx.doi.org/10.1007/s10482-010-9508-3] [PMID: 20924674]

[50] Anapi GR, Aba RPM, Gabriel AA. Screening for heat-resistant reference yeast isolate in orange juice. Food Microbiol 2021; 94: 103639. [http://dx.doi.org/10.1016/j.fm.2020.103639] [PMID: 33279065]

[51] Gava AJ, Da Silva CAB, Frias JRG. Tecnologia de alimentos - Princípios e aplicações. São Paulo: Nobel 2008; p. 116.

[52] Bevilacqua A, Corbo MR, Sinigaglia M. Use of essential oils to inhibit *alicyclobacillus acidoterrestris*: a short overview of the literature. Front Microbiol 2011; 2: 195. [http://dx.doi.org/10.3389/fmicb.2011.00195] [PMID: 21991262]

[53] Aneja KR, Dhiman R, Aggarwal NK, Kumar V, Kaur M. Microbes associated with freshly prepared juices of citrus and carrots. Int J Food Sci 2014; 2014: 408085. [http://dx.doi.org/10.1155/2014/408085] [PMID: 26904628]

[54] Barnett JA, Payne RW, Yarrow D. Yeasts: characteristics and identification. 3rd ed. Cambrige: University Press, ed. 2000.

[55] Bakir M, Cerikcioğlu N, Tirtir A, Berrak S, Ozek E, Canpolat C. *Pichia anomala* fungaemia in immunocompromised children. Mycoses 2004; 47(5-6): 231-5. [http://dx.doi.org/10.1111/j.1439-0507.2004.00962.x] [PMID: 15189190]

[56] Otag F, Kuyucu N, Erturan Z, Sen S, Emekdas G, Sugita T. An outbreak of Pichia ohmeri infection in the paediatric intensive care unit: case reports and review of the literature. Mycoses 2005; 48(4): 265-9. [http://dx.doi.org/10.1111/j.1439-0507.2005.01126.x] [PMID: 15982209]

[57] Daniel HM, Vrancken G, Takrama JF, Camu N, De Vos P, De Vuyst L. Yeast diversity of Ghanaian cocoa bean heap fermentations. FEMS Yeast Res 2009; 9(5): 774-83. [http://dx.doi.org/10.1111/j.1567-1364.2009.00520.x] [PMID: 19473277]

[58] Araújo MAM. Isolamento e seleção de leveduras para produção de enzimas de interesse industrial a partir de frutos do cerrado [master's thesis]. Campo Grande: Universidade Católica De Bom Gosto; 2015; 67.

[59] Jutakanoke R, Tanasupawat S, Akaracharanya A. Characterization and ethanol fermentation of *Pichia* and *Torulaspora strains*. J Appl Pharm Sci 2014; 4(4): 52-6.

[60] Lima R, Lima R, Martins C. Isolamento e caracterização de leveduras de caldo de cana de uma indústria de fermentação alcoólica no nordeste brasileiro. Enciclop biosf 2015; 11(22): 3019-5.

[61] Poojary A, Sapre G. *Kodamaea ohmeri* infection in a neonate. Indian Pediatr 2009; 46(7): 629-31. [PMID: 19638663]

[62] Distasi MA, Del Gaudio T, Pellegrino G, Pirronti A, Passera M, Farina C. Fungemia due to *Kodamaea ohmeri*: First isolating in Italy. Case report and review of literature. J Mycol Med 2015; 25(4): 310-6. [http://dx.doi.org/10.1016/j.mycmed.2015.08.002] [PMID: 26404421]

[63]   Vivas R, Beltran C, Munera MI, Trujillo M, Restrepo A, Garcés C. Fungemia due to *Kodamaea ohmeri* in a young infant and review of the literature. Med Mycol Case Rep 2016; 13: 5-8.
        [http://dx.doi.org/10.1016/j.mmcr.2016.06.001] [PMID: 27630816]

[64]   Faber KN, Harder W, Ab G, Veenhuis M. Review: methylotrophic yeasts as factories for the production of foreign proteins. Yeast 1995; 11(14): 1331-44.
        [http://dx.doi.org/10.1002/yea.320111402] [PMID: 8585317]

[65]   Veenhuís M, van der Klei IJ. Peroxisomes: surprisingly versatile organelles. Biochim Biophys Acta 2002; 1555(1-3): 44-7.
        [http://dx.doi.org/10.1016/S0005-2728(02)00252-9] [PMID: 12206889]

[66]   Gould SJ, Valle D. Peroxisome biogenesis disorders: genetics and cell biology. Trends Genet 2000; 16(8): 340-5.
        [http://dx.doi.org/10.1016/S0168-9525(00)02056-4] [PMID: 10904262]

[67]   Choi MH, Park YH. Growth of *Pichia guilliermondii* A9, an osmotolerant yeast, in waste brine generated from kimchi production. Bioresour Technol 1999; 70: 231-6.
        [http://dx.doi.org/10.1016/S0960-8524(99)00049-8]

[68]   Silva VC, Peres MFS, Gattás EAL. Application of methylotrophic yeast *Pichia pastoris* in the field of food industry – A review. J Food Agric Environ 2009; 7(2): 268-73.

[69]   Kurzman CP, Fell JW. The yeasts: a taxonomic study 4. New York: Elsevier 1998.

[70]   Baptista AS, Horii J, Baptista AS. Fatores físico-químicos e biológicos ligados à produção de micotoxinas. Boletim CEPPA 2004; 22(1): 1-14.
        [http://dx.doi.org/10.5380/cep.v22i1.1175]

[71]   Droby S, Hofstein R, Wilson CL, Wisniewski M, Fridlender B, Cohen L, *et al.* Pilot testing of *Pichia guilliermondii*: a biological agent of postharvest diseases of citrus fruit. Biol Control 1993; 3: 47-52.
        [http://dx.doi.org/10.1006/bcon.1993.1008]

[72]   Saligkarias ID, Gravanis FT, Epton HAS. Biological control of *Botrytis cinerea* on tomato plants by the use of epiphytic yeasts *Candida guilliermondii* strains 101 and US 7 and *Candida oleophila* strain I-182: I. *in vivo* studies. Biol Control 2002; 25: 143-50.
        [http://dx.doi.org/10.1016/S1049-9644(02)00051-8]

[73]   van der Heijden AM, van Hoek P, Kaliterna J, van Dijken JP, van Rantwijk F, Pronk JT. Use of the yeast *Hansenula polymorpha* (*Pichia angusta*) to remove contaminating sugars from ethyl beta--fructofuranoside produced during sucrose ethanolysis catalysed by invertase. J Biosci Bioeng 1999; 87(1): 82-6.
        [http://dx.doi.org/10.1016/S1389-1723(99)80012-7] [PMID: 16232429]

[74]   Passoth V, Fredlund E, Druvefors UA, Schnürer J. Biotechnology, physiology and genetics of the yeast *Pichia anomala.* FEMS Yeast Res 2006; 6(1): 3-13.
        [http://dx.doi.org/10.1111/j.1567-1364.2005.00004.x] [PMID: 16423066]

[75]   Masih EI, Alie I, Paul B. Can the grey mould disease of the grape-vine be controlled by yeast? FEMS Microbiol Lett 2000; 189(2): 233-7.
        [http://dx.doi.org/10.1111/j.1574-6968.2000.tb09236.x] [PMID: 10930744]

[76]   Petersson S, Jonsson N, Schnürer J. *Pichia anomala* as a biocontrol agent during storage of high-moisture feed grain under airtight conditions. Postharvest Biol Technol 1999; 15(2): 175-84.
        [http://dx.doi.org/10.1016/S0925-5214(98)00081-7]

[77]   Druvefors U, Jonsson N, Boysen ME, Schnürer J. Efficacy of the biocontrol yeast *Pichia anomala* during long-term storage of moist feed grain under different oxygen and carbon dioxide regimens. FEMS Yeast Res 2002; 2(3): 389-94.
        [PMID: 12702289]

[78]   Passoth V, Schnürer J. Non-conventional yeasts in antifungal application. Berlin, Functional Genetics

of Industrial Yeasts (DeWinde JH ed) 2003;. 2003; pp. 297-329.
[http://dx.doi.org/10.1007/3-540-37003-X_10]

[79] Vohra A, Kaur P, Satyanarayana T. Production, characteristics and applications of the cell-bound phytase of *Pichia anomala*. Antonie van Leeuwenhoek 2011; 99(1): 51-5.
[http://dx.doi.org/10.1007/s10482-010-9498-1] [PMID: 20730601]

[80] Kurtzman CP, Fell JW, Boekhout T, Robert V. Methods for isolation, phenotypic characterization and maintenance of yeasts The yeasts, a taxonomic study. 5th ed. Amsterdam: Elsevier 2011; pp. 87-110.

[81] Wang J, Wang HY, Xia XM, Li PP, Wang KY. Synergistic effect of *Lentinula edodes* and *Pichia membranefaciens* on inhibition of *Penicillium expansum* infections. Postharvest Biol Technol 2013; 81: 7-12.
[http://dx.doi.org/10.1016/j.postharvbio.2013.02.002]

[82] Zhou YH, Ming J, Deng LL, Zeng KF. Effect of *Pichia membranaefaciens* in combination with salicylic acid on postharvest blue and green mold decay in citrus fruits. Biol Control 2014; 74: 21-9.
[http://dx.doi.org/10.1016/j.biocontrol.2014.03.007]

[83] Niu X, Deng L, Zhou Y, Wang W, Yao S, Zeng K. Optimization of a protective medium for freeze-dried *Pichia membranifaciens* and application of this biocontrol agent on citrus fruit. J Appl Microbiol 2016; 121(1): 234-43.
[http://dx.doi.org/10.1111/jam.13129] [PMID: 26972894]

[84] Gramiscia BR, Lutz MC, Lopes CA, Sangorrina MP. Enhancing the efficacy of yeast biocontrol agents against postharvest pathogens through nutrient profiling and the use of other additives. Biol Control 2018; 121: 151-8.
[http://dx.doi.org/10.1016/j.biocontrol.2018.03.001]

[85] Wu Q, Chen L, Xu Y. Yeast community associated with the solid state fermentation of traditional Chinese Maotai-flavor liquor. Int J Food Microbiol 2013; 166(2): 323-30.
[http://dx.doi.org/10.1016/j.ijfoodmicro.2013.07.003] [PMID: 23978339]

[86] You L, Wang S, Zhou R, Hu X, Chu Y, Wang T. Characteristics of yeast flora in Chinese strong□flavoured liquor fermentation in the Yibin region of China. J Inst Brew 2016; 122: 517-23.
[http://dx.doi.org/10.1002/jib.352]

[87] Šuranská H, Vránová D, Omelková J, Vadkertiová R. Monitoring of yeast population isolated during spontaneous fermentation of Moravian wine. Chem Pap 2012; 66: 861-8.
[http://dx.doi.org/10.2478/s11696-012-0198-3]

[88] Stefanini I, Albanese D, Cavazza A, *et al.* Dynamic changes in microbiota and mycobiota during spontaneous 'Vino Santo Trentino' fermentation. Microb Biotechnol 2016; 9(2): 195-208.
[http://dx.doi.org/10.1111/1751-7915.12337] [PMID: 26780037]

[89] Pereira EL, Ramalhosa E, Borges A, Pereira JÁ, Baptista P. Yeast dynamics during the natural fermentation process of table olives (Negrinha de Freixo cv.). Food Microbiol 2015; 46: 582-6.
[http://dx.doi.org/10.1016/j.fm.2014.10.003] [PMID: 25475331]

[90] Grounta A, Doulgeraki AI, Nychas GJ, Panagou EZ. Biofilm formation on Conservolea natural black olives during single and combined inoculation with a functional *Lactobacillus pentosus* starter culture. Food Microbiol 2016; 56: 35-44.
[http://dx.doi.org/10.1016/j.fm.2015.12.002] [PMID: 26919816]

[91] Benítez-Cabello A, Romero-Gil V, Rodríguez-Gómez F, Garrido-Fernández A, Jiménez-Díaz R, Arroyo-López FN. Evaluation and identification of poly-microbial biofilms on natural green Gordal table olives. Antonie van Leeuwenhoek 2015; 108(3): 597-610.
[http://dx.doi.org/10.1007/s10482-015-0515-2] [PMID: 26115883]

[92] Baumann K, Adelantado N, Lang C, Mattanovich D, Ferrer P. Protein trafficking, ergosterol biosynthesis and membrane physics impact recombinant protein secretion in *Pichia pastoris*. Microb Cell Fact 2011; 10: 93.

[http://dx.doi.org/10.1186/1475-2859-10-93] [PMID: 22050768]

[93]   Spohner SC, Müller H, Quitmann H, Czermak P. Expression of enzymes for the usage in food and feed industry with *Pichia pastoris*. J Biotechnol 2015; 202: 118-34.
        [http://dx.doi.org/10.1016/j.jbiotec.2015.01.027] [PMID: 25687104]

[94]   Zahrl RJ, Peña DA, Mattanovich D, Gasser B. Systems biotechnology for protein production in *Pichia pastoris*. FEMS Yeast Res 2017; 17(7)
        [http://dx.doi.org/10.1093/femsyr/fox068] [PMID: 28934418]

[95]   Patra P, Das M, Kundu P, Ghosh A. Recent advances in systems and synthetic biology approaches for developing novel cell-factories in non-conventional yeasts. Biotechnol Adv 2021; 47: 107695.
        [http://dx.doi.org/10.1016/j.biotechadv.2021.107695] [PMID: 33465474]

[96]   Chauhan AK, Arora D, Khanna N. A novel feeding strategy for enhanced protein production by fed-batch fermentation in recombinant *Pichia pastoris*. Process Biochem 1999; 34: 139-45.
        [http://dx.doi.org/10.1016/S0032-9592(98)00080-6]

[97]   Li P, Anumanthan A, Gao XG, *et al*. Expression of recombinant proteins in *Pichia pastoris*. Appl Biochem Biotechnol 2007; 142(2): 105-24.
        [http://dx.doi.org/10.1007/s12010-007-0003-x] [PMID: 18025573]

[98]   Potvin G, Ahmad A, Zhang Z. Bioprocess engineering aspects of heterologous protein production in *Pichia pastoris*: A review. Biochem Eng J 2012; 64: 91-105.
        [http://dx.doi.org/10.1016/j.bej.2010.07.017]

[99]   Pennell CA, Eldin P. *In vitro* production of recombinant antibody fragments in *Pichia pastoris*. Res Immunol 1998; 149(6): 599-603.
        [http://dx.doi.org/10.1016/S0923-2494(98)80012-6] [PMID: 9835424]

[100]   Daly R, Hearn MTW. Expression of heterologous proteins in *Pichia pastoris*: a useful experimental tool in protein engineering and production. J Mol Recognit 2005; 18(2): 119-38.
        [http://dx.doi.org/10.1002/jmr.687] [PMID: 15565717]

[101]   Spadiut O, Capone S, Krainer F, Glieder A, Herwig C. Microbials for the production of monoclonal antibodies and antibody fragments. Trends Biotechnol 2014; 32(1): 54-60.
        [http://dx.doi.org/10.1016/j.tibtech.2013.10.002] [PMID: 24183828]

[102]   Vogl T, Hartner FS, Glieder A. New opportunities by synthetic biology for biopharmaceutical production in *Pichia pastoris*. Curr Opin Biotechnol 2013; 24(6): 1094-101.
        [http://dx.doi.org/10.1016/j.copbio.2013.02.024] [PMID: 23522654]

[103]   Malpied LP, Díaz CA, Nerli B, Pessoa A. Single-chain antibody fragments: Purification methodologies. Process Biochem 2013; 48(8): 1242-51.
        [http://dx.doi.org/10.1016/j.procbio.2013.06.008]

[104]   Serrano-Rivero Y, Marrero-Domínguez K, Fando-Calzada R. *Pichia pastoris*: una plataforma para la producción de proteínas heterólogas. Rev CENIC 2015; 47(2): 67-77.

[105]   Weinacker D, Rabert C, Zepeda AB, Figueroa CA, Pessoa A, Farías JG. Applications of recombinant *Pichia pastoris* in the healthcare industry. Braz J Microbiol 2014; 44(4): 1043-8.
        [http://dx.doi.org/10.1590/S1517-83822013000400004] [PMID: 24688491]

[106]   Guerreiro DJLM. Produção de biofármacos pela levedura Pichia pastoris [master's thesis]. Caparica: Instituto Superior De Ciências Da Saúde Egas Moniz; 2016; 52.

[107]   Kwon WT, Lee WS, Park PJ, Park TK, Kang H. Protective immunity of *Pichia pastoris*-expressed recombinant envelope protein of Japanese encephalitis virus. J Microbiol Biotechnol 2012; 22(11): 1580-7.
        [http://dx.doi.org/10.4014/jmb.1205.05047] [PMID: 23124351]

[108]   Löbs AK, Schwartz C, Wheeldon I. Genome and metabolic engineering in non-conventional yeasts: Current advances and applications. Synth Syst Biotechnol 2017; 2(3): 198-207.

[http://dx.doi.org/10.1016/j.synbio.2017.08.002] [PMID: 29318200]

[109] Lünsdorf H, Gurramkonda C, Adnan A, Khanna N, Rinas U. Virus-like particle production with yeast: ultrastructural and immunocytochemical insights into *Pichia pastoris* producing high levels of the hepatitis B surface antigen. Microb Cell Fact 2011; 10: 48.
       [http://dx.doi.org/10.1186/1475-2859-10-48] [PMID: 21703024]

[110] Holguera J, Villar E, Muñoz-Barroso I. Identification of cellular proteins that interact with Newcastle Disease Virus and human Respiratory Syncytial Virus by a two-dimensional virus overlay protein binding assay (VOPBA). Virus Res 2014; 191: 138-42.
       [http://dx.doi.org/10.1016/j.virusres.2014.07.031] [PMID: 25109545]

[111] Rossey I, McLellan JS, Saelens X, Schepens B. Clinical potential of prefusion RSV F-specific antibodies. Trends Microbiol 2018; 26(3): 209-19.
       [http://dx.doi.org/10.1016/j.tim.2017.09.009] [PMID: 29054341]

[112] Adivitiya Babbal, S Mohanty, YP Khasa. Nitrogen supplementation ameliorates product quality and quantity during high cell density bioreactor studies of *Pichia pastoris:* A case study with proteolysis prone streptokinase. Int J Biol Macromol 2021; 0141-8130.

[113] Pichia. Pichia Produced Products on the Market. [cited 2021 Mar 26]. Available from: http://www.pichia.com/science-center/commercialized-products/

[114] Fickers P. Pichia pastoris: a workhorse for recombinant protein production. Curr Res Microbiol Biotechnol 2014; 2(3): 354-63.

[115] Walsh G. Biopharmaceutical benchmarks 2014. Nat Biotechnol 2014; 32(10): 992-1000.
       [http://dx.doi.org/10.1038/nbt.3040] [PMID: 25299917]

[116] Zhu W, Xu R, Gong G, Xu L, Hu Y, Xie L. Medium optimization for high yield production of human serum albumin in *Pichia pastoris* and its efficient purification. Protein Expr Purif 2021; 181: 105831.
       [http://dx.doi.org/10.1016/j.pep.2021.105831] [PMID: 33508474]

[117] Zhu W, Gong G, Pan J, *et al.* High level expression and purification of recombinant human serum albumin in *Pichia pastoris.* Protein Expr Purif 2018; 147: 61-8.
       [http://dx.doi.org/10.1016/j.pep.2018.02.003] [PMID: 29518537]

[118] Liu YK, Li YT, Lu CF, Huang LF. Enhancement of recombinant human serum albumin in transgenic rice cell culture system by cultivation strategy. N Biotechnol 2015; 32(3): 328-34.
       [http://dx.doi.org/10.1016/j.nbt.2015.03.001] [PMID: 25765580]

[119] Mallem M, Warburton S, Li F, *et al.* Maximizing recombinant human serum albumin production in a Mut(s) Pichia pastoris strain. Biotechnol Prog 2014; 30(6): 1488-96.
       [http://dx.doi.org/10.1002/btpr.1990] [PMID: 25196297]

[120] Dong Y, Zhang F, Wang Z, *et al.* Extraction and purification of recombinant human serum albumin from *Pichia pastoris* broths using aqueous two-phase system combined with hydrophobic interaction chromatography. J Chromatogr A 2012; 1245: 143-9.
       [http://dx.doi.org/10.1016/j.chroma.2012.05.041] [PMID: 22658659]

[121] Lei J, Guan B, Li B, *et al.* Expression, purification and characterization of recombinant human interleukin-2-serum albumin (rhIL-2-HSA) fusion protein in *Pichia pastoris.* Protein Expr Purif 2012; 84(1): 154-60.
       [http://dx.doi.org/10.1016/j.pep.2012.05.003] [PMID: 22609631]

[122] Guan B, Chen F, Lei J, *et al.* Constitutive expression of a rhIL-2-HSA fusion protein in *Pichia pastoris* using glucose as carbon source. Appl Biochem Biotechnol 2013; 171(7): 1792-804.
       [http://dx.doi.org/10.1007/s12010-013-0423-8] [PMID: 23999737]

[123] Duman-Özdamar ZE, Binay B. Production of industrial enzymes *via* Pichia pastoris as a cell factory in bioreactor: current status and future aspects. Protein J 2021. print.

[124] Zhang L, Liu WQ, Li J. Establishing a eukaryotic *Pichia pastoris* cell-free protein synthesis system.

Front Bioeng Biotechnol 2020; 8: 536.
[http://dx.doi.org/10.3389/fbioe.2020.00536] [PMID: 32626695]

[125] Tavasoli T, Arjmand S, Ranaei Siadat SO, Shojaosadati SA, Sahebghadam Lotfi A. Enhancement of alpha 1-antitrypsin production in *Pichia pastoris* by designing and optimizing medium using elemental analysis. Iran J Biotechnol 2017; 15(4): 224-31.
[http://dx.doi.org/10.15171/ijb.1808] [PMID: 29845074]

[126] Robert JM, Lattari FS, Machado AC, *et al.* Production of recombinant lipase B from Candida Antarctica in *Pichia pastoris* under control of the promoter PGK using crude glycerol from biodiesel production as carbon source. Biochem Eng J 2017; 118: 123-31.
[http://dx.doi.org/10.1016/j.bej.2016.11.018]

[127] Fang Z, Xu L, Pan D, Jiao L, Liu Z, Yan Y. Enhanced production of *Thermomyces lanuginosus* lipase in *Pichia pastoris*via genetic and fermentation strategies. J Ind Microbiol Biotechnol 2014; 41(10): 1541-51.
[http://dx.doi.org/10.1007/s10295-014-1491-7] [PMID: 25074457]

[128] Sha C, Yu XW, Zhang M, Xu Y. Efficient secretion of lipase r27RCL in *Pichia pastoris* by enhancing the disulfide bond formation pathway in the endoplasmic reticulum. J Ind Microbiol Biotechnol 2013; 40(11): 1241-9.
[http://dx.doi.org/10.1007/s10295-013-1328-9] [PMID: 23990169]

[129] Wang JY, Lu L, Feng FJ. Combined strategies for improving production of a thermo-alkali stable laccase in *Pichia pastoris*. Electron J Biotechnol 2017; 28: 7-13.
[http://dx.doi.org/10.1016/j.ejbt.2017.04.002]

[130] Visioli F, Artaria C. Astaxanthin in cardiovascular health and disease: mechanisms of action, therapeutic merits, and knowledge gaps. Food Funct 2017; 8(1): 39-63.
[http://dx.doi.org/10.1039/C6FO01721E] [PMID: 27924978]

[131] Zeni J, Colet R, Cence K, *et al.* Screening of microorganisms for production of carotenoids. CYTA J Food 2011; 9: 160-6.
[http://dx.doi.org/10.1080/19476337.2010.499570]

[132] Gayeski L, Colet R, Urnau L, Burkert JFM, Steffens C, Valduga E. Semi-continuous carotenoid production in bioreactor from *Phaffia rhodozyma* using agro-industrial residues. Biointerface Res Appl Chem 2012; 11: 7501-10.

[133] Urnau L, Colet R, Gayeski L, *et al.* Fed-batch carotenoid production by *Phaffia rhodozyma* Y-17268 using agroindustrial substrates. Biointerface Res Appl Chem 2020; 10: 5348-54.
[http://dx.doi.org/10.33263/BRIAC103.348354]

[134] Colet R, Urnau L, De Souza Hassemer G, *et al.* Kinetic Parameters of Fed-Batch Production of Carotenoids by *Sporidiobolus salmonicolor* using low-cost agro-industrial substrates. Ind Biotechnol (New Rochelle NY) 2019; 15: 311-21.
[http://dx.doi.org/10.1089/ind.2019.0015]

[135] Lange N, Steinbüchel A. β-Carotene production by *Saccharomyces cerevisiae* with regard to plasmid stability and culture media. Appl Microbiol Biotechnol 2011; 91(6): 1611-22.
[http://dx.doi.org/10.1007/s00253-011-3315-2] [PMID: 21573686]

[136] Miura Y, Kondo K, Saito T, Shimada H, Fraser PD, Misawa N. Production of the carotenoids lycopene, beta-carotene, and astaxanthin in the food yeast *Candida utilis*. Appl Environ Microbiol 1998; 64(4): 1226-9.
[http://dx.doi.org/10.1128/AEM.64.4.1226-1229.1998] [PMID: 9546156]

[137] Johns AM, Love J, Aves SJ. Four inducible promoters for controlled gene expression in the oleaginous yeast *Rhodotorula toruloides*. Front Microbiol 2016; 7: 1666.
[http://dx.doi.org/10.3389/fmicb.2016.01666] [PMID: 27818654]

[138] Papp T, Velayos A, Bartók T, Eslava AP, Vágvölgyi C, Iturriaga EA. Heterologous expression of

astaxanthin biosynthesis genes in *Mucor circinelloides*. Appl Microbiol Biotechnol 2006; 69(5): 526-31.
[http://dx.doi.org/10.1007/s00253-005-0026-6] [PMID: 16034557]

[139] Veiga-Crespo P, Araya-Garay JM, Villa TG. Engineering Pichia pastoris for the production of carotenoids: Methods and Protocols. Microbial Carotenoids 2018; pp. 311-26.
[http://dx.doi.org/10.1007/978-1-4939-8742-9_19]

[140] Araya-Garay JM, Feijoo-Siota L, Rosa-dos-Santos F, Veiga-Crespo P, Villa TG. Construction of new Pichia pastoris X-33 strains for production of lycopene and β-carotene. Appl Microbiol Biotechnol 2012; 93(6): 2483-92.
[http://dx.doi.org/10.1007/s00253-011-3764-7] [PMID: 22159890]

[141] Otero DM, Bulsing BA, Huerta KM, *et al.* Carotenoid-producing yeasts in the brazilian biodiversity: isolation, identification and cultivation in agroindustrial waste. Braz J Chem Eng 2019; 36(1): 117-29.
[http://dx.doi.org/10.1590/0104-6632.20190361s20170433]

[142] Cipolatti EP, Remedi RD, Sá CS, *et al.* Use of agroindustrial byproducts as substrate for production of carotenoids with antioxidant potential by wild yeasts. Biocatal Agric Biotechnol 2019; 20: 101208.
[http://dx.doi.org/10.1016/j.bcab.2019.101208]

[143] Sene L, Arruda PV, Oliveira SMM, Felipe MGA. Evaluation of sorghum straw hemicellulosic hydrolysate for biotechnological production of xylitol by *Candida guilliermondii*. Braz J Microbiol 2011; 42(3): 1141-6.
[http://dx.doi.org/10.1590/S1517-83822011000300036] [PMID: 24031733]

[144] Ling H, Cheng K, Ge J, Ping W. Statistical optimization of xylitol production from corncob hemicellulose hydrolysate by *Candida tropicalis* HDY-02. N Biotechnol 2011; 28(6): 673-8.
[http://dx.doi.org/10.1016/j.nbt.2010.05.004] [PMID: 20466087]

[145] Zou YZ, Qi K, Chen X, Miao XL, Zhong JJ. Favorable effect of very low initial K(L)a value on xylitol production from xylose by a self-isolated strain of *Pichia guilliermondii*. J Biosci Bioeng 2010; 109(2): 149-52.
[http://dx.doi.org/10.1016/j.jbiosc.2009.07.013] [PMID: 20129099]

[146] Pardillo TB, Batista GG, Garriga LM, Suárez EG, Galiano EC. Estrategia para evaluar las alternativas de uso de la xilosa para la obtención de xilitol o etanol. Cent Azúcar 2011; 38(3): 71-6.

[147] Canettieri EV, Almeida SJB, Almeida-Felipe MG. Obtenção biotecnológica de xilitol a partir de cavacos de eucalipto. Rev Bras Cienc Farm 2002; 38(3): 323-31.
[http://dx.doi.org/10.1590/S1516-93322002000300008]

[148] Barbosa MFS, Medeiros MB, Mancilha IM, Schneider H, Lee H. Screening of yeasts for production of xylitol from d-xylose and some factors which affect xylitol yield in *Candida guilliermondii*. J Ind Microbiol 1988; 3: 241-51.
[http://dx.doi.org/10.1007/BF01569582]

[149] Jeffries TW. Utilization of xylose by bacteria, yeasts, and fungi. Adv Biochem Eng Biotechnol 1983; 27: 1-32.
[http://dx.doi.org/10.1007/BFb0009101] [PMID: 6437152]

[150] Slininger PJ, Bolen PL, Kurtzman CP. *Pachysolen tannophilus*: properties and process consideration for ethanol production from D-xylose. Enzyme Microb Technol 1987; 9: 5-15.
[http://dx.doi.org/10.1016/0141-0229(87)90043-3]

[151] Nolleau V, Preziosi-Belloy L, Delgenes JP, Delgenes JM. Xylitol production from xylose by two yeast strains: Sugar tolerance. Curr Microbiol 1993; 27: 191-7.
[http://dx.doi.org/10.1007/BF01692875]

[152] Silva CJ, Roberto IC. Improvement of xylitol production by *Candida guilliermondii* FTI 20037 previously adapted to rice straw hemicellulosic hydrolysate. Lett Appl Microbiol 2001; 32(4): 248-52.
[http://dx.doi.org/10.1046/j.1472-765X.2001.00899.x] [PMID: 11298935]

[153] Rodrigues RDCLB, Jeffries TW. Applying response surface methodology to kinetic parameter to improve xylitol production by *Pichia stipitis* d-xylulokinase mutant on mixed sugar model. N Biotechnol 2009; 25: 266-7.
[http://dx.doi.org/10.1016/j.nbt.2009.06.596]

[154] Menezes CR, Durrant LR. Xilooligossacarídeos: produção, aplicações e efeitos na saúde humana. Cienc Rural 2008; 38(2): 587-92.
[http://dx.doi.org/10.1590/S0103-84782008000200050]

[155] Nascimento CEO. Produção de xilo-oligossacarídeos a partir do bagaço de cana-de-açúcar pela ação de xilanase GH10 de Thermoascus aurantiacus com atividade expressa em Pichia pastoris e aplicação [master's thesis]. São José do Rio Preto: IBILCE Universidade Estadual Paulista Júlio de Mesquita Filho 2019; 78.

[156] Martins M, Ávila PF, Paim de Andrade CC, Goldbeck R. Synergic recombinant enzyme association to optimize xylo-oligosaccharides production from agricultural waste. Biocatal Agric Biotechnol 2020; 28: 101747.
[http://dx.doi.org/10.1016/j.bcab.2020.101747]

[157] Tao Y, Yang L, Yin L, *et al.* Novel approach to produce biomass-derived oligosaccharides simultaneously by recombinant endoglucanase from *Trichoderma reesei.* Enzyme Microb Technol 2020; 134: 109481.
[http://dx.doi.org/10.1016/j.enzmictec.2019.109481] [PMID: 32044028]

[158] Trujillo LE, Arrieta JG, Dafhnis F, *et al.* Fructo-oligosaccharides production by the *Gluconacetobacter diazotrophicus* levansucrase expressed in the methylotrophic yeast *Pichia pastoris.* Enz Microb Tech 2001; 28(2-3): 139-44.
[http://dx.doi.org/10.1016/S0141-0229(00)00290-8] [PMID: 11166804]

[159] Kalidas NR, Saminathan M, Ismail IS, *et al.* Structural characterization and evaluation of prebiotic activity of oil palm kernel cake mannanoligosaccharides. Food Chem 2017; 234: 348-55.
[http://dx.doi.org/10.1016/j.foodchem.2017.04.159] [PMID: 28551246]

[160] St-Onge MP, Salinardi T, Herron-Rubin K, Black RM. A weight-loss diet including coffee-derived mannooligosaccharides enhances adipose tissue loss in overweight men but not women. Obesity (Silver Spring) 2012; 20(2): 343-8.
[http://dx.doi.org/10.1038/oby.2011.289] [PMID: 21938072]

[161] Wang Y, Vilaplana F, Brumer H, Aspeborg H. Enzymatic characterization of a glycoside hydrolase family 5 subfamily 7 (GH5_7) mannanase from *Arabidopsis thaliana.* Planta 2014; 239(3): 653-65.
[http://dx.doi.org/10.1007/s00425-013-2005-y] [PMID: 24327260]

[162] Li YX, Yi P, Liu J, Yan QJ, Jiang ZQ. High-level expression of an engineered β-mannanase (mRmMan5A) in *Pichia pastoris* for manno-oligosaccharide production using steam explosion pretreated palm kernel cake. Bioresour Technol 2018; 256: 30-7.
[http://dx.doi.org/10.1016/j.biortech.2018.01.138] [PMID: 29428611]

[163] Li YX, Liu HJ, Shi YQ, Yan QJ, You X, Jiang ZQ. Preparation, characterization, and prebiotic activity of manno-oligosaccharides produced from cassia gum by a glycoside hydrolase family 134 β-mannanase. Food Chem 2020; 309: 125709.
[http://dx.doi.org/10.1016/j.foodchem.2019.125709] [PMID: 31708343]

[164] Michalak M, Thomassen LV, Roytio H, Ouwehand AC, Meyer AS, Mikkelsen JD. Expression and characterization of an endo-1,4-β-galactanase from Emericella nidulans in Pichia pastoris for enzymatic design of potentially prebiotic oligosaccharides from potato galactans. Enzyme Microb Technol 2012; 50(2): 121-9.
[http://dx.doi.org/10.1016/j.enzmictec.2011.11.001] [PMID: 22226198]

[165] Srivastava PK, Kapoor M. Production, properties, and applications of endo-β-mannanases. Biotechnol Adv 2017; 35(1): 1-19.
[http://dx.doi.org/10.1016/j.biotechadv.2016.11.001] [PMID: 27836790]

[166]   Wang NN, Liu J, Li YX, Ma JW, Yan QJ, Jiang ZQ. High-level expression of a glycoside hydrolase family 26 β-mannanase from *Aspergillus niger* in *Pichia pastoris* for production of partially hydrolysed fenugreek gum. Process Biochem 2021; 100: 90-7.
[http://dx.doi.org/10.1016/j.procbio.2020.09.034]

[167]   Limtong S, Sringiew C, Yongmanitchai W. Production of fuel ethanol at high temperature from sugar cane juice by a newly isolated *Kluyveromyces marxianus*. Bioresour Technol 2007; 98(17): 3367-74.
[http://dx.doi.org/10.1016/j.biortech.2006.10.044] [PMID: 17537627]

[168]   Christensen AD, Kádár Z, Oleskowicz-Popiel P, Thomsen MH. Production of bioethanol from organic whey using *Kluyveromyces marxianus*. J Ind Microbiol Biotechnol 2011; 38(2): 283-9.
[http://dx.doi.org/10.1007/s10295-010-0771-0] [PMID: 20632200]

[169]   Charoensopharat K, Thanonkeo P, Thanonkeo S, Yamada M. Ethanol production from Jerusalem artichoke tubers at high temperature by newly isolated thermotolerant inulin-utilizing yeast *Kluyveromyces marxianus* using consolidated bioprocessing. Antonie van Leeuwenhoek 2015; 108(1): 173-90.
[http://dx.doi.org/10.1007/s10482-015-0476-5] [PMID: 25980834]

[170]   Dhaliwal SS, Oberoi HS, Sandhu SK, Nanda D, Kumar D, Uppal SK. Enhanced ethanol production from sugarcane juice by galactose adaptation of a newly isolated thermotolerant strain of *Pichia kudriavzevii*. Bioresour Technol 2011; 102(10): 5968-75.
[http://dx.doi.org/10.1016/j.biortech.2011.02.015] [PMID: 21398115]

[171]   Gallardo JCM, Souza CS, Cicarelli RMB, Oliveira KF, Morais MR, Laluce C. Enrichment of a continuous culture of Saccharomyces cerevisiae with the yeast *Issatchenkia orientalis* in the production of ethanol at increasing temperatures. J Ind Microbiol Biotechnol 2011; 38(3): 405-14.
[http://dx.doi.org/10.1007/s10295-010-0783-9] [PMID: 20697927]

[172]   Kwon YJ, Ma AZ, Li Q, Wang F, Zhuang GQ, Liu CZ. Effect of lignocellulosic inhibitory compounds on growth and ethanol fermentation of newly-isolated thermotolerant *Issatchenkia orientalis*. Bioresour Technol 2011; 102(17): 8099-104.
[http://dx.doi.org/10.1016/j.biortech.2011.06.035] [PMID: 21737262]

[173]   Kaewkrajay C, Dethoup T, Limtong S. Ethanol production from cassava using a newly isolated thermotolerant yeast strain. Sci Asia 2014; 40: 268-77.
[http://dx.doi.org/10.2306/scienceasia1513-1874.2014.40.268]

[174]   Isono N, Hayakawa H, Usami A, Mishima T, Hisamatsu M. A comparative study of ethanol production by *Issatchenkia orientalis* strains under stress conditions. J Biosci Bioeng 2012; 113(1): 76-8.
[http://dx.doi.org/10.1016/j.jbiosc.2011.09.004] [PMID: 22018735]

[175]   Yuangsaard N, Yongmanitchai W, Yamada M, Limtong S. Selection and characterization of a newly isolated thermotolerant *Pichia kudriavzevii* strain for ethanol production at high temperature from cassava starch hydrolysate. Antonie van Leeuwenhoek 2013; 103(3): 577-88.
[http://dx.doi.org/10.1007/s10482-012-9842-8] [PMID: 23132277]

[176]   Hisamatsu M, Furubayashi T, Karita S, Mishima T, Isono N. Isolation and identification of a novel yeast fermenting ethanol under acidic conditions. J Appl Glycosci 2006; 53: 111-3.
[http://dx.doi.org/10.5458/jag.53.111]

[177]   Kitagawa T, Tokuhiro K, Sugiyama H, *et al.* Construction of a β-glucosidase expression system using the multistress-tolerant yeast *Issatchenkia orientalis*. Appl Microbiol Biotechnol 2010; 87(5): 1841-53.
[http://dx.doi.org/10.1007/s00253-010-2629-9] [PMID: 20467739]

[178]   Chi Z, Arnebory N. *Saccharomyces cerevisiae* strains with different degrees of ethanol tolerance exhibit different adaptive responses to produced ethanol. J Ind Microbiol Biotechnol 2000; 24: 75-8.
[http://dx.doi.org/10.1038/sj.jim.2900769]

[179]   Edgardo A, Carolina P, Manuel R, Juanita F, Baeza J. Selection of thermotolerant yeast strains

Saccharomyces cerevisiae for bioethanol production. Enzyme Microb Technol 2008; 43: 120-3.
[http://dx.doi.org/10.1016/j.enzmictec.2008.02.007]

[180] Hahn-Hägerdal B, Karhumaa K, Fonseca C, Spencer-Martins I, Gorwa-Grauslund MF. Towards industrial pentose-fermenting yeast strains. Appl Microbiol Biotechnol 2007; 74(5): 937-53.
[http://dx.doi.org/10.1007/s00253-006-0827-2] [PMID: 17294186]

[181] Ribeiro N, Freita L, Tralli L, Silva A, *et al.* Otimização das condições fermentativas de *Pichia membranifaciens* para produção de etanol de segunda geração. Quim Nova 2019; 42: 720-8.
[http://dx.doi.org/10.21577/0100-4042.20170385]

[182] Sheetal KR, Prasad S, Renjith PS. Effect of cultivar variation and *Pichia stipitis* NCIM 3498 on cellulosic ethanol production from rice straw. Biomass Bioenergy 2019; 127: 105253.
[http://dx.doi.org/10.1016/j.biombioe.2019.105253]

[183] Nosrati-Ghods N, Harrison STL, Isafiade AJ, Tai SL. Mathematical Modelling of Bioethanol Fermentation From Glucose, Xylose or Their Combination – A Review. Chem Bio Eng 2020; 7(3): 68-88.
[http://dx.doi.org/10.1002/cben.201900024]

[184] Raina N, Slathia PS, Sharma P. Experimental optimization of thermochemical pretreatment of sal (*Shorea robusta*) sawdust by Central Composite Design study for bioethanol production by co-fermentation using *Saccharomyces cerevisiae* (MTCC-36) and *Pichia stipitis* (NCIM-3498). Biomass Bioenergy 2020; 143: 105819.
[http://dx.doi.org/10.1016/j.biombioe.2020.105819]

[185] Burini JA, Eizaguirre JI, Loviso C, Libkind D. Levaduras no convencionales como herramientas de innovación y diferenciación en la producción de cerveza, Rev Argent Microb 2021; 0325-7541.

[186] Holt S, Mukherjee V, Lievens B, Verstrepen KJ, Thevelein JM. Bioflavoring by non-conventional yeasts in sequential beer fermentations. Food Microbiol 2018; 72: 55-66.
[http://dx.doi.org/10.1016/j.fm.2017.11.008] [PMID: 29407405]

[187] Padilla B, Gil JV, Manzanares P. Past and future of non-*Saccharomyces* yeasts: 650 from spoilage microorganisms to biotechnological tools for improving wine aroma 651 complexity. Front Microbiol 2016; 7: 411.
[http://dx.doi.org/10.3389/fmicb.2016.00411] [PMID: 27065975]

[188] Zhong W, Chen T, Yang H, Li E. Isolation and Selection of Non-*Saccharomyces* yeasts being capable of degrading citric acid and evaluation its effect on kiwifruit wine fermentation. Ferment 2020; 6: 25.
[http://dx.doi.org/10.3390/fermentation6010025]

[189] Lee PR, Ong YL, Yu B, Curran P, Liu SQ. Evolution of volatile compounds in papaya wine fermented with three *Williopsis saturnus* yeasts. Int J Food Sci Technol 2010; 45(10): 2032-41.
[http://dx.doi.org/10.1111/j.1365-2621.2010.02369.x]

[190] Contreras A, Hidalgo C, Henschke PA, Chambers PJ, Curtin C, Varela C. Evaluation of non-*Saccharomyces* yeasts for the reduction of alcohol content in wine. Appl Environ Microbiol 2014; 80(5): 1670-8.
[http://dx.doi.org/10.1128/AEM.03780-13] [PMID: 24375129]

[191] Agbogbo FK, Coward-Kelly G. Cellulosic ethanol production using the naturally occurring xylose-fermenting yeast, *Pichia stipitis*. Biotechnol Lett 2008; 30(9): 1515-24.
[http://dx.doi.org/10.1007/s10529-008-9728-z] [PMID: 18431677]

[192] Caputo L, Quintieri L, Baruzzi F, Borcakli M, Morea M. Molecular and phenotypic characterization of *Pichia fermentans* strains found among Boza yeasts. Food Res Int 2012; 48: 755-62.
[http://dx.doi.org/10.1016/j.foodres.2012.06.022]

[193] Božič JT, Butinar L, Albreht A, Vovk I, Korte D, Vodopivec BM. The impact of *Saccharomyces* and non-*Saccharomyces* yeasts on wine colour: A laboratory study of vinylphenolic pyranoanthocyanin formation and anthocyanin cell wall adsorption. Lebensm Wiss Technol 2020; 123: 109072.

[http://dx.doi.org/10.1016/j.lwt.2020.109072]

[194] Zhong W, Liu S, Yang H, Li E. Effect of selected yeast on physicochemical and oenological properties of blueberry wine fermented with citrate-degrading *Pichia* fermentans. Lebensm Wiss Technol 2021; 111261.
[http://dx.doi.org/10.1016/j.lwt.2021.111261]

[195] Mingorance-Cazorla L, Clemente-Jiménez JM, Martínez-Rodríguez S, Heras-Vázquez FJ, Rodríguez-Vico F. Contribution of different natural yeasts to the aroma of two alcoholic beverages. World J Microbiol Biotechnol 2003; 19: 297-04.
[http://dx.doi.org/10.1023/A:1023662409828]

[196] Méndez-Zamora A, Gutiérrez-Avendaño DO, Arellano-Plaza M, *et al.* The non-*Saccharomyces* yeast *Pichia kluyveri* for the production of aromatic volatile compounds in alcoholic fermentation. FEMS Yeast Res 2021; 20(8): 67.
[http://dx.doi.org/10.1093/femsyr/foaa067] [PMID: 33316048]

[197] Hansen CHR. https://www.chr-hansen.com/pt/food-cultures-and-enzymes/fermented-beverages/cards/product-cards/frootzen-first-ever-pichia-kluyveri-yeast

[198] Portugal CB, de Silva AP, Bortoletto AM, Alcarde AR. How native yeasts may influence the chemical profile of the Brazilian spirit, cachaça? Food Res Int 2017; 91: 18-25.
[http://dx.doi.org/10.1016/j.foodres.2016.11.022] [PMID: 28290322]

[199] Oliveira ES, Rosa CA, Morgano MA, Serra GE. Fermentation characteristics as criteria for selection of cachaça yeast. World J Microbiol Biotechnol 2004; 20: 19-24.
[http://dx.doi.org/10.1023/B:WIBI.0000013286.30695.4e]

[200] Portugal CB, Alcarde AR, Bortoletto AM, Silva AP. The role of spontaneous fermentation for the production of cachaça: a study of case. Eur Food Res Technol 2016; 242: 1587-97. [Internet].
[http://dx.doi.org/10.1007/s00217-016-2659-3]

[201] Domizio P, Romani C, Lencioni L, *et al.* Outlining a future for non-*Saccharomyces* yeasts: selection of putative spoilage wine strains to be used in association with *Saccharomyces cerevisiae* for grape juice fermentation. Int J Food Microbiol 2011; 147(3): 170-80.
[http://dx.doi.org/10.1016/j.ijfoodmicro.2011.03.020] [PMID: 21531033]

[202] Chung H Jr, Lee SL, Chou CC. Production and molar yield of 2-phenylethanol by *Pichia fermentans* L-5 as affected by some medium components. J Biosci Bioeng 2000; 90(2): 142-7.
[http://dx.doi.org/10.1016/S1389-1723(00)80101-2] [PMID: 16232833]

[203] Jolly NP, Varela C, Pretorius IS. Not your ordinary yeast: non-*Saccharomyces* yeasts in wine production uncovered. FEMS Yeast Res 2014; 14(2): 215-37.
[http://dx.doi.org/10.1111/1567-1364.12111] [PMID: 24164726]

[204] Wang C, Sun J, Lassabliere B, Yu B, Liu SQ. Coffee flavour modification through controlled fermentations of green coffee beans by *Saccharomyces cerevisiae* and *Pichia kluyveri*: Part I. Effects from individual yeasts. Food Res Int 2020; 136: 109588.
[http://dx.doi.org/10.1016/j.foodres.2020.109588] [PMID: 32846616]

[205] da Silva Vale A, de Melo Pereira GV. de Carvalho Neto DP, Rodrigues C, Pagnoncelli MGB, Soccol CR. Effect of Co-Inoculation with *Pichia fermentans* and *Pediococcus acidilactici* on metabolite produced during fermentation and volatile composition of coffee beans. Ferment 2019; 5(3): 67.
[http://dx.doi.org/10.3390/fermentation5030067]

[206] de Carvalho Neto DP, de Melo Pereira G, Tanobe V, *et al.* Yeast diversity and physicochemical characteristics associated with coffee bean fermentation from the Brazilian Cerrado Mineiro region. Ferment 2017; 3(1): 11.
[http://dx.doi.org/10.3390/fermentation3010011]

[207] Gonzalez-Rios O, Suarez-Quiroz ML, Boulanger R, *et al.* Impact of "ecological" post-harvest processing on the volatile fraction of coffee beans: I. Green coffee. J Food Compos Anal 2007;

20(3–4): 289-96.
[http://dx.doi.org/10.1016/j.jfca.2006.07.009]

[208]   Vilela DM, Pereira GV, Silva CF, Batista LR, Schwan RF. Molecular ecology and polyphasic characterization of the microbiota associated with semi-dry processed coffee (Coffea arabica L.). Food Microbiol 2010; 27(8): 1128-35.
[http://dx.doi.org/10.1016/j.fm.2010.07.024] [PMID: 20832694]

[209]   Haile M, Kang WH. The role of microbes in coffee fermentation and theirimpact on coffee quality. J Food Qual 2019; 12.

[210]   De Bruyn F, Zhang SJ, Pothakos V, *et al.* Exploring the impacts of postharvest processing on the microbiota and metabolite profiles during green coffee bean production. Appl Environ Microbiol 2016; 83(1): e02398-16.
[PMID: 27793826]

[211]   Evangelista SR, Miguel MGCP, Silva CF, Pinheiro ACM, Schwan RF. Microbiological diversity associated with the spontaneous wet method of coffee fermentation. Int J Food Microbiol 2015; 210: 102-12.
[http://dx.doi.org/10.1016/j.ijfoodmicro.2015.06.008] [PMID: 26119187]

[212]   Zhang SJ, De Bruyn F, Pothakos V, *et al.* Following coffee production from cherries to cup: Microbiological and metabolomic analysis of wet processing of Coffea Arabica. Appl Environ Microbiol 2019; 85(6): e02635-18.
[http://dx.doi.org/10.1128/AEM.02635-18] [PMID: 30709820]

[213]   Bressani APP, Martinez SJ, Evangelista SR, Dias DR, Schwan RF. Characteristics of fermented coffee inoculated with yeast starter cultures using different inoculation methods. Lebensm Wiss Technol 2018; 92: 212-9.
[http://dx.doi.org/10.1016/j.lwt.2018.02.029]

[214]   Martinez SJ, Bressani APP, Miguel MGDCP, Dias DR, Schwan RF. Different inoculation methods for semi-dry processed coffee using yeasts as starter cultures. Food Res Int 2017; 102: 333-40.
[http://dx.doi.org/10.1016/j.foodres.2017.09.096] [PMID: 29195956]

[215]   de Melo Pereira GV, Neto E, Soccol VT, Medeiros ABP, Woiciechowski AL, Soccol CR. Conducting starter culture-controlled fermentations of coffee beans during on-farm wet processing: Growth, metabolic analyses and sensorial effects. Food Res Int 2015; 75: 348-56.
[http://dx.doi.org/10.1016/j.foodres.2015.06.027] [PMID: 28454966]

[216]   Ribeiro LS, Ribeiro DE, Evangelista SR. Controlled fermentation of semi-dry coffee (Coffea arabica) using starter cultures: A sensory perspective. Lebensm Wiss Technol 2017; 82: 32-8.
[http://dx.doi.org/10.1016/j.lwt.2017.04.008]

[217]   Saerens S, Swiegers JH. Enhancement of coffee quality and flavor by using pichia kluyveri yeast starter culture for coffee fermentation US Patent: US20160058028A1 2016.

[218]   Takahashi K, Minami Y, Kanabuchi Y, Togami K, Mitsuhashi M. Method of processing green coffee US Patent: US20070190207A1 2007.

[219]   Owusu M, Petersen MA, Heimdal H. Effect of fermentation method, roasting and conching conditions on the aroma volatiles of dark chocolate. J Food Process Preserv 2012; 36: 446-56.
[http://dx.doi.org/10.1111/j.1745-4549.2011.00602.x]

[220]   Ho VTT, Zhao J, Fleet G. Yeasts are essential for cocoa bean fermentation. Int J Food Microbiol 2014; 174: 72-87.
[http://dx.doi.org/10.1016/j.ijfoodmicro.2013.12.014] [PMID: 24462702]

[221]   Pereira GVM, Alvarez JP, Carvalho Neto DP, Soccol VT, Tanobe VOA, Rogez H, *et al.* Great intraspecies diversity of *Pichia kudriavzevii* in cocoa fermentation highlights the importance of yeast strain selection for flavor modulation of cocoa beans. Lebensm Wiss Technol 2017; 84: 290-7.
[http://dx.doi.org/10.1016/j.lwt.2017.05.073]

[222]   Ooi TS, Ting ASY, Siow LF. Influence of selected native yeast starter cultures on the antioxidant activities, fermentation index and total soluble solids of *Malaysia cocoa* beans: A simulation study. Lebensm Wiss Technol 2020; 122: 108977.
[http://dx.doi.org/10.1016/j.lwt.2019.108977]

[223]   Chagas Junior GCA, Ferreira NR, Gloria MBA, Martins LHDS, Lopes AS. Chemical implications and time reduction of on-farm cocoa fermentation by *Saccharomyces cerevisiae* and *Pichia kudriavzevii.* Food Chem 2021; 338: 127834. a
[http://dx.doi.org/10.1016/j.foodchem.2020.127834] [PMID: 32810810]

[224]   Junior GCAC, Ferreira NR, Andrade EHDA. Nascimento LDd, Siqueira FCd, Lopes AS. Profile of volatile compounds of on-farm fermented and dried cocoa beans inoculated with *Saccharomyces cerevisiae* KY794742 and *Pichia kudriavzevii* KY794725. Molecules 2021; 26(2): 344.
[http://dx.doi.org/10.3390/molecules26020344]

[225]   Santos DS, Rezende RP, Santos TF, Marques ELS, Ferreira ACR, Silva ABC, *et al.* Fermentation in fine cocoa type Scavina: Change in standard quality as the effect of use of starters yeast in fermentation. Food Chemist 202; 328: 127110.

[226]   Fernández Maura Y, Balzarini T, Clapé Borges P, Evrard P, De Vuyst L, Daniel HM. The environmental and intrinsic yeast diversity of Cuban cocoa bean heap fermentations. Int J Food Microbiol 2016; 233: 34-43.
[http://dx.doi.org/10.1016/j.ijfoodmicro.2016.06.012] [PMID: 27322722]

[227]   Romo-Sánchez S, Alves-Baffi M, Arévalo-Villena M, Ubeda-Iranzo J, Briones-Pérez A. Yeast biodiversity from oleic ecosystems: study of their biotechnological properties. Food Microbiol 2010; 27(4): 487-92.
[http://dx.doi.org/10.1016/j.fm.2009.12.009] [PMID: 20417397]

[228]   Montaño A, Cortés-Delgado A, Sánchez AH, Ruiz-Barba JL. Production of volatile compounds by wild-type yeasts in a natural olive-derived culture medium. Food Microbiol 2021; 98: 103788.
[http://dx.doi.org/10.1016/j.fm.2021.103788] [PMID: 33875216]

[229]   Lucena-Padrós H, Ruiz-Barba JL. Microbial biogeography of Spanish-style green olive fermentations in the province of Seville, Spain. Food Microbiol 2019; 82: 259-68.
[http://dx.doi.org/10.1016/j.fm.2019.02.004] [PMID: 31027782]

[230]   Jafari N, Soudi MR, Kasra-Kermanshahi R. Biodecolorization of textile azo dyes by isolates yeast from activated sludge: *Issatchenkia orientalis* JKS6. Ann Microbiol 2014; 64: 475-82.
[http://dx.doi.org/10.1007/s13213-013-0677-y]

[231]   Pajot HF, Martorell MM, Figueroa LIC. Ecology of dye decolorizing yeasts. In book Bioremediation in Latin America 2014; 223-40.
[http://dx.doi.org/10.1007/978-3-319-05738-5_14]

[232]   Tan L, He M, Song L, Fu X, Shi S. Aerobic decolorization, degradation and detoxification of azo dyes by a newly isolated salt-tolerant yeast *Scheffersomyces spartinae* TLHS-SF1. Bioresour Technol 2016; 203: 287-94.
[http://dx.doi.org/10.1016/j.biortech.2015.12.058] [PMID: 26744802]

[233]   Delane EI, Evangelista-Barreto NS, Cazetta ML. Levedura *Pichia kudriavzevii* SD5 como biocatalizador na descoloração do corante Preto Reativo 5. Eng Sanit Ambient 2020; 25(2): 361-9.
[http://dx.doi.org/10.1590/s1413-41522020182911]

[234]   Saravanan P, Kumaran S, Bharathi S, Sivakumar P, Sivakumar P, Pugazhvendan SR, *et al.* Bioremediation of synthetic textile dyes using live yeast *Pichia pastoris.* Environ Technol Innovation 2021; 22: 101442.
[http://dx.doi.org/10.1016/j.eti.2021.101442]

[235]   Eze FN, Nwabor OF. Valorization of *Pichia* spent medium *via* one-pot synthesis of biocompatible silver nanoparticles with potent antioxidant, antimicrobial, tyrosinase inhibitory and reusable catalytic

activities. Mater Sci Eng C 2020; 115: 111104.
[http://dx.doi.org/10.1016/j.msec.2020.111104] [PMID: 32600707]

[236] Ramos DMB, Silva CF, Batista LR, Schwan RF. Inibição *in vitro* de fungos toxigênicos por *Pichia sp.* e *Debaryomyces sp.* isoladas de frutos de café (Coffea arabica). Acta Sci Agron 2010; 32(3): 397-02.

[237] El-Ghaouth A, Smilanick JL, Wisniewski M, Wilson CL. Improved control of apple and citrus fruit decay with a combination of *Candida saitoana* and 2- deoxy-D-glucose. Plant Dis 2000; 84(3): 249-53.
[http://dx.doi.org/10.1094/PDIS.2000.84.3.249] [PMID: 30841237]

[238] Luo Y, Zeng KF, Ming J. Control of blue and green mold decay of citrus fruit by *Pichia membranefaciens* and induction of defense responses. Sci Hortic (Amsterdam) 2012; 135: 120-7.
[http://dx.doi.org/10.1016/j.scienta.2011.11.031]

[239] Luo Y, Zhou YH, Zeng KF. Effect of *Pichia membranaefaciens* on ROS metabolism and postharvest disease control in citrus fruit. Crop Prot 2013; 53: 96-102.
[http://dx.doi.org/10.1016/j.cropro.2013.06.015]

[240] Zhang X, Zong Y, Li Z, *et al.* Postharvest *Pichia guilliermondii* treatment promotes wound healing of apple fruits. Postharvest Biol Technol 2020; 167: 111228.
[http://dx.doi.org/10.1016/j.postharvbio.2020.111228]

[241] Zhao Y, Tu K, Shao XF, Jing W, Su ZP. Effects of the yeast *Pichia guilliermondii* against *Rhizopus nigricans* on tomato fruit. Postharvest Biol Technol 2008; 49: 113-20.
[http://dx.doi.org/10.1016/j.postharvbio.2008.01.001]

[242] Jijakli MH, Lepoivre P. Characterization of an exo-b-1, 3-glucanase produced by *Pichia anomala* strain K, antagonist of *Botrytis cinerea* on apples. Phytopathology 1998; 88(4): 335-43.
[http://dx.doi.org/10.1094/PHYTO.1998.88.4.335] [PMID: 18944957]

[243] Lassois L, de Bellaire L, Jijakli MH. Biological control of crown rot of bananas with *Pichia anomala* strain K and *Candida oleophila* strain O. Biol Control 2008; 45(3): 410-8.
[http://dx.doi.org/10.1016/j.biocontrol.2008.01.013]

[244] Lima G, Ippolito A, Nigro F, Salerno M. Effectiveness of *Aureobasidium pullulans* and *Candida oleophila* against postharvest strawberry rots. Postharvest Biol Technol 1997; 10(2): 169-78.
[http://dx.doi.org/10.1016/S0925-5214(96)01302-6]

[245] Sharma RR, Singh D, Singh R. Biological control of postharvest diseases of fruits and vegetables by microbial antagonists: A review. Biol Control 2009; 50(3): 205-21.
[http://dx.doi.org/10.1016/j.biocontrol.2009.05.001]

[246] Arras G, de-Cicco V, Arru S, Lima G. Biocontrol by yeasts of blue mold of citrus fruits and the mode of action of an isolate of *Pichia guilliermondii*. J Hortic Sci Biotechnol 1998; 73: 413-8.
[http://dx.doi.org/10.1080/14620316.1998.11510993]

[247] Janisiewicz WJ, Korsten L. Biological control of postharvest diseases of fruits. Annu Rev Phytopathol 2002; 40: 411-41.
[http://dx.doi.org/10.1146/annurev.phyto.40.120401.130158] [PMID: 12147766]

[248] Massart S, Jijakli HM. Use of molecular techniques to elucidate the mechanisms of action of fungal biocontrol agents: a review. J Microbiol Methods 2007; 69(2): 229-41.
[http://dx.doi.org/10.1016/j.mimet.2006.09.010] [PMID: 17084929]

[249] Yu T, Zhang H, Li X, Zheng X. Biocontrol of *Botrytis cinerea* in apple fruit by *Cryptococcus laurentii* and indole-3-acetic acid. Biol Control 2008; 46(2): 171-7.
[http://dx.doi.org/10.1016/j.biocontrol.2008.04.008]

[250] Srinivasan R, Chaitanyakumar A, Subramanian P, *et al.* Recombinant engineered phage-derived enzybiotic in *Pichia pastoris* X-33 as whole cell biocatalyst for effective biocontrol of *Vibrio parahaemolyticus* in aquaculture. Int J Biol Macromol 2020; 154: 1576-85.
[http://dx.doi.org/10.1016/j.ijbiomac.2019.11.042] [PMID: 31715237]

[251] Mukherjee PK, Chandra J, Retuerto M, *et al.* Oral mycobiome analysis of HIV-infected patients: identification of *Pichia* as an antagonist of opportunistic fungi. PLoS Pathog 2014; 10(3): e1003996.
[http://dx.doi.org/10.1371/journal.ppat.1003996] [PMID: 24626467]

[252] Kurtzman CP. Pichia EC: Hansen (1904) The Yeasts. Elsevier 2011; Vol. 2: pp. 685-707.
[http://dx.doi.org/10.1016/B978-0-444-52149-1.00057-4]

[253] Hansen EC. Grundlinien zur Systematik der Saccharomyceten. Zentralbl Bakteriol Parasitenk 1904; 2: 529-38.

[254] Kurtzman CP, Robnett CJ. Identification and phylogeny of ascomycetous yeasts from analysis of nuclear large subunit (26S) ribosomal DNA partial sequences. Antonie van Leeuwenhoek 1998; 73(4): 331-71.
[http://dx.doi.org/10.1023/A:1001761008817] [PMID: 9850420]

[255] Kurtzman CP, Robnett CJ, Basehoar-Powers E. Phylogenetic relationships among species of Pichia, Issatchenkia and Williopsis determined from multigene sequence analysis, and the proposal of Barnettozyma gen. nov., Lindnera gen. nov. and Wickerhamomyces gen. nov. FEMS Yeast Res 2008; 8(6): 939-54.
[http://dx.doi.org/10.1111/j.1567-1364.2008.00419.x] [PMID: 18671746]

[256] Casaregola S, Weiss S, Morel G. New perspectives in hemiascomycetous yeast taxonomy. C R Biol 2011; 334(8-9): 590-8.
[http://dx.doi.org/10.1016/j.crvi.2011.05.006] [PMID: 21819939]

[257] Kurtzman CP, Robnett CJ. Systematics of methanol assimilating yeasts and neighboring taxa from multigene sequence analysis and the proposal of Peterozyma gen. nov., a new member of the Saccharomycetales. FEMS Yeast Res 2010; 10(3): 353-61.
[http://dx.doi.org/10.1111/j.1567-1364.2010.00625.x] [PMID: 20522116]

[258] Kurtzman CP. Synonomy of the yeast genera Hansenula and Pichia demonstrated through comparisons of deoxyribonucleic acid relatedness. Antonie van Leeuwenhoek 1984; 50(3): 209-17.
[http://dx.doi.org/10.1007/BF02342132] [PMID: 6486768]

[259] Kurtzman CP. Wickerhamomyces Kurtzman, Robnett & Basehoar-Powers (2008) The Yeasts. Elsevier 2011; Vol. 2: pp. 899-917.

[260] Kurtzman CP. Phylogeny of the ascomycetous yeasts and the renaming of Pichia anomala to Wickerhamomyces anomalus. Antonie van Leeuwenhoek 2011; 99(1): 13-23.
[http://dx.doi.org/10.1007/s10482-010-9505-6] [PMID: 20838888]

[261] Kurtzman CP, Robnett CJ. Multigene phylogenetic analysis of the Trichomonascus, Wickerhamiella and Zygoascus yeast clades, and the proposal of Sugiyamaella gen. nov. and 14 new species combinations. FEMS Yeast Res 2007; 7(1): 141-51.
[http://dx.doi.org/10.1111/j.1567-1364.2006.00157.x] [PMID: 17311592]

[262] Yamada Y, Maeda K, Mikata K. The phylogenetic relationships of the hat-shaped ascospore-forming, nitrate-assimilating Pichia species, formerly classified in the genus Hansenula Sydow et Sydow, based on the partial sequences of 18S and 26S ribosomal RNAs (Saccharomycetaceae): the proposals of three new genera, Ogataea, Kuraishia, and Nakazawaea. Biosci Biotechnol Biochem 1994; 58(7): 1245-57.
[http://dx.doi.org/10.1271/bbb.58.1245] [PMID: 7765249]

[263] Suh S-O, Zhou JJ. Methylotrophic yeasts near Ogataea (Hansenula) polymorpha: a proposal of *Ogataea angusta* comb. nov. and *Candida parapolymorpha* sp. nov. FEMS Yeast Res 2010; 10(5): 631-8. [no-no.].
[http://dx.doi.org/10.1111/j.1567-1364.2010.00634.x] [PMID: 20491937]

[264] Yamada Y, Kawasaki H, Nagatsuka Y, Mikata K, Seki T. The phylogeny of the cactophilic yeasts based on the 18S ribosomal RNA gene sequences: the proposals of Phaffomyces antillensis and Starmera caribaea, new combinations. Biosci Biotechnol Biochem 1999; 63(5): 827-32.

[http://dx.doi.org/10.1271/bbb.63.827] [PMID: 10380625]

[265] Nagatsuka Y, Saito S, Sugiyama J. Ogataea neopini sp. nov. and O. corticis sp. nov., with the emendation of the ascomycete yeast genus Ogataea, and transfer of Pichia zsoltii, P. dorogensis, and P. trehaloabstinens to it. J Gen Appl Microbiol 2008; 54(6): 353-65.
[http://dx.doi.org/10.2323/jgam.54.353] [PMID: 19164878]

[266] Kurtzman CP, Suzuki M. Phylogenetic analysis of ascomycete yeasts that form coenzyme Q-9 and the proposal of the new genera Babjeviella, Meyerozyma, Millerozyma, Priceomyces, and Scheffersomyces. Mycoscience 2010; 51: 2-14.
[http://dx.doi.org/10.1007/S10267-009-0011-5]

[267] Mallet S, Weiss S, Jacques N, Leh-Louis V, Sacerdot C, Casaregola S. Insights into the life cycle of yeasts from the CTG clade revealed by the analysis of the Millerozyma (Pichia) farinosa species complex. PLoS One 2012; 7(5): e35842.
[http://dx.doi.org/10.1371/journal.pone.0035842] [PMID: 22574125]

[268] Tammawong S, Ninomiya S, Kawasaki H, Boonchird C, Sumpradit T. Millerozyma phetchabunensis sp. nov., a novel ascomycetous yeast species isolated from Nam Nao forest soil in Thailand, and the transfer of Pichia koratensis to the genus Millerozyma. J Gen Appl Microbiol 2010; 56(1): 37-42.
[http://dx.doi.org/10.2323/jgam.56.37] [PMID: 20339218]

[269] Yamada Y, Matsuda M, Maeda K, Mikata K. The phylogenetic relationships of methanol-assimilating yeasts based on the partial sequences of 18S and 26S ribosomal RNAs: the proposal of Komagataella gen. nov. (Saccharomycetaceae). Biosci Biotechnol Biochem 1995; 59(3): 439-44.
[http://dx.doi.org/10.1271/bbb.59.439] [PMID: 7766181]

[270] Kurtzman CP. Description of Komagataella phaffii sp. nov. and the transfer of Pichia pseudopastoris to the methylotrophic yeast genus Komagataella. Int J Syst Evol Microbiol 2005; 55(Pt 2): 973-6.
[http://dx.doi.org/10.1099/ijs.0.63491-0] [PMID: 15774694]

[271] Limtong S, Srisuk N, Yongmanitchai W, Yurimoto H, Nakase T. Ogataea chonburiensis sp. nov. and Ogataea nakhonphanomensis sp. nov., thermotolerant, methylotrophic yeast species isolated in Thailand, and transfer of Pichia siamensis and Pichia thermomethanolica to the genus Ogataea. Int J Syst Evol Microbiol 2008; 58(Pt 1): 302-7.
[http://dx.doi.org/10.1099/ijs.0.65380-0] [PMID: 18175726]

[272] Kurtzman CP. Scheffersomyces Kurtzman & M. Suzuki (2010). Yeasts 2011; 2: 773-7.
[http://dx.doi.org/10.1016/B978-0-444-52149-1.00065-3]

CHAPTER 7

Brettanomyces: Diversity and Potential Applications in Industrial Fermentation

Manoela Martins[1], **Maria Paula Jiménez Castro**[1], **Marcus Bruno Soares Forte**[1] and **Rosana Goldbeck**[1,*]

[1] *Bioprocess and Metabolic Engineering Laboratory (LEMEB), Department of Food Engineering, Faculty of Food Engineering, University of Campinas (UNICAMP), Campinas - SP, 13083-970, Brazil*

Abstract: Although mainly known for their role in wine spoilage, *Brettanomyces* yeasts have been increasingly recognized as having beneficial effects on fermented beverages. These microorganisms can, for instance, increase flavor complexity, a property that can be controlled by understanding the physiological, genetic, and biochemical traits of *Brettanomyces* species in fermentation processes. Moreover, their genetic diversity, exceptional stress and low-pH tolerance, and peculiar metabolism suggest great potential for bioethanol production. This chapter summarizes the most notable features of *Brettanomyces*, briefly highlights recent insights into their genetic characteristics, and discusses potential applications in industrial fermentation processes, such as for the production of specialty beers, wines, and bioethanol.

Keywords: Acetic acid, Aroma, Beer, Bioethanol, Bioprocess, Crabtree effect, Custers effect, Fermentation, Oak barrel, Off-flavor, Spoilage yeast, Volatile phenol, Wine.

INTRODUCTION

Usually considered no more than a spoilage yeast, *Brettanomyces* isolated from ale beer was in fact, the first microorganism to be patented in history. Unique flavors produced by the fungus have become associated with British beers, hence the genus name *Brettanomyces*, derived from the Greek words brettano (British) and myces (fungus). Since the first description in 1904 [1], *Brettanomyces* species have been isolated in wineries and breweries all over the world [2]. The taxonomy of the genus has gone through several reclassifications over the years. In 1940, Custers [3] performed the first systematic study of *Brettanomyces* yeasts. Initially, the classification was based solely on asexually reproducing (anamorphic)

---

* **Corresponding author Rosana Goldbeck:** Bioprocess and Metabolic Engineering Laboratory (LEMEB), Department of Food Engineering, Faculty of Food Engineering, University of Campinas (UNICAMP), Campinas - SP, 13083-970, Brazil; Tel: +55 19 981040716; E-mails: goldbeck@unicamp.br

Sérgio Luiz Alves Júnior, Helen Treichel, Thiago Olitta Basso and Boris Ugarte Stambuk (Eds.)
All rights reserved-© 2022 Bentham Science Publishers

variants. In 1960, the genus *Dekkera* was proposed as a teleomorphic (sexually reproducing) counterpart of *Brettanomyces* after the observation of ascospores in some strains. More recent molecular DNA techniques revealed no differences between anamorphic and teleomorphic forms, and currently, there is no separation between these groups. Both terms (*Brettanomyces* and *Dekkera*) are commonly used in wine research, but the term *Brettanomyces* is preferred in industrial settings [4 - 6].

Molecular analysis of the genus identified five species, the anamorphs *B. bruxellensis*, *B. anomalus*, *B. custersianus*, *B. naardenensis*, and *B. nanus*, the first two of which also occur as teleomorphs, known as *Dekkera bruxellensis* and *Dekkera anomala* [5]. The species primarily associated with winemaking is *B. bruxellensis* (or *D. bruxellensis*), although recent wine-related investigations often include *D. anomala* along with *D. bruxellensis*, as current methods have had difficulty in differentiating between these two species [6].

Brettanomyces is a controversial yeast that has gained increasing attention in recent years because of its association with wine spoilage and the production of ethyl phenols. The yeast grows slowly; therefore, it usually imparts intense flavors (volatile phenols) in aged beverages, the so-called Brett flavors and odors, described as strong, smoky, or aromatic [7, 8]. Such descriptors may be either negative or positive depending on compound concentration and consumer expectation [9, 10]. This chapter presents a summary of the major phenotypic characteristics, growth patterns, and roles of *Brettanomyces* in fermentation processes, encompassing from beverage off-flavors to future perspectives in the production of spontaneously fermented beer, wine, and bioethanol.

DIVERSITY OF *BRETTANOMYCES* HABITATS

Brettanomyces is ubiquitous in nature. The yeast can be isolated from fermented food products, particularly during post-fermentation processing and aging of alcoholic beverages such as wine, beer, and cider, and is scarcely found outside these environments [6]. *Brettanomyces* niches comprise spontaneous alcoholic fermentation media with high ethanol concentrations, low pH, absence of readily fermentable nitrogen and carbon sources, and low oxygen [11]. Once alcoholic fermentation is completed, the remaining traces of residual sugars are sufficient for the proliferation of this slow-growing yeast [4]. *Brettanomyces* shows a high preference for fermented media, but it usually occurs at low concentrations and is, therefore, not considered a contaminant. Contamination only occurs when other microorganisms have been inhibited, evidence of the yeast's exceptional resistance to low-nutrient conditions, which allows it to adapt to harsh environments and outcompete other microorganisms [6, 11]. Malolactic fermentation and aging in used oak barrels are recognized as the most critical stages of wine production for *Brettanomyces* contamination. Low concentrations of free sulfur dioxide (SO_2) and residual sugars, yeast autolysis with the nutrient release, presence of cellobiose (the main disaccharide in wood), and difficulty in sanitizing used barrels are factors that favor *Brettanomyces* growth and wine contamination [4].

Microbial contamination is an inevitable, undesired, complex event. Knowledge of yeast growth requirements and awareness of natural occurrence in raw ingredients may facilitate the identification of unwanted spoilage microorganisms. *Brettanomyces* is the most monitored yeast, particularly in winemaking [12]. The prevalence of *Brettanomyces* on grape skins is remarkably low, as previously assessed with the aid of an enrichment medium [13]. *Brettanomyces* has not yet been detected in the air during the first stages of harvesting [12], but it was identified in air samples of crush pads, tanks, barrels, and bottling rooms of a winery [14, 15]. Dweck *et al.* [16] observed that flies, known to be vectors of *Brettanomyces,* react to the yeast's smell. As expected, it was not difficult to detect the yeast in washing water and winery equipment used at advanced stages of vinification, oak aging, and bottle aging [12, 17, 18]. Differences in yeast population levels throughout winemaking can be attributed to production stage, time of year, degree of cleanliness, and cleaning protocols [12, 17]. Barrel sanitation is a very difficult task, and the use of sanitizing agents, such as SO_2, may contribute to the development of tolerant strains, as will be discussed in the following paragraphs [4]. Chemicals, ozone, biofilms, and sonication are alternative methods recommended for barrel and equipment sanitation to decrease *Brettanomyces* populations and volatile phenol production, but the effectiveness of such methods is debatable [4, 19]. Fig. (1) shows some factors associated with the presence of *Brettanomyces* during winemaking.

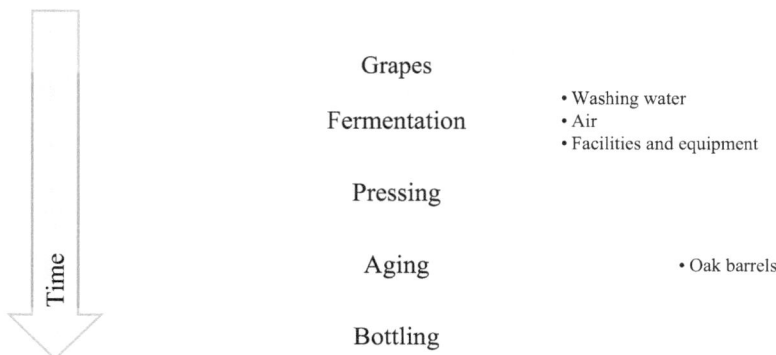

Fig. (1). Factors promoting the growth of Brettanomyces populations during winemaking.

Given the slow growth and low occurrence of *Brettanomyces*, isolation is difficult and requires long incubation times (up to 2 weeks). The incubation periods used for other yeasts (3 to 6 days) are inadequate for routine microbiological screenings during winemaking. Several possibilities have been investigated to detect *Brettanomyces* spp., including the use of selective media or DNA-based techniques. Rodrigues *et al.* [20] were the first to develop a selective or differential medium specifically for *Brettanomyces* spp. isolation. The medium was named DBDM (*Dekkera/Brettanomyces* differential medium) and was shown to recover less than 1% of the target yeasts from the total microbial population. DBDM contains yeast nitrogen base, ethanol and cycloheximide as antimicrobial agents, bromocresol green as an indicator of medium acidification, and *p*-coumaric acid as substrate, whose degradation results in a distinct phenolic off-odor indicative of *Brettanomyces* activity. To overcome the lack of optimal isolation media and poor detection limits, Renouf and Lonvaud-Funel [21] developed a new enrichment medium that successfully detected *B. bruxellensis*, even on the surface of grape berries.

A more rapid and reliable identification method emerged with the development of molecular DNA-based techniques involving the use of ribosomal RNA probes. This indirect method enables the visualization of fluorescent cells after hybridization [22] and polymerase chain reaction (PCR) [23]. One of the concerns about direct PCR methods is that sensitivity depends on the level of contamination [6], and only a high detection limit ($\geq 10^4$ CFU mL^{-1}) may provide a positive result [24]. Several studies have since been performed to identify different strains using amplified fragment length polymorphism [25], mitochondrial DNA restriction analysis and randomly amplified polymorphic DNA PCR with OPA-primers [26], and restriction enzyme analysis [27]. The relevance of genetic strain characterization in the wine industry lies in the connection between different strains, their geographic origin, their stress tolerance, and their effects on the phenolic profile of wine [8, 10].

With affordable sequencing technologies, it is possible to explore the phenotypic variability of industrially relevant species, providing a better view of the evolutionary history of genetic variations underlying the phenotypic diversity of *Brettanomyces* species. Gounot *et al.* [28] performed the first deep investigation of genetic variants of a large population and reported the presence of at least two hybridization events. Such events are one of the main factors involved in the generation of subpopulations within a species and are probably a driving mechanism of *Brettanomyces* adaptation to harsh environments. Furthermore, significant genomic differences can also be found within subpopulations, suggesting the presence of genomic adaptations specific to each subpopulation. Several aneuploidies, segmental duplications, and copy-number variants were also found in the whole population, particularly in triploid hybrids, indicating that these genome dynamics favor structural variations. Similar observations have been reported for *Saccharomyces pastorianus*, an interspecies hybrid of *Saccharomyces cerevisiae* and *Saccharomyces eubayanus*, whose subpopulations show extensive chromosome and heterozygosity loss events [29]. An example is seen in the nitrate assimilation gene: such mutations may result in identical alleles (haplotyped sequences) and inability to utilize nitrate as a nitrogen source [30].

A study compared genome sequences from *Brettanomyces* strains of the five recognized species isolated from different food-related sources and geographic areas to those of beer and wine strains using LSU rRNA gene sequences and a number of established DNA fingerprinting techniques [30]. The results indicated a stronger correlation of genetic profiles with isolation sources than with geographic origin or year of isolation, suggesting niche adaptation, which is uncommon for *Saccharomyces* species, for which geography seems more important than ecology in shaping population structure [31]. The association between genotype and niche was supported by evidence that strains isolated from similar niches in different locations clustered together, whereas isolates obtained from the same geographic region but different niches did not (Fig. **2**). In agreement with these findings, Avramova *et al.* [32] observed that an allopolyploid strain with high SO_2 tolerance was specifically adapted to environments with high SO_2 concentrations compared with other *B. bruxellensis* wine strains, indicating a potential correlation between allotriploidization origin and environmental adaptation.

A recent study confirmed the assumption that *B. bruxellensis* populations are composed of strains with different ploidy levels. The polyploid state has a high fitness cost for eukaryotic cells given the difficulty in maintaining an imbalanced number of chromosomes during cell division and enlargement as well as nucleus effects. Therefore, the authors presumed that a stable polyploid or aneuploid state is only maintained if it confers a survival advantage to cells under particular conditions. Aneuploidy and polyploidy contribute to genome plasticity and confer selective and fitness advantages under extreme conditions [9]. One example is the development of a triploid state for SO_2 tolerance, pushing *Brettanomyces* wine-related populations toward this characteristic. SO_2 is an antimicrobial agent widely used in winemaking, exerting constant selective pressure on *B. bruxellensis* wine populations, thereby leading to the establishment of more tolerant genotypes that are closely related to human activity [32]. High levels of SO_2 cause various deleterious effects on cells, such as membrane damage, increased permeability, and disruption of the synthesis of binding molecules (*e.g.*, acetaldehyde) that remove free SO_2 from the extracellular environment and reduce molecular stress on cells [33]. As noted by Capozzi *et al.* [2], this pressure for

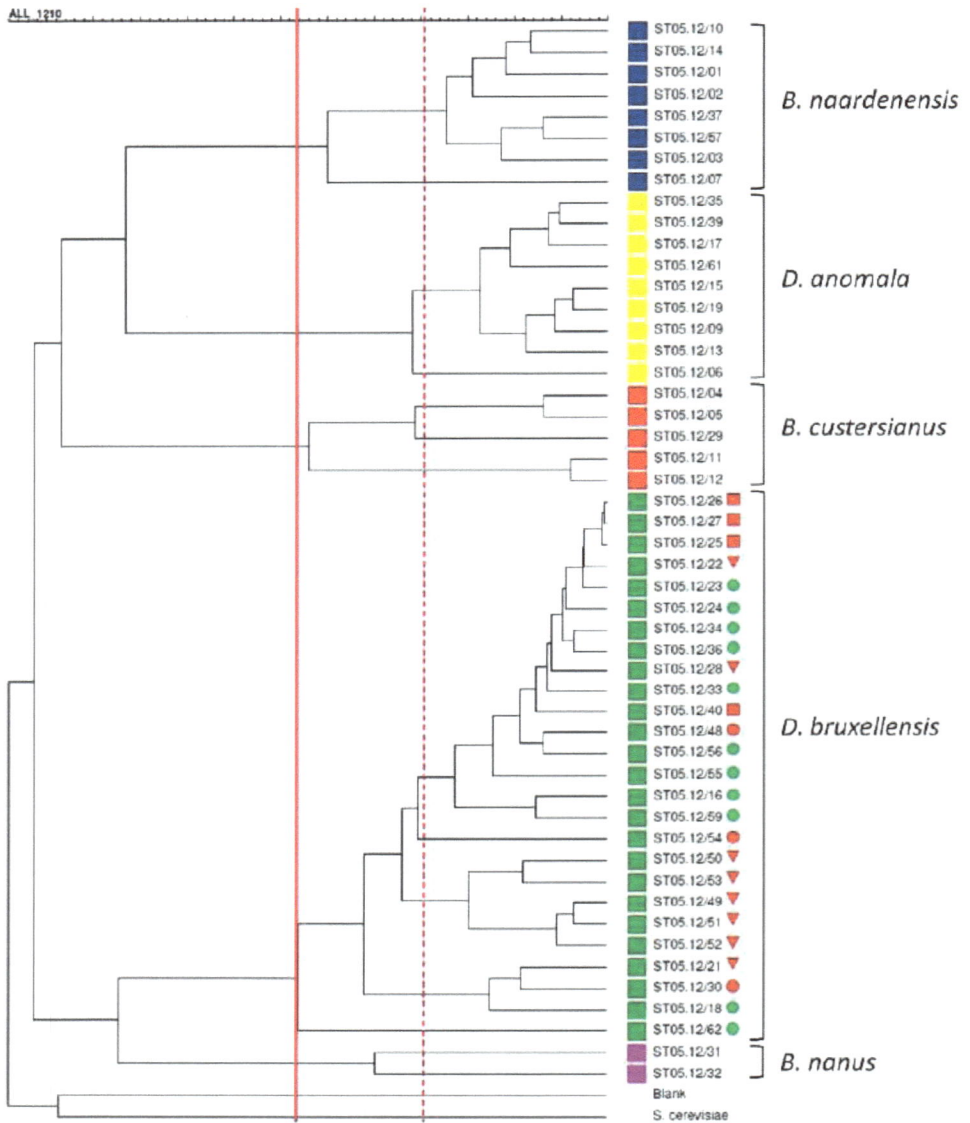

Fig. (2). Dendrogram of combined fingerprinting data sets for all Brettanomyces strains from the study of Crauwels *et al.* [30]. Thirteen clusters were identified (marked by the dotted red line); B. bruxellensis subclusters generally represent strains from a similar environment. Sterile distilled water was used as blank (negative control). B. bruxellensis strains marked with a circle were shown to carry the complete nitrate assimilation gene cluster, consisting of genes encoding nitrate reductase, nitrite reductase, and nitrate transporter. Isolates marked with a square lost the genes encoding nitrate reductase and nitrite reductase. Isolates marked with a triangle lost the complete nitrate assimilation gene cluster. B. bruxellensis strains that were able or unable to utilize nitrate as nitrogen source are indicated with a green or red mark, respectively. Isolates ST05.12/30 and ST05.12/54 were negative for ammonium and nitrate.

natural adaptation is likely the cause of genomic diversity and, consequently, phenotypic variability between *Brettanomyces* strains. Such characteristics make it difficult to predict spoilage potential and are a major concern for winemaking industries [9]. Chromosomal translocation and ability to restore functional permeability constitute adaptation mechanisms in *B. bruxellensis,* reflecting in a longer lag phase for more tolerant strains [34]. Tolerant triploid strains were able to grow in sweet wine containing 0.6 mg L^{-1} SO_2 [35]. It is also hypothesized that the capacity to metabolize part of the added SO_2 through sulfur amino acid biosynthesis could be a resistance mechanism for the species [33, 35]. Most SO_2-tolerant strains belong to specific genetic groups that can be detected by microsatellite genotyping; although this technique is expensive and time-consuming for routine analysis, it may provide relevant information to guide microbial control strategies and prevent spoilage [34]. An alternative analysis based on PCR and classical gel electrophoresis has been patented for the identification of *B. bruxellensis* species and SO_2 tolerance [36].

One of the most important phenotypes that determine the habitat of *Brettanomyces* is associated with the Crabtree effect [37], seen in strains with the ability to ferment sugars to ethanol even under aerobic conditions and high sugar concentrations. These yeasts first produce and accumulate ethanol and acetic acid to prevent the growth of competing microbes, withstanding the resulting low-pH environment. After depletion of glucose, yeast cells are able to metabolize ethanol and acetate. Additionally, in a nutrient-poor aerobic or semi-aerobic environment, cells can use ethanol or acetic acid as the sole carbon source, causing a strong redox imbalance with excessive production of NADH [11, 38]. Under stress conditions, *B. bruxellensis* may store compounds such as organic acids and sugars (arabinose, fructose, glucose). These storage reserves seem to contribute to the production of ethanol and biomass after the exponential cell growth phase, being readily consumed as soon as other nutrients become unavailable, evidence of the ability of *B. bruxellensis* to use limited nutrients in a stressful environment [11, 39]. However, under anaerobic conditions, *B. bruxellensis* strains are not able to metabolize ethanol, acetic acid, or even glycerol, and therefore show reduced growth [39].

Another interesting characteristic is the efficient degradation of cellobiose observed in some strains that cannot utilize maltose as substrate. Low maltose assimilation, combined with the production of off-flavors such as H_2S, prevent *B. custersianus* and *B. naardenensis* from being applied in brewing, but these species hold potential for the production of low-alcohol beers and as secondary fermentation strains. Moreover, strains with high β-glucosidase activity display a highly flocculent phenotype, which is desired to facilitate downstream processing steps [2]. β-Glucosidases are able to breakdown cellobiose, explaining the long survival of β-glucosidase-producing species in wooden casks. β-Glucosidase activity also has a major impact on beer flavors. During beer fermentation, β-glucosidases act on hops to release previously odorless and nonvolatile glucosides from sugar molecules. These enzymes can result in a significant increase in several volatile compounds in beer, including terpenes such as linalool [38, 40]. Yeasts can further convert monoterpenes into β-citronellol or α-terpineol, enhancing the floral citrusy flavor of beer. *Brettanomyces* strains with high β-glucosidase activity have been investigated for their potential as bioflavoring agents [41] and for the production of resveratrol, an important antioxidant and antiaging compound [42].

Brettanomyces species have developed a strategy to survive in nitrogen-poor environments. Even though glutamine is their preferred nitrogen source [43], *Brettanomyces* can assimilate nitrogen from other compounds, such as nitrate. This adaptation increases fermentation efficiency by bypassing the Custers effect, which inhibits fermentation under anaerobic conditions. Nitrate assimilation allows cells to replenish the NAD(P)H pool through reduction of nitrate to ammonium. However, the nitrate assimilation gene cluster (nitrate reductase, nitrite reductase, and nitrate transporter genes) is not present in all *Brettanomyces* strains [2, 38].

BRETTANOMYCES YEASTS IN BIOPROCESSES

Brettanomyces species are mainly found in wine and beer production processes. These microorganisms have been regarded by winemakers as undesirable agents that produce unpleasant flavors and aromas. In contrast to these views, researchers have described various positive effects for these microorganisms, attracting the attention of academic and industrial communities to the possibility of increasing the efficiency of fermentation processes and improving the organoleptic profile of final products [44]. Currently, the most discussed topics related to *Brettanomyces* yeasts are beverage spoilage and off-flavor production, techniques to eliminate or reduce *Brettanomyces* populations, optimal growth and fermentation conditions, and production of bioethanol and special beverages.

Spoilage and Off-Flavors

Phenolic off-flavor production is one of the major concerns related to *Brettanomyces* yeasts in winemaking, being associated with large economic losses. Metabolic products responsible for wine spoilage include mainly volatile phenols, especially vinyl- and ethylphenols [8], but also acetic acid and tetrahydropyridines. *Brettanomyces* produces volatile phenols through enzymatic decarboxylation of hydroxycinnamic acids by phenolic acid decarboxylase, forming hydroxystyrenes (step 1). *p*-Coumaric acid is converted to 4-vinylphenol, ferulic acid to 4-vinylguaiacol, and caffeic acid to 4-vinylcatechol. These compounds are reduced to ethylphenol, ethylguaiacol, and ethylcatechol, respectively, by vinylphenol reductase (step 2) [7, 10, 45] (Fig. 3). It is interesting to note that hydroxycinnamic acids, which are naturally present at high concentrations in grapes and wine, are inhibitory toward many microorganisms. *Brettanomyces*, however, is capable of overcoming this toxicity by converting hydroxycinnamic acids to volatile phenols [4]. Ethyl forms remain free and volatile, and can be detected by most wine consumers at concentrations above 400 μg L^{-1} [10].

4-Ethylphenol and 4-ethylguaiacol production is a determining factor of the influence of *B. bruxellensis* on sensory properties, particularly in wines [46]. At high concentrations, these compounds may negatively impact aroma quality and displease consumers. Sensory studies [20, 47, 48] showed that consumers perceive these aromas as "phenolic," "leathery," "horse sweat," "horse stable," or "varnish." By contrast, at low concentrations, these compounds may impart a pleasant aroma [49]. Of note, some *Brettanomyces* strains possess high amylase activity and are being investigated for their potential to produce new flavors and aromas in alcoholic beverages [50].

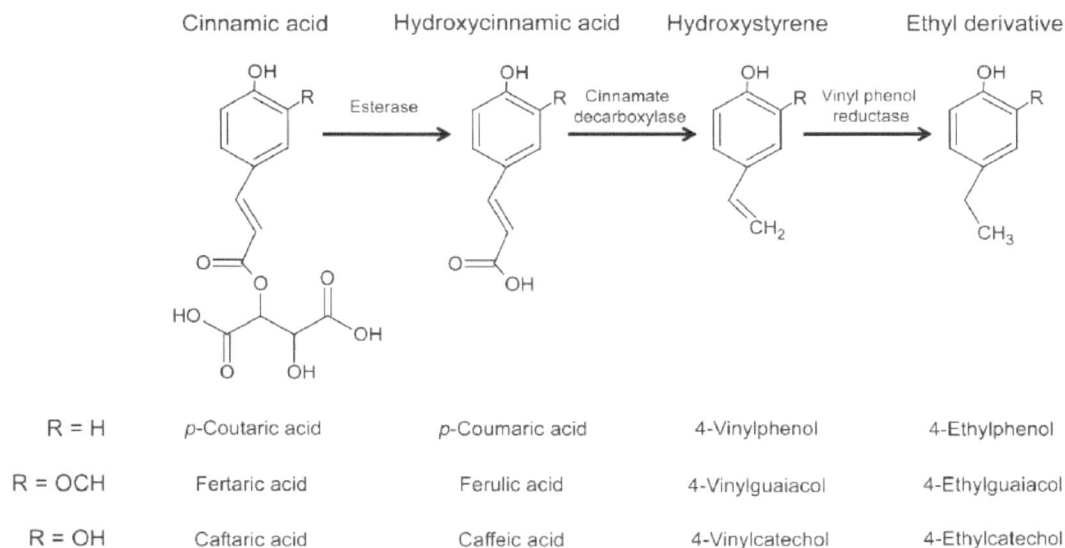

R = H	*p*-Coutaric acid	*p*-Coumaric acid	4-Vinylphenol	4-Ethylphenol
R = OCH	Fertaric acid	Ferulic acid	4-Vinylguaiacol	4-Ethylguaiacol
R = OH	Caftaric acid	Caffeic acid	4-Vinylcatechol	4-Ethylcatechol

Fig. (3). Production pathway of volatile phenols.

Cibrario *et al*. [10] reported that the production of volatile phenols is ubiquitous within the species *B. bruxellensis*, which shows high genetic diversity. Volatile phenol production is independent of yeast growth profile and not affected by growth substrate or level of constraints. The rate of volatile phenol production raises by increasing aeration but is little affected by pH decrease until 3.0 or by ethanol concentration increase up to 12% vol. A combination of constraints (for instance, pH < 3.0 and ethanol ≥ 14% vol.) may decrease but not cease volatile phenol production. Other volatile fatty acids synthesized by *Brettanomyces* that can negatively affect wine sensory quality are isovaleric acid (associated with a rancid olfactory note [51]), 2-methylbutyric acid, isobutyric acid, and acetic acid, which constitutes more than 90% of wine volatile acidity [52]. Acetic acid imparts a vinegary taste [53] and may halt the fermentation process [54].

Brettanomyces can also indirectly contribute to the natural flavors of beer through its high β-glucosidase activity. Enzymatic hydrolysis *via* β-glucosidases is the preferred industrial strategy to release aglycones, volatile aromas separated from glycones. Vervoot *et al.* [41] were the first to identify β-glucosidase-encoding genes in *B. anomalus* and *B. bruxellensis* and apply the recombinant enzyme from these yeasts, thereby avoiding off-flavor production. Although these yeast species are generally considered safe for food production, their use is often not possible because of their spoilage effects. In the referred study, beers with *B. anomalus* β-glucosidase contained more eugenol (clove/honey aroma) and benzyl alcohol (sweet/flower aroma) than untreated beers. *B. anomalus* β-glucosidase released more eugenol and less linalool (citral/flower aroma), and forest fruit milk beverages with *B. anomalus* β-glucosidase contained more methyl salicylate (peppermint/wintergreen aroma), benzyl alcohol, and linalool than untreated beverages.

Methods to Eliminate or Reduce *Brettanomyces* Populations

Over the past years, many techniques have been developed to reduce *Brettanomyces* levels in fermentation processes, thereby avoiding yeast proliferation and generation of undesirable flavors and odors in fermentation products. Table **1** lists some of the traditional methods.

Table 1. Methods traditionally used to control *Brettanomyces* populations.

Method	Description	Refs.
Sulfur dioxide	The molecule exerts antioxidant, antimicrobial, and antiseptic action	[35]
Fining agents	Fining with casein, potassium, or liquid gelatin at a dosage of 0.6 mL L^{-1}	[55]
Filtration	Effective removal of yeast cells using membranes with a pore size smaller than 0.45 μm	[56]

Emerging technologies have been proposed to reduce the use of SO_2 and overcome some of the disadvantages of traditional methods, thereby enhancing final product quality (Table **2**).

Table 2. Emerging methods to reduce the use of antimicrobial agents for *Brettanomyces* control.

Method	Description	Refs.
Polysaccharide application	Chitosan at 3–6 g L^{-1} inhibits *Brettanomyces* cells	[57]
High-pressure treatment	Spoilage yeasts such as *Brettanomyces* are controlled by subjecting wine to 100–200 MPa or 400 MPa treatment for 10 min	[58]
Biological control	Control is achieved using antimicrobial agents such as zymocins or yeast killer toxins produced by *Pichia anomala* (DBVPG 3003) and *Kluyveromyces wickerhamii* (DBVPG 6077)	[59, 60]
Ultrasonication	Ultrasonication at 400 W, 24 kHz, and 100 μm amplitude in continuous flow reduced *Brettanomyces* spp. populations by 89.1–99.7%	[61]
UV radiation	Different UV-C dosages (459 to 3672 J L^{-1}) can inactivate *Brettanomyces*	[62]
Electrolyzed water	Electrolysis of chlorinated salt solutions (400 ppm free chlorine) can reduce yeast cell viability	[63]
Pulsed electric fields	Application of micropulses of high voltage electric fields during short periods	[64]
Metal nanoparticles	Silver nanoparticles show potential to completely or partially replace SO_2 in winemaking for their antimicrobial activity	[65]

Moreover, some methods have been developed to decrease volatile phenol production (Table **3**).

Table 3. Methods to reduce the production of volatile phenols in fermented beverages [53].

Method	Description	Refs.
Reverse osmosis	Reverse osmosis system with tangential flow filtration and a hydrophobic absorbent resin, resulting in a 77% reduction of total ethylphenols	[66]
Absorption	Polyvinylpolypyrrolidone and charcoal are used to decrease ethylphenol levels	[55]
Biosorption	Active dry wine yeast (lees) can be used as biosorbent	[67]
Ozone treatment	Gaseous ozone (ca. 30 ppm) provides a high level of yeast inactivation	[63]

Growth and Fermentation

Brettanomyces occurs mainly in beer, bioethanol, and wine production, particularly in wine stored for long periods in oak barrels, whose cellobiose content serves as a sugar source and whose porous microstructure allows the passage of oxygen [53]. The yeast has gained increasing importance in fermentative processes because of its differentiated flavor profile (which may contribute to the development of new alcoholic beverages) [50], tolerance to low pH, high ethanol-producing capacity, and growth under oxygen-limited conditions. In the following sections, we will discuss important factors that influence yeast growth and development in fermentation media, such as oxygen level, nitrogen source, carbon source, temperature, and pH.

Oxygen

This parameter influences *Brettanomyces* metabolism because the yeast is sensitive to oxygen availability. *Brettanomyces* species (although not all phenotypes) are known as facultative anaerobic and Crabtree-positive organisms. As previously mentioned, these characteristics allow *Brettanomyce*s to produce ethanol and acetic acid in the presence of oxygen but only when sugar concentrations are sufficiently high [11]. For instance, *B. bruxellensis* tends to ferment sugars to ethanol to prevent the growth of other microorganisms; when glucose is depleted, it consumes ethanol [68]. On the other hand, *B. naardenensis* is unable to grow in the absence of oxygen [37].

Dijken *et al.* [69] reported that *B. bruxellensis* produced ethanol in the presence of oxygen and detected pyruvate decarboxylase and alcohol dehydrogenase activities under both aerobic and oxygen-limited conditions, as expected for a Crabtree-positive yeast. This phenomenon can be attributed to the higher ATP production rate in fermentation than in respiration, although the yield of ATP is at least one order of magnitude higher by respiration, explaining why yeasts prefer respiration under aerobic low-sugar conditions [11].

On the other hand, when sugar concentrations are low, some species, such as *B. bruxellensis*, block sugar fermentation to ethanol under anaerobic conditions [70]. This phenomenon was, at first, referred to as the negative Pasteur effect [3]; nowadays, it is called the Custers effect. As shown in Fig. (**4a**), the redox balance in *B. bruxellensis* is responsible for the Custers effect and explains why this yeast is not able to restore or maintain its internal redox balance in the absence of oxygen. The species is capable of producing acetic acid under aerobic conditions *via* NAD^+ aldehyde dehydrogenase, affording NADH. When the environment changes from aerobic to anaerobic, the lack of NAD^+ generated by acetic acid production induces a blockage of glycolysis, and, consequently, glycerol is not produced [11]. Such an inability to produce glycerol decreases yeast growth rate and generates a lag phase, during which cells transition from aerobic to anaerobic metabolism, as shown in (Fig. **4b**).

A

B

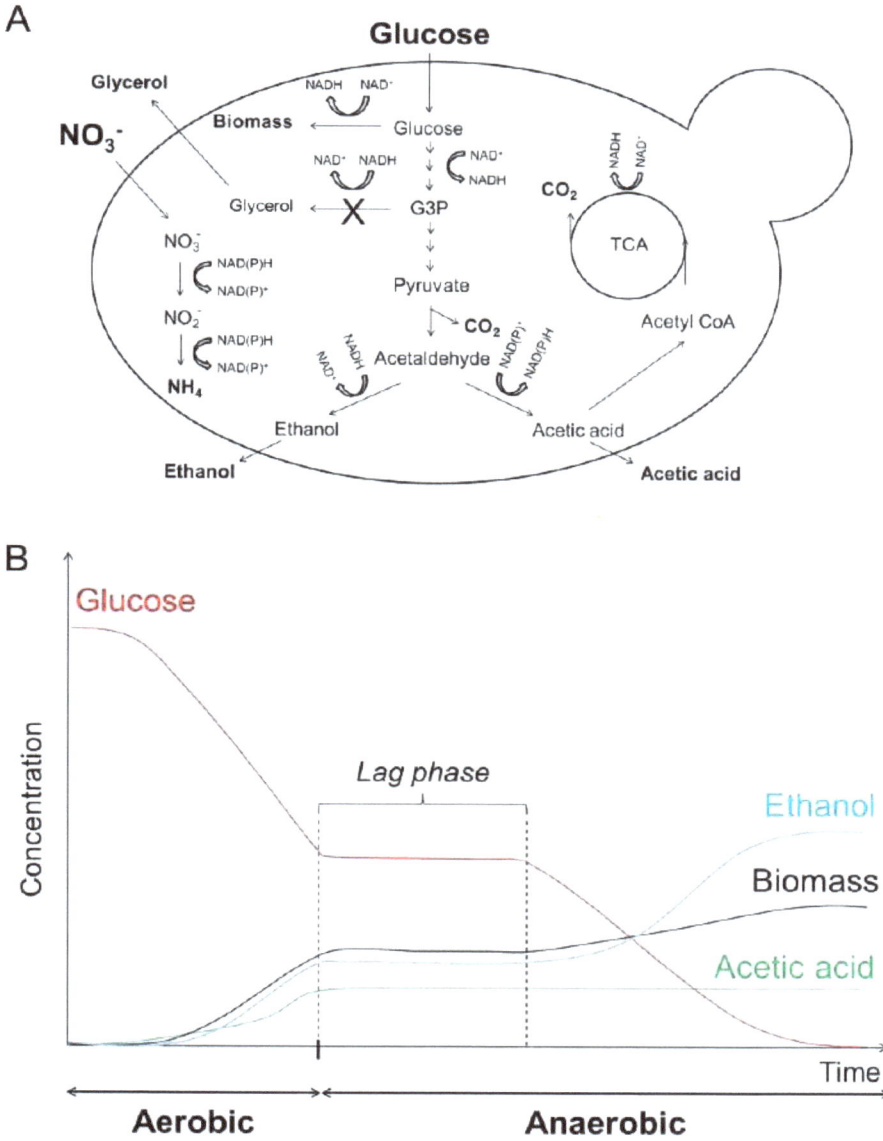

Fig. (4). Custers effect in *Brettanomyces.***(A)** Schematic overview of redox balance, the main factor responsible for the Custers effect. Glucose is fermented to ethanol more rapidly under aerobic than anaerobic conditions. **(B)** Effect of the transition from aerobic to anaerobic culture conditions on growth kinetics. The lag phase is caused by the blockage of glycolysis resulting from a lack of NAD$^+$ [11].

Nitrogen

Brettanomyces can use different nutrients, not depending on those that are limiting to fermentation. *B. bruxellensis* GDB 248 was shown to have a higher growth rate in ammonia/nitrate media than *S. cerevisiae* JP1 [71]. It is important to note that *B. bruxellensis* can use nitrate as the sole carbon source, unlike other ethanol-producing strains, such as *S. cerevisiae* [72]. Because nitrate is not generally used as a nutrient, *B. bruxellensis* may have an important advantage in terms of ethanol production, given that the most common raw material (sugarcane) can contain significant nitrate levels, derived from soil bacteria metabolism and/or field fertilizers [71].

Carbon Source

One of the most striking characteristics of *B. bruxellensis* is its ability to degrade complex sugars, differently from *Saccharomyces* spp. For instance, the species can degrade cellobiose, a disaccharide formed by the hydrolysis of lignocellulosic materials in wood. Cellobiose, found in oak barrels used for wine production, is hydrolyzed by β-glucosidase. Many *Brettanomyces* strains contain this enzyme [15]. Dextrins such as maltotetraose and maltopentaose are residual sugars produced during beer fermentation. These compounds are also hydrolyzed by *Brettanomyces* through the action of β-glucosidase, generating over-fermented beer [73].

Other carbon sources used by some *Brettanomyces* strains are pentoses (*e.g.*, xylose) and hexoses (*e.g.*, glucose) [74]. Galafassi *et al.* [75] tested the effects of different carbon sources, including pentoses (D-xylose, xylitol, L-arabinose), hexoses (D-glucose, D-galactose, D-mannose), disaccharides (sucrose, maltose, cellobiose), and polysaccharide (starch), on ethanol production by *Brettanomyces* strains. *B. naardenensis* was shown to produce ethanol from xylose. According to previous studies, few yeasts use pentoses for ethanol production, namely *Pichia stipitis*, *Candida shehatae*, and *Pachysolen tannophilus* [76, 77]. Galafassi *et al.* [75] also showed that several *B. anomala* strains, one *B. custersianus* strain, and one *B. nanus* strain were able to use agar as the sole carbon source, although the highest ethanol yield was achieved in the presence of glucose. These findings demonstrate that *Brettanomyces* yeasts can use different carbon sources under oxygen-limited and low pH conditions (commonly used in industrial processes) and produce ethanol yields similar to those of *S. cerevisiae* (0.44 g g^{-1} glucose) [75].

Another study reported the ability of *B. bruxellensis* CBS 11269 to assimilate and ferment cellobiose into ethanol in oxygen-limited cultures, albeit at a slower rate and less effectively than with glucose. The ethanol yield from cellobiose was 0.29 ± 0.10 g g^{-1} cellobiose, whereas the yield from glucose was 0.41 ± 0.00 g g^{-1} glucose under the same conditions [78]. Spindler *et al.* [79] investigated the efficiency of *B. custersii* strains in medium containing 150 g L^{-1} cellobiose. The final ethanol concentration and yield were 60.3 g L^{-1} and 0.40 g g^{-1} cellobiose, respectively, much closer to the ethanol yield obtained from glucose in the previous study.

Temperature

Temperature is an important parameter that directly influences fermentation and yeast metabolism kinetics. The optimum temperature range of *Brettanomyces* is 25 to 28 °C [80]. According to Bradman *et al*. [81], the optimum temperatures for ethanol and acetic acid production are 32 and 25 °C, respectively. However, the authors reported that, at 15–32 °C, the final yeast biomass concentration was very close to the maximum value (5.6 g L^{-1}). Interestingly, the maximal biomass concentration was achieved after 220 h at 15 °C and after 50 h at 32 °C, confirming that temperature influences *B. bruxellensis* cell viability. At 35 °C, cell death rates were high, and biomass yield decreased by 65% (to 2 g L^{-1}) compared with yields obtained at other temperatures. Blomqvist *et al*. [78] observed that *B. bruxellensis* CBS 11269 was temperature-dependent in the 25–37 °C range. It is well-known that ethanol yield and yeast growth are strongly influenced by temperature. The wide temperature range of *Brettanomyces* contributes to the yeast's robustness to rapid changes in fermentation conditions [11, 78, 82].

pH

Given that *Brettanomyces* strains are mostly found in alcoholic fermentation, their ideal pH is low, ranging from pH 2 to 5, as reported by Blomqvist *et al*. [78]. The authors cultivated *B. bruxellensis* CBS 2499 and CBS 2796 under uncontrolled pH conditions, resulting in a decrease in pH, from 4.5 to 2.5. Both strains converted glucose into ethanol with good yields, and differences between strains were related to differences in pH values. Romano *et al*. [51] observed that, by lowering the pH, the production of vinylphenol and ethylphenol was inhibited. As discussed above, these volatile phenols are responsible for unpleasant aromas and flavors in wines.

Bioethanol Production

Studies on bioethanol production by *B. bruxellensis* have increased greatly in recent years, stemming from the yeast's ability to ferment sugars to ethanol under aerobic conditions, even when respiration is theoretically possible (Crabtree effect) [11]. At first, Brazilian, Swedish, and Canadian researchers believed the yeast to be a contaminant in ethanol production [83 - 85]. However, recent discoveries showed that *Brettanomyces* strains are capable of fermenting sugars to ethanol, with the added benefit of adapting easily to the harsh conditions of ethanol fermentation tanks [11, 78, 84]. Furthermore, as discussed by Steensels *et al.* [11] in their review, *B. bruxellensis* can outcompete inoculated *S. cerevisiae* strains, without affecting ethanol yield. *B. bruxellensis* also has the ability to co-consume nitrate and use other carbon sources instead of glucose, indicative of high industrial potential. Galafassi *et al.* [75] studied the ability of 50 *Brettanomyces* strains to produce ethanol from different carbon sources and under similar conditions to those used in industrial fermentation, such as limited oxygen and low pH. *B. bruxellensis* produced 0.44 g g^{-1} glucose, a yield comparable to that of *S. cerevisiae*. The authors also observed that *B. naardenensis* produced ethanol from xylose.

Blomqvist *et al.* [78] investigated the factors influencing the fermentation, such as pH, temperature, yeast growth rate, and ethanol production. The industrial strain *B. bruxellensis* CBS 11269, two *B. bruxellensis* strains, and two industrial *S. cerevisiae* strains were used in shake flasks and bioreactors. *B. bruxellensis* and *S. cerevisiae* generated similar ethanol yields, but the former had higher biomass yields and lower glycerol yields, indicating that *B. bruxellensis* is more energy-efficient than *S. cerevisiae*. The lower growth rate of *B. bruxellensis*, however, suggests that it might not be competitive in batch fermentations. Nevertheless, according to Blomqvist *et al.* [78], in a continuous process in which growth rate can be controlled by technical parameters, efficient substrate use and high tolerance to inhibitors may be more important determinant factors. These findings have attracted much attention from the industry because of the interest in using strains that are capable of resisting environmental changes and using different carbon sources.

Brettanomyces has the ability to produce acetic acid under aerobic conditions and withstand low pH environments. Currently, acetic acid production is mainly achieved by petrochemical routes. Thus, *Brettanomyces* strains are interesting candidates for sustainable acetic acid production. Freer *et al.* [86] reported that *D. intermedia* NRRL YB-4553 and *B. bruxellensis* NRRL Y-17525 are the two most promising yeasts for acetic acid production when glucose is used as a carbon source. The optimum pH was found to be 5.5, at which acetic acid was afforded in yields of 57.5 and 65.1 g acetic acid L^{-1} from an initial glucose concentration of 150 and 200 g L^{-1}, respectively.

Production of Beer and Wine with Unique Aromas

Brettanomyces has, for many years, been known to be responsible for unpleasant smells and flavors in wine, attributed to the presence of 4-ethylphenol (phenolic/medicinal aroma) and 4-ethylguaiacol (clove/holiday spice aroma) [55]. For some styles of wine, these aromas are appreciated, as they generate positive and special organoleptic characteristics and give an aged flavor to some young red wines [87, 88].

Brettanomyces has β-glucosidase, an important enzyme that can release natural flavors and aromas from different sources and hydrolyze nonvolatile, chemically bound aroma compounds [50]. This characteristic makes the yeast attractive to the bioflavor industry, which is constantly searching for new, natural sources of differentiated aroma profiles for food and beverages. Esters generate a fruity taste and/or smell; ethyl esters (*e.g.*, ethyl acetate, hexanoate, and octanoate) contribute to tropical fruit and pineapple-like flavors [75]. For this reason, *Brettanomyces* yeasts are currently being used in the production of craft beers and natural wines. *Brettanomyces* is crucial for the characteristic flavor of Belgian lambic and Gueuze, produced by spontaneous fermentation [89]. In the United States of America, some breweries have developed differentiated products, such as American coolship ales [90].

More studies on the potential applications of *Brettanomyces* are expected in the near future. Investigations of the large-scale production of fermented beverages and bioethanol should be carried out to confirm the advantages of *Brettanomyces* and promote the discovery of new fermentative processes. Furthermore, it would be interesting to perform a technical-economic analysis and a life cycle assessment of *Brettanomyces* to evaluate the environmental impacts of all process stages.

CONCLUDING REMARKS

Exploring the potential of nonconventional yeasts, such as *Brettanomyces* (teleomorph *Dekkera*) for industrial applications is highly attractive for the development of novel fermented beverages. The yeast's high adaptability to extreme conditions, as confirmed by phenotypic, metabolome, and transcriptome characterizations, allows its use for primary, secondary, and bottle fermentation. *Brettanomyces* gives fermented products a peculiar flavor profile with great potential, provided that undesirable aromas are avoided and fermentation patterns are optimized. The metabolic pathways for flavor production have been elucidated by studies aiming to propel the industrial use of *Brettanomyces* species for generating attractive flavor compositions. Furthermore, the tolerance of *Brettanomyces* to low pH, high ethanol concentration, and overall stressful conditions underscore the economical relevance of the yeast as a starter culture in second-generation bioethanol production. Research on metabolic control and genetic modification could improve the abilities of *Brettanomyces* strains and turn them into the protagonists of fermentation processes.

CONSENT FOR PUBLICATION

Not applicable.

CONFLICT OF INTERESTS

The authors declare no conflict of interest, financial or otherwise.

ACKNOWLEDGEMENTS

The authors acknowledge the São Paulo State Research Foundation (FAPESP, grant no. 2017/24520-4 and 2019/08542-3), and the Coordenação de Aperfeiçoamento de Pessoal de Nível Superior (CAPES, Finance Code 001).

REFERENCES

[1]   Claussen NH. On a Method for the Application of Hansen's Pure Yeast System in the Manufacturing of Well-Conditioned English Stock Beers. J Inst Brew 1904; 10: 308-31.
[http://dx.doi.org/10.1002/j.2050-0416.1904.tb04656.x]

[2]   Colomer MS, Chailyan A, Fennessy RT, *et al.* Assessing Population Diversity of *Brettanomyces* Yeast Species and Identification of Strains for Brewing Applications. Front Microbiol 2020; 11: 637.
[http://dx.doi.org/10.3389/fmicb.2020.00637] [PMID: 32373090]

[3]   Custers MTJ. Onderzoekingen over het Gistgeslacht Brettanomyces. TUDelft 1940.

[4]   Oelofse A, Pretorius IS, Toit M. Brettanomyces review 2008. S Afr J Enol Vitic 2008; 29: 128-44.

[5]   Spencer JFT. Yeasts: Characteristics and identification. Food Microbiol 1984; 1: 349-50.
[http://dx.doi.org/10.1016/0740-0020(84)90068-6]

[6]   Loureiro V, Malfeito-Ferreira M. Dekkera/Brettanomyces spp. In: Blackburn Cd W Food Spoilage Microrganisms Woodhead Publishing Series in Food Science, Technology and Nutrition. 2006; pp. 354-98.
[http://dx.doi.org/10.1533/9781845691417.3.354]

[7]   Edlin DAN, Narbad A, Gasson MJ, Dickinson JR, Lloyd D. Purification and characterization of hydroxycinnamate decarboxylase from Brettanomyces anomalus. Enzyme Microb Technol 1998; 22: 232-9.
[http://dx.doi.org/10.1016/S0141-0229(97)00169-5]

[8]   Silva LR, Andrade PB, Valentão P, Seabra RM, Trujillo ME, Velázquez E. Analysis of non-coloured phenolics in red wine: Effect of Dekkera bruxellensis yeast. Food Chem 2005; 89: 185-9.
[http://dx.doi.org/10.1016/j.foodchem.2004.02.019]

[9]   Avramova M, Cibrario A, Peltier E, *et al.* Brettanomyces bruxellensis population survey reveals a diploid-triploid complex structured according to substrate of isolation and geographical distribution. Sci Rep 2018; 8(1): 4136.
[http://dx.doi.org/10.1038/s41598-018-22580-7] [PMID: 29515178]

[10]   Cibrario A, Miot-Sertier C, Paulin M, *et al.* Brettanomyces bruxellensis phenotypic diversity, tolerance to wine stress and wine spoilage ability. Food Microbiol 2020; 87: 103379.
[http://dx.doi.org/10.1016/j.fm.2019.103379] [PMID: 31948620]

[11]   Steensels J, Daenen L, Malcorps P, Derdelinckx G, Verachtert H, Verstrepen KJ. Brettanomyces yeasts--From spoilage organisms to valuable contributors to industrial fermentations. Int J Food Microbiol 2015; 206: 24-38.
[http://dx.doi.org/10.1016/j.ijfoodmicro.2015.04.005] [PMID: 25916511]

[12] Hernández A, Pérez-Nevado F, Ruiz-Moyano S, *et al.* Spoilage yeasts: What are the sources of contamination of foods and beverages? Int J Food Microbiol 2018; 286: 98-110.
[http://dx.doi.org/10.1016/j.ijfoodmicro.2018.07.031] [PMID: 30056262]

[13] Renouf V, Lonvaud-Funel A. Development of an enrichment medium to detect Dekkera/Brettanomyces bruxellensis, a spoilage wine yeast, on the surface of grape berries. Microbiol Res 2007; 162(2): 154-67.
[http://dx.doi.org/10.1016/j.micres.2006.02.006] [PMID: 16595174]

[14] Connell L, Stender H, Edwards CG. Rapid detection and identification of brettanomyces from winery air samples based on peptide nucleic acid analysis. Am J Enol Vitic 2002; 53: 322-4.

[15] Ocón E, Garijo P, Sanz S, Olarte C, López R, Santamaría P, *et al.* Screening of yeast mycoflora in winery air samples and their risk of wine contamination. Food Control 2013; 34: 261-7.
[http://dx.doi.org/10.1016/j.foodcont.2013.04.044]

[16] Dweck HKM, Ebrahim SAM, Farhan A, Hansson BS, Stensmyr MC. Olfactory proxy detection of dietary antioxidants in Drosophila. Curr Biol 2015; 25(4): 455-66.
[http://dx.doi.org/10.1016/j.cub.2014.11.062] [PMID: 25619769]

[17] Garijo P, González-Arenzana L, López-Alfaro I, *et al.* Analysis of grapes and the first stages of the vinification process in wine contamination with Brettanomyces bruxellensis. Eur Food Res Technol 2015; 240: 525-32.
[http://dx.doi.org/10.1007/s00217-014-2351-4]

[18] Wedral D, Shewfelt R, Frank J. The challenge of Brettanomyces in wine. Lebensm Wiss Technol 2010; 43: 1474-9.
[http://dx.doi.org/10.1016/j.lwt.2010.06.010]

[19] Stadler E, Schmarr H-G, Fischer U. Influence of physical and chemical barrel sanitization treatments on the volatile composition of toasted oak wood. Eur Food Res Technol 2020; 246: 497-511.
[http://dx.doi.org/10.1007/s00217-019-03417-7]

[20] Rodrigues N, Gonçalves G, Pereira-da-Silva S, Malfeito-Ferreira M, Loureiro V. Development and use of a new medium to detect yeasts of the genera Dekkera/Brettanomyces. J Appl Microbiol 2001; 90(4): 588-99.
[http://dx.doi.org/10.1046/j.1365-2672.2001.01275.x] [PMID: 11309071]

[21] Renouf V, Falcou M, Miot-Sertier C, Perello MC, De Revel G, Lonvaud-Funel A. Interactions between *Brettanomyces bruxellensis* and other yeast species during the initial stages of winemaking. J Appl Microbiol 2006; 100(6): 1208-19.
[http://dx.doi.org/10.1111/j.1365-2672.2006.02959.x] [PMID: 16696668]

[22] Stender H, Kurtzman C, Hyldig-Nielsen JJ, *et al.* Identification of *Dekkera bruxellensis* (Brettanomyces) from wine by fluorescence *in situ* hybridization using peptide nucleic acid probes. Appl Environ Microbiol 2001; 67(2): 938-41.
[http://dx.doi.org/10.1128/AEM.67.2.938-941.2001] [PMID: 11157265]

[23] Cocolin L, Rantsiou K, Iacumin L, Zironi R, Comi G. Molecular detection and identification of *Brettanomyces/Dekkera bruxellensis* and Brettanomyces/*Dekkera anomalus* in spoiled wines. Appl Environ Microbiol 2004; 70(3): 1347-55.
[http://dx.doi.org/10.1128/AEM.70.3.1347-1355.2004] [PMID: 15006752]

[24] Delaherche A, Claisse O, Lonvaud-Funel A. Detection and quantification of *Brettanomyces bruxellensis* and 'ropy' Pediococcus damnosus strains in wine by real-time polymerase chain reaction. J Appl Microbiol 2004; 97(5): 910-5.
[http://dx.doi.org/10.1111/j.1365-2672.2004.02334.x] [PMID: 15479405]

[25] Curtin CD, Bellon JR, Henschke PA, Godden PW, de Barros Lopes MA. Genetic diversity of *Dekkera bruxellensis* yeasts isolated from Australian wineries. FEMS Yeast Res 2007; 7(3): 471-81.
[http://dx.doi.org/10.1111/j.1567-1364.2006.00183.x] [PMID: 17233769]

[26]   Martorell P, Barata A, Malfeito-Ferreira M, Fernández-Espinar MT, Loureiro V, Querol A. Molecular typing of the yeast species *Dekkera bruxellensis* and *Pichia guilliermondii* recovered from wine related sources. Int J Food Microbiol 2006; 106(1): 79-84.
[http://dx.doi.org/10.1016/j.ijfoodmicro.2005.05.014] [PMID: 16229917]

[27]   Miot-Sertier C, Lonvaud-Funel A. Development of a molecular method for the typing of *Brettanomyces bruxellensis (Dekkera bruxellensis)* at the strain level. J Appl Microbiol 2007; 102(2): 555-62.
[http://dx.doi.org/10.1111/j.1365-2672.2006.03069.x] [PMID: 17241362]

[28]   Gounot J-S, Neuvéglise C, Freel KC, *et al.* High complexity and degree of genetic variation in brettanomyces bruxellensis population. Genome Biol Evol 2020; 12(6): 795-807.
[http://dx.doi.org/10.1093/gbe/evaa077] [PMID: 32302403]

[29]   Okuno M, Kajitani R, Ryusui R, Morimoto H, Kodama Y, Itoh T. Next-generation sequencing analysis of lager brewing yeast strains reveals the evolutionary history of interspecies hybridization. DNA Res 2016; 23(1): 67-80.
[http://dx.doi.org/10.1093/dnares/dsv037] [PMID: 26732986]

[30]   Crauwels S, Zhu B, Steensels J, *et al.* Assessing genetic diversity among Brettanomyces yeasts by DNA fingerprinting and whole-genome sequencing. Appl Environ Microbiol 2014; 80(14): 4398-413.
[http://dx.doi.org/10.1128/AEM.00601-14] [PMID: 24814796]

[31]   Liti G, Carter DM, Moses AM, *et al.* Population genomics of domestic and wild yeasts. Nature 2009; 458(7236): 337-41.
[http://dx.doi.org/10.1038/nature07743] [PMID: 19212322]

[32]   Avramova M, Grbin P, Borneman A, Albertin W, Masneuf-Pomarède I, Varela C. Competition experiments between *Brettanomyces bruxellensis* strains reveal specific adaptation to sulfur dioxide and complex interactions at intraspecies level. FEMS Yeast Res 2019; 19(3): 10.
[http://dx.doi.org/10.1093/femsyr/foz010] [PMID: 30721945]

[33]   Divol B, du Toit M, Duckitt E. Surviving in the presence of sulphur dioxide: strategies developed by wine yeasts. Appl Microbiol Biotechnol 2012; 95(3): 601-13.
[http://dx.doi.org/10.1007/s00253-012-4186-x] [PMID: 22669635]

[34]   Avramova M, Vallet-Courbin A, Maupeu J, Masneuf-Pomarède I, Albertin W. Molecular diagnosis of brettanomyces bruxellensis' sulfur dioxide sensitivity through genotype specific method. Front Microbiol 2018; 9: 1260.
[http://dx.doi.org/10.3389/fmicb.2018.01260] [PMID: 29942296]

[35]   Dimopoulou M, Hatzikamari M, Masneuf-Pomarede I, Albertin W. Sulfur dioxide response of brettanomyces bruxellensis strains isolated from Greek wine. Food Microbiol 2019; 78: 155-63.
[http://dx.doi.org/10.1016/j.fm.2018.10.013] [PMID: 30497597]

[36]   Albertin W, Masneuf-Pomarede I, Peltier E. Method for analysing a sample to detect the presence of sulphite-resistant yeasts of the. Brettanomyces Bruxellensis 2017.

[37]   Rozpędowska E, Hellborg L, Ishchuk OP, *et al.* Parallel evolution of the make-accumulate-consume strategy in Saccharomyces and Dekkera yeasts. Nat Commun 2011; 2: 302.
[http://dx.doi.org/10.1038/ncomms1305] [PMID: 21556056]

[38]   Serra Colomer M, Funch B, Forster J. The raise of Brettanomyces yeast species for beer production. Curr Opin Biotechnol 2019; 56: 30-5.
[http://dx.doi.org/10.1016/j.copbio.2018.07.009] [PMID: 30173102]

[39]   Smith BD, Divol B. The carbon consumption pattern of the spoilage yeast *Brettanomyces bruxellensis* in synthetic wine-like medium. Food Microbiol 2018; 73: 39-48.
[http://dx.doi.org/10.1016/j.fm.2017.12.011] [PMID: 29526225]

[40]   Haslbeck K, Jerebic S, Zarnkow M. Characterization of the unfertilized and fertilized hop varieties progress and hallertauer tradition – analysis of free and glycosidic-bound flavor compounds and β-

glucosidase activity. BrewingScience 70:148–58. 2017; 70: 148-58 .https://doi.org/10.23763/ BrSc17-15haslbeck

[41]   Vervoort Y, Herrera-Malaver B, Mertens S, *et al*. Characterization of the recombinant Brettanomyces anomalus β-glucosidase and its potential for bioflavouring. J Appl Microbiol 2016; 121(3): 721-33.
[http://dx.doi.org/10.1111/jam.13200] [PMID: 27277532]

[42]   Kuo H-P, Wang R, Huang C-Y, Lai J-T, Lo Y-C, Huang S-T. Characterization of an extracellular β-glucosidase from *Dekkera bruxellensis* for resveratrol production. J Food Drug Anal 2018; 26(1): 163-71.
[http://dx.doi.org/10.1016/j.jfda.2016.12.016] [PMID: 29389552]

[43]   Parente DC, Cajueiro DBB, Moreno ICP, Leite FCB, De Barros Pita W, De Morais MA Jr. On the catabolism of amino acids in the yeast Dekkera bruxellensis and the implications for industrial fermentation processes. Yeast 2018; 35(3): 299-309.
[http://dx.doi.org/10.1002/yea.3290] [PMID: 29065215]

[44]   Steensels J, Verstrepen KJ. Taming wild yeast: potential of conventional and nonconventional yeasts in industrial fermentations. Annu Rev Microbiol 2014; 68: 61-80.
[http://dx.doi.org/10.1146/annurev-micro-091213-113025] [PMID: 24773331]

[45]   Heresztyn T. Metabolism of volatile phenolic compounds from hydroxycinnamic acids byBrettanomyces yeast. Arch Microbiol 1986; 146: 96-8.
[http://dx.doi.org/10.1007/BF00690165]

[46]   Vigentini I, Romano A, Compagno C, *et al*. Physiological and oenological traits of different *Dekkera/Brettanomyces bruxellensis* strains under wine-model conditions. FEMS Yeast Res 2008; 8(7): 1087-96.
[http://dx.doi.org/10.1111/j.1567-1364.2008.00395.x] [PMID: 18565109]

[47]   Chatonnet P, Dubourdieu D, Boidron JN, Lavigne V. Synthesis of volatile phenols by *Saccharomyces cerevisiae* in wines. J Sci Food Agric 1993; 62: 191-202.
[http://dx.doi.org/10.1002/jsfa.2740620213]

[48]   Chatonnet P, Dubourdie D, Boidron J, Pons M. The origin of ethylphenols in wines. J Sci Food Agric 1992; 60: 165-78. https://doi.org/https://doi.org/10.1002/jsfa.2740600205
[http://dx.doi.org/10.1002/jsfa.2740600205]

[49]   Licker J, Henick-Kling T, Acree T. Impact of Brettanomyces Yeast on Wine Flavor: Sensory Description of Wines with Different Brett Aroma Character. Proc 12th Int Enol Symp. 31 May - 2 June 1999; 1999; pp. 218-40.

[50]   Daenen L, Saison D, De Schutter DP, De Cooman L, Verstrepen KJ, Delvaux FR, *et al*. Bioflavoring of beer through fermentation, refermentation and plant parts addition. Beer Heal Dis Prev 2008; pp. 33-48.

[51]   Romano A, Perello MC, de Revel G, Lonvaud-Funel A. Growth and volatile compound production by *Brettanomyces/Dekkera bruxellensis* in red wine. J Appl Microbiol 2008; 104(6): 1577-85.https://doi.org/https://doi.org/10.1111/j.1365-2672.2007.03693.x
[http://dx.doi.org/10.1111/j.1365-2672.2007.03693.x] [PMID: 18194246]

[52]   Van Der Walt JP, Van Kerken AE. The wine yeasts of the cape. Antonie van Leeuwenhoek 1958; 24: 239-52.
[http://dx.doi.org/10.1007/BF02548451]

[53]   Oelofse A, Pretorius IS, du Toit M. Significance of Brettanomyces and Dekkera during Winemaking: A Synoptic Review. S Afr J Enol Vitic 2008; 29: 128-44.
[http://dx.doi.org/10.21548/29-2-1445]

[54]   Licker JL, Acree TE, Henick-Kling T. What Is "Brett" (Brettanomyces) Flavor?: A Preliminary Investigation Chem Wine Flavor. American Chemical Society 1998; Vol. 714: pp. 8-96.
https://doi.org/doi:10.1021/bk-1998-0714.ch008

[55]   Suárez R, Suárez-Lepe JA, Morata A, Calderón F. The production of ethylphenols in wine by yeasts of the genera Brettanomyces and Dekkera: A review. Food Chem 2007; 102: 10-21.
[http://dx.doi.org/10.1016/j.foodchem.2006.03.030]

[56]   Barrado AM, Uthurry CA, Fernández FC, Lepe JAS. Aplicaciones de la ultrafiltración en la industria enológica: últimos avances tecnológicos. Tecnol Del Vino Trat y Equipos Para Vitic y Enol 2004; pp. 49-54.

[57]   Gómez-Rivas L, Escudero-Abarca BI, Aguilar-Uscanga MG, Hayward-Jones PM, Mendoza P, Ramírez M. Selective antimicrobial action of chitosan against spoilage yeasts in mixed culture fermentations. J Ind Microbiol Biotechnol 2004; 31(1): 16-22.
[http://dx.doi.org/10.1007/s10295-004-0112-2] [PMID: 14747932]

[58]   Morata A, Loira I, Vejarano R, González C, Callejo MJ, Suárez-Lepe JA. Emerging preservation technologies in grapes for winemaking. Trends Food Sci Technol 2017; 67: 36-43.https://doi.org/https://doi.org/10.1016/j.tifs.2017.06.014
[http://dx.doi.org/10.1016/j.tifs.2017.06.014]

[59]   Comitini F, De Ingeniis J, Pepe L, Mannazzu I, Ciani M. *Pichia anomala* and *Kluyveromyces wickerhamii* killer toxins as new tools against Dekkera/Brettanomyces spoilage yeasts. FEMS Microbiol Lett 2004; 238(1): 235-40.
[http://dx.doi.org/10.1111/j.1574-6968.2004.tb09761.x] [PMID: 15336427]

[60]   du Toit M, Pretorius IS. Microbial Spoilage and Preservation of Wine: Using Weapons from Nature's Own Arsenal -A Review. S Afr J Enol Vitic 2019; 21
[http://dx.doi.org/10.21548/21-1-3559]

[61]   Gracin L, Jambrak AR, Juretić H, *et al.* Influence of high power ultrasound on Brettanomyces and lactic acid bacteria in wine in continuous flow treatment. Appl Acoust 2016; 103: 143-7.
[http://dx.doi.org/10.1016/j.apacoust.2015.05.005]

[62]   Fredericks IN, du Toit M, Krügel M. Efficacy of ultraviolet radiation as an alternative technology to inactivate microorganisms in grape juices and wines. Food Microbiol 2011; 28(3): 510-7.
https://doi.org/https://doi.org/10.1016/j.fm.2010.10.018
[http://dx.doi.org/10.1016/j.fm.2010.10.018] [PMID: 21356459]

[63]   Cravero F, Englezos V, Rantsiou K, *et al.* Control of Brettanomyces bruxellensis on wine grapes by post-harvest treatments with electrolyzed water, ozonated water and gaseous ozone. Innov Food Sci Emerg Technol 2018; 47: 309-16.https://doi.org/https://doi.org/10.1016/j.ifset.2018.03.017
[http://dx.doi.org/10.1016/j.ifset.2018.03.017]

[64]   González-Arenzana L, Reinares JP, López N, Santamaría P, Garde-Cerdán T, Gutiérrez AR, *et al.* Impact of pulsed electric field treatment on must and wine quality. Handb Electroporation Springer; Berlin, Ger 2017; 2391-406.
[http://dx.doi.org/10.1007/978-3-319-32886-7_174]

[65]   Izquierdo-Cañas PM, López-Martín R, García-Romero E, *et al.* Effect of kaolin silver complex on the control of populations of *Brettanomyces* and acetic acid bacteria in wine. J Food Sci Technol 2018; 55(5): 1823-31.
[http://dx.doi.org/10.1007/s13197-018-3097-y] [PMID: 29666535]

[66]   Ugarte P, Agosin E, Bordeu E, Villalobos JI. Reduction of 4-ethylphenol and 4-ethylguaiacol concentration in red wines using reverse osmosis and adsorption. Am J Enol Vitic 2005; 56: 30-6.

[67]   Chassagne D, Guilloux-Benatier M, Alexandre H, Voilley A. Sorption of wine volatile phenols by yeast lees. Food Chem 2005; 91: 39-44.
[http://dx.doi.org/10.1016/j.foodchem.2004.05.044]

[68]   De Deken RH. The Crabtree effect: a regulatory system in yeast. J Gen Microbiol 1966; 44(2): 149-56.https://doi.org/https://doi.org/10.1099/00221287-44-2-149.
[http://dx.doi.org/10.1099/00221287-44-2-149] [PMID: 5969497]

[69] Van Dijken JP, Scheffers WA. Redox balances in the metabolism of sugars by yeasts. FEMS Microbiol Rev 1986; 1: 199-224.
[http://dx.doi.org/10.1111/j.1574-6968.1986.tb01194.x]

[70] Barnett JA, Entian KD. A history of research on yeasts 9: regulation of sugar metabolism. Yeast 2005; 22(11): 835-94.
[http://dx.doi.org/10.1002/yea.1249] [PMID: 16134093]

[71] de Barros Pita W, Leite FCB, de Souza Liberal AT, Simões DA, de Morais MA Jr. The ability to use nitrate confers advantage to *Dekkera bruxellensis* over *S. cerevisiae* and can explain its adaptation to industrial fermentation processes. Antonie van Leeuwenhoek 2011; 100(1): 99-107.
[http://dx.doi.org/10.1007/s10482-011-9568-z] [PMID: 21350883]

[72] Conterno L, Joseph CML, Arvik TJ, Henick-Kling T, Bisson LF. Genetic and Physiological Characterization of *Brettanomyces bruxellensis* Strains Isolated from Wines. Am J Enol Vitic 2006; 57: 139: 147.

[73] Kumara HMCS, De Cort S, Verachtert H. Localization and characterization of α-glucosidase activity in *Brettanomyces lambicus*. Appl Environ Microbiol 1993; 59(8): 2352-8.
[http://dx.doi.org/10.1128/aem.59.8.2352-2358.1993] [PMID: 16349005]

[74] Parekh SR, Parekh RS, Wayman M. Fermentation of xylose and cellobiose by *Pichia stipitis* and *Brettanomyces clausenii*. Appl Biochem Biotechnol 1988; 18: 325-38.
[http://dx.doi.org/10.1007/BF02930836]

[75] Galafassi S, Merico A, Pizza F, *et al.* Dekkera/Brettanomyces yeasts for ethanol production from renewable sources under oxygen-limited and low-pH conditions. J Ind Microbiol Biotechnol 2011; 38(8): 1079-88.
[http://dx.doi.org/10.1007/s10295-010-0885-4] [PMID: 20936422]

[76] Agbogbo FK, Coward-Kelly G, Torry-Smith M, Wenger K, Jeffries TW. The effect of initial cell concentration on xylose fermentation by Pichia stipitis Appl Biochem Biotecnol. Springer 2007; pp. 653-62.

[77] Perego P, Converti A, Palazzi E, Del Borghi M, Ferraiolo G. Fermentation of hardwood hemicellulose hydrolysate by *Pachysolen tannophilus, Candida shehatae* and *Pichia stipitis*. J Ind Microbiol Biotechnol 1990; 6: 157-64.

[78] Blomqvist J, Eberhard T, Schnürer J, Passoth V. Fermentation characteristics of *Dekkera bruxellensis* strains. Appl Microbiol Biotechnol 2010; 87(4): 1487-97.
[http://dx.doi.org/10.1007/s00253-010-2619-y] [PMID: 20437232]

[79] Spindler DD, Wyman CE, Grohmann K, Philippidis GP. Evaluation of the cellobiose-fermenting yeast *Brettanomyces custersii* in the simultaneous saccharification and fermentation of cellulose. Biotechnol Lett 1992; 14: 403-7.
[http://dx.doi.org/10.1007/BF01021255]

[80] Zuehlke JM, Edwards CG. Impact of sulfur dioxide and temperature on culturability and viability of *Brettanomyces bruxellensis* in Wine. J Food Prot 2013; 76(12): 2024-30.
[http://dx.doi.org/10.4315/0362-028X.JFP-13-243R] [PMID: 24290676]

[81] Brandam C, Castro-Martínez C, Délia M-L, Ramón-Portugal F, Strehaiano P. Effect of temperature on *Brettanomyces bruxellensis*: metabolic and kinetic aspects. Can J Microbiol 2008; 54(1): 11-8.
[http://dx.doi.org/10.1139/W07-126] [PMID: 18388967]

[82] Yakobson C. Pure culture fermentation characteristics of brettanomyces yeast species and their use in the brewing industry. Masters Degree Heriot-Watt Univ Edinburgh 2010.

[83] de Souza Liberal AT, Basílio ACM, do Monte Resende A, *et al.* Identification of *Dekkera bruxellensis* as a major contaminant yeast in continuous fuel ethanol fermentation. J Appl Microbiol 2007; 102(2): 538-47.
[http://dx.doi.org/10.1111/j.1365-2672.2006.03082.x] [PMID: 17241360]

[84]   Passoth V, Blomqvist J, Schnürer J. *Dekkera bruxellensis* and *Lactobacillus vini* form a stable ethanol-producing consortium in a commercial alcohol production process. Appl Environ Microbiol 2007; 73(13): 4354-6.
[http://dx.doi.org/10.1128/AEM.00437-07] [PMID: 17483277]

[85]   Abbott DA, Hynes SH, Ingledew WM. Growth rates of Dekkera/Brettanomyces yeasts hinder their ability to compete with *Saccharomyces cerevisiae* in batch corn mash fermentations. Appl Microbiol Biotechnol 2005; 66(6): 641-7.
[http://dx.doi.org/10.1007/s00253-004-1769-1] [PMID: 15538553]

[86]   Freer SN, Dien B, Matsuda S. Production of acetic acid by Dekkera/Brettanomyces yeasts under conditions of constant pH. World J Microbiol Biotechnol 2003; 19: 101-5.
[http://dx.doi.org/10.1023/A:1022592810405]

[87]   Delfini C, Formica JV. Wine Microbiology. Рипол Классик 2001.
[http://dx.doi.org/10.1201/9781482294644]

[88]   Loureiro V, Malfeito-Ferreira M. Spoilage yeasts in the wine industry. Int J Food Microbiol 2003; 86(1-2): 23-50.
[http://dx.doi.org/10.1016/S0168-1605(03)00246-0] [PMID: 12892920]

[89]   Spitaels F, Wieme AD, Janssens M, *et al.* The microbial diversity of traditional spontaneously fermented lambic beer. PLoS One 2014; 9(4): e95384.
[http://dx.doi.org/10.1371/journal.pone.0095384] [PMID: 24748344]

[90]   Bokulich NA, Bamforth CW, Mills DA. Brewhouse-resident microbiota are responsible for multi-stage fermentation of American coolship ale. PLoS One 2012; 7(4): e35507.
[http://dx.doi.org/10.1371/journal.pone.0035507] [PMID: 22530036]

CHAPTER 8

Spathaspora and *Scheffersomyces*: Promising Roles in Biorefineries

Thamarys Scapini[1], Aline F. Camargo[1], Jéssica Mulinari[2], Camila E. Hollas[3], Charline Bonatto[1], Bruno Venturin[3], Alan Rempel[4], Sérgio L. Alves Jr.[5] and **Helen Treichel[1,*]**

[1] *Laboratory of Microbiology and Bioprocess, Federal University of Fronteira Sul (UFFS), Erechim/RS, Brazil*

[2] *Laboratory of Membrane Processes, Department of Chemical Engineering and Food Engineering, Federal University of Santa Catarina (UFSC), Florianópolis SC, Brazil*

[3] *Center for Exact and Technological Sciences, Graduate Program in Agricultural Engineering, Western Paraná State University (UNIOESTE), Cascavel PR, Brazil*

[4] *Graduate Program in Environmental and Civil Engineering, University of Passo Fundo (UPF), Passo Fundo/RS, Brazil*

[5] *Laboratory of Biochemistry and Genetics, Federal University of Fronteira Sul, Chapecó/SC, Brazil*

Abstract: Currently, biotechnologies that aim to optimize the residual lignocellulosic biomass are receiving widespread attention, mainly when it comes to developing integrated systems that allow the generation of multi-products in industrial plants, especially for ethanol production. One of the main bottlenecks for efficient conversion of lignocellulosic biomass into ethanol is the limitation of *Saccharomyces cerevisiae*, the most widely used yeast in bioethanol production, in metabolizing xylose. This pentose is the main constituent of the hemicellulose fractions in plant cell walls and the second most abundant monosaccharide in lignocellulosic biomass. This challenge is being overcome by the isolation and intense molecular evaluation of new yeast species, mainly members of the genera *Spathaspora* and *Scheffersomyces*, since they have shown high capacities for xylose assimilation, which has been corroborated through studies aimed at improving ethanol production and other products *via* the association of these yeasts with improved fermentation capacity. In this sense, this chapter addresses the recent advances in the identification of novel isolates of the genera *Spathaspora* and *Scheffersomyces*, particularly emphasizing the applications of these genera in ethanol and xylitol production.

* **Corresponding author Helen Treichel:** Laboratory of Microbiology and Bioprocess, Federal University of Fronteira Sul (UFFS), Erechim/RS, Brazil; Tel: +55 54 981126456; E-mail: helentreichel@gmail.com

Sérgio Luiz Alves Júnior, Helen Treichel, Thiago Olitta Basso and Boris Ugarte Stambuk (Eds.)
All rights reserved-© 2022 Bentham Science Publishers

Keywords: Atlantic rainforest, Biofuel, Biotechnology, *Candida*, Cellobiose, Ethanol, Fermentation, Fungi, Genomic analysis, Gut of insects, Hexoses, Lignocellulosic biomass, Pentoses, *Pichia*, Rotting wood, *Sp. passalidarum*, *Sc. stipitis*, Sugarcane hydrolysate, Xylitol, Xylose.

INTRODUCTION

The current model of society is characterized by increased consumption of biomass, which puts the environmental systems under pressure to meet the growing demand. Unchecked development and limited resources make waste management a key component to sustaining the current consumption pattern. The current linear thinking model, "take-make-dispose," is no longer sustainable, necessitating the shift to a circular economy, based on the use of waste as raw materials to acquire new products [1, 2].

Due to the necessity to change production systems, the biorefinery concept has been highlighted as a more holistic view of the exploration of biomass, ensuring the maximum use of the structure of the raw material by generating different products in the same industrial plant. This configuration enables the lignocellulosic biomass conversion in energy, chemicals, or biomaterials, thus adding value to waste, minimizing the impacts of production activities, and ensuring the sustainability of these activities [3].

The conversion of lignocellulosic biomass is dependent on fractionating the complex structure formed by different polymers, such as cellulose, hemicellulose, lignin, and pectin, in varying concentrations depending on the type of biomass [4]. Among the sugars released from the lignocellulosic biomass, glucose is released from cellulose. This hexose is a high-affinity substrate for various microorganisms, showing high conversion rates into ethanol. The primary sugar released from hemicellulose hydrolysis is xylose, which is the most abundant in xylan. However, unlike glucose, xylose offers low conversion efficiency by microorganisms [5]. The successful conversion of lignocellulosic biomass into bioproducts depends on the efficient use of the released sugars, such as glucose, D-xylose, cellobiose, galactose, rhamnose, arabinose, and mannose, by the microorganisms suitable to produce value-added products [6, 7].

Second-generation ethanol is one of the leading products resulting from the biotechnological processes involving the use of lignocellulosic biomass, and is a fundamental fuel to meet the demand and change the current energy matrix. Besides ethanol, other products of high added value can be obtained from these residues by biological routes, such as xylitol, a sugar-alcohol with great industrial applicability (*e.g.,* pharmaceutical and food), which is currently obtained *via* expensive chemical processes. This further emphasizes the importance of finding economically and environmentally viable alternatives to get these products as well as the importance of the biorefinery approach [8].

Traditionally, yeasts of *Saccharomyces cerevisiae* and *Zymomonas mobilis* species are used industrially, but they either metabolize the pentoses found in the hydrolysates with low efficiency or do not metabolize them at all [9]. New approaches to prospect strains capable of acting widely in the fermentation process of complex substrates using different sugars have gained strength, presenting underutilized and unexploited bioresources with relevant characteristics for industrial exploration in biorefineries. In this scenario, yeasts of the genera *Spathaspora* and *Scheffersomyces* are highlighted as some species that can convert hexose and pentose sugars to ethanol and xylitol, which are isolated from different environments and which have shown a wide spectrum of industrial applications.

The performance evaluation of these yeast species is being explored as a valuable bioresource for the configurations of multi-product industrial plants. The metabolic capacity for pentoses and hexoses and the possibility of changes in the fermentative conditions to obtain different products have been intensely explored. In this scenario, this chapter aims to address the characteristics of yeasts of the genera *Spathaspora* and *Scheffersomyces*, emphasizing the environments from which they can be isolated and the primary studies on their application in fermentation systems, emphasizing the production of ethanol and xylitol.

SCHEFFERSOMYCES AND *SPATHASPORA* PHYLOGENY AND TAXONOMY

Scheffersomyces and *Spathaspora* yeasts have also demonstrated the potential for converting mixed sugars, which is interesting for biorefineries as products, such as xylitol, enzymes, and ethanol, can also be obtained [7, 10]. The relationship between these genera was described by Kurtzman e Suzuki [11] as being relatively closely related due to the fermentative capacity of D-xylose present in the species of both genera.

The genus *Spathaspora* was first described in 2006 by Nguyen and co-workers [12] to accommodate the single species *Sp. passalidarum*. The yeast was isolated from the intestine of *Odontotaenius disjunctus* (beetle) in Louisiana (USA) and highlighted for its characteristics of fermenting xylose as well as its distinct morphology, containing a single ascospore with curved ends, unlike any other known yeast [12, 13]. *Spathaspora* means a type of wide sword (*spatha*) and seed (*spora*), which was also named in honor of Joseph W. Spatafora for his contributions to the field of insect-fungus interactions, which were extremely relevant to the discovery of this yeast [12].

Sp. passalidarum was the first species described within the genus that occurred in a small clade of xylose-fermenting yeasts, which comprised three *Candida* species, including *C. jeffriesii*, *C. insectamans*, and *C. lyxosophila* [12, 14]. The production of elongated ascospores with curved ends was a feature highlighted by the researchers that isolated *Sp. passalidarum* [14]. This same characteristic was observed in the *Sp. arborariae* species isolated from rotting wood, and this is the second species of this genus. *Sp. arborariae* also assimilates and ferments xylose efficiently and was differentiated from *Sp. passalidarum* based on L-sorbose assimilation and by growing with 50% w v^{-1} glucose concentration [15].

Some species were later introduced into this genus and are identified as *Sp. brasiliensis, Sp. xylofermentans, Sp. Roraimanensis*, and *Sp. Suhii*, all isolated in rotting wood and have presented the capacity to ferment xylose [16]. These species differ from the first two isolated species (*Sp. passalidarum* and *Sp. arborariae*) as they cannot assimilate D-ribose. According to the authors, *Sp. roraimanesis* and *Sp. xylofermentans* can grow in the presence of 0.01% cycloheximide and share the unique ascospore morphology of this group, which distinguishes these two species from all other species. Concerning *Sp.brasiliensis* and *Sp. suhii*, the authors suggest that the sequencing of D1/D2 domains is essential to differentiate these species as both have an identical growth profile [16, 17].

Many additional *Spathaspora* species have since been described, with most of them being associated with insects or rotting wood. Wang and co-workers [18] reported the species *Sp. allomyrinae*, isolated from the gut of the beetles *Allomyrina dichotoma*, and highlighted the formation of two ascospores surrounded by a rather circular membrane enclosing the two spores in an ascus, differing from the previously described species of the *Spathaspora* genus [18]. Another additional species is *Sp. boniae*, which can metabolize cellobiose and has good efficiency for converting D-xylose to xylitol [19], and *Sp. piracicabensis*, isolated from rotting wood, demonstrated the potential to produce a higher ethanol titer than xylitol from D-xylose [6].

Three species were described by Lopes and co-workers [20] isolated from rotting wood that ferment D-xylose, including *Sp. girioi*, *Sp. Hagerdaliae*, and *Sp. gorwiae*. More recently, five new species of the genus have been described: *Sp. elongata, Sp. mengyangensis, Sp. jiuxiensis, Sp. Parajiuxiensis*, and *Sp. rosae*, the latter was named as a tribute by Chinese authors to the Brazilian researcher Carlos Rosa for his contributions in yeast taxonomy [21]. Likewise, the species *Sp. boniae* was also named as a tribute to the Brazilian researcher Elba Pinto da Silva Bon [19] for her studies using lignocellulosic materials for ethanol production, and *Sp. girioi* was named in honor of Francisco Gírio for his contribution to the study of ethanol and xylitol production [20].

The proposition of species *Sp. boniae* expanded the discussions of the character of the paraphilia of the genus *Spathaspora*. The analysis of bardocode sequences and other regions in the rRNA gene cluster positioned the species in the *C. albicans/Lodderomyces* clade [19]. The relationship between the *Spathaspora* and *Candida* genera was again discussed by Varize and co-workers [6] in the study of isolation and identification of the species *Sp. piracicabensis*, which presented phylogenetic placement with proximity to several species of *Spathaspora* and two species of the genus *Candida* (*C. materiae* and *C. jeffriesii*). In other previous studies of identification of the *Spathaspora* species, the proximity relation, mainly of these two *Candida* species, had already been reported, as reported by Nguyen and co-workers [12] who, when defining the genus *Spathaspora*, related the species *C. jeffriesii* as the sister taxon.

With the expansion of discussions based on the paraphilia of the *Spathaspora* genus, in addition to the identification of five new species (*Sp. elongata, Sp. jiuxiensis, Sp. mengyangensis, Sp. parajiuxiensis,* and *Sp. rosae*), Lv and co-workers [21] also proposed the relocation of two species of the *Candida* genus (*C. materiae* and *C. jeffriesii*) for the *Spathaspora* genus, due to the phylogenetic analysis of the combined ITS and nuc 28S rDNA sequences due to their phylogenetic placement within that genus [21].

The paraphyletic character of the genus has not been entirely resolved, and the asymmetric taxonomy divergence will depend on the isolation of new species and the sequencing of the complete genome of the members. For Lv and co-workers [21], the *Spathaspora* genus needs to be comprising the type-species *Sp. passalidarum*, with the other members being possible representatives of another genus. However, the authors reaffirm that only complete genome sequencing and isolation of new species can clarify the heterogeneity of this genus [21].

In addition to discussions on phylogenetic positions, the fermentative characteristics of the genus are highlighted, mainly by the ability of most species of this genus to ferment xylose. Species assigned to *Spathaspora* are explored primarily in ethanol and xylitol production due to their metabolic capacity evolved for xylose consumption, which is considered an excellent choice for biotechnological applications [13, 22]. The species *Sp. passalidarum* stands out within the genus due to its high metabolic capacity for pentoses, such as xylose, under aerobic or strictly anaerobic conditions [23].

The genus *Scheffersomyces* was described four years after *Spathaspora* in 2010 by Kurtzman and Suzuki [11] based on phylogenetic analysis from the combined sequences of the D1/D2 domain of the large subunit (LSU) and the nearly complete small subunit (SSU) rRNA genes. The authors studied species from the genus *Pichia* and, based on their results, three species were transferred to the new genus *Scheffersomyces* (*Sc. stipitis, Sc. segobiensis,* and *Sc. spartiniae*) [11, 24]. According to the authors, *Scheffersomyces* yeasts can ferment glucose and other sugars. Some species, for example, can metabolize D-xylose, acting as suitable candidates for biofuel production through biological routes. Besides sugars, polyols and organic acids can also be assimilated by the species of *Scheffersomyces* [11]. Based on the genus' ability to ferment different sugars, the name *Scheffersomyces* was chosen in honor of Professor W. Alexander Scheffers (Delft University of Technology, The Netherlands) due to his contributions to yeast physiology and biotechnology, particularly regarding D-xylose fermentation [11].

Two years later, the first mention of the genus *Scheffersomyces* was made by Urbina and Blackwell [25], based on a multilocus phylogenetic analysis, nucleotide differences in the rRNA markers (SSU, LSU, and internal transcribed spacers – ITS), *RPB1* and *XYL1*, and biochemical differences concerning their closest relatives, and they included three new species to the *Scheffersomyces* genus: *Sc. illinoinensis, Sc. quercinus,* and *Sc. virginianus*. These species were named after the substrates from which they were isolated: *Carya illinoinensis* (pecan), *Quercus nigra* (water oak), and *Quercus virginiana* (live oak) rotted wood, respectively. In addition, the authors also proposed seven asexual species, previously assigned to the polyphyletic genus *Candida*, in the *Scheffersomyces* genus: *Sc. coipomoensis, Sc. gosingicus, Sc. insectosa, Sc. lignicola, Sc. lignosus, Sc. queiroziae,* and *Sc. shehatae* [25]. Therefore, the genus was composed of 13 species with phylogenetic similarities (three included by Kurtzman and Suzuki in 2010 [11] and ten by Urbina and Blackwell in 2012 [25] that probably evolved from a common ancestor capable of fermenting xylose, which makes *Scheffersomyces* an independent clade based on the multilocus phylogenetic analysis [24, 25].

Moreover, Urbina and Blackwell [25] proposed a subdivision of the *Scheffersomyces* clade into three subclades: 1) the early-diverging *Sc. spartinae* (the only member that cannot ferment both xylose and cellobiose) and *Sc. gosingicus* (cellobiose-fermenting yeast); 2) cellobiose-fermenting subclade composed of *Sc. coipomoensis, Sc. lignicola, Sc. queiroziae, Sc. Ergatensis,* and *Sc. amazonensis* (the last two were isolated by Santa María [26] and Cadete and co-workers [27], respectively, and firstly assigned to the *Candida* genus, being later included in the *Scheffersomyces* clade as reported by Urbina and Blackwell [28]); and 3) xylose-fermenting subclade, including *Sc. stipitis, Sc. lignosus, Sc. illinoinensis, Sc. insectosa, Sc. quercinus, Sc. segobiensis, Sc. shehatae,* and *Sc.*

virginianus [25]. The authors' phylogenetic analyses suggest that the ability to ferment xylose might have played a central role in the xylose-fermenting subclade's speciation process as xylose reductase (XR) was detected in all *Scheffersomyces* clade members. Furthermore, the same tree topology of the multilocus analysis was obtained using either XR or XYL1 [25].

With the insertion of new isolates and transference of species to the genus *Scheffersomyces*, it was observed that the gene datasets result in several positions in the phylogenetic tree due to genetic diversity. The polyphyletic character of the genus *Scheffersomyces* was first observed by Kutzman e Robnett [29]. It was shown that *Sc. spatinae* was allocated to a different clade of the *Sc. stipitis* and in the same clade as the yeast *Sp. passalidarum*. Suh and co-workers [30] added to this discussion by comparing the multilocus DNA sequence and suggested the existence of at least four phylogenetically distinct groups in the phylogenetic tree of the *Scheffersomyces* genus: 1) the species, *Sc. stipis* and *Sc. shehatae*, and other species xylose fermenters; 2) *Sc. coipomoensis* and *Sc. ergatensis*; 3) *Sc. lignicola* and *Sc. queiroziae*; 4) *Sc. spartiniae* and *Sc. gosingicus*. The authors suggest that the genus *Scheffersomyces* should be limited based on the monophyletic group of yeasts capable of fermenting xylose circumscribed to the type-species *Sc. stipitis* and other groups may become representatives of new genera based on the studies of the genetic relationships of the isolates [30].

The polyphyletic character of the genus *Scheffersomyces* was again reported by Liu and co-workers [31], that in the same study described the species *Sc. titanus*. In line with what was proposed in other studies that observed polyphilia in the genus, the authors suggested that there are two phylogenetically distinct groups, one of which exhibits a monophyletic character, comprising the species surrounding the yeast *Sc. stipitis*, and a second group that comprises the other representatives of the genus [31]. More recently, Jia and co-workers [32] highlighted that, based on the species described until now, a change in the proposed phylogenetic groups could be suggested, demonstrating through the analysis of genes and composition of the phylogenetic tree the existence of three distinct groups: 1) *Sc. stipitis*, *Sc. coipomoensis*, *Sc. shehatae* and species-related; 2) *Sc. stambukii* and *Sc. anoplophorae* (new species described in the article [32]); 3) *Sc. gosingicus* and *Sc. spartiniae* and the species *C. thasaenensis* (phylogenetically related to the group). In this same study, the authors isolated three new species of the genus *Scheffersomyces*: *Sc. jinghongensis*, *Sc. paraergatensis,* and *Sc. anoplophorae*, that share morphological similarities with *Sc. titanus*, *Sc. ergatensis,* and *Sc. stambukii*, respectively, and according to the separation described by the authors belong to the phylogenetically distinct groups one, two, and three, respectively [32].

Together, studies on the phylogenetic relation and the polyphyletic character of the *Scheffersomyces* suggest the separation of individuals, making the genus monophyletic based on the type-species *Sc. stipitis* and the other related xylose-fermenting species. However, further clarification is needed about the heterogeneity of the *Scheffersomyces* genus and the phylogenetic relations of the individuals it comprises, resulting in the emergence of new genera that allocate the members of some species currently related to the genus *Scheffersomyces* [29 - 32].

WHERE HAVE *SPATHASPORA* AND *SCHEFFERSOMYCES* YEASTS BEEN FOUND?

Each environment is home to a multitude of biodiversity, from which several new yeast species can be isolated. Isolation and identification of new yeasts can help fill some gaps to better understand the phylogenetic relationships since it is estimated that approximately 90% of the diversity of yeasts with potential for use in biotechnological processes remains unknown [10]. Different species of *Spathaspora* and *Scheffersomyces* have been explored to obtain robust strains with unique performance characteristics. Table **1** presents several examples of species of *Spathaspora* and *Scheffersomyces* isolated from different environments.

Table 1. *Spathaspora* and *Scheffersomyces* species isolated from different substrates and environments.

Species	Substrate	Environmental	References
Spathaspora			
Sp. passalidarum	Gut of beetle (*Odontotaenius disjunctus*)	Southeastern Louisiana (USA)	[12, 14]
Sp. arborariae	Rotting wood	Atlantic Rainforest (Brazil)	[15, 40]
Sp. brasiliensis	Rotting wood	Amazonian Florest (Brazil)	[16]
Sp. roraimanensis	Rotting wood	Amazonian Florest (Brazil)	[16]
Sp. suhii	Rotting wood	Amazonian Florest (Brazil)	[16]
Sp. xylofermentans	Rotting wood	Amazonian Florest (Brazil)	[16]
Sp. allomyrinae	Gut of beetle (*Allomyrina dichotoma*)	Baotianman National Nature Reserve (China)	[18]
Sp. girioi	Rotting wood	Atlantic Rainforest (Brazil)	[20]
Sp. hagerdaliae	Rotting wood	Atlantic Rainforest (Brazil)	[20]
Sp. gorwiae	Rotting wood	Atlantic Rainforest (Brazil)	[20]
Sp. boniae	Rotting wood	Atlantic Rainforest (Brazil)	[10, 19]
Sp. piracicabensis	Rotting wood	Atlantic Rainforest (Brazil)	[6]

(Table 1) cont.....

Species	Substrate	Environmental	References
Sp. elongate	Rotting wood	Jinghong City, Yunnan Province (China)	[21]
Sp. jiuxiensis	Rotting wood	Honghe Prefecture, Yunnan Province, (China)	[21]
Sp. mengyangensis	Rotting wood	Jinghong City, Yunnan Province (China)	[21]
Sp. parajiuxiensis	Rotting wood	Honghe Prefecture, Yunnan Province, (China)	[21]
Sp. rosae	Rotting wood	Jinghong City, Yunnan Province (China)	[21]
Sp. (Candida) materiae	Rotting wood	Atlantic Rainforest (Brazil)	[21, 45]
Sp. (Candida) jeffriesi	Gut of wood-boring bleetle	Chiriqui (Panamá)	[21, 46]
Scheffersomyces			
Sc. (Pichia) stipitis	Rotting wood; Larvae of insects *Cetonia* sp., *Dorcus parallelipidus* and *Laphria*; Gut of beetles	Atlantic Rainforest (Brazil); Guatemala	[10, 11, 44, 47]
Sc. (Pichia) segobiensis	Larvae of *Calcophora mariana massiliensis* and insect frass; Saccate organ of an insect	Honshu (Japan)	[11, 47, 48]
Sc. (Pichia) spartiniae	Oyster grass marshes	Louisiana (USA)	[11, 47]
Sc. illinoinensis	Rotting wood (*Carya illinoinensis*); Rotting wood and Gut of beetle	Louisiana (USA); Atlantic Rainforest (Brazil)	[25, 40]
Sc. quercinus	Rotting wood (*Quercus niger*)	Louisiana (USA)	[25]
Sc. virginianus	Rotting wood (*Quercus virginiana*)	Louisiana (USA)	[25]
Sc. (Candida) coipomoensis	Rotting wood and larvae of cerambicid beetle	Chile	[25, 46]
Sc. (Candida gosingica) gosingicus	Soil	Nanto (Taiwan)	[25, 49]
Sc. (Candida) insectosa	Longhorn beetle (*Leptura maculicornis*)	-	[25, 46]
Sc. (Candida) lignicola	Insect frass	Khao-Yai National Park (Thailand)	[25, 50]
Sc. (Candida lignosa) lignosus	Rotting wood	Chile	[25, 46]
Sc. (Candida) queiroziae	Rotting wood and Wood-boring insect	Atlantic Rain Forest (Brazil)	[25, 35, 39]

(Table 1) cont.....

Species	Substrate	Environmental	References
Sc. (Candida) shehatae	Gut of beetle; Soil	Guatemala; Munich (Germany); Kyoto (Japan)	[44, 51, 52]
Sc. stambukii	Rotting wood	Amazonian forest (Brazil)	[36]
Sc. anoplophorae	Gut of beetle (*Anoplophora leechi*)	Nanyang City, Henan Province (China)	[32]
Sc. jinghongensis	Rotting wood	Jinghong City, Yunnan Province (China)	[32]
Sc. paraergatensis	Rotting wood	Nanyang City, Henan Province (China)	[32]
Sc. (Candida) ergatensis	Larvae of *Ergates faber*	Madrid (Spain)	[26, 28]
Sc. titanus	Gut of beetle (*Dorsus titanus*)	Baotianman Mountain, Henan Province, China	[31]
Sc. henanensis	Rotting wood	Baotianman Nature Reserve, Henan Province (China)	[24]
Sc. xylosifermentans	Insect tunnel on a rotten log	Bull Run Mountain, Broad Run, Virginia (USA)	[30]
Sc. parashehatae	Gut of *Odontotaenius disjunctus* larvae; Rotting wood	Bull Run Mountain, Broad Run, Virginia (USA); Atlantic Rainforest (Brazil)	[30, 40]
Sc. cryptocercus	Gut of wood roach (*Cryptocercus* sp.)	Appalachian Trail at Newfound Gap (USA)	[53]
Sc. (Candida) amazonensis	Rotting wood	Amazonian forest (Brazil)	[27, 28]

Tropical and humid environments are considered as biodiversity hotspots of these species. The regions belonging to the Atlantic Forest and the Amazon Forest, for example, present valuable ecosystems for the development and isolation of new yeast species of the genera *Spathaspora* and *Scheffersomyces*, some capable of producing xylanases and fermenting D-xylose. Yeasts are easily obtained in these places as they are considered decomposing microorganisms, acting on substrates rich in nutrients, such as plants and wood, carrying out the process of organic matter degradation [6, 10, 22, 33 - 35]. Despite comprising about 30% of the world's primary tropical forests; playing key roles in biodiversity, the carbon cycle, rain regime, and global climate; and being constantly used in various research studies, the Amazonian environment, specifically, still has a lot to offer in terms of biotechnological inputs, in terms of both new raw materials for the most diverse areas as well as new species of efficient microorganisms [33, 36].

In these tropical and humid ecosystems, rotting wood samples are valuable habitats of *Spathaspora* and *Scheffersomyces*. The physiological traces associated with the assimilation of xylose and cellobiose are related to the environments where these yeasts are usually isolated and the requirement to use the available lignocellulosic biomass [32].

Yeasts are highlighted for their ability to decompose and are among the primary colonizers of nutrient-rich substrates, such as soil and plant tissues. Moreover, they play a key role in relationships in habitats with other organisms, including mutualism, competition, parasitism, and pathogenicity [37]. As it is a complex tissue composed of a lignocellulosic structure, the survival of these organisms will depend on their ability to act in the biodegradation of the polysaccharide complex. This process is mediated by xylanases and cellulases, which act as agents to convert the structural complexes into di- and mono-saccharides. It will also depend on the yeast's ability to assimilate and metabolize pentoses and hexoses efficiently. And in this scenario, the species of yeasts *Spathaspora* and *Scheffersomyces* are always highlighted, given their ability to ferment pentoses [38].

The species of *Spathaspora* and *Scheffersomyces* vary in fermentation efficiency, which may be linked to local factors related to climate, landscape history, and especially the quality of the biomass in which the species develops and evolves [38]. They are generally isolated from rotting wood in tropical and high humidity environments such as tropical forests (see Table **1**). Morais and co-workers [35] identified the yeast *Sc. queiroziae* frequently in samples of rotting wood in the Atlantic Forest (Brazil), which corroborates the study by Santos and co-workers [39] that identified the same species in samples of rotting wood and wood-boring insect in two different locations of the Atlantic Forest (Brazil), suggesting that the species *Sc. queiroziae* inhabits this place as its natural habitat. Interestingly, the species *Sc. queiroziae* can assimilate cellobiose and other compounds related to rotting wood [39], and is just one example of the genus associated with habitats, such as rotting wood. *Sp. passalidarum*, one of the most highlighted species of the *Spathaspora* genus is known for its ability to ferment xylose, which also has as its primary habitat rotting wood in tropical forests, in addition to the gut of wood-feeding insects [12, 33]. The presence in habitats with complex substrates defines the physiological specialties for the essential needs of the organism [37], such as the enzymatic production of specialized ligninolytic enzymes and the consumption of disaccharides and pentoses.

Producing ethanol from cellobiose and xylose is rare among microorganisms, but it frequently occurs within members of the *Scheffersomyces* and *Spathaspora* genera. Still, it varies within different genera and species [16, 27, 30, 40]. The emphasis given to these yeasts is their relevance to technological advances in second-generation biofuels, considering the fermentation rates and yields of some species of *Spathaspora* and *Scheffersomyces*, such as the fermentation capacity of *Sp. hagerdaliae* to efficiently convert xylose into ethanol without reprogramming due to the presence of glucose [41].

Another habitat frequently reported is the gut of insects, in which the food base is composed basically of rotting wood. Yeasts depend on habitats with assimilable carbon sources, making herbivorous insects a perfect niche because these invertebrates feed on carbohydrate-rich biomasses [42]. Furthermore, the greater efficiency in metabolizing pentoses and cellobiose than other yeast species may result from this symbiosis with insects [25].

Yeasts isolated from the guts of insects are likely to be efficient in digesting the components of the host's diet. To survive in the environment, these organisms must resist toxic secondary metabolites and adapt to the intestinal physiological environment (low concentrations of oxygen, high concentrations of carbon dioxide, and extreme pH variation), giving it unique characteristics due to the selective pressure. Consequently, in each host generation, symbiotic yeasts can be exposed to different environments and selective pressures, which can favor the performance of the yeast with improved characteristics over others [25].

Although the role of yeasts in the gut of insects is not fully understood, there is evidence that the enzymes produced by these microorganisms aid in the digestion and detoxification of the host's diet, providing essential nutrients for direct conversion by insects. Therefore, insects that feed on woody biomasses are an excellent habitat for isolating microorganisms capable of assimilating lignocellulosic structures [30, 43]. An example of this is the description of *Spathaspora* and *Scheffersomyces* in the beetle's gut highlighted by Urbina and co-workers [44], corresponding to 76.5% of isolated yeast species, with the most abundant members being *Sc. shehatae* and *Sc. stipitis*.

Isolating and identifying pentose fermenting yeasts in environments, such as guts of insects and rotting wood, in different regions and climates will contribute to the expansion of environmental and industrial biotechnology by identifying valuable biological and genetic resources for the development of biorefineries from lignocellulosic biomasses [32].

THE INDUSTRIAL POTENTIAL OF THE SPECIES OF *SPATHASPORA* AND *SCHEFFERSOMYCES*

Most microorganisms, including native xylose-fermenting yeasts, cannot efficiently convert xylose to ethanol under industrially relevant conditions, such as strict anaerobic environment, high osmotic stress, high ethanol concentrations, and presence of inhibitors [54]. This has been one of the issues involved in the use of lignocellulosic hydrolysates in biorefineries since xylose is the second most abundant sugar present. Its non-utilization in the fermentative process affects the fermentation efficiency and consequently the economic viability of the bioprocess.

The most used yeast in the fermentation processes for ethanol production in distilleries is *S. cerevisiae*, but this species has a limited capacity to metabolize xylose [6]. The difficulty in assimilating and fermenting xylose reduces the efficiency in producing 2G ethanol, significantly increasing economic costs and reducing market attractiveness. In contrast, the *Spathapora* and *Scheffersomyces* genera are described as having species that exhibit effective xylose assimilation and fermentation capacities, converting this sugar into ethanol and xylitol [6]. Thus, besides being a valuable biological resource that is isolated from conserved natural habitats (mainly forests and parks), it can offer other products with high added value and lead to other novel scientific discoveries.

Several studies have reported the possibility of using these yeast genera for large-scale ethanol and xylitol production. What draws more attention, besides their ability to utilize xylose and other sugars in the metabolic conversion of these organisms, is their ability to produce ethanol and xylitol together in the same production process, which according to Unrean and Ketsub [55], can improve the profitability of the system by up to 2.3 times compared to producing only ethanol.

Table **2** shows some species of the genera *Scheffersomyces* and *Spathaspora*, producers of ethanol and/or xylitol from different substrates.

Table 2. Different species, substrates, and respective ethanol and xylitol yields.

Species	Carbon Source	Product	Yield (g g^{-1})	Productivity (g L^{-1} h^{-1})	References
Spathaspora					
Sp. arborariae UFMG-HM19.1A	Rice hull hydrolysates	Ethanol	0.45	0.16	[57]
		Xylitol	0.33	-	
Sp. arborariae UFMG-CM-Y352	Sugarcane bagasse hydrolysate	Ethanol	0.16	0.034	[40]
Sp. arborariae NRRL Y-48658	Soybean hull hydrolysate	Ethanol	0.38	-	[59]
		Xylitol	0.25	-	
Sp. arborariae UFMG-CM-Y352	Synthetic medium YPX (Yeast extract, peptone, xylose)	Ethanol	0.32	-	[54]
		Xylitol	0.18	-	
Sp. passalidarum NRRL Y-27907	Sugarcane bagasse hydrolysate	Ethanol	-	0.38-0.81	[56]

(Table 2) cont.....

Species	Carbon Source	Product	Yield (g g⁻¹)	Productivity (g L⁻¹ h⁻¹)	References
Sp. passalidarum UFMG-HDM-14.1	Sugarcane bagasse hydrolysate	Ethanol	0.32	0.34	[68]
Sp. passalidarum NRRL-27907	Sugarcane bagasse hydrolysate	Ethanol	0.25	0.16	[10]
		Xylitol	0.03	0.02	
Sp. passalidarum UFMG-CM-Y469	Synthetic medium YPX (Yeast extract, peptone, xylose)	Ethanol	0.48	-	[54]
		Xylitol	0.03	-	
Sp. boniae UFMG-CM-306	Sugarcane bagasse hydrolysate	Ethanol	0.03	0.02	[10]
		Xylitol	0.71	0.46	
Sp. brasiliensis UFMG-CM-Y353	Synthetic medium YPX (Yeast extract, peptone, xylose)	Ethanol	0.16	-	[54]
		Xylitol	0.47	-	
Sp. roraimanensis UFMG-CM-Y477	Synthetic medium YPX (Yeast extract, peptone, xylose)	Ethanol	0.06	-	[54]
		Xylitol	0.56	-	
Sp. suhii UFMG-CM-Y475	Synthetic medium YPX (Yeast extract, peptone, xylose)	Ethanol	0	-	[54]
		Xylitol	0.92	-	
Sp. xylofermentans UFMG-CM-Y478	Synthetic medium YPX (Yeast extract, peptone, xylose)	Ethanol	0.08	-	[54]
		Xylitol	0.51	-	
Sp. xylofermentans UFMG-CM-Y479	Mineral medium xylose	Ethanol	0.34	0.07	[68]
Scheffersomyces					
Sc. shehatae UFMG-HM 52.2	Sugarcane bagasse hydrolysate	Ethanol	0.38	0.19	[64]
Sc. amazonensis UFMG-CM-Y493	Rice hulls hydrolysate	Ethanol	> 0.22	-	[63]
		Xylitol	1.04	0.05	
Sc. (Candida) shehatae HM 52.2	Acid–enzymatic soybean hull hydrolysate	Ethanol	0.47	-	[59]
		Xylitol	0.23	-	
Sc. stipitis NRRL Y-7124	Sugarcane bagasse hydrolysate	Ethanol	-	0.35-0.40	[56]
Sc. stipitis NRRL Y-7124	Sugarcane bagasse hydrolysate	Ethanol	0.25	0.10	[40]
Sc. stipitis UFMG-CM-2303	Sugarcane bagasse hydrolysate	Ethanol	0.32	0.25	[10]
		Xylitol	0	0	
Sc. parashehatae UFMG-CM-Y507	Sugarcane bagasse hydrolysate	Ethanol	0.25	0.14	[40]
Sc. parashehatae UFMG-CM-Y507	Mineral medium xylose	Ethanol	0.25	0.16	[68]

(Table 2) cont.....

Species	Carbon Source	Product	Yield (g g^{-1})	Productivity (g L^{-1} h^{-1})	References
Sc. illinoinensis UFMG-CM-Y513	Sugarcane bagasse hydrolysate	Ethanol	0.23	0.15	[40]
Sc. lignose CBS 4705	Sugarcane bagasse hydrolysate	Ethanol	0.16	0.048	[40]

The yeast strain plays a crucial role in ethanol and xylitol production performance, directly affecting the productivity and yield. The utilization of different sugars, cellulosic and hemicellulosic, becomes an essential characteristic of the technological processes. Nakanishi and co-workers [56] showed that the yeast *Sp. passalidarum* can metabolize different sugars, such as glucose, xylose, and cellobiose, for the ethanol production in a fed-batch process with cell recycle, consuming all the glucose and more than 90% of the xylose when grown on sugarcane bagasse hydrolysate after 4th recycle, which was also observed for the yeast *Sc. stipitis*, but with 30% lower yield in ethanol production.

Using hydrolyzate with different fermentation processes allowed Cunha-Pereira and co-workers [57] to observe the diauxic effect on *Sp. arborariae* in the rice husk hydrolyzate xylose fermented only after the complete depletion of glucose. In addition, after the complete glucose depletion, ethanol was also metabolized. According to the authors, the conversion of xylose into ethanol is carried out until the xylose concentration reaches 10 g L^{-1}, a phase in which the cells start to consume ethanol concomitantly if microaerophilic conditions are maintained [57, 58]. However, this fact was not observed after the immobilization of the yeasts *Sp. arborariae* and *Sc. shehatae*, wherein soybean husk hydrolyzate medium was not catabolic repression in the presence of glucose, and the xylose consumption after 24 hours of fermentation was started concomitantly with the glucose consumption [59]. Still, as in the study by Cunha-Pereira and co-workers [57], the diauxic effect was observed for ethanol consumption after glucose depletion [59].

Cadete and co-workers [54] studied a group of xylose-fermenting *Spathaspora* species. The study revealed that *Sp. passalidarum* and *Sp. arborariae* species are ethanol-producing yeasts and *Sp. brasiliensis*, *Sp. roraimanensis*, *Sp. suhii* and *Sp. xylofermentanos* xylitol producers. Among the ethanol producers, *Sp. passalidarum* obtained the highest yields (0.48 g g^{-1}) under severe oxygen-limiting conditions (oxygen transfer rate (OTR) of approximately 1–2 mmol L^{-1} min^{-1}), making it industrially relevant. Although aeration rates are low, it is a requirement of great importance to achieve high process efficiency in an industrial environment [54].

Furthermore, these xylose-fermenting species can be applied directly for ethanol and xylitol production or to provide genes, enzymes, or sugar transporters to develop new industrial pentose fermentation strains [18, 54].

Use of *Scheffersomyces* and *Spathaspora* yeasts in large-scale biotechnological processes and for ethanol or xylitol production are influenced by several factors. Besides the specific yeast species, factors such as pH, oxygen availability, temperature, nutrient content, or the presence of inhibitors, for example, can drastically affect the productivity and yield of xylose fermentation [8].

According to Carneiro and co-workers [60], who analyzed the performance of the strain *Spathaspora* sp. JA1 isolated from rotting wood [61], acetic acid is one of the most critical inhibitors of xylose fermentation. This carboxylic acid is common to all hydrolysis processes, being formed by deacetylation of hemicelluloses. Concentrations of 2–5 g L^{-1} can already be considered inhibitory for most xylose-positive yeasts. Acetic acid permeates the cell, dissociates to its ionic form (acetate), and releases protons, decreasing cytosolic pH. This intracellular acidification may bring damage to the cell and inhibit yeast xylose metabolism [62]. Even after hydrolyzate detoxification, the presence of 3.2 g L^{-1} of acetic acid reduced the fermentation parameters of *Sc. parashehatae*, *Sc. illinoinensis*, *Sp. arborariae*, *Sc. stipitis*, and *Sc. lignosa* [40].

Besides acetic acid, other inhibitory compounds can originate from the hydrolysis step, such as furfural, hydroxymethylfurfural, and phenolic compounds, for example, with characteristics that vary depending on the substrate and type of hydrolysis performed [63]. In a study by Antunes and co-workers [64], the use of sugarcane bagasse hydrolyzate for ethanol production demonstrated the need for detoxification of the medium to increase the fermentative yield of *Sc. shehatae*. The authors used a neutralization and activated charcoal strategy with significant reduction of inhibitors, particularly hydroxymethylfurfural and furfural concentration from $1.0 \cdot 10^{-3}$ g L^{-1} to $5.0 \cdot 10^{-4}$ g L^{-1} and from $2.0 \cdot 10^{-3}$ g L^{-1} to $1.0 \cdot 10^{-3}$ g L^{-1}, respectively.

The operational processes also emerge as key mechanisms in the bioconversion of xylose. Santos and co-workers [65] demonstrated that batch operation combined with cell recycling and gradual temperature reduction (30 to 26 °C) throughout the batches is a viable and beneficial practice for ethanol production by *Sc. stipitis* from lignocellulosic hydrolysates. These operating conditions promoted a six-time increase in xylitol reductase enzyme activity, which increased the xylose utilization efficiency from 27–81%, resulting in an ethanol productivity of 1.53 g L^{-1} h^{-1}. Furthermore, the operation in batches and cell recycling minimizes the inhibitory effects. It allows higher yeast adaptation due to the possibility of slow assimilation of inhibitors by microorganisms, which decreases their inhibitory effects [56, 63].

Another key point for converting sugars (pentoses) into ethanol is avoiding the accumulation of xylitol (when it is not the product of interest), which can be achieved *via* controlling aeration since the regeneration of the NAD^+ cofactor is performed under these circumstances. In converting xylose into ethanol, this sugar is reduced to xylitol by the enzyme XR and then oxidized to xylulose by the enzyme xylitol dehydrogenase (XDH). Subsequently, xylulose is phosphorylated and enters the pentose phosphate pathway, where it is converted into glyceraldehyde-3-P and reduced to ethanol after glycolysis [8, 13]. According to Hou [23], most XRs have dual specificity, using NADPH and NADH as a cofactor, whereas XDHs use NAD+ as a cofactor time. Thus, the absence of this cofactor may bring about the accumulation of xylitol, inhibiting the conversion of xylulose into glyceraldehyde-3-P and the subsequent reduction to ethanol, which makes the oxygen supply fundamental to this process [61, 66]. However, excessive aeration favors aerobic cell growth resulting in low ethanol yields [13, 23].

Sp. passalidarum has stood out as one of the ethanol producers with higher yields and productivity under conditions of oxygen limitation, as demonstrated by Su and co-workers [66]. These authors found higher ethanol yields (0.45 g g^{-1}) for this yeast under more limited oxygenation conditions. Cadete and co-workers [63]. showed that *Sc. amazonensis* under severe oxygen limiting conditions resulted in higher xylitol productivity due to the higher XR activity and the shortage of NAD^+ needed by XDH. Thus, taking into account an industrial scale, the precise control of oxygen supply is a complex and expensive operation, which has encouraged research on the conversion of xylose to ethanol/xylitol under anaerobic conditions or severe oxygen limitation, being the product (ethanol/xylitol or both) a particularity of each metabolism [36, 54].

Adjustment of the operating conditions of the system is crucial to obtain the products of interest. *Sp. hagerdaliae* UFMG-CM-Y303 showed the ability to convert xylose into ethanol and concomitant consumption of xylose and glucose under anaerobic conditions, suggesting the presence of differentiated enzymatic and metabolic transport. Moreover, the suitability of the operating system for microaeration conditions (limited oxygen concentration) and complete aeration of the medium increased xylitol production, which approached the concentrations obtained for ethanol in the system [41].

Co-fermentation strategies can also be used to increase the fermentation yield and design multi-product systems. The simultaneous saccharification and co-fermentation process under oxygen limitation conditions was studied for *S. cerevisiae* (excellent converter of hexose to ethanol) and *Sp. arborariae* (assimilates and metabolizes pentose and hexose to ethanol and xylitol). Rice hull hydrolyzate was used, and an enzymatic complex developed for saccharification of lignocellulosic biomasses was added to the medium. The results demonstrated that cell co-culture combined with the enzyme complex improved the system's performance, increasing the consumption of xylose and arabinose and ethanol production [67].

By adjusting the system conditions, these yeasts can be used at the industrial level to produce ethanol, xylitol, or both. However, further research is still needed to elucidate all the factors that influence the conversion of sugars into these products and the possibilities of using these yeasts for genetic improvement to further favor various biotechnological processes.

WHOLE-GENOME SEQUENCED SPECIES OF *SPATHASPORA* AND *SCHEFFERSOMYCES*

During the last few years, due to the advances in the isolations and transfers within the genera *Scheffersomyces* and *Spathaspora*, genomic analysis has been suggested to allocate species in monophyletic structures and identify interesting genes for biotechnological processes of ethanol and xylitol production. Furthermore, the conversion of lignocellulosic biomass is dependent on a well-established microbiological system. For this, it is necessary to understand the genes involved in the conversion of the sugars. Thus, different yeasts of the *Spathaspora* and *Scheffersomyces* genera have already been sequenced by genomic analysis, and some of these studies are described below.

Lopes and co-workers [20] performed the genomic analysis of three species of the genus *Spathaspora*: *Sp. hagerdaliae* f.a., sp.nov. UFMG-CM-Y303[T] (=CBS 13475), *Sp. girioi* sp.nov. UFMG-CM-Y302[T] (=CBS 13476) and *Sp.gorwiae* f.a., sp. nov. UFMG-CM-Y312[T] (=CBS13472). These species were isolated from rotting wood collected from Atlantic Rainforest in Brazil. The genome analysis of the species had strong support for description within the genus *Spathapora*, maintaining a monophyletic structure. *Sp. girioi* was allocated as a sister species of *Sp. arborariae* and both were associated with *Sp. passalidarum* by a common ancestor. *Sp. gorwiae* and *Sp. hagerdaliae* formed a separate subclade, still associated with the other species of the genus.

Sp. girioi sp.nov. UFMG-CM-Y302[T] (=CBS 13476) can assimilate D-ribose and glycerol, and it cannot grow in the presence of ethanol, sorbitol, and hexadecane. This yeast is still able to grow on citrate and D-gluconate. These characteristics differ from closely related species, such as *Sp. materiae* and *Sp. brasiliensis. Sp.*

hagerdaliae f.a., sp.nov. UFMG-CM-Y303[T] (=CBS 13475) and *Sp. girioi* sp.nov. UFMG-CM-Y302[T] (=CBS 13476) can grow on D-gluconate, hexadecane, and L-arabinose and differ from the closest relative *C. lyxosophila* in these characteristics and is not growing on glycerol and soluble starch [20]. Also, according to the authors, *Sp. hagerdaliae* f.a., sp.nov. UFMG-CM-Y303[T] (=CBS 13475) is not able to assimilate L-sorbose, which may be associated with the absence of a second copy of the gene encoding sorbose reductase (*SOU1, K17742*, EC 1.1.1.289) [20].

Genomic analysis of the three *Spathaspora* species by Lopes and co-workers [20] enabled the identification of different genes related to the metabolic capacity of the species. According to the authors, one of the genes identified was responsible for coding xylan 1,4-beta-xylosidase (*XYL4, K15920*, EC 3.2.1.37), which according to the authors, is present in the genus *Spathaspora* and performs hydrolysis of (1→4)-beta-D-xylans and remove residues of D-xylose from the non-reducing termini. The presence of this enzyme corroborates with the environment in which the isolations were carried out (rotting wood) as well as the fermentation medium based on xylan as the only carbon source, which provides a competitive advantage for these species that can assimilate and metabolize D-xylose [20].

Another relevant point identified by Lopes and co-workers [20] was that the yeasts *Sp. hagerdaliae* f.a., sp.nov. UFMG-CM-Y303[T] (=CBS 13475) and *Sp. girioi* sp.nov. UFMG-CM-Y302[T] (=CBS 13476) exhibited a general loss of the urate/allantoin degradation pathway, which may be related to the environmental adaptation of anaerobic/fermentative pathways since the urate/allantoin degradation pathway uses a substantial amount of oxygen.

The isolation, identification, and genomic analysis of yeast *Spathaspora boniae* sp. nov. UFMG-CM-Y306[T] (=CBS 13262T) was also performed. This strain was isolated in the Atlantic Rainforest (Brazil). This yeast can assimilate cellobiose and salicin as the only carbon sources. Genomic analysis identified the presence of genes involved in the D-xylose conversion. Furthermore, the enzyme activity involved in xylose metabolism was determined, and a preference for XR activity was observed, which was detected with both cofactors (NADPH and NADH), although it was higher with NADPH. The choice of XR activity in *Sp. boniae* species suggests that xylitol is the main product during xylose fermentation, as XR is involved in converting D-xylose to xylitol [10, 19].

Scheffersomyces shehatae ATY839, isolated from Kyoto (Japan) soil samples, is characterized by its ability to ferment xylose to ethanol at 37 °C. When performing the genomic analysis, high-temperature tolerance was associated with

25 genes encoding heat shock proteins, which promote inducible tolerance in other microorganisms and are associated with survival at high temperatures. In addition, genes related to xylose and starch fermentation have also been reported [69].

The strategy of genome sequencing was also performed for *Spathaspora arborariae* UFMG-HM19.1AT (CBS 11463 = NRRL Y-48658) isolated from rotting wood in Atlantic Rainforest and Cerrado ecosystems in Brazil [15] and described by Lobo and co-workers [70]. Genes coding for sugar transport and enzymes responsible for converting D-xylose to D-xylose-5P have been found, which provides an explanation for the xylose fermentation assimilation capacity by the species [70]. Recently, several genes encoding xylose reductase and xylitol dehydrogenase (and combinations of them) from *Sc. passalidarum* and *Sc. arborariae* were expressed in *S. cerevisiae* for the efficient xylitol fermentative production from xylose [71].

Lopes and co-workers [72] sequenced the genome of the species of *Spathaspora xylofermentans* UFMG-HMD23.3 (=CBS 12681), isolated from rotting wood in Amazonian Forest (Brazil). This yeast can ferment xylose, and genes for the conversion of this sugar were identified: for the conversion of D-xylose to D-xylulose (XYL1 and XYL2), xylulokinase for incorporation of D-xylose-5P in the pentose phosphate pathway. Other species were also sequenced, such as *Sc. stambukii* [36], *Sc. stipitis* [73], and *Sp. passalidarum* [74].

The genomic analysis of the species that make up the genera *Spathaspora* and *Scheffersomyces* may aid in positioning the phylogenetic tree and determining the groups with a unique common ancestor to define the monophyletic structure of the genera. In addition, genomic sequencing may be performed to identify the genes of biotechnological interest.

CONCLUSION

The isolation and identification of new xylose-fermenting yeasts widen the horizon for the development of highly efficient multi-product industries. The genera *Scheffersomyces* and *Spathaspora* are valuable biotechnological resources for application in 2G ethanol conversion processes or for the identification of the genes and enzymes of interest. In humid and forest environments, yeasts are widely associated with the rotting wood or gut of the wood-eating insect. Genes related to xylose conversion present in the genera's main species demonstrate a strong correlation between the habitat, in which these organisms are isolated, and the ability to assimilate and metabolize a range of different sugars, which is of interest for processes based on the lignocellulosic biomass.

The main challenge for biorefineries is associated with the scaling of processes that produce different products from the same biomass with high efficiency *via* efficient pre-treatment methods involving the fractionation of structures (Fig. **1**). In this scenario, this chapter presented the yeasts of the genera *Scheffersomyces* and *Spathaspora* as alternatives to conventional yeasts that do not ferment pentoses. Thus, advances have already been made in the identification of new species as well as their applications in ethanol and xylitol conversion.

Future perspectives indicate the necessity for further studies on the taxonomic relationship of these genera, which may advance with the identification of new species and genomic analysis for alignment of the phylogenetic tree. Still, it is necessary to elucidate their mechanisms and fermentation capacities against inhibitors and other factors found on an industrial scale. The exploration of isolates in lignocellulosic biomass hydrolysates in different process configurations, co-cultures, and the production of multi-products will significantly advance this scenario.

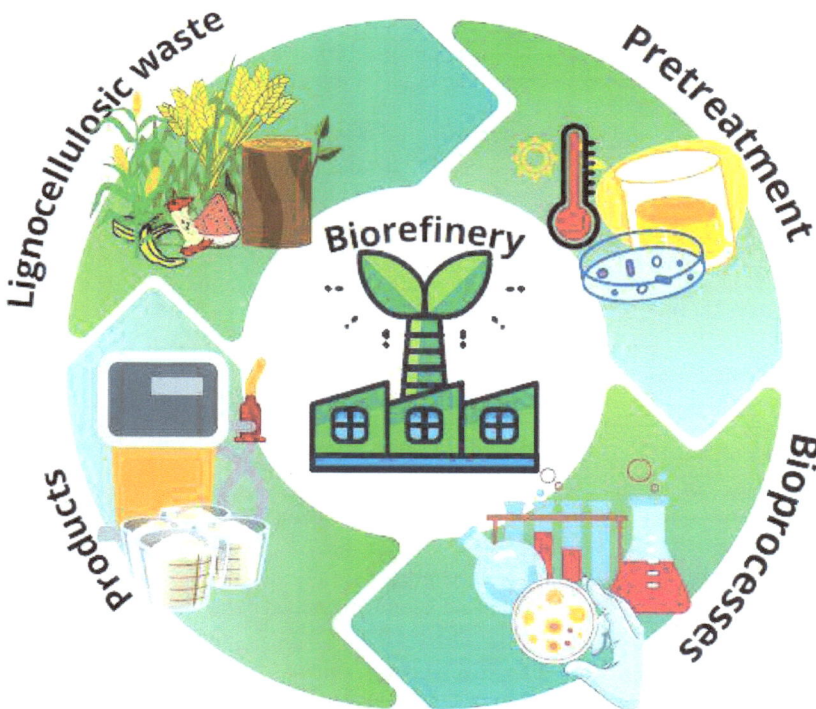

Fig. (1). Integration of different processes (waste recovery, pretreatment, efficient use of microorganisms, enzymes, and multiproducts) for the success of biorefineries based on lignocellulosic matrices towards the circularity of the production chains.

CONSENT FOR PUBLICATION

Not applicable.

CONFLICT OF INTEREST

The authors declare no conflict of interest, financial or otherwise.

ACKNOWLEDGEMENT

Declared none.

REFERENCES

[1]   Clark J, Deswarte F. The biorefinery concept. In: Clark J, Deswarte F, Eds. Introd to Chem from Biomass. 2nd ed. Wiley 2015; pp. 1-29.
       [http://dx.doi.org/10.1002/9781118714478.ch1]

[2]   Kumar B, Bhardwaj N, Agrawal K, Chaturvedi V, Verma P. Current perspective on pretreatment technologies using lignocellulosic biomass: An emerging biorefinery concept. Fuel Process Technol 2020; 199: 106244.
       [http://dx.doi.org/10.1016/j.fuproc.2019.106244]

[3]   Ruiz HA, Conrad M, Sun SN, *et al.* Engineering aspects of hydrothermal pretreatment: From batch to continuous operation, scale-up and pilot reactor under biorefinery concept. Bioresour Technol 2020; 299: 122685.
       [http://dx.doi.org/10.1016/j.biortech.2019.122685] [PMID: 31918970]

[4]   Treichel H, Fongaro G, Scapini T, Camargo AF, Stefanski FS, Venturin B, Eds. Utilising Biomass in Biotechnology. 1st ed., Cham: Springer International Publishing 2020.
       [http://dx.doi.org/10.1007/978-3-030-22853-8]

[5]   Rodrussamee N, Sattayawat P, Yamada M. Highly efficient conversion of xylose to ethanol without glucose repression by newly isolated thermotolerant *Spathaspora passalidarum* CMUWF1-2. BMC Microbiol 2018; 18(1): 73.
       [http://dx.doi.org/10.1186/s12866-018-1218-4] [PMID: 30005621]

[6]   Varize CS, Cadete RM, Lopes LD, *et al. Spathaspora piracicabensis* f. a., sp. nov., a D-xylos--fermenting yeast species isolated from rotting wood in Brazil. Antonie van Leeuwenhoek 2018; 111(4): 525-31.
       [http://dx.doi.org/10.1007/s10482-017-0974-8] [PMID: 29124467]

[7]   Selim KA, Easa SM, El-Diwany AI. The xylose metabolizing yeast *Spathaspora passalidarum* is a promising genetic treasure for improving bioethanol production. Fermentation (Basel) 2020; 6: 33-44.
       [http://dx.doi.org/10.3390/fermentation6010033]

[8]   Antunes FAF, Thomé LC, Santos JC, *et al.* Multi-scale study of the integrated use of the carbohydrate fractions of sugarcane bagasse for ethanol and xylitol production. Renew Energy 2021; 163: 1343-55.
       [http://dx.doi.org/10.1016/j.renene.2020.08.020]

[9]   Soares LB, Bonan CIDG, Biazi LE, *et al.* Investigation of hemicellulosic hydrolysate inhibitor resistance and fermentation strategies to overcome inhibition in non-saccharomyces species. Biomass Bioenergy 2020; 137: 105549.
       [http://dx.doi.org/10.1016/j.biombioe.2020.105549]

[10]   Morais CG, Sena LMF, Lopes MR, *et al.* Production of ethanol and xylanolytic enzymes by yeasts inhabiting rotting wood isolated in sugarcane bagasse hydrolysate. Fungal Biol 2020; 124(7): 639-47.
       [http://dx.doi.org/10.1016/j.funbio.2020.03.005] [PMID: 32540187]

[11] Kurtzman CP, Suzuki M. Phylogenetic analysis of ascomycete yeasts that form coenzyme Q-9 and the proposal of the new genera *Babjeviella, Meyerozyma, Millerozyma, Priceomyces*, and *Scheffersomyces*. Mycoscience 2010; 51: 2-14.
[http://dx.doi.org/10.1007/S10267-009-0011-5]

[12] Nguyen NH, Suh S-O, Marshall CJ, Blackwell M. Morphological and ecological similarities: wood-boring beetles associated with novel xylose-fermenting yeasts, *Spathaspora passalidarum* gen. sp. nov. and *Candida jeffriesii* sp. nov. Mycol Res 2006; 110(Pt 10): 1232-41.
[http://dx.doi.org/10.1016/j.mycres.2006.07.002] [PMID: 17011177]

[13] Cadete RM, Rosa CA. The yeasts of the genus *Spathaspora*: potential candidates for second-generation biofuel production. Yeast 2018; 35(2): 191-9.
[http://dx.doi.org/10.1002/yea.3279] [PMID: 28892565]

[14] Nguyen NH, Suh SO, Blackwell M. *Spathaspora* N.H. Nguyen, SO Suh & M. Blackwell (2006). In: Kurtzman CP, Fell JW, Boekhout T, Eds. The Yeasts. 5th ed. Elsevier 2011; pp. 795-7.
[http://dx.doi.org/10.1016/B978-0-444-52149-1.00068-9]

[15] Cadete RM, Santos RO, Melo MA, *et al. Spathaspora arborariae* sp. nov., a d-xylose-fermenting yeast species isolated from rotting wood in Brazil. FEMS Yeast Res 2009; 9(8): 1338-42.
[http://dx.doi.org/10.1111/j.1567-1364.2009.00582.x] [PMID: 19840117]

[16] Cadete RM, Melo MA, Zilli JE, *et al. Spathaspora brasiliensis* sp. nov., *Spathaspora suhii* sp. nov., *Spathaspora roraimanensis* sp. nov. and *Spathaspora xylofermentans* sp. nov., four novel (D)-xylos--fermenting yeast species from Brazilian Amazonian forest. Antonie van Leeuwenhoek 2013; 103(2): 421-31.
[http://dx.doi.org/10.1007/s10482-012-9822-z] [PMID: 23053696]

[17] Daniel HM, Lachance MA, Kurtzman CP. On the reclassification of species assigned to *Candida* and other anamorphic ascomycetous yeast genera based on phylogenetic circumscription. Antonie van Leeuwenhoek 2014; 106(1): 67-84.
[http://dx.doi.org/10.1007/s10482-014-0170-z] [PMID: 24748333]

[18] Wang Y, Ren YC, Zhang ZT, Ke T, Hui FL. *Spathaspora allomyrinae* sp. nov., a d-xylose-fermenting yeast species isolated from a scarabeid beetle *Allomyrina dichotoma*. Int J Syst Evol Microbiol 2016; 66(5): 2008-12.
[http://dx.doi.org/10.1099/ijsem.0.000979] [PMID: 26895992]

[19] Morais CG, Batista TM, Kominek J, *et al. Spathaspora boniae* sp. nov., a D-xylose-fermenting species in the *Candida albicans/Lodderomyces* clade. Int J Syst Evol Microbiol 2017; 67(10): 3798-805.
[http://dx.doi.org/10.1099/ijsem.0.002186] [PMID: 28884677]

[20] Lopes MR, Morais CG, Kominek J, *et al.* Genomic analysis and D-xylose fermentation of three novel *Spathaspora* species: *Spathaspora girioi* sp. nov., *Spathaspora hagerdaliae* f. a., sp. nov. and *Spathaspora gorwiae* f. a., sp. nov. FEMS Yeast Res 2016; 16(4): fow044.
[http://dx.doi.org/10.1093/femsyr/fow044] [PMID: 27188884]

[21] Lv SL, Chai CY, Wang Y, Yan ZL, Hui FL. Five new additions to the genus *Spathaspora* (Saccharomycetales, Debaryomycetaceae) from southwest China. MycoKeys 2020; 75: 31-49.
[http://dx.doi.org/10.3897/mycokeys.75.57192] [PMID: 33223920]

[22] Lopes MR, Lara CA, Moura MEF, *et al.* Characterisation of the diversity and physiology of cellobiose-fermenting yeasts isolated from rotting wood in Brazilian ecosystems. Fungal Biol 2018; 122(7): 668-76.
[http://dx.doi.org/10.1016/j.funbio.2018.03.008] [PMID: 29880202]

[23] Hou X. Anaerobic xylose fermentation by *Spathaspora passalidarum*. Appl Microbiol Biotechnol 2012; 94(1): 205-14.
[http://dx.doi.org/10.1007/s00253-011-3694-4] [PMID: 22124720]

[24] Ren Y, Chen L, Niu Q, Hui F. Description of *Scheffersomyces henanensis* sp. nov., a new D-xylos-

-fermenting yeast species isolated from rotten wood. PLoS One 2014; 9(3): e92315.
[http://dx.doi.org/10.1371/journal.pone.0092315] [PMID: 24647466]

[25]   Urbina H, Blackwell M. Multilocus phylogenetic study of the *Scheffersomyces* yeast clade and characterization of the N-terminal region of xylose reductase gene. PLoS One 2012; 7(6): e39128.
[http://dx.doi.org/10.1371/journal.pone.0039128] [PMID: 22720049]

[26]   Santa María J. *Candida ergatensis* nov. spec. An Del Inst Nac Investig Agrar 1971; 1: 85-8.

[27]   Cadete RM, Melo MA, Lopes MR, *et al. Candida amazonensis* sp. nov., an ascomycetous yeast isolated from rotting wood in the Amazonian forest. Int J Syst Evol Microbiol 2012; 62(Pt 6): 1438-40.
[http://dx.doi.org/10.1099/ijs.0.036715-0] [PMID: 21856981]

[28]   Urbina H, Blackwell M. New combinations, *Scheffersomyces amazonensis* and *S. ergatensis.* Mycotaxon 2013; 123: 233-4.
[http://dx.doi.org/10.5248/123.233]

[29]   Kurtzman CP, Robnett CJ. Relationships among genera of the Saccharomycotina (Ascomycota) from multigene phylogenetic analysis of type species. FEMS Yeast Res 2013; 13(1): 23-33.
[http://dx.doi.org/10.1111/1567-1364.12006] [PMID: 22978764]

[30]   Suh SO, Houseknecht JL, Gujjari P, Zhou JJ. Scheffersomyces parashehatae f.a., sp. nov., Scheffersomyces xylosifermentans f.a., sp. nov., Candida broadrunensis sp. nov. and Candida manassasensis sp. nov., novel yeasts associated with wood-ingesting insects, and their ecological and biofuel implications. Int J Syst Evol Microbiol 2013; 63(Pt 11): 4330-9.
[http://dx.doi.org/10.1099/ijs.0.053009-0] [PMID: 24014624]

[31]   Liu XJ, Cao WN, Ren YC, *et al.* Taxonomy and physiological characterisation of *Scheffersomyces titanus* sp. nov., a new D-xylose-fermenting yeast species from China. Sci Rep 2016; 6: 32181.
[http://dx.doi.org/10.1038/srep32181] [PMID: 27558134]

[32]   Jia RR, Lv SL, Chai CY, Hui FL. Three new *Scheffersomyces* species associated with insects and rotting wood in China. MycoKeys 2020; 71: 87-99.
[http://dx.doi.org/10.3897/mycokeys.71.56168] [PMID: 32855604]

[33]   Cadete RM, Melo MA, Dussán KJ, *et al.* Diversity and physiological characterization of D-xylos--fermenting yeasts isolated from the Brazilian Amazonian Forest. PLoS One 2012; 7(8): e43135.
[http://dx.doi.org/10.1371/journal.pone.0043135] [PMID: 22912807]

[34]   Cadete RM, Lopes MR, Rosa CA. Yeasts associated with decomposing plant material and rotting wood. In: Buzzini P, Lachance M-A, Yurkov A, Eds. Yeasts Nat Ecosyst Divers. 1st ed. Cham: Springer International Publishing 2017; pp. 265-92.
[http://dx.doi.org/10.1007/978-3-319-62683-3_9]

[35]   Morais CG, Cadete RM, Uetanabaro APT, Rosa LH, Lachance MA, Rosa CA. D-xylose-fermenting and xylanase-producing yeast species from rotting wood of two Atlantic Rainforest habitats in Brazil. Fungal Genet Biol 2013; 60: 19-28.
[http://dx.doi.org/10.1016/j.fgb.2013.07.003] [PMID: 23872280]

[36]   Lopes MR, Batista TM, Franco GR, *et al. Scheffersomyces stambukii* f.a., sp. nov., a d-xylos--fermenting species isolated from rotting wood. Int J Syst Evol Microbiol 2018; 68(7): 2306-12.
[http://dx.doi.org/10.1099/ijsem.0.002834] [PMID: 29786499]

[37]   Starmer WT, Lachance MA. Yeast Ecology. The Yeasts. Elsevier 2011; pp. 65-83.
[http://dx.doi.org/10.1016/B978-0-444-52149-1.00006-9]

[38]   Abrego N, Christensen M, Bässler C, Ainsworth AM, Heilmann-Clausen J. Understanding the distribution of wood-inhabiting fungi in European beech reserves from species-specific habitat models. Fungal Ecol 2017; 27: 168-74.
[http://dx.doi.org/10.1016/j.funeco.2016.07.006]

[39]   Santos RO, Cadete RM, Badotti F, *et al. Candida queiroziae* sp. nov., a cellobiose-fermenting yeast

species isolated from rotting wood in Atlantic Rain Forest. Antonie van Leeuwenhoek 2011; 99(3): 635-42.
[http://dx.doi.org/10.1007/s10482-010-9536-z] [PMID: 21136162]

[40] Cadete RM, Melo-Cheab MA, Dussán KJ, *et al.* Production of bioethanol in sugarcane bagasse hemicellulosic hydrolysate by *Scheffersomyces parashehatae, Scheffersomyces illinoinensis* and *Spathaspora arborariae* isolated from Brazilian ecosystems. J Appl Microbiol 2017; 123(5): 1203-13.
[http://dx.doi.org/10.1111/jam.13559] [PMID: 28799253]

[41] Dall Cortivo PR, Hickert LR, Rosa CA, Ayub MAZ. Conversion of fermentable sugars from hydrolysates of soybean and oat hulls into ethanol and xylitol by *Spathaspora hagerdaliae* UFMG-CM-Y303. Ind Crops Prod 2020; 146: 112218.
[http://dx.doi.org/10.1016/j.indcrop.2020.112218]

[42] Alves SL, Müller C, Bonatto C, *et al.* Bioprospection of enzymes and microorganisms in insects to improve second-generation ethanol production. Ind Biotechnol (New Rochelle NY) 2019; 15: 336-49.
[http://dx.doi.org/10.1089/ind.2019.0019]

[43] Vega FE, Dowd PE. The role of yeast as insect endosymbionts. In: Vega FE, Blackwell M, Eds. Insect-Fungal Assoc Ecol Evol. New York: Oxford University Press 2005; pp. 211-43.

[44] Urbina H, Schuster J, Blackwell M. The gut of Guatemalan passalid beetles: a habitat colonized by cellobiose- and xylose-fermenting yeasts. Fungal Ecol 2013; 6: 339-55.
[http://dx.doi.org/10.1016/j.funeco.2013.06.005]

[45] Barbosa AC, Cadete RM, Gomes FCO, Lachance M-A, Rosa CA. *Candida materiae* sp. nov., a yeast species isolated from rotting wood in the Atlantic Rain Forest. Int J Syst Evol Microbiol 2009; 59(Pt 8): 2104-6.
[http://dx.doi.org/10.1099/ijs.0.009175-0] [PMID: 19605715]

[46] Lachance MA, Boekhout T, Scorzetti G, Fell JW, Kurtzman CP. *Candida* Berkhout (1923). In: Kurtzman CP, Fell JW, Boekhout T, Eds. The Yeasts. 5th ed. Elsevier 2011; pp. 987-1278.
[http://dx.doi.org/10.1016/B978-0-444-52149-1.00090-2]

[47] Kurtzman CP. *Pichia* E.C. Hansen emend. Kurtzman. In: Kurtzman CP, Fell J, Eds. Yeasts A Taxon study. 4th ed. Amsterdam: Elsevier Science B.V. 1998; pp. 273-352.
[http://dx.doi.org/10.1016/B978-044481312-1/50046-0]

[48] Tanahashi M, Kubota K, Matsushita N, Togashi K. Discovery of mycangia and the associated xylose-fermenting yeasts in stag beetles (Coleoptera: Lucanidae). Naturwissenschaften 2010; 97(3): 311-7.
[http://dx.doi.org/10.1007/s00114-009-0643-5] [PMID: 20107974]

[49] Chang CF, Yao CH, Young SS, *et al. Candida gosingica* sp. nov., an anamorphic ascomycetous yeast closely related to *Scheffersomyces spartinae*. Int J Syst Evol Microbiol 2011; 61(Pt 3): 690-4.
[http://dx.doi.org/10.1099/ijs.0.020511-0] [PMID: 20382788]

[50] Jindamorakot S, Limtong S, Yongmanitchai W, *et al.* Two new anamorphic yeasts, *Candida thailandica* sp. nov. and *Candida lignicola* sp. nov., isolated from insect frass in Thailand. FEMS Yeast Res 2007; 7(8): 1409-14.
[http://dx.doi.org/10.1111/j.1567-1364.2007.00305.x] [PMID: 17854399]

[51] Grünwald S, Pilhofer M, Höll W. Microbial associations in gut systems of wood- and bark-inhabiting longhorned beetles [Coleoptera: Cerambycidae]. Syst Appl Microbiol 2010; 33(1): 25-34.
[http://dx.doi.org/10.1016/j.syapm.2009.10.002] [PMID: 19962263]

[52] Tanimura A, Nakamura T, Watanabe I, Ogawa J, Shima J. Isolation of a novel strain of *Candida shehatae* for ethanol production at elevated temperature. Springerplus 2012; 1: 27.
[http://dx.doi.org/10.1186/2193-1801-1-27] [PMID: 23961357]

[53] Urbina H, Frank R, Blackwell M. *Scheffersomyces cryptocercus*: a new xylose-fermenting yeast associated with the gut of wood roaches and new combinations in the *Sugiyamaella* yeast clade. Mycologia 2013; 105(3): 650-60.

[http://dx.doi.org/10.3852/12-094] [PMID: 23233509]

[54]   Cadete RM, de Las Heras AM, Sandström AG, *et al.* Exploring xylose metabolism in *Spathaspora* species: XYL1.2 from *Spathaspora passalidarum* as the key for efficient anaerobic xylose fermentation in metabolic engineered *Saccharomyces cerevisiae.* Biotechnol Biofuels 2016; 9: 167.
        [http://dx.doi.org/10.1186/s13068-016-0570-6] [PMID: 27499810]

[55]   Unrean P, Ketsub N. Integrated lignocellulosic bioprocess for co-production of ethanol and xylitol from sugarcane bagasse. Ind Crops Prod 2018; 123: 238-46.
        [http://dx.doi.org/10.1016/j.indcrop.2018.06.071]

[56]   Nakanishi SC, Soares LB, Biazi LE, *et al.* Fermentation strategy for second generation ethanol production from sugarcane bagasse hydrolyzate by *Spathaspora passalidarum* and *Scheffersomyces stipitis.* Biotechnol Bioeng 2017; 114(10): 2211-21.
        [http://dx.doi.org/10.1002/bit.26357] [PMID: 28627711]

[57]   da Cunha-Pereira F, Hickert LR, Sehnem NT, de Souza-Cruz PB, Rosa CA, Ayub MAZ. Conversion of sugars present in rice hull hydrolysates into ethanol by *Spathaspora arborariae, Saccharomyces cerevisiae,* and their co-fermentations. Bioresour Technol 2011; 102(5): 4218-25.
        [http://dx.doi.org/10.1016/j.biortech.2010.12.060] [PMID: 21220201]

[58]   Guo C, He P, Lu D, Shen A, Jiang N. Cloning and molecular characterization of a gene coding D-xylulokinase (CmXYL3) from *Candida maltosa.* J Appl Microbiol 2006; 101(1): 139-50.
        [http://dx.doi.org/10.1111/j.1365-2672.2006.02915.x] [PMID: 16834601]

[59]   Hickert LR, Cruz MM, Dillon AJP, Fontana RC, Rosa CA, Ayub MAZ. Fermentation kinetics of acid–enzymatic soybean hull hydrolysate in immobilized-cell bioreactors of *Saccharomyces cerevisiae, Candida shehatae, Spathaspora arborariae,* and their co-cultivations. Biochem Eng J 2014; 88: 61-7.
        [http://dx.doi.org/10.1016/j.bej.2014.04.004]

[60]   Carneiro CVGC, de Paula E Silva FC, Almeida JRM. Silva FC de P e, Almeida JRM. Xylitol production: Identification and comparison of new producing yeasts. Microorganisms 2019; 7(11): 484.
        [http://dx.doi.org/10.3390/microorganisms7110484] [PMID: 31652879]

[61]   Trichez D, Steindorff AS, Soares CEVF, Formighieri EF, Almeida JRM. Physiological and comparative genomic analysis of new isolated yeasts *Spathaspora* sp. JA1 and *Meyerozyma caribbica* JA9 reveal insights into xylitol production. FEMS Yeast Res 2019; 19(4): foz034.
        [http://dx.doi.org/10.1093/femsyr/foz034] [PMID: 31073598]

[62]   Bonatto C, Venturin B, Mayer DA, *et al.* Experimental data and modelling of 2G ethanol production by *Wickerhamomyces* sp. UFFS-CE-3.1.2. Renew Energy 2020; 145: 2445-50.
        [http://dx.doi.org/10.1016/j.renene.2019.08.010]

[63]   Cadete RM, Melo-Cheab MA, Viana AL, Oliveira ES, Fonseca C, Rosa CA. The yeast *Scheffersomyces amazonensis* is an efficient xylitol producer. World J Microbiol Biotechnol 2016; 32(12): 207.
        [http://dx.doi.org/10.1007/s11274-016-2166-5] [PMID: 27807756]

[64]   Antunes FAF, Chandel AK, Milessi TSS, Santos JC, Rosa CA, da Silva SS. Bioethanol production from sugarcane bagasse by a novel brazilian pentose fermenting yeast *Scheffersomyces shehatae* UFMG-HM 52.2: Evaluation of fermentation medium. Int J Chem Eng 2014; 2014: 1-8.
        [http://dx.doi.org/10.1155/2014/180681]

[65]   Santos SC, de Sousa AS, Dionísio SR, *et al.* Bioethanol production by recycled *Scheffersomyces stipitis* in sequential batch fermentations with high cell density using xylose and glucose mixture. Bioresour Technol 2016; 219: 319-29.
        [http://dx.doi.org/10.1016/j.biortech.2016.07.102] [PMID: 27498013]

[66]   Su YK, Willis LB, Jeffries TW. Effects of aeration on growth, ethanol and polyol accumulation by *Spathaspora passalidarum* NRRL Y-27907 and *Scheffersomyces stipitis* NRRL Y-7124. Biotechnol Bioeng 2015; 112(3): 457-69.

[http://dx.doi.org/10.1002/bit.25445] [PMID: 25164099]

[67] Hickert LR, de Souza-Cruz PB, Rosa CA, Ayub MAZ. Simultaneous saccharification and co-fermentation of un-detoxified rice hull hydrolysate by *Saccharomyces cerevisiae* ICV D254 and *Spathaspora arborariae* NRRL Y-48658 for the production of ethanol and xylitol. Bioresour Technol 2013; 143: 112-6.
[http://dx.doi.org/10.1016/j.biortech.2013.05.123] [PMID: 23792660]

[68] Souza RFR, Dutra ED, Leite FCB, *et al.* Production of ethanol fuel from enzyme-treated sugarcane bagasse hydrolysate using D-xylose-fermenting wild yeast isolated from Brazilian biomes. 3 Biotech 2018; 8: 312.
[http://dx.doi.org/10.1007/s13205-018-1340-x]

[69] Okada N, Tanimura A, Hirakawa H, Takashima M, Ogawa J, Shima J. Draft genome sequences of the xylose-fermenting yeast *Scheffersomyces shehatae* NBRC 1983T and a thermotolerant isolate of *S. shehatae* ATY839 (JCM 18690). Genome Announc 2017; 5(20): e00347-17.
[http://dx.doi.org/10.1128/genomeA.00347-17] [PMID: 28522710]

[70] Lobo FP, Gonçalves DL, Alves SL Jr, *et al.* Draft genome sequence of the D-xylose-fermenting yeast *Spathaspora arborariae* UFMG-HM19.1A. Genome Announc 2014; 2(1): e01163-13.
[http://dx.doi.org/10.1128/genomeA.01163-13] [PMID: 24435867]

[71] Mouro A, dos Santos AA, Agnolo DD, *et al.* Combining xylose reductase from *Spathaspora arborariae* with xylitol dehydrogenase from *Spathaspora passalidarum* to promote xylose consumption and fermentation into xylitol by *Saccharomyces cerevisiae*. Fermentation (Basel) 2020; 6: 72.
[http://dx.doi.org/10.3390/fermentation6030072]

[72] Lopes DD, Cibulski SP, Mayer FQ, *et al.* Draft genome sequence of the D-xylose-fermenting yeast *Spathaspora xylofermentans* UFMG-HMD23.3. Genome Announc 2017; 5(33): e00815-17.
[http://dx.doi.org/10.1128/genomeA.00815-17] [PMID: 28818907]

[73] Jeffries TW, Grigoriev IV, Grimwood J, *et al.* Genome sequence of the lignocellulose-bioconverting and xylose-fermenting yeast *Pichia stipitis*. Nat Biotechnol 2007; 25(3): 319-26.
[http://dx.doi.org/10.1038/nbt1290] [PMID: 17334359]

[74] Wohlbach DJ, Kuo A, Sato TK, *et al.* Comparative genomics of xylose-fermenting fungi for enhanced biofuel production. Proc Natl Acad Sci USA 2011; 108(32): 13212-7.
[http://dx.doi.org/10.1073/pnas.1103039108] [PMID: 21788494]

CHAPTER 9

Engineered *Saccharomyces* or Prospected non-*Saccharomyces*: Is There Only One Good Choice for Biorefineries?

Sérgio L. Alves Jr[1]**, Thamarys Scapini**[2]**, Andressa Warken**[2]**, Natalia Klanovicz**[2,3]**, Dielle P. Procópio**[4]**, Viviani Tadioto**[1]**, Boris U. Stambuk**[5]**, Thiago O. Basso**[4] **and Helen Treichel**[2,*]

[1] *Laboratory of Biochemistry and Genetics, Federal University of Fronteira Sul, Chapecó/SC, Brazil*

[2] *Laboratory of Microbiology and Bioprocesses, Federal University of Fronteira Sul, Erechim/RS, Brazil*

[3] *Research Group in Advanced Oxidation Processes (AdOx), Department of Chemical Engineering, Escola Politécnica, University of São Paulo, São Paulo SP, Brazil*

[4] *Department of Chemical Engineering, University of São Paulo, São Paulo SP, Brazil*

[5] *Department of Biochemistry, Federal University of Santa Catarina, Florianópolis SC, Brazil*

Abstract: Biorefineries require residual biomass as a raw material for their processes. Among all the possible products, 2G ethanol is undoubtedly the most studied and is probably the most desired in environmental terms. Carbohydrate-rich feedstocks used in biorefineries are mainly composed of polysaccharides, cellulose and hemicellulose (xylan), which initially require the action of hydrolytic enzymes to release their constituent monosaccharides, mostly glucose (from cellulose) and xylose (from hemicellulose). The conversion of glucose into ethanol is carried out by the yeast *Saccharomyces cerevisiae* with an efficiency close to the theoretical maximum yield (> 90%). Although it is the most widely used yeast in alcoholic fermentation processes, *S. cerevisiae* cannot metabolize xylose unless it undergoes genetic or evolutionary engineering. However, in recent decades, wild yeasts with an innate capacity to ferment this pentose and even hydrolyze the polysaccharides from lignocellulosic biomasses have been isolated and characterized from natural environments. Facing this duality, we conducted a major literature review and presented the data both in favor of engineering *S. cerevisiae* and the prospective use of wild yeasts in this chapter. To analyze the strengths of each strategy, this chapter also highlights the applications of integrated hydrolysis and fermentation processes and the possibility of simultaneously generating xylitol as the second product in biorefineries.

Corresponding author Helen Treichel: Laboratory of Microbiology and Bioprocess, Federal University of Fronteira Sul (UFFS), Erechim/RS, Brazil; Tel: +55 54 981126456; E-mail: helentreichel@gmail.com

Sérgio Luiz Alves Júnior, Helen Treichel, Thiago Olitta Basso and Boris Ugarte Stambuk (Eds.)
All rights reserved-© 2022 Bentham Science Publishers

Keywords: Bioprospection, *Candida*, Ethanol, Evolutionary engineering, Fermentation, Hydrolysis, Lignocellulosic biomass, *Meyerozyma*, *Pichia*, *Saccharomyces*, *Scheffersomyces*, *Spathaspora*, *Sugiyamaella*, *Wickerhamomyces*, Xylan, Xylanases, Xylitol, Xylooligosaccharide, Xylose.

INTRODUCTION

In terms of maximum production volume, fuel ethanol processing is the most extensive process to employ yeast as a fermenting microorganism. The world production of biofuel exceeds 100 billion liters per year. In this scenario, the USA and Brazil are the two largest producers, with ~60 and ~30 billion liters/year, respectively. These data place both countries at the forefront of bioethanol, although they have significantly different first-generation processes. In US production, yeasts ferment corn starch hydrolysates, whereas in Brazilian production, these microorganisms primarily ferment sucrose from the juice and molasses obtained from the milling of sugarcane [1].

Almost all ethanol production in both countries relies on the yeast *Saccharomyces cerevisiae*. This species is one of the best-studied eukaryotes, and its presence in fermentation processes dates back to the Neolithic revolution. Over thousands of years of coexistence with humanity, this yeast suffered different selective pressures that ended up domesticating it and generating a true workhorse microbe for the fermentation industry [1]. Although strains of *S. cerevisiae* differ genetically and phenotypically depending on the industrial sector in which they are found, some common characteristics make the species the preferred one in alcoholic fermentation, such as the ability to ferment sugars efficiently even in the presence of oxygen and to tolerate: (i) high concentrations of ethanol in the final stages of fermentation, (ii) the low pH levels of the medium, (iii) the osmotic stress caused by the high concentrations of sugars, and (iv) the hydrostatic pressure caused by the large volume of liquid contained in the fermentation tanks [2 - 4]. This yeast also stands out for being among the best glucose fermenters [5], which is the most abundant sugar in lignocellulosic residues, a biomass rich in cellulose and hemicellulose used as raw material in biorefineries [6 - 8]. From these residues, biorefineries can, separately or concomitantly, produce different fermentation products, including xylitol and second-generation ethanol (2G ethanol) [9 - 13]. Taking Brazil as an example, and considering only the sugarcane residue from the first-generation production of the fuel, it would be possible to increase the volume of ethanol produced in Brazil by up to 50%. To this end, however, it would be necessary for alcoholic fermentation to occur with a degree of efficiency of ~90%, similar to what already occurs for 1G ethanol [14, 15].

However, to achieve the production increase mentioned above, the fermenting microorganism must convert into ethanol the second-most abundant monosaccharide in lignocellulosic residue hydrolysates: xylose [16, 17]. In the fermentation of this pentose, one of the main obstacles of 2G ethanol is found, given that wild and industrial strains of *S. cerevisiae* are incapable of fermenting it [1, 15]. As a result, second-generation production is still in its infancy, representing less than 1% of the total volume of ethanol produced annually worldwide [18]. Thus, especially in the last two decades, research groups worldwide have made efforts to overcome the xylose-fermentation obstacle, either by engineering industrial strains of *S. cerevisiae* or by bioprospecting of wild non-*Saccharomyces* yeasts. In the present chapter, we address these different approaches to verify if there is a better choice for biorefineries.

FEEDSTOCK STRUCTURE AND FERMENTATION CHALLENGES

As an abundant source and for not serving as food for animals and humans, the great effort from scientists to replace the output of oil with ethanol is not unexpected [19]. The need for a transition from fossil fuels to renewable energy sources is evident on account of the climatic changes caused by modern times, mainly due to undesirable effects on atmospheric carbon balance and its disastrous effects on global warming. The development of transportation, for example, has influenced the environment in which emissions from internal combustion engines used in automobiles are the major source of air pollution in many urban areas [20]. The reduction of CO_2 emissions significantly contributes to minimizing environmental impact, and the use of a sugarcane ethanol system, for instance (like the Brazilian one), may offset 86% of CO_2 emissions compared to oil use [14].

Lignocellulosic biomass is found in several raw materials, ranging from urban and industrial waste, wood, and agricultural residues such as corn straw, wheat straw, rice straw, and sugarcane bagasse [21]. This material is derived from the cell wall of plants and is a rich source of inspiration for biotechnology, biofuels, and industrial biomaterials. The plant cell wall is a structure characterized by a mesh of polysaccharides, structural proteins, and phenolic compounds that protect the plant cell against external attacks and provide structural and mechanical support to the plant tissue, making it highly compact and treatment-resistant structure. It consists essentially of cellulose microfibrils, representing 30–60% of the total composition, as well as hemicellulose and lignin, representing between 20–40% and 10–20%, respectively. The chemical composition variation is due to various factors, such as climatic variability [7, 8, 22].

Cellulose is a linear homopolysaccharide composed of glucose units joined by β-1,4-glycosidic bonds. The hemicellulose fraction is essentially composed of xylan, whose structure is represented by complex polysaccharides. These polymers consist of a backbone of β-1,4-linked xylopyranoside, partially substituted with acetyl, glucuronosyl, and arabinosyl side chains [23]. It is structurally amorphous and variable, composed mainly of hexoses (such as glucose, galactose, and mannose), pentoses (such as xylose and arabinose), and uranic acids (such as galacturonic, glucuronic, and methylgalacturonic acid). The main hemicellulose chain contains mainly β-1,4-xylose (approximately 90%) and arabinose (10%) [24]. It is found in the walls of plant cells, associated with cellulose, and binds non-covalently to the surface of the cellulose fibrils by keeping them in place [25].

Before fermentation takes place, the lignocellulosic biomass must be pretreated and hydrolyzed to convert cellulose and hemicellulose into fermentable sugars. Based on the polysaccharide compositions mentioned above, we see that glucose and xylose are the first- and second-most abundant monosaccharides in lignocellulosic hydrolysates, thus making them the two most important sugars to be fermented by yeast cells in a biorefinery. However, as previously stated, *S. cerevisiae*, the most employed yeast in alcoholic fermentation processes, cannot ferment xylose, which discourages 2G ethanol production. This led the scientific community to find ways to make this fermentation feasible.

Xylose may reach up to 40% of lignocellulosic biomass [26]. Although wild-type *S. cerevisiae* strains cannot use xylose as a carbon source, they can slowly and aerobically metabolize xylulose [27, 28]. This indicates that xylulokinase (XK) encoded by the *S. cerevisiae XKS1* gene can assist the introduction of xylulose into the pentose-phosphate pathway (PPP) at low rates. Furthermore, the sugar transporters encoded by the *HXT4*, *HXT5*, *HXT7*, and *GAL2* genes are able to transport xylose, albeit with poor affinity [29, 30]. To confer xylose-fermenting capabilities to *S. cerevisiae*, two metabolic pathways from other microorganisms have been introduced into this yeast, xylose isomerase (XI) or xylose reductase/xylitol dehydrogenase (XR/XDH) [31].

While XI is mainly retrieved from bacteria [32, 33], genes encoding XR and XDH are mostly cloned from filamentous fungi and xylose-fermenting yeasts [34]. In the first approach, xylose is directly isomerized to xylulose. In the second one, both XR and XDH, in two subsequent reactions, will reduce xylose to xylitol and then oxidize it into xylulose. Finally, either through XI or XDH/XR, XK will phosphorylate xylulose into xylulose-5P, which is then sent to PPP (Fig. **1**). However, for the XR/XDH approach to improve xylose utilization, XR and XDH should be capable of recycling the same coenzyme in both oxide-reduction reactions conducted by them (NADH/NAD⁺), thus avoiding a possible redox imbalance inside the cells, which tends to accumulate xylitol to the detriment of ethanol production [1, 15]. This is due to the dual specificity of XR for both NADPH and NADH. The activity of an NADPH-specific XR in line with an NAD-specific XDH gives rise to an excess of NADH and a lack of NAD⁺, making XDH activity unfeasible. This makes it impossible to convert xylose into ethanol anaerobically (see Fig. **1**) [35]. Indeed, this causes some yeast species to end up only generating xylitol from xylose, which also presents itself as a desirable biotechnological product – see below [36 - 38].

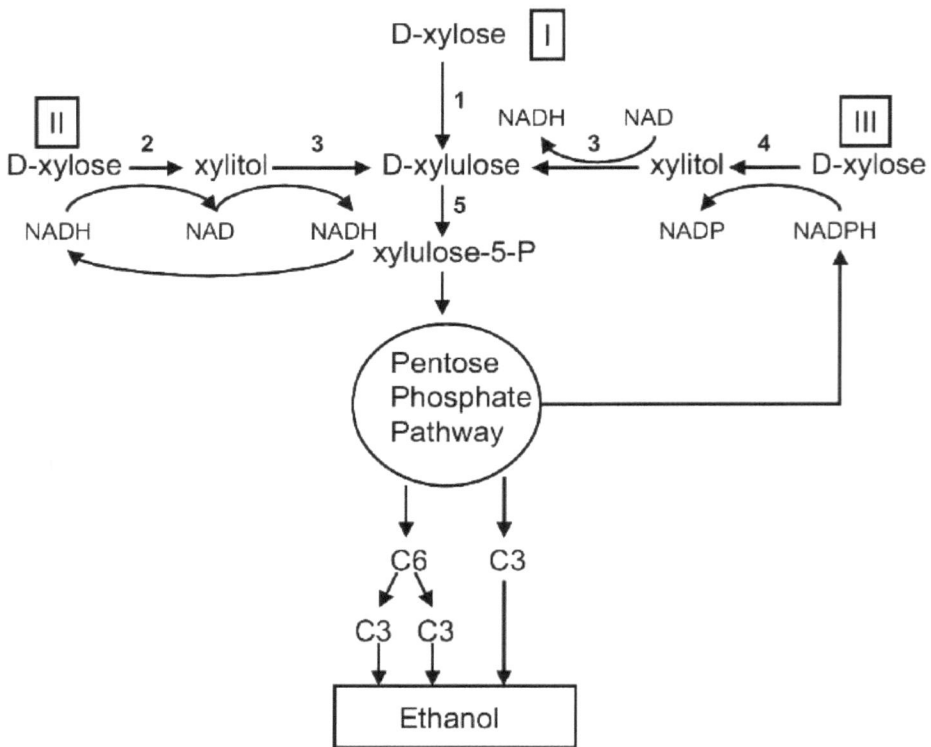

Fig. (1). Conversion of xylose into ethanol *via* XI (the route I) and XR/XDH (route II and III). 1, Xylose isomerase. 2, NADH-linked xylose reductase. 3, NAD-specific xylitol dehydrogenase. 4, NADPH-linked xylose reductase. 5, xyluloquinase. Adapted from Kuyper *et al.* [39].

This fermentative capacity implementation on yeasts has been the subject of many scientific studies. In addition to the redox-imbalance issue mentioned above, some studies have reported two other bottlenecks that must be overcome to enable *S. cerevisiae* for industrial applications: (i) the inefficient simultaneous co-fermentation of all sugars in the hydrolysate [40] and (ii) the lower ethanol productivity often observed in strains containing the XI pathway as compared to the XR/XDH pathway [41, 42]. In fact, to improve xylose fermentation, the literature reports the requirement of additional mutations in *S. cerevisiae*'s genetic background [39, 43 - 45]. Bracher *et al.* [45] verified that the deletion of the *GRE3* aldose-reductase gene and the overexpression of genes encoding xylulokinase (*XKS1*) and non-oxidative pentose phosphate pathway enzymes (*RKI1, RPE1, TAL1, TKL1*) enabled xylose metabolism under aerobiosis by a CEN.PK strain genetically engineered with xylose isomerase gene on xylose. In contrast, Verhoeven *et al.* [46] reported that XI-based *S. cerevisiae* requires an adaptation period before growth on xylose in anaerobic bioreactor cultures. These adapted mutants are known to carry mutations in *PMR1*, which encodes a Golgi Mn^{2+}/Ca^{2+} ATPase responsible for transporting Mn^{2+}/Ca^{2+} ions into the cell. These mutations implied an increase in the cellular content of Mn^{2+}, the preferred metal cofactor of XI [47].

Alternative metabolic engineering of XR/XDH-based *S. cerevisiae* also addressed the overexpression of the *TAL1* gene encoding transaldolase to increase the flux through the pentose phosphate pathway [48]. *TAL1* converts sedoheptulose-7-phosphate and glyceraldehyde-3-phosphate to erythrose-4-phosphate and fructose-6-phosphate. The intermediates glyceraldehyde-3-phosphate and fructose-6-phosphate are shared with glycolysis. Walfridsson *et al.* [48] mentioned that the bottleneck in PPP might be due to competition for glyceraldehyde-3-phosphate. Moreover, Hasunuma *et al.* [49, 50] demonstrated, in their work with laboratory strains of *S. cerevisiae*, an increase in tolerance to inhibitors (acetic acid and formic acid) and, consequently, an increase in ethanol yields in fermentations carried out at different concentrations of weak acids. In this case, besides the *TAL1* gene, the authors also overexpressed the *TKL1* gene, which encodes the PPP enzyme transketolase. These results were corroborated by other authors, who combined the overexpression of *TAL1* with the deletion of the *PHO13* gene (which encodes a *p*-nitrophenyl phosphatase) and verified an increase in fermentation performance in media containing weak acids as fermentation inhibitors [51, 52].

This last combination (*TAL1* overexpression and *PHO13* deletion), in particular, has been consistently pointed out in the literature as one of the most effective for enhancing the xylose-fermenting capabilities of *S. cerevisiae* [32, 53 - 55]. Indeed, by catalyzing the dephosphorylation of seduheptolose-7P in PPP, Pho13p indirectly represses the activity of transaldolase Tal1p and, therefore, prevents the formation of glycolytic pathway intermediates and, consequently, fermentation [32]. Furthermore, it has been shown that the transcription of non-oxidative PPP genes is increased in *S. cerevisiae pho13Δ* strains, intensifying the metabolic flux of xylose [54 - 56]. Kim *et al.* [54] showed that the transcriptional changes induced by *pho13Δ* cells required the transcription factor Stb5p, specifically activated under NADPH-limiting conditions. Therefore, as a response to oxidative stress, PPP genes and NADPH-producing genes are upregulated, suggesting that Pho13p may also be involved in cellular redox maintenance.

In addition to the positive effects of *PHO13* deletion on xylose fermentation, a novel knockout target for the improvement of lignocellulosic ethanol production was found: the global transcriptional factor *GCR2*. The investigators showed that *GCR2* deletion led to the upregulation of PPP genes and downregulation of glycolytic genes [57]. Those authors found that the *gcr2Δ* strain displayed an ethanol yield 3.8 times higher than the control strain. However, the *gcr2Δ* performance was not as good as that displayed by the *pho13Δ* strain, which showed a yield 35% higher. The combination of both deletions did not improve the fermentation either [57].

In contrast, the loss of xylose alcoholic fermentation favoring the accumulation of xylitol can also be desirable because this sugar alcohol is one of the top 12 bio-products. According to the United States Department of Energy, it is also among the leading products derived from biomass [58, 59]. Xylitol production in biorefineries is advantageous, with relatively high values in a growing global market with strong demand [60, 61]. It can be used as a sweetener, a functional food additive with proposed anticancerogenic effects [10], and as a drug carrier because of its high permeability and non-toxic nature. Moreover, xylitol prevents demineralization of teeth and bones, otitis media, and respiratory tract infections [58]. Due to this diversity of applications, *S. cerevisiae* strains have been engineered to optimize the production of this sugar alcohol [62 - 65]. In fact, xylitol production is in agreement with the biorefinery concept, as it provides opportunities for the production of different bio-products from residual biomass [9]. Interestingly, during xylitol accumulation in engineered yeasts, xylose is entirely channeled to the formation of this bio-product, that is, without being diverted to cell growth. In this sense, other carbon sources need to support the growth and metabolism of yeasts, which can be achieved by using the other carbohydrates present in lignocellulosic hydrolysates used as feedstocks in biorefineries [63].

Metabolism of Xylooligosaccharides by *S. cerevisiae*

Similar to xylose, xylooligosaccharides derived from plant cell walls are not catabolized by non-engineered *S. cerevisiae* [66]. In current industrial methods, in addition to harsh pretreatment of biomass, large quantities of cellulase and hemicellulose enzymes are required to release monosaccharides from plant cell wall polymers, posing unsolved economic and logistical challenges [67 - 69]. Enabling *S. cerevisiae* to metabolize oligosaccharides (such as xylobiose, xylotriose, and xylotetraose), instead of monomers only, would reduce costs, thus making the process more viable. In this sense, integrated technologies are estimated to significantly reduce the cost of converting lignocellulose into useful bioproducts [70]. The two combined approaches mostly addressed in the literature are: (i) simultaneous saccharification and fermentation (SSF), which integrates the enzyme responsible for substrate hydrolysis and the yeast responsible for the fermentation of monosaccharides [71]; and (ii) consolidated bioprocessing (CBP), which relies on the fact that one single microorganism accounts for the hydrolysis of polysaccharides as well as for the fermentation of sugars resulting from this hydrolysis [72].

The xylooligosaccharides are derived from hydrolyzed xylan by the action of the endo-β-xylanase (EC 3.2.1.8) enzyme, followed by β-D-xylosidase (EC 3.2.1.37) hydrolyzes of xylooligosaccharides to xylose [73, 74]. Some reports suggest that many fungi consume xylooligosaccharides derived from plant cell walls [75, 76]. Furthermore, as with intracellular β-glucosidases [77, 78], intracellular β-xylosidases are also widespread in fungi [75, 79]. Many fungal species can grow on plants because of their capacity to break down cellulose and hemicellulose into simple sugars.

After expressing a xylooligosaccharide transport and consumption pathway from the mold *Neurospora crassa* in *S. cerevisiae* coupled to the XR/XDH pathway, Li *et al.* [75] showed that the yeast was able to uptake xylobiose, xylotriose, and xylotetraose, and to hydrolyze these oligomers inside the cells, suggesting that the ability of *S. cerevisiae* to break down hemicellulose has the potential to improve the efficiency of biofuel production. However, the authors found that this conversion (xylobiose/xylotriose into ethanol) takes a considerable amount of time, which implies that improvements may be required before making it feasible for the industrial environment. In fact, the same expressive requirement of time had been previously seen by Fujii *et al.* [80], whose data showed that yeast cells took at least 72 h to consume half of the amount of xylotriose.

Nevertheless, the desired improvement seems to be nearly achieved. Recently, CBP assays carried out in corn cob liquor with a potential 32 g of xylose per liter afforded a yield of 0.328 g of ethanol per gram of potential sugar when the liquor was inoculated with the genetically modified *S. cerevisiae* ER-X-2P [81]. In this yeast, derived from the commercial bioethanol strain Ethanol Red, the investigators heterologously expressed genes encoding for *Aspergillus aculeatus* β-glucosidase 1 (*BGL1*), *A. oryzae* β-xylosidase A (*XYLA*), *Trichoderma reesei* endoxylanase II (*XYN*), *Clostridium phytofermentans* xylose isomerase (*xylA*) and

Scheffersomyces stipitis xylose reductase (*XYL1*) and xylitol dehydrogenase (*XYL2*). Moreover, Cunha *et al.* [81] also overexpressed *S. cerevisiae* xylulokinase and transaldolase genes in the strain ER-X-2P. With this genetically improved strain under CPB, the authors found a whole-process yield of 102.8 kg of ethanol from 1 ton of corn cob with no exogenous enzymes. It is worth noting that this amount represents, in terms of mass balance, a higher yield than that found in first-generation production, which is estimated to be ~63 kg of ethanol per ton of sugarcane [15].

CRISPR*ING* BIOREFINERIES IN HOPES OF A (BIO)SAFE CIRCULAR ECONOMY

Employing genetically modified organisms in the industry may be considered a biosafety matter [82]. The necessary precaution in this context is primarily due to the heterologous genes in many genetically modified organisms (GMOs). In such cases, the fermenting microorganism is considered transgenic, as it has been transformed with a DNA sequence coming from an unrelated species. Among these foreign sequences, antibiotic-resistance genes are usually required as transformant selective markers. Thus, considering (i) the large fermentation vats in biorefineries, (ii) the eventuality of transgenic microorganisms escaping into nature, (iii) their fitness advantage over indigenous microbes, (iv) the risk of horizontal gene transfer, and (v) the frequent appearance of novel microbial resistance against antibiotics, the large-scale industrial use of GM microorganisms is commonly avoided [83 - 88].

In contrast, especially in the last decade, the growing development of a novel genome-editing technology has brought to the table efficient alternative approaches for the generation of non-transgenic genetically engineered microorganisms. The so-called CRISPR-Cas system (an abbreviation for <u>C</u>lustered <u>R</u>egularly <u>I</u>nterspaced <u>S</u>hort <u>P</u>alindromic <u>R</u>epeats and <u>C</u>RISPR-<u>as</u>sociated genes) is a natural adaptable prokaryotic defense against exogenous DNA, such as bacteriophage's, that works as an "immune memory" in some Bacteria and Archaea, conferring resistance to eventually repeated infections caused by viruses [89 - 91]. On every new contact with phages, a CRISPR-containing prokaryote is able to integrate a DNA sequence from the invading pathogen as a new spacer sequence into its CRISPR locus [91]. In case the cell is reinfected by the same kind of virus (or a related one), a CRISPR RNA (crRNA) is transcribed from the above-mentioned phage-complementary spacer sequence and targets the invading DNA with the guidance of a *trans*-encoded small RNA (tracrRNA — which has complementary sequences to the repeat regions of crRNA precursor transcripts). Then, the Cas protein that assembles the

CRISPR system with these two RNA molecules (crRNA and tracrRNA) disarms the virus by cleaving its DNA [92 - 97].

Knowing the operational mode of the CRISPR-Cas system in prokaryotes, the 2020 Nobel laureate in chemistry [98], Emmanuelle Charpentier and Jennifer Doudna, proposed that both crRNA and tracrRNA can be fused together to form an active, chimeric single-guide RNA molecule (sgRNA). With this achievement, a simple two-component endonuclease system (sgRNA plus a Cas protein) was proven to be universally programmable to cleave any desired DNA sequence, as sgRNA assumes the pre-ordered shape [96, 99 - 105]. Indeed, this Nobel-awarded method has been increasingly used for biotechnological purposes, allowing, at the most unpretentious setting, at least the construction of *S. cerevisiae* industrial strains [106 - 113] and non-conventional yeasts [114, 115] without selective markers. Besides, these genetic editions enable the manipulation of multiple traits with one single transformation step [116 - 119].

Considering that many fermentative improvements can be achieved with gene knockouts, self-cloning (cisgenic transformation), or site-directed mutagenesis (single- or a few nucleotide mutations), the markerless CRISPR-Cas techniques consequently allows the construction of GM yeasts with no foreign DNA [120 - 122]. It is worth noting that sending the remaining yeasts (collected at the end of the fermentation processes) to animal feeding as single-cell protein (SCP) is imperative to increase profits to the biorefineries. While non-transgenic GM yeasts are accepted as SCP in many countries, those with foreign genes are not [88].

Although limited to the transference of genes from one yeast species to the same or to another closely related one, cisgenic transformations have proven to be an effective alternative for the food and beverage industry [122, 123]. Unfortunately, despite this well-played role in industrial sectors where transgenics are commonly frowned upon, the literature lacks articles that apply non-transgenic genome-editing techniques for biorefinery purposes. Nonetheless, the closer humanity gets to the UN 2030 Agenda deadline [124], the more it realizes how faster it needs to work to make up for the lost time. New bioprocesses that fit into the circular economy approach are thus mandatory. For all intents and purposes, humans may be putting biodiversity and the one-health concept [125] at risk by trying to achieve sustainability through the use of transgenic yeasts. Therefore, in this somewhat controversial scenario, CRISPR emerges as a promising tool not only to meet the biorefineries' urges, but also the environmental safety requirements.

WHAT ABOUT EVOLUTIONARY ENGINEERING?

Evolutionary engineering has become one of the most prominent and promising strain improvement methodologies owing to advances in "-omic" technologies. Also known as adaptive laboratory evolution (ALE), this strategy follows nature's engineering. It is based on exploring the plasticity of microbial genomes by designing and imposing cultivation regimes that confer an appropriate selective pressure, thus forcing the studied microorganism into the desired phenotype [126, 127].

The ALE tools may be either serial (repeated batch culture approach) or chemostats (continuous culture approach), depending on the study objectives. The fact is that a population of microorganisms, after several generations, ends up undergoing mutations and, within an environment that confers selective pressure, will also pass through selection for individuals with better conditions of development in this environment. In this way, the less adapted cells will be eliminated from these cropping systems; in repeated batches, each time the medium is renewed, and in the chemostats, continuously by the medium withdrawal system [128].

Evolutionary engineering can also be classified by the goal of the industrially relevant trait, which is a complementary strain improvement strategy. Improvement of substrate utilization and product formation, and improvement of stress resistance, such as high temperatures in the fermentation process, are examples of goals often required to provide an industrially efficient yeast strain [127]. In the biorefineries context, the most appropriate flowchart to follow to achieve the goal, from the wild-type to the evolved strain, is presented in Fig. (**2**). In sequence, we will present the basic concept of evolutionary engineering, the possibilities of operational approaches, and the latest studies and results applying this tool for ethanol production.

Basic Concept of Evolutionary Engineering

As mentioned above and shown in Fig. (**2**), strain adaptive evolution is considered a complementary tool and often requires an upstream process. Again, the understanding of the basic evolutionary engineering concept can indicate why.

Based on Darwin's theory of the natural evolution of species, the desired phenotype in ALE can provide both positive, neutral, and deleterious mutations in strains. In fact, these tools assume the existence of a natural evolutionary path, which we can explore to find a beneficial mutation, resulting in a desired phenotypic change in the strain performance. Mutations are caused by sudden changes in the extracellular environment, perturbing the intracellular environment, and provoking changes in biochemical reactions, a natural response of living organisms to survive [129].

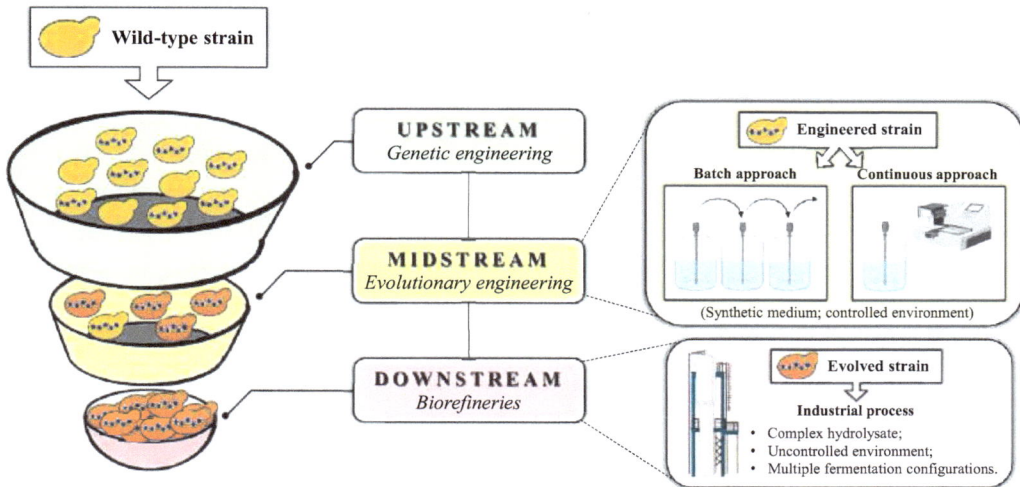

Fig. (2). Flowchart of a sustainable industrial practice of evolutionary engineering, from wild-type strain to evolved strain, in biorefineries context.

The desired genotype has already been developed when genetic engineering is applied to ALE. From this point on, the evolution tool becomes more powerful, since genes capable of providing beneficial mutations have already been selected and overexpressed. In addition, the enormous number of evolutionary possibilities is reduced, and step-wise improvements in the strain's phenotype are facilitated, considering that even modest genetic changes in cells provide great adaptations to the environment [130, 131].

Nespolo *et al.* [132] reanalyzed data from a representative yeast phylogenetic diversity library, including *Saccharomyces* spp., and tested whether multiple or a single event explains their fermentative capacity. Considering glycerol production, ethanol yield, and respiratory quotient, the study concluded that a single evolutionary episode did provide a phenotypic change in yeasts, instead of a serial genomic rearrangement. Thus, a single event appears to be the key factor behind fermentative yeast diversification and can be critical for creating novel mutations in strains through evolutionary engineering.

In this sense, it has already been recognized that end-point analyses are important in evolution experiments because continuous recombination and evolution of large populations reach the maximum positive response in a certain number of generations. Moreover, immediate selection or screening to recover only positive mutations is a successful way to achieve the study goal [127, 130].

Screening for mutations is difficult in heterogeneous and large populations, so the search for individual clones is not recommended to identify beneficial phenotypes. Instead, strains are often characterized as cultures, and the desired change is defined by their ability to grow in a certain environment and assessed by visual inspection. In the OD_{600} method, the optical density in a spectrophotometer at 600 nm — for example, the wild-type strain has an already known OD_{600}. The culture selection is made by measuring the value of each batch or time interval, representing the culture's ability to grow in media under unfavorable conditions. In this method, species survival is a statistical process through many generations or cycles [130].

However, screening through growth capability can become a trade-off when other aspects of yeast physiology are neglected. For example, the balance between substrate assimilation and product formation is important considering the context of biorefineries, and preferential resource allocation to grow is not considered a beneficial mutation, since the main objective is to produce ethanol. It has also been reported that after the desired phenotype is acquired and selective pressure is alleviated, the yeast stops expressing the beneficial characteristic [127]. To overcome these trade-offs, several operational approaches for evolving strains have been developed in evolutionary engineering experiments, as presented below.

Operational Approaches Used in Adaptive Laboratory Evolution

As mentioned above, the operational regime for evolving strains can be divided into batch and continuous cultivation. However, as Mans *et al.* [78] reported, several manipulations can be performed to induce more selective pressure on the strain culture and maintain the desired phenotype even after pressure alleviation.

When adopting the serial flask cultivation strategy, the strain was inoculated and maintained in a constant environment with selective pressure for a batch, and in the next cycle, it is transferred to a flask with the same environment or increasing pressure. This operational approach remains powerful because it is cheap, simple, and compatible with robotization. However, the desired phenotype can be lost after pressure alleviation, and the dynamic selection pressure strategy becomes an interesting alternative in these cases. This approach is based on batch intercalation under different selective pressures or batches with and without stress induction. Thus, the strain evolves under environmental fluctuations, and its phenotype will adapt to adverse conditions [127].

Continuous chemostat cultivation also enables yeast evolution under environmental fluctuations by selecting clones by substrate affinity. This approach enables investigations into the effects of individual environmental parameters on sugar transport in yeast, for example. In sugar-limited chemostat cultures, yeasts adapt their sugar transport systems to cope with low residual sugar concentrations, which are often in the micromolar range. Under these conditions, yeasts with high-affinity proton symport carriers have a competitive advantage over yeasts that transport sugars *via* facilitated-diffusion carriers [127, 130].

Chemostat cultivation offers unique possibilities for studying the energetic consequences of sugar transport in growing cells. For example, anaerobic sugar-limited chemostat cultivation has been used to quantify the energy requirement for the maltose-proton symport in *S. cerevisiae*. In addition, this operational approach has been used to produce ethanol using mixed sugars (xylose and arabinose) [133], to improve strain resistance under inhibitory compounds of lignocellulosic biomass [134], and to improve *S. cerevisiae* tolerance to ethanol [135] or even sucrose fermentation [136]. Although these studies successfully applied chemostat cultivation in yeast adaptive laboratory evolution, in recent years, the most used approach has been batch cultivation, as presented in the next section of this chapter.

Evolutionary Engineering with Ethanol Production Purpose

In addition to its excellent characteristics for a broad range of sugar assimilation in ethanol production, *S. cerevisiae* has exceptional biotechnological characteristics for adaptive evolution, such as high transformation efficiency, compatibility with a broad repertoire of episomal vectors, easy adaptation to demands of high-throughput screening, and production of numerous minor and major intermediates and metabolites, among others [131, 137].

Even robust industrial strains highly used in first-generation ethanol production, such as *S. cerevisiae* PE-2 and *S. cerevisiae* SA-1, have the plasticity to be explored by laboratory evolution, as reported by de Melo *et al.* [138]. In their study, thermotolerance was improved through a consecutive batch fermentation approach, where the strains were grown under increasing temperatures for up to 183 cycles in yeast extract-peptone-dextrose medium containing 20 or 60 g L^{-1} of glucose. Compared to the parental strain, the evolved *S. cerevisiae* PE-2 and SA-1 had 61% and 63% higher cell growth at 40 °C, respectively. Besides that, the evolved strains grew faster and had higher ethanol productivity (~2.5 g L^{-1} h^{-1}) than the parental strain with increased glucose concentration, with an ethanol yield of around 0.44 g ethanol per g sugar.

Although wild-type *S. cerevisiae* has the potential to evolve, there is a predominance in recently published studies applying this tool in engineered strains, following the flowchart in Fig. (**2**). Ko *et al.* [139] successfully applied this strategy to evolve the engineered *S. cerevisiae* XUSE, harboring the xylose isomerase pathway, with the aim of improving lignocellulosic biomass conversion under inhibitory conditions (acetic acid stress). After 13 cycles of serial subculturing in yeast synthetic medium containing 20 g L^{-1} of xylose, complete supplement mixture, and yeast nitrogen base, in the presence of increasing concentrations of acetic acid, the adapted strain showed 2.5-fold higher sugar consumption efficiency than the parental strain, even when the pH was not controlled. When the evolved strain was subjected to sugarcane bagasse hydrolysate medium (containing 34 g L^{-1} of glucose, 32 g L^{-1} of xylose, 3.1 g L^{-1} of acetic acid, and 0.7 g L^{-1} of phenolics), it showed better fermentation performance with a higher specific xylose consumption rate (0.88 g L^{-1} hr^{-1}) and ethanol yield (0.49 g ethanol per g sugars) than the not evolved strain. In addition, it is important to emphasize that the acetic acid concentrations in the growth media reached up to 6 g L^{-1} as byproducts of the reactions, and the evolved strain maintained high fermentation performance, highlighting its robustness under adverse conditions.

Also aiming at improving inhibitors tolerance, but adding the goals of improve xylose utilization and obtain a robust strain, Wei *et al.* [140] applied atmospheric and room temperature plasma mutagenesis, followed by consecutive batch fermentations under increasing concentrations of pretreated corn straw leachate to the engineered industrial strain *S. cerevisiae* LF1. The yeast already had superior co-fermentation capacity of glucose and xylose, but after the evolution experiment under yeast extract-peptone with increasing diluted leachate, the co-fermentation performance was increased, with ethanol productivity of 0.525 g g^{-1} h^{-1} in high-level mixed sugars (80 g L^{-1} glucose and 40 g L^{-1} xylose) in the absence of inhibitors. In addition, the fermentation time was shortened by 8 h compared to the parental strain. When non-detoxified hydrolysate was used in the fermentation process, containing inhibitors (acetic acid, HMF, furfural, and phenolics), the xylose conversion reached 94%, and ethanol yield was 0.43 g per g sugars.

Xie *et al.* [141] also improved xylose utilization by the engineered industrial strain *S. cerevisiae* HX57D after evolution through consecutive batch fermentations under increasing yeast nitrogen base and xylose concentrations (20 g L^{-1} or 40 g L^{-1}). After 56 cycles under selective pressure, the evolved strain reached xylose consumption rates around 1.02 g L^{-1} h^{-1} (up to 7 times higher than the parental strain), and ethanol yield of ~0.33 g ethanol per g xylose. When fermenting pretreated wheat straw slurry containing cellulose (81.3 g L^{-1}), glucose (4.2 g L^{-1}), xylose (14.8 g L^{-1}), formic acid (5.3 g L^{-1}), acetic acid (4.0 g L^{-1}), and phenolics (3.4 g L^{-1}), the xylose conversion was 47%, with a consumption rate of 0.10 g L^{-1} h^{-1}. The ethanol yield reached 0.38 g ethanol per g sugars, even with increasing acetic acid concentration. It is also worth emphasizing the higher xylitol yield and lower glycerol yield of the evolved strain compared with the parental strain.

In contrast to *S. cerevisiae*, non-conventional strains can be very sensitive to fermentation inhibitors and high ethanol concentrations, presenting more obstacles to be inserted in industrial processes but presenting an amazing resource to metabolize unusual substrates [142]. In some cases, these strains were isolated from coculture with industrial *S. cerevisiae* from adaptive laboratory evolution cultures, as reported by Moreno *et al.* [143]. In their study, after 400 generations of repetitive batch cultivation in a medium containing xylose and increasing wheat straw hydrolysate concentrations in the presence of *S. cerevisiae* and *Candida intermedia*, clones were isolated and their xylose conversion capability was evaluated. In a fermentation process containing 20 g L^{-1} of xylose and hydrolysate with glucose and inhibitors (acetic acid, formic acid, furfural, and 5-HMF), the *C. intermedia* clones were able to convert 42% of xylose, with an ethanol yield of 0.30 g ethanol per g sugars. According to the authors, *C. intermedia* may have prevailed in the coculture because of its superior capacity to utilize xylose, rather than a high inhibitor tolerance, and showed an interesting performance when fermenting mixed sugars.

In the studies mentioned above, simple evolutionary engineering approaches were used to achieve the pre-defined goals in the fermentative process to produce ethanol. In addition, the synergism between genetic engineering and adaptive laboratory evolution is a frequently used strategy to improve strain performance under adverse conditions. In this sense, a comprehensive understanding of natural microbial evolution and genome, combined with experimental exploration, testing hypotheses, and approaches, is necessary to develop beneficial phenotypes in strains, based on scientific knowledge and evolutionary engineering [130, 132].

Evolutionary Engineering in Biorefineries Context

Biorefineries show promising results in biomass processing and valuation, making them relevant for the circular economy and the generation of value-added products, especially biofuels [144]. However, their processes have limitations, and evolutionary engineering appears to be an interesting alternative for selecting characteristics of industrial interest in strains. Through techniques of identification and molecular engineering mechanisms, the performance in fuel production or other products can be significantly improved, guaranteeing economic viability [127, 145].

In addition to the metabolism of pentoses from hemicellulose, other significant problems in ethanol generation *via* alcoholic fermentation with wild-type strains are acetic acid release during hydrolysis and pretreatment processes that inhibit anaerobic growth. Moreover, during biomass processing, innumerable inhibitors derived from furan are generated, affecting glycolysis, fermentation, and the yeast tricarboxylic acid cycle during the stages of reducing polymerization degree, thermal treatments for breaking the crystalline cellulose, and hydrolysis [18]. Therefore, the selection of characteristics to cease or hinder inhibitor action during the various stages of biorefineries is of enormous importance in increasing productivity. Furthermore, evolutionary techniques to improve the strain tolerance to temperature and pH flocculation are of great interest.

In this regard, in-depth studies of evolutionary engineering are relevant in improving the conversion rate of sugars to ethanol or other products of interest, providing greater economic viability to the biorefineries associated with the valuation of agricultural by-products. The impact of evolutionary engineering in industrial processes has yet to be explored, but it currently promises to accelerate *in vivo* mutation rates and reduce strain developmental problems [127].

Evolution tools have the potential to fill gaps between upstream and downstream processes, because the controlled environment and synthetic medium, often used in genetic engineering, are not the reality of industrial processes, as shown in Fig. (**2**). Also, developing techniques to rapidly sequence, analyze and edit yeast genomes, while the adaptive evolution experiments are conducted, have the power to transform simply approaches, as batch cultivation, in a valuable cell factory to biorefineries. As small increases in sugar conversion rates result in a significant improvement in industrial-scale processes, evolved strains are promising for optimizing biorefinery performance, presenting the possibility of automation and scaling up, aspects that still need to be explored. This opens up opportunities for future research in partnership with the industrial sector.

EXPERIENCES WITH NON-*SACCHAROMYCES* YEASTS

S. cerevisiae strains have been the most used in initial cultures to produce 1G ethanol. This species became a model organism and was the first eukaryotic microbial species from which the entire genome was sequenced [146, 147]. However, as mentioned above, ethanol production from lignocellulosic raw material presents challenges for the yeast *S. cerevisiae*, linked mainly to the low tolerance to compounds with cellular inhibition capacity and limitations in the metabolism of pentoses. These technical challenges encourage the search for strategies to reduce ethanol production costs and increase fermentative yield by exploiting microorganisms capable of efficiently metabolizing hexoses and pentoses, and which tolerate the presence of inhibitory compounds in the hydrolysate [148]. In addition, the search for greater economic viability in 2G ethanol production pressures the exploration of alternatives beyond *Saccharo-myces* [149].

More than 2000 species of yeast have been isolated from different habitats and have industrially relevant characteristics, such as the ability to metabolize complex nutrients, tolerance to fermentation inhibitors, and various stresses, including osmotic, temperature, and ethanol concentration [150 - 152]. The need for survival in high-stress environments has driven the development of specific metabolic pathways in these microorganisms. This evolution of most species occurred in a manner adapted to the environment in which they evolved; therefore, most of them have new and unique mechanisms, which are not present in the yeast *S. cerevisiae* [151].

Several studies have reported strains of non-*Saccharomyces* yeasts, such as *Candida membranifaciens* [153], *Wickerhamomyces anomalus* [154], *Scheffer-somyces spitis* [155], *Spathaspora passalidarum* [156], *Zygosaccharomyces bailii* [113], and *Pichia kudriavzevii* [158], with promising fermentative characteristics and a wide spectrum of biotechnological applications in biorefineries. These bioresources are still in the phase of intense exploration and can harbor interesting genetic characteristics, mainly related to identifying characteristics of tolerance to environmental stress that are rarely presented by natural or industrial strains of *S. cerevisiae*. These new strategies can enable the development of superior *S. cerevisiae* strains or allow the direct use of non-*Saccharomyces* yeasts in biorefineries of multiple products (Table **1**).

Table 1. Non-*Saccharomyces* yeast with a remarkable feature for ethanol production.

Specie	Isolation Site	Remarkable Feature	References
Candida membranifaciens M2	Marine sources (dried or wet seaweed, marine organisms, sand, or driftwood)	Tolerance to inhibitor (acetic acid = 95.2 mM). Faster rates of fermentation than terrestrial yeast.	[153]
Candida sp.	Marine sediments	Tolerance to salt present (9% NaCl; 10% KCl; 10% CaCl$_2$). Wide range of pH tolerance (4.0 to 9.0). Consumed galactose.	[159]
Kluyveromyces marxianus	Soil	Thermotolerant yeast (> 37 °C).	[160]
Pichia caribbica UM-5	Rotten bagasse of *Agave tequilana*	Fermentation capacity of xylose in ethanol.	[161]

(Table 1) cont.....

Specie	Isolation Site	Remarkable Feature	References
Pichia kudriavzevii	Sugarcane molasses	Thermotolerant yeast (40 °C) Cell viability in a recycling system. High-ethanol and sugar tolerance. Consumed galactose.	[162]
Spathaspora passalidarum NRRL Y-27907	Rotting wood	Produce ethanol using xylose. Recycle of cells increase the xylose consumed. Consumed pentoses and hexoses in short time.	[156, 163]
Scheffersomyces parashehatae	Rotting wood and beetle larvae	Produce ethanol using xylose.	[164]
Wickerhamomyces sp. UFFS-CE-3.1.2	Rotting wood (Chapecó, Brazil)	Tolerance to inhibitor (acetic acid > 2 g L^{-1}); Consumed xylose in controlled pH (= 7,0) Fermentative capacity is not affected in the presence of saline water and shrimp wastewater	[154, 165, 166]
Wickerhamomyces anomalus M15	Marine sources (dried or wet seaweed, marine organisms, sand, or driftwood)	Tolerance a salt present (twice more than terrestrial yeast). Faster rates of fermentation than terrestrial yeast.	[153]
Wickerhamomyces anomalus WA-HF5.5	Rice hulls	Xylitol-producer capability when cultivated under oxygen-limiting.	[167]
Wickerhamomyces anomalus X19	Rooting wood (Tunísia)	Assimilation capacity of D-glucose, D-fructose, D-mannose, sucrose, D-xylose e L-arabinose. Tolerance to low pH values. Fermentation capacity of xylose, arabinose, and galactose.	[168, 169]
Zygosaccharomyces bailii MTCC 8177	Tea fungal mat of *Medusomyces gisevii*	Produce ethanol in seawater-based system (tolerance a salt stress). Tolerance a polyphenol present in tea waste (substrate).	[170, 171]

The isolation of yeasts in natural ecosystems can increase the possibility of obtaining genetic diversity among strains with desirable phenotypes for ethanol production and that thrive in intense stress environments while maintaining high yields. The exploration of these yeasts can lead to the identification of genes that provide tolerance responses to the chemical components of lignocellulosic biomass [153]. Similar to the case of the yeast *Zygosaccharomyces bailii*, which is tolerant to high concentrations of acetic acid, one of the main inhibitory compounds generated in 2G ethanol production [157, 172], and *Wickerhamo-*

myces strains, which have been isolated from different environments and have the ability to produce ethanol in systems based on seawater and in the presence of inhibitors [154, 165].

The ability of yeast to sustain the fermentation process in industrial systems, in an environment with high concentrations of sugars and ethanol, is a challenge for the adaptation of the cells and has been widely studied over the decades in non-*Saccharomyces* yeasts. Osmoadaptation is an important phenotype in yeast cells and results from the development of mechanisms to adjust metabolism, within certain limits, to high external osmolarity. These processes occur actively, based on detecting changes in molarity and stimulating dynamic cellular responses aimed at maintaining cellular activity [173]. This is the case of *Zygosaccharomyces rouxii*, which is recognized for its extreme osmotolerance, showing growth in concentrations of up to 90% (m v^{-1}) of sugars, tolerance to low pH, fast hexose fermentation, as well as tolerance to temperature increases and high concentrations of NaCl [172, 174, 175]. Other non-*Saccharomyces* yeasts isolated from contaminated food and beverages showed a growth capacity of 70% (m v^{-1}) of glucose. They were identified as *Candida bombi, Candida metapsilosis, Candida parapsilosis, Citeromyces matritensis, Kodamaea ohmeri, Metschnikowia pulcherrima, Schizosaccharomyces pombe, Torulaspora delbrueckii,* and *Zygosaccharomyces mellis* [176].

The ability to tolerate high-salt environments is an interesting feature, especially considering the recent discussions about the need to reduce freshwater consumption in industrial biofuel production and the search for alternatives, such as substitution for seawater or wastewater [165, 171, 177, 178]. Moreover, the great challenge of these systems is related to the fact that most of the isolated yeasts for ethanol production are terrestrial and have limited capacity to ferment in environments with a high salt content, which makes it an interesting alternative to the screening of yeasts in marine environments.

In marine environments, microorganisms adapt due to constant exposure to extreme conditions that force the evolution of protection and survival mechanisms that can override terrestrial strains [153]. This is corroborated by isolates of *S. cerevisiae* from marine environments, as demonstrated in the study by Zaky *et al.* [179], where the strain *S. cerevisiae* AZ65 showed fermentative yield greater than 73% using sugarcane molasses prepared in seawater and showed greater osmotic tolerance than the terrestrial yeast *S. cerevisiae* isolated from distillery. Metabolism adapted to extreme environmental conditions can give marine yeasts, in addition to the improved ability to tolerate salt stress, tolerance to inhibitory compounds (acetic acid, formic acid, hydroxymethylfurfural, and furfural), and better growth profile compared to terrestrial yeasts [153].

Tolerance to inhibitory compounds higher than that seen in *S. cerevisiae* has been reported for other non-*Saccharomyces* yeast species, such as *Wickerhamomyces* species that have already been isolated from different environments, from marine environments to rotting wood [9, 24]; they are able to metabolize a wide spectrum of carbon sources, act at low pH, high osmotic pressure, in addition to presenting physiological plasticity adapting to stress environments, which is a remarkable feature for biotechnological applications [180, 181]. Various strains of *Wickerhamomyces* have already demonstrated potential for application in ethanol production from glycerol [182], lignocellulosic hydrolysates, algae [154, 168], and starch substrates [180], as well as improved fermentation capacity in co-culture systems with *S. cerevisiae* producing ethanol and xylitol [167] and absence of inhibition in systems based on seawater and wastewater [165].

Some *Kluyveromyces* species also show excellent fermentative performance, and some strains show a broad spectrum of substrate assimilation, being able to produce ethanol from hexoses (glucose, galactose), pentoses (xylose, arabinose), and disaccharides (sucrose, cellobiose, lactose) [183, 184]. Besides, it is widely known that the species *K. marxianus* can tolerate and ferment at high temperatures [160]. The fermentative capacity at high temperatures is interesting for 2G ethanol production, considering that the process is carried out through SSF (see above). Since the optimum temperature for (hemi)cellulolytic enzymes (responsible for the enzymatic hydrolysis of biomass) occurs at approximately 50 °C, it is desirable that the fermenting microorganism can act well at high temperatures [185]. Moreover, compared to *S. cerevisiae*, these yeasts tolerate higher concentrations of inhibitors such as furans and acetic acid, showing superior growth and higher ethanol production [186].

A critical aspect of the isolation of non-*Saccharomyces* yeasts for ethanol production is the prospecting species that ferment high-yielding hemicellulosic sugars, mainly xylose. As mentioned above, xylose is the primary pentose in hemicellulose structures and the second most abundant monosaccharide in nature [164]. Yeasts that produce ethanol using pentose as a substrate have already been isolated from different environments, mainly from rotting wood and insect gut, and generally stand out for species of *Spathaspora* and *Scheffersomyces* (see Chapter 8 of this book).

Among the different species of *Spathaspora* capable of fermenting xylose, three are potential candidates for ethanol production: *Sp. gorwiae*, *Sp. hagerdaliae*, and *Sp. passalidarum*, the latter being the best ethanol producer, presenting high yields and productivity when grown in xylose medium under anaerobic conditions or with oxygen limitation [36, 156, 187, 188]. Another potential candidate for second-generation fuels is *Scheffersomyces* (*Pichia*) *stipitis*, which has an excellent ability to metabolize xylose and convert it into ethanol – up to 0.44 g of ethanol per g of xylose [189], that is, 86% of the maximum theoretical yield. The same holds true for *Scheffersomyces* (*Candida*) *shehatae*, a microorganism normally found to be associated with decaying lignocellulosic biomass. For decades, it has been considered one of the best xylose-fermenting yeast species [190 - 194]. Indeed, *Sc. shehatae* has been shown to ferment this pentose with yields of up to 0.44 g g^{-1} in media with xylose as the sole carbon source [194]. It is worth noting that *Sc. shehatae* performed ~95% fermentation efficiency when inoculated in wood hydrolysates [195] and ~96% in rice straw hydrolysate [196]. A drawback of these yeasts is their low tolerance to the ethanol produced, especially at high sugar concentrations, and also if the cells need to be recycled from one fermentation to the next.

In addition to ethanol production, in multi-product biorefineries, non-*Saccharomyces* yeasts may have an even more interesting potential because they are capable of producing other compounds in addition to ethanol, such as sugar alcohols, with an emphasis on xylitol [59]. In this sense, the search for non-*Saccharomyces* yeasts that can produce different products with potential applications for incorporation in biorefineries has been widely explored. *Hansenula polymorpha* is a yeast species known for its ability to tolerate high concentrations of ethanol, ferment at high temperatures, and mainly for its ability to ferment a wide spectrum of sugars and produce xylitol as the main product in the presence of glucose and xylose [61, 197].

In addition to *H. polymorpha*, the species of the genera *Spathaspora* and *Scheffersomyces* are excellent candidates for xylitol production, with high yield strains that can positively contribute to multiproduct biorefineries, such as strains of *Sp. boniae* isolated from rotting wood [198]. In addition to these species, the yeasts of the genera *Candida*, *Pichia*, and *Pachylosen* are also excellent candidates, such as *Candida amazonensis*, which can efficiently consume xylose and show high yields of xylitol [188].

Table **2** shows recent studies using non-*Saccharomyces* yeast in simulated processes using lignocellulosic biomass hydrolysates, demonstrating the relevance of these investigations for scientific advances, mainly for the development of biorefineries.

Although studies using non-*Saccharomyces* yeasts in ethanol production and multi-product systems have increasingly expanded, large-scale operation is still a major challenge, mainly because the production performance depends on the yeast strain employed. Moreover, challenges linked to low conversion rates, ethanol tolerance, system sensitivity to oxygen, and problems with the presence of inhibitors in the hydrolysate still need to be overcome [191]. Nevertheless, the advances related to isolation and batch experiments are relevant to these challenges, and non-*Saccharomyces* yeasts can be potential candidates to accelerate and enhance the development of biorefineries with multiple high-value-added products.

Table 2. Recent studies using non-*Saccharomyces* yeast in simulated processes using lignocellulosic biomass.

Specie	Raw Material	Fermentation Process	Ethanol Yield (%)	Ethanol Productivity (g L^{-1} h^{-1})	References
Candida tropicalis Y-27290	Sugarcane bagasse	Two-stage sequential enzymatic hydrolysis of pretreated biomass: 1st: pre-treatment 5 L reactor; NH$_4$OH 20% (w v^{-1}); 50°C and 48 h 2nd: hydrolysis with commercial xylanases and xylitol production 3rd: hydrolysis with commercial cellulase and ethanol production (SHF): supplementation with yeast extract, malt extract and peptone; 30 °C.	94.65 ± 1.2	1.54 ± 0.03	[199]
Scheffersomyces shehatae UFMG-HM 52.2	Sugarcane bagasse	Pre-treatment: stirred tak reactor; NaOH 0.5 M and 4 h. Enzymatic hydrolysis: Cellulases; 50 °C, 0.2 vvm and 72 h. Fermentation: stirred tank reactor; 0.2 vvm, 30 °C and 48 h	66.54	0.18 ± 0.02	[200]

Specie	Raw Material	Fermentation Process	Ethanol Yield (%)	Ethanol Productivity (g L^{-1} h^{-1})	References
Spathaspora passalidarum NRRL Y-27907	Sugarcane bagasse	Pre-treatment: 350 L reactor; NaOH solution (1.5% w v^{-1}) 130 °C Enzymatic hydrolysis: 15 L reactor; Celluclast 10 FPU g^{-1} and β-glucosidase 20 IU g^{-1}; 50 °C and 72 h Fermentation: 1.4 L reactor; NH$_4$Cl supplementation; recycle of cells (4 times); 200 rpm and 0.1 vvm	90 in 4th fed batch	0.81	[163]
Scheffersomyces stipitis NRRL Y-7124		Pre-treatment: 350 L reactor; NaOH solution (1.5% w v^{-1}) 130 °C Enzymatic hydrolysis: 15 L reactor; Celluclast 10 FPU g^{-1} and β-glucosidase 20 IU g^{-1}; 50 °C and 72 h Fermentation: 1.4 L reactor; recycle of cells (4 times); 200 rpm and no aeration	56 in 4th fed batch	0.36	
Wickerhamomyces anomalus WA-HF5.5 (rice hulls)	Soybean hull hydrolysate	Soybean hull pretreated with diluted acid (H$_2$SO$_4$) in autoclave. Bioreactor 2 L Co-culture: *S. cerevisiae* P6H9 Aeration rate of 0.33 vvm 30 °C and 180 rpm	99.8	4.28 ± 0.03	[167]

Xylanolytic Non-*Saccharomyces* Yeasts

As mentioned earlier in this chapter, it is highly desirable for biorefinery processes that the hydrolysis step of lignocellulosic-biomass polysaccharides is simultaneously carried out during the fermentation stage, either by means of SSF (combining hydrolytic enzymes and fermenting microorganisms) or by CBP (in which the same yeast is responsible for both hydrolysis and fermentation). In this sense, the literature points out some wild yeast species that present the required potential for a process to be conducted in the CBP format.

Lara *et al.* [193] isolated yeasts from decaying lignocellulosic materials collected from different Brazilian biomes and found three yeast strains that displayed high growth rates in xylan and high xylanase activity: *Cryptococcus laurentii* UFMG-HB-48, *Sugiyamaella smithiae* UFMG-HM-80.1, and *Sc. shehatae* UFMG-H-9.1a. As we have already mentioned in the present chapter, the last of these strains is also widely acknowledged for being an efficient xylose-fermenting yeast, making it a suitable choice for lignocellulosic-based bioprocesses through CBP. Furthermore, in a previous study by the same group, the authors also detected xylanase activity in *Sc. stipitis*, another potential yeast for 2G ethanol [201].

Xylanases are hydrolases indeed found in many genera of yeast. Besides *Cryptococcus*, *Sugiyamaella*, and *Scheffersomyces*, these enzymes have been reported in *Candida*, *Meyerozyma*, *Trichosporon*, *Moesziomyces*, *Pseudozyma*, *Hannaella*, *Kodamaea*, *Papiliotrema*, *Cystobasidium*, *Wickerhamomyces*, *Saturnispora*, *Saitozyma*, *Aureobasidium*, *Cyberlindnera*, *Pichia*, *Spencermartinsiella*, *Galactomyces*, *Spencermartinsiella*, *Lindnera*, *Dioszegia*, *Enteroramus*, and *Bulleromyces* [193, 198, 201 - 206].

Although decaying wood may be considered the main isolation substrate for xylanolytic yeasts, some of them have been found in insect guts (especially in beetles and termites), where they help those invertebrates to digest xylan [207, 208]. It is worth noting that, once again, the two species that stand out the most in the dual-role performance of xylan hydrolysis and xylose fermentation, *Sc. shehatae* and *Sc. stipitis,* are also among the most found in the insect microbiota [209 - 211]. Interestingly, it seems that xylanolytic yeasts are transient microorganisms in the gut of these animals swallowed during feeding. In this context, lignocellulosic biomass, whose decomposition process depends on microorganisms, is the primary reason for the symbiosis between yeast and insects, working as a carbon source for both parties [211 - 213]. The analysis of these ecological niches retraces humanity's path since the Neolithic revolution, but now the prospecting process has been done consciously and in a more accurate and accelerated way. When faced with the difficulties presented by *S. cerevisiae* — the yeast that has accompanied us the most —, humans now return to nature to take new yeasts to the laboratory and bioprocesses.

CONCLUSION

The literature points out three ways — mainly genetic engineering, evolutionary engineering, and prospecting for wild non-*Saccharomyces* yeasts — to optimize the fermentation process by yeasts in biorefineries, which may all be combined. Numerous studies, as discussed here, have indicated the success or potential of each strategy, which makes it difficult to choose one among them and makes it clear that there is not one specific choice that is better for biorefineries. In addition to this remarkable potential, the environmental and energy gains with the production of 2G ethanol have been mathematically proven. Nevertheless, the industrial sector does not seem to be convinced of the existence of an economic gain, given the asymmetry between the production volume of first- and second-generation ethanol. At the laboratory scale, several simulations of industrial conditions have already been tested; however, not much work has been carried out on semi-industrial and industrial scales. Perhaps this scaling up is what it takes to convince the industry.

CONSENT FOR PUBLICATION

Not applicable.

CONFLICT OF INTEREST

The authors declare no conflict of interest, financial or otherwise.

ACKNOWLEDGEMENTS

The authors would like to acknowledge the following Brazilian funding agencies: Conselho Nacional de Desenvolvimento Científico e Tecnológico – CNPq (grant numbers 454215/2014-2, 305258/2018-4, 429029/2018-7, and 308389/2019-0), Coordenação de Aperfeiçoamento de Pessoal de Nível Superior – CAPES (grant number 359/14), Financiadora de Estudos e Projetos – FINEP (grant number 01.09.0566.00/1421-08), Fundação de Amparo à Pesquisa do Estado de São Paulo – FAPESP (grant numbers 2018/17172-2 and 2018/01759-4), Fundação de Amparo à Pesquisa do Estado do Rio Grande do Sul – FAPERGS (grant number 17/2551-0001086-8), and Fundação de Amparo à Pesquisa do Estado de Santa Catarina – FAPESC (grant numbers 17293/2009-6 and 749/2016 T.O. 2016TR2188).

REFERENCES

[1]   Eliodório KP, de G e. Cunha GC, Müller C, Lucaroni AC, Giudici R, Walker GM. Advances in yeast alcoholic fermentations for the production of bioethanol, beer and wine. Adv Appl Microbiol. Academic Press Inc. 2019; 109: pp. 61-119.
        [http://dx.doi.org/10.1016/bs.aambs.2019.10.002]

[2]   Lagunas R. Energetic irrelevance of aerobiosis for *S. cerevisiae* growing on sugars. Mol Cell Biochem 1979; 27(3): 139-46.
        [http://dx.doi.org/10.1007/BF00215362] [PMID: 390364]

[3]   Lagunas R. Misconceptions about the energy metabolism of *Saccharomyces cerevisiae*. Yeast 1986; 2(4): 221-8.
        [http://dx.doi.org/10.1002/yea.320020403] [PMID: 3333454]

[4]   Landry CR, Townsend JP, Hartl DL, Cavalieri D. Ecological and evolutionary genomics of *Saccharomyces cerevisiae*. Mol Ecol 2006; 15(3): 575-91.
        [http://dx.doi.org/10.1111/j.1365-294X.2006.02778.x] [PMID: 16499686]

[5]   Merico A, Sulo P, Piškur J, Compagno C. Fermentative lifestyle in yeasts belonging to the Saccharomyces complex. FEBS J 2007; 274(4): 976-89.
        [http://dx.doi.org/10.1111/j.1742-4658.2007.05645.x] [PMID: 17239085]

[6]   Somerville C, Bauer S, Brininstool G, *et al.* Toward a systems approach to understanding plant cell walls. Science (80-) 2004; 306: 2206-11.
        [http://dx.doi.org/10.1126/science.1102765]

[7]   Murphy JD, McCarthy K. Ethanol production from energy crops and wastes for use as a transport fuel in Ireland. Appl Energy 2005; 82: 148-66.
        [http://dx.doi.org/10.1016/j.apenergy.2004.10.004]

[8]   Shen D, Xiao R, Gu S, Zhang H. The overview of thermal decomposition of cellulose in lignocellulosic biomass. Cellul. - Biomass Convers., InTech 2013.
        [http://dx.doi.org/10.5772/51883]

[9]   Bonatto C, Camargo AF, Scapini T, *et al.* Biomass to bioenergy research: current and future trends for biofuels Recent Dev Bioenergy Res. Elsevier 2020; pp. 1-17.
        [http://dx.doi.org/10.1016/B978-0-12-819597-0.00001-5]

[10]   Bertels LK, Fernández Murillo L, Heinisch JJ. The pentose phosphate pathway in yeasts–more than a poor cousin of glycolysis. Biomolecules 2021; 11(5): 725.
        [http://dx.doi.org/10.3390/biom11050725] [PMID: 34065948]

[11]   Monteiro de Oliveira P, Aborneva D, Bonturi N, Lahtvee P-J. Screening and growth characterization of non-conventional yeasts in a hemicellulosic hydrolysate. Front Bioeng Biotechnol 2021; 9: 659472.
        [http://dx.doi.org/10.3389/fbioe.2021.659472] [PMID: 33996782]

[12]   Ning P, Yang G, Hu L, *et al.* Recent advances in the valorization of plant biomass. Biotechnol Biofuels 2021; 14(1): 102.
        [http://dx.doi.org/10.1186/s13068-021-01949-3] [PMID: 33892780]

[13]   Rosales-Calderon O, Arantes V. A review on commercial-scale high-value products that can be produced alongside cellulosic ethanol. Biotechnol Biofuels 2019; 12: 240.
        [http://dx.doi.org/10.1186/s13068-019-1529-1] [PMID: 31624502]

[14]   Jaiswal D, De Souza AP, Larsen S, *et al.* Brazilian sugarcane ethanol as an expandable green alternative to crude oil use. Nat Clim Chang 2017; 7: 788-92.
        [http://dx.doi.org/10.1038/nclimate3410]

[15]   Stambuk BU, Eleutherio ECA, Florez-Pardo LM, Souto-Maior AM, Bon EPS. Brazilian potential for biomass ethanol: Challenge of using hexose and pentose cofermenting yeast strains. J Sci Ind Res (India) 2008; 67: 918-26.

[16]   Hans M, Garg S, Pellegrini VOA, *et al.* Liquid ammonia pretreatment optimization for improved release of fermentable sugars from sugarcane bagasse. J Clean Prod 2021; 281
        [http://dx.doi.org/10.1016/j.jclepro.2020.123922]

[17]   Ávila PF, Forte MBS, Goldbeck R. Evaluation of the chemical composition of a mixture of sugarcane bagasse and straw after different pretreatments and their effects on commercial enzyme combinations for the production of fermentable sugars. Biomass Bioenergy 2018; 116: 180-8.
        [http://dx.doi.org/10.1016/j.biombioe.2018.06.015]

[18]   Jansen MLA, Bracher JM, Papapetridis I, *et al.* *Saccharomyces cerevisiae* strains for second-generation ethanol production: from academic exploration to industrial implementation. FEMS Yeast Res 2017; 17(5): 44.
        [http://dx.doi.org/10.1093/femsyr/fox044] [PMID: 28899031]

[19] Aristidou A, Penttilä M. Metabolic engineering applications to renewable resource utilization. Curr Opin Biotechnol 2000; 11(2): 187-98.
[http://dx.doi.org/10.1016/S0958-1669(00)00085-9] [PMID: 10753763]

[20] Manahan SE. Industrial ecology: Environmental chemistry and hazardous waste. Lewis Publishers 2017.
[http://dx.doi.org/10.1201/9780203751091]

[21] Cardona CA, Sánchez ÓJ. Fuel ethanol production: process design trends and integration opportunities. Bioresour Technol 2007; 98(12): 2415-57.
[http://dx.doi.org/10.1016/j.biortech.2007.01.002] [PMID: 17336061]

[22] Pettersen RC. The Chemical Composition of Wood. 1984; pp. 57-126.
[http://dx.doi.org/10.1021/ba-1984-0207.ch002]

[23] Katahira S, Fujita Y, Mizuike A, Fukuda H, Kondo A. Construction of a xylan-fermenting yeast strain through codisplay of xylanolytic enzymes on the surface of xylose-utilizing *Saccharomyces cerevisiae* cells. Appl Environ Microbiol 2004; 70(9): 5407-14.
[http://dx.doi.org/10.1128/AEM.70.9.5407-5414.2004] [PMID: 15345427]

[24] McMillan JD. Pretreatment of Lignocellulosic Biomass. 1994; pp. 292-324.
[http://dx.doi.org/10.1021/bk-1994-0566.ch015]

[25] Timell TE. Recent progress in the chemistry of wood hemicelluloses. Wood Sci Technol 1967; 1: 45-70.
[http://dx.doi.org/10.1007/BF00592255]

[26] Ji X-J, Huang H, Nie Z-K, Qu L, Xu Q, Tsao GT. Fuels and chemicals from hemicellulose sugars. Adv Biochem Eng Biotechnol 2012; 128: 199-224.
[http://dx.doi.org/10.1007/10_2011_124] [PMID: 22249365]

[27] Eliasson A, Christensson C, Wahlbom CF, Hahn-Hägerdal B. Anaerobic xylose fermentation by recombinant *Saccharomyces cerevisiae* carrying XYL1, XYL2, and XKS1 in mineral medium chemostat cultures. Appl Environ Microbiol 2000; 66(8): 3381-6.
[http://dx.doi.org/10.1128/AEM.66.8.3381-3386.2000] [PMID: 10919795]

[28] Jeffries TW. Utilization of xylose by bacteria, yeasts, and fungi. Pentoses and Lignin. Berlin, Heidelberg: Springer-Verlag 1983; pp. 1-32.
[http://dx.doi.org/10.1007/BFb0009101]

[29] Hamacher T, Becker J, Gárdonyi M, Hahn-Hägerdal B, Boles E. Characterization of the xylose-transporting properties of yeast hexose transporters and their influence on xylose utilization. Microbiology 2002; 148(Pt 9): 2783-8.
[http://dx.doi.org/10.1099/00221287-148-9-2783] [PMID: 12213924]

[30] Patiño MA, Ortiz JP, Velásquez M, Stambuk BU. d-Xylose consumption by nonrecombinant *Saccharomyces cerevisiae*: A review. Yeast 2019; 36: 541-56.
[http://dx.doi.org/10.1002/yea.3429] [PMID: 31254359]

[31] Wisselink HW, Toirkens MJ, del Rosario Franco Berriel M, *et al.* Engineering of *Saccharomyces cerevisiae* for efficient anaerobic alcoholic fermentation of L-arabinose. Appl Environ Microbiol 2007; 73(15): 4881-91.
[http://dx.doi.org/10.1128/AEM.00177-07] [PMID: 17545317]

[32] Kwak S, Jin YS. Production of fuels and chemicals from xylose by engineered *Saccharomyces cerevisiae*: a review and perspective. Microb Cell Fact 2017; 16(1): 82.
[http://dx.doi.org/10.1186/s12934-017-0694-9] [PMID: 28494761]

[33] Cunha JT, Soares PO, Romaní A, Thevelein JM, Domingues L. Xylose fermentation efficiency of industrial *Saccharomyces cerevisiae* yeast with separate or combined xylose reductase/xylitol dehydrogenase and xylose isomerase pathways. Biotechnol Biofuels 2019; 12: 20.
[http://dx.doi.org/10.1186/s13068-019-1360-8] [PMID: 30705706]

[34]   Matsushika A, Inoue H, Kodaki T, Sawayama S. Ethanol production from xylose in engineered *Saccharomyces cerevisiae* strains: current state and perspectives. Appl Microbiol Biotechnol 2009; 84(1): 37-53.
[http://dx.doi.org/10.1007/s00253-009-2101-x] [PMID: 19572128]

[35]   Kuyper M, Winkler AA, van Dijken JP, Pronk JT. Minimal metabolic engineering of *Saccharomyces cerevisiae* for efficient anaerobic xylose fermentation: a proof of principle. FEMS Yeast Res 2004; 4(6): 655-64.
[http://dx.doi.org/10.1016/j.femsyr.2004.01.003] [PMID: 15040955]

[36]   Cadete RM, Rosa CA. The yeasts of the genus Spathaspora: potential candidates for second-generation biofuel production. Yeast 2018; 35(2): 191-9.
[http://dx.doi.org/10.1002/yea.3279] [PMID: 28892565]

[37]   Carneiro CVGC, de Paula E Silva FC, Almeida JRM. E Silva FC de P, Almeida JRM. Xylitol production: Identification and comparison of new producing yeasts. Microorganisms 2019; 7(11): E484.
[http://dx.doi.org/10.3390/microorganisms7110484] [PMID: 31652879]

[38]   Trichez D, Steindorff AS, Soares CEVF, Formighieri EF, Almeida JRM. Physiological and comparative genomic analysis of new isolated yeasts Spathaspora sp. JA1 and *Meyerozyma caribbica* JA9 reveal insights into xylitol production. FEMS Yeast Res 2019; 19(4): foz034.
[http://dx.doi.org/10.1093/femsyr/foz034] [PMID: 31073598]

[39]   Kuyper M, Harhangi HR, Stave AK, *et al.* High-level functional expression of a fungal xylose isomerase: the key to efficient ethanolic fermentation of xylose by *Saccharomyces cerevisiae?* FEMS Yeast Res 2003; 4(1): 69-78.
[http://dx.doi.org/10.1016/S1567-1356(03)00141-7] [PMID: 14554198]

[40]   Chakrabortee S, Byers JS, Jones S, *et al.* Intrinsically Disordered Proteins Drive Emergence and Inheritance of Biological Traits Intrinsically disordered proteins act in a prion-like manner to create protein-based molecular memories that drive the emergence of new, frequently adaptive traits. Int Cell 2016; 167(2): 369-381.e12.
[http://dx.doi.org/10.1016/j.cell.2016.09.017] [PMID: 27693355]

[41]   Li X, Park A, Estrela R, Kim S-R, Jin YS, Cate JH. Comparison of xylose fermentation by two high-performance engineered strains of *Saccharomyces cerevisiae*. Biotechnol Rep (Amst) 2016; 9: 53-6.
[http://dx.doi.org/10.1016/j.btre.2016.01.003] [PMID: 28352592]

[42]   Karhumaa K, Garcia Sanchez R, Hahn-Hägerdal B, *et al.* Comparison of the xylose reductase-xylitol dehydrogenase and the xylose isomerase pathways for xylose fermentation by recombinant *Saccharomyces cerevisiae*. Microb Cell Fact 2007; 6: 5.
[http://dx.doi.org/10.1186/1475-2859-6-5] [PMID: 17280608]

[43]   Van Vleet JH, Jeffries TW. Yeast metabolic engineering for hemicellulosic ethanol production. Curr Opin Biotechnol 2009; 20(3): 300-6.
[http://dx.doi.org/10.1016/j.copbio.2009.06.001] [PMID: 19545992]

[44]   Dos Santos LV, Carazzolle MF, Nagamatsu ST, *et al.* Unraveling the genetic basis of xylose consumption in engineered *Saccharomyces cerevisiae* strains. Sci Rep 2016; 6: 38676.
[http://dx.doi.org/10.1038/srep38676] [PMID: 28000736]

[45]   Bracher JM, Martinez-Rodriguez OA, Dekker WJC, Verhoeven MD, van Maris AJA, Pronk JT. Reassessment of requirements for anaerobic xylose fermentation by engineered, non-evolved *Saccharomyces cerevisiae* strains. FEMS Yeast Res 2019; 19(1)
[http://dx.doi.org/10.1093/femsyr/foy104] [PMID: 30252062]

[46]   Verhoeven MD, Lee M, Kamoen L, *et al.* Mutations in PMR1 stimulate xylose isomerase activity and anaerobic growth on xylose of engineered *Saccharomyces cerevisiae* by influencing manganese homeostasis. Sci Rep 2017; 7: 46155.
[http://dx.doi.org/10.1038/srep46155] [PMID: 28401919]

[47]   Lee M, Rozeboom HJ, de Waal PP, de Jong RM, Dudek HM, Janssen DB. Metal dependence of the xylose isomerase from *piromyces* sp. E2 explored by activity profiling and protein crystallography. Biochemistry 2017; 56(45): 5991-6005.
       [http://dx.doi.org/10.1021/acs.biochem.7b00777] [PMID: 29045784]

[48]   Walfridsson M, Hallborn J, Penttilä M, Keränen S, Hahn-Hägerdal B. Xylose-metabolizing Saccharomyces cerevisiae strains overexpressing the TKL1 and TAL1 genes encoding the pentose phosphate pathway enzymes transketolase and transaldolase. Appl Environ Microbiol 1995; 61(12): 4184-90.
       [http://dx.doi.org/10.1128/aem.61.12.4184-4190.1995] [PMID: 8534086]

[49]   Hasunuma T, Sung KM, Sanda T, Yoshimura K, Matsuda F, Kondo A. Efficient fermentation of xylose to ethanol at high formic acid concentrations by metabolically engineered Saccharomyces cerevisiae. Appl Microbiol Biotechnol 2011; 90(3): 997-1004.
       [http://dx.doi.org/10.1007/s00253-011-3085-x] [PMID: 21246355]

[50]   Hasunuma T, Ismail KSK, Nambu Y, Kondo A. Co-expression of TAL1 and ADH1 in recombinant xylose-fermenting *Saccharomyces cerevisiae* improves ethanol production from lignocellulosic hydrolysates in the presence of furfural. J Biosci Bioeng 2014; 117(2): 165-9.
       [http://dx.doi.org/10.1016/j.jbiosc.2013.07.007] [PMID: 23916856]

[51]   Li Y-C, Gou Z-X, Liu Z-S, Tang Y-Q, Akamatsu T, Kida K. Synergistic effects of TAL1 over-expression and PHO13 deletion on the weak acid inhibition of xylose fermentation by industrial *Saccharomyces cerevisiae* strain. Biotechnol Lett 2014; 36(10): 2011-21.
       [http://dx.doi.org/10.1007/s10529-014-1581-7] [PMID: 24966040]

[52]   Fujitomi K, Sanda T, Hasunuma T, Kondo A. Deletion of the PHO13 gene in *Saccharomyces cerevisiae* improves ethanol production from lignocellulosic hydrolysate in the presence of acetic and formic acids, and furfural. Bioresour Technol 2012; 111: 161-6.
       [http://dx.doi.org/10.1016/j.biortech.2012.01.161] [PMID: 22357292]

[53]   Kim SR, Skerker JM, Kang W, *et al.* Rational and evolutionary engineering approaches uncover a small set of genetic changes efficient for rapid xylose fermentation in *Saccharomyces cerevisiae*. PLoS One 2013; 8(2): e57048.
       [http://dx.doi.org/10.1371/journal.pone.0057048] [PMID: 23468911]

[54]   Kim SR, Xu H, Lesmana A, *et al.* Deletion of PHO13, encoding haloacid dehalogenase type IIA phosphatase, results in upregulation of the pentose phosphate pathway in *Saccharomyces cerevisiae*. Appl Environ Microbiol 2015; 81(5): 1601-9.
       [http://dx.doi.org/10.1128/AEM.03474-14] [PMID: 25527558]

[55]   Xu H, Kim S, Sorek H, *et al.* PHO13 deletion-induced transcriptional activation prevents sedoheptulose accumulation during xylose metabolism in engineered *Saccharomyces cerevisiae*. Metab Eng 2016; 34: 88-96.
       [http://dx.doi.org/10.1016/j.ymben.2015.12.007] [PMID: 26724864]

[56]   Ye S, Jeong D, Shon JC, *et al.* Deletion of PHO13 improves aerobic L-arabinose fermentation in engineered *Saccharomyces cerevisiae*. J Ind Microbiol Biotechnol 2019; 46(12): 1725-31.
       [http://dx.doi.org/10.1007/s10295-019-02233-y] [PMID: 31501960]

[57]   Shin M, Park H, Kim S, *et al.* Transcriptomic changes induced by deletion of transcriptional regulator *GCR2* on pentose sugar metabolism in *saccharomyces cerevisiae*. Front Bioeng Biotechnol 2021; 9: 654177.
       [http://dx.doi.org/10.3389/fbioe.2021.654177] [PMID: 33842449]

[58]   Ahuja V, Macho M, Ewe D, Singh M, Saha S, Saurav K. Biological and pharmacological potential of xylitol: A molecular insight of unique metabolism. Foods 2020; 9(11): E1592.
       [http://dx.doi.org/10.3390/foods9111592] [PMID: 33147854]

[59]   Werpy T, Petersen G. 2004.Top value added chemicals from biomass: Volume I -- Results of screening for potential candidates from sugars and synthesis gas. Golden, CO (United States):

[http://dx.doi.org/10.2172/15008859]

[60]   de Albuquerque TL, da Silva IJ, de Macedo GR, Rocha MVP. Biotechnological production of xylitol from lignocellulosic wastes: A review. Process Biochem 2014; 49: 1779-89.
[http://dx.doi.org/10.1016/j.procbio.2014.07.010]

[61]   Yamakawa CK, Kastell L, Mahler MR, Martinez JL, Mussatto SI. Exploiting new biorefinery models using non-conventional yeasts and their implications for sustainability. Bioresour Technol 2020; 309: 123374.
[http://dx.doi.org/10.1016/j.biortech.2020.123374] [PMID: 32320924]

[62]   Reshamwala SMS, Lali AM. Exploiting the NADPH pool for xylitol production using recombinant Saccharomyces cerevisiae. Biotechnol Prog 2020; 36(3): e2972.
[http://dx.doi.org/10.1002/btpr.2972] [PMID: 31990139]

[63]   Lane S, Dong J, Jin YS. Value-added biotransformation of cellulosic sugars by engineered *Saccharomyces cerevisiae*. Bioresour Technol 2018; 260: 380-94.
[http://dx.doi.org/10.1016/j.biortech.2018.04.013] [PMID: 29655899]

[64]   Jo JH, Oh SY, Lee HS, Park YC, Seo JH. Dual utilization of NADPH and NADH cofactors enhances xylitol production in engineered Saccharomyces cerevisiae. Biotechnol J. 10(12): 1935-43.
[http://dx.doi.org/10.1002/biot.201500068]

[65]   Mouro A, dos Santos AA, Agnolo DD, *et al*. Combining Xylose Reductase from Spathaspora arborariae with Xylitol Dehydrogenase from Spathaspora passalidarum to Promote Xylose Consumption and Fermentation into Xylitol by Saccharomyces cerevisiae. Fermentation (Basel) 2020; 6: 72.
[http://dx.doi.org/10.3390/fermentation6030072]

[66]   Young E, Lee S-M, Alper H. Optimizing pentose utilization in yeast: the need for novel tools and approaches. Biotechnol Biofuels 2010; 3: 24. [http://dx.doi.org/10.1186/1754-6834-3-24] [PMID: 21080929]

[67]   Lynd LR, Weimer PJ, van Zyl WH, Pretorius IS. Microbial cellulose utilization: fundamentals and biotechnology. Microbiol Mol Biol Rev 2002; 66(3): 506-77.
[http://dx.doi.org/10.1128/MMBR.66.3.506-577.2002] [PMID: 12209002]

[68]   Himmel ME, Ding S-Y, Johnson DK, *et al*. Biomass Recalcitrance: Engineering Plants and Enzymes for Biofuels Production. Science (80-) 2007; 315: 804-7.
[http://dx.doi.org/10.1126/science.1137016]

[69]   Jarboe LR, Zhang X, Wang X, Moore JC, Shanmugam KT, Ingram LO. Metabolic engineering for production of biorenewable fuels and chemicals: contributions of synthetic biology. J Biomed Biotechnol 2010; 2010: 761042.
[http://dx.doi.org/10.1155/2010/761042] [PMID: 20414363]

[70]   Lynd LR, van Zyl WH, McBride JE, Laser M. Consolidated bioprocessing of cellulosic biomass: an update. Curr Opin Biotechnol 2005; 16(5): 577-83.
[http://dx.doi.org/10.1016/j.copbio.2005.08.009] [PMID: 16154338]

[71]   Sewsynker-Sukai Y, Gueguim Kana EB. Simultaneous saccharification and bioethanol production from corn cobs: Process optimization and kinetic studies. Bioresour Technol 2018; 262: 32-41.
[http://dx.doi.org/10.1016/j.biortech.2018.04.056] [PMID: 29689438]

[72]   Robak K, Balcerek M. Review of second generation bioethanol production from residual biomass. Food Technol Biotechnol 2018; 56(2): 174-87.
[http://dx.doi.org/10.17113/ftb.56.02.18.5428] [PMID: 30228792]

[73]   Biely P. Microbial xylanolytic systems. Trends Biotechnol 1985; 3: 286-90.
[http://dx.doi.org/10.1016/0167-7799(85)90004-6]

[74]   Kulkarni N, Shendye A, Rao M. Molecular and biotechnological aspects of xylanases. FEMS Microbiol Rev 1999; 23(4): 411-56.

[http://dx.doi.org/10.1111/j.1574-6976.1999.tb00407.x] [PMID: 10422261]

[75] Li X, Yu VY, Lin Y, *et al.* Expanding xylose metabolism in yeast for plant cell wall conversion to biofuels. eLife 2015; 4
[http://dx.doi.org/10.7554/eLife.05896] [PMID: 25647728]

[76] Olson DG, McBride JE, Shaw AJ, Lynd LR. Recent progress in consolidated bioprocessing. Curr Opin Biotechnol 2012; 23(3): 396-405.
[http://dx.doi.org/10.1016/j.copbio.2011.11.026] [PMID: 22176748]

[77] Galazka JM, Tian C, Beeson WT, Martinez B, Glass NL, Cate JHD. Cellodextrin transport in yeast for improved biofuel production. Science (80-) 330: 84-6.2010;
[http://dx.doi.org/10.1126/science.1192838]

[78] Goldbeck R, Filho FM. Screening, characterization, and biocatalytic capacity of lipases producing wild yeasts from Brazil biomes. Food Sci Biotechnol 2013; 22: 79-87.
[http://dx.doi.org/10.1007/s10068-013-0052-6]

[79] Sun J, Tian C, Diamond S, Glass NL. Deciphering transcriptional regulatory mechanisms associated with hemicellulose degradation in Neurospora crassa. Eukaryot Cell 2012; 11(4): 482-93.
[http://dx.doi.org/10.1128/EC.05327-11] [PMID: 22345350]

[80] Fujii T, Yu G, Matsushika A, *et al.* Ethanol production from xylo-oligosaccharides by xylose-fermenting Saccharomyces cerevisiae expressing β-xylosidase. Biosci Biotechnol Biochem 2011; 75(6): 1140-6.
[http://dx.doi.org/10.1271/bbb.110043] [PMID: 21670522]

[81] Cunha JT, Romaní A, Inokuma K, *et al.* Consolidated bioprocessing of corn cob-derived hemicellulose: engineered industrial *Saccharomyces cerevisiae* as efficient whole cell biocatalysts. Biotechnol Biofuels 2020; 13: 138.
[http://dx.doi.org/10.1186/s13068-020-01780-2] [PMID: 32782474]

[82] Tolin SA, Vidaver AK. Genetically modified organisms: guidelines and regulations for research Ref Modul Biomed Sci. Elsevier 2018; pp. 378-89.
[http://dx.doi.org/10.1016/B978-0-12-801238-3.02357-6]

[83] Stirling F, Silver PA. Controlling the implementation of transgenic microbes: are we ready for what synthetic biology has to offer? Mol Cell 2020; 78(4): 614-23.
[http://dx.doi.org/10.1016/j.molcel.2020.03.034] [PMID: 32442504]

[84] Beacham TA, Sweet JB, Allen MJ. Large scale cultivation of genetically modified microalgae: A new era for environmental risk assessment. Algal Res 2017; 25: 90-100.
[http://dx.doi.org/10.1016/j.algal.2017.04.028]

[85] Glass DJ. Government regulation of the uses of genetically modified algae and other microorganisms in biofuel and bio-based chemical production. Algal Biorefineries. Cham: Springer International Publishing 2015; pp. 23-60.
[http://dx.doi.org/10.1007/978-3-319-20200-6_2]

[86] Andrade PP, Melo MA, Kido EA. Post-release monitoring: the Brazilian system, its aims and requirements for information. Transgenic Res 2014; 23(6): 1043-7.
[http://dx.doi.org/10.1007/s11248-014-9787-y] [PMID: 24659218]

[87] Gorter de Vries AR, Pronk JT, Daran JG. Lager-brewing yeasts in the era of modern genetics. FEMS Yeast Res 2019; 19(7): 63.
[http://dx.doi.org/10.1093/femsyr/foz063] [PMID: 31553794]

[88] Ritala A, Häkkinen ST, Toivari M, Wiebe MG. Single Cell Protein—State-of-the-Art, Industrial Landscape and Patents 2001–2016. Front Microbiol 2017.
[http://dx.doi.org/10.3389/fmicb.2017.02009]

[89] Mojica FJM, Díez-Villaseñor C, García-Martínez J, Soria E. Intervening sequences of regularly spaced prokaryotic repeats derive from foreign genetic elements. J Mol Evol 2005; 60(2): 174-82.

[http://dx.doi.org/10.1007/s00239-004-0046-3] [PMID: 15791728]

[90] Pourcel C, Salvignol G, Vergnaud G. CRISPR elements in *Yersinia pestis* acquire new repeats by preferential uptake of bacteriophage DNA, and provide additional tools for evolutionary studies. Microbiology 2005; 151(Pt 3): 653-63.
[http://dx.doi.org/10.1099/mic.0.27437-0] [PMID: 15758212]

[91] Barrangou R, Fremaux C, Deveau H, *et al.* CRISPR Provides Acquired Resistance Against Viruses in Prokaryotes. Science (80-) 2007; 315: 1709-2.
[http://dx.doi.org/10.1126/science.1138140]

[92] Brouns SJJ, Jore MM, Lundgren M, *et al.* Small CRISPR RNAs Guide Antiviral Defense in Prokaryotes. Science (80-) 2008; 315: 1709-2.
[http://dx.doi.org/10.1126/science.1159689]

[93] Deltcheva E, Chylinski K, Sharma CM, *et al.* CRISPR RNA maturation by trans-encoded small RNA and host factor RNase III. Nature 2011; 471(7340): 602-7.
[http://dx.doi.org/10.1038/nature09886] [PMID: 21455174]

[94] Bhaya D, Davison M, Barrangou R. CRISPR-Cas systems in bacteria and archaea: versatile small RNAs for adaptive defense and regulation. Annu Rev Genet 2011; 45: 273-97.
[http://dx.doi.org/10.1146/annurev-genet-110410-132430] [PMID: 22060043]

[95] Terns MP, Terns RM. CRISPR-based adaptive immune systems. Curr Opin Microbiol 2011; 14(3): 321-7.
[http://dx.doi.org/10.1016/j.mib.2011.03.005] [PMID: 21531607]

[96] Jinek M, Chylinski K, Fonfara I, Hauer M, Doudna JA, Charpentier E. A Programmable Dual-RN--Guided DNA Endonuclease in Adaptive Bacterial Immunity. Science (80-) 2012; 337: 816-21.
[http://dx.doi.org/10.1126/science.1225829]

[97] Gasiunas G, Barrangou R, Horvath P, Siksnys V. Cas9-crRNA ribonucleoprotein complex mediates specific DNA cleavage for adaptive immunity in bacteria. Proc Natl Acad Sci USA 2012; 109(39): E2579-86.
[http://dx.doi.org/10.1073/pnas.1208507109] [PMID: 22949671]

[98] The Nobel Prize. Press Release: The Nobel Prize in Chemistry 2020. Nobel Prize Outreach 2020; pp. 1-4.

[99] Charpentier E, Doudna JA. Biotechnology: Rewriting a genome. Nature 2013; 495(7439): 50-1.
[http://dx.doi.org/10.1038/495050a] [PMID: 23467164]

[100] Charpentier E, Marraffini LA. Harnessing CRISPR-Cas9 immunity for genetic engineering. Curr Opin Microbiol 2014; 19: 114-9.
[http://dx.doi.org/10.1016/j.mib.2014.07.001] [PMID: 25048165]

[101] Doudna JA, Charpentier E. The new frontier of genome engineering with CRISPR-Cas9. Science (80-) 2014; 346.
[http://dx.doi.org/10.1126/science.1258096]

[102] Barrangou R, Doudna JA. Applications of CRISPR technologies in research and beyond. Nat Biotechnol 2016; 34(9): 933-41.
[http://dx.doi.org/10.1038/nbt.3659] [PMID: 27606440]

[103] Jiang F, Doudna JA. CRISPR-Cas9 Structures and Mechanisms. Annu Rev Biophys 2017; 46: 505-29.
[http://dx.doi.org/10.1146/annurev-biophys-062215-010822] [PMID: 28375731]

[104] Doudna JA, Gersbach CA. Genome editing: the end of the beginning. Genome Biol 2015; 16: 292.
[http://dx.doi.org/10.1186/s13059-015-0860-5] [PMID: 26700220]

[105] Jiang F, Zhou K, Ma L, Gressel S, Doudna JA A. Cas9-guide RNA complex preorganized for target DNA recognition. Science (80-) 2015; 348: 1477-81.
[http://dx.doi.org/10.1126/science.aab1452]

[106] Cámara E, Lenitz I, Nygård Y. A CRISPR activation and interference toolkit for industrial Saccharomyces cerevisiae strain KE6-12. Sci Rep 2020; 10(1): 14605.
[http://dx.doi.org/10.1038/s41598-020-71648-w] [PMID: 32884066]

[107] Lee ME, DeLoache WC, Cervantes B, Dueber JE. A highly characterized yeast toolkit for modular, multipart assembly. ACS Synth Biol 2015; 4(9): 975-86.
[http://dx.doi.org/10.1021/sb500366v] [PMID: 25871405]

[108] Stovicek V, Borodina I, Forster J. CRISPR-Cas system enables fast and simple genome editing of industrial *Saccharomyces cerevisiae* strains. Metab Eng Commun 2015; 2: 13-22.
[http://dx.doi.org/10.1016/j.meteno.2015.03.001] [PMID: 34150504]

[109] Lian J, Bao Z, Hu S, Zhao H. Engineered CRISPR/Cas9 system for multiplex genome engineering of polyploid industrial yeast strains. Biotechnol Bioeng 2018; 115(6): 1630-5.
[http://dx.doi.org/10.1002/bit.26569] [PMID: 29460422]

[110] Shi S, Liang Y, Zhang MM, Ang EL, Zhao H. A highly efficient single-step, markerless strategy for multi-copy chromosomal integration of large biochemical pathways in *Saccharomyces cerevisiae*. Metab Eng 2016; 33: 19-27.
[http://dx.doi.org/10.1016/j.ymben.2015.10.011] [PMID: 26546089]

[111] Shi S, Qi N, Nielsen J. Microbial production of chemicals driven by CRISPR-Cas systems. Curr Opin Biotechnol 2021; 73: 34-42.
[http://dx.doi.org/10.1016/j.copbio.2021.07.002] [PMID: 34303184]

[112] Jessop-Fabre MM, Jakočiūnas T, Stovicek V, *et al.* EasyClone-MarkerFree: A vector toolkit for marker-less integration of genes into *Saccharomyces cerevisiae via* CRISPR-Cas9. Biotechnol J 2016; 11(8): 1110-7.
[http://dx.doi.org/10.1002/biot.201600147] [PMID: 27166612]

[113] Ryan OW, Cate JHD. Multiplex Engineering of Industrial Yeast Genomes Using CRISPRm. Methods Enzymol. Academic Press 2014; Vol. 546: pp. 473-89.
[http://dx.doi.org/10.1016/B978-0-12-801185-0.00023-4]

[114] Patra P, Das M, Kundu P, Ghosh A. Recent advances in systems and synthetic biology approaches for developing novel cell-factories in non-conventional yeasts. Biotechnol Adv 2021; 47: 107695.
[http://dx.doi.org/10.1016/j.biotechadv.2021.107695] [PMID: 33465474]

[115] Schwartz C, Shabbir-Hussain M, Frogue K, Blenner M, Wheeldon I. Standardized markerless gene integration for pathway engineering in *yarrowia lipolytica*. ACS Synth Biol 2017; 6(3): 402-9.
[http://dx.doi.org/10.1021/acssynbio.6b00285] [PMID: 27989123]

[116] Utomo JC, Hodgins CL, Ro D-K. Multiplex genome editing in yeast by CRISPR/Cas9 - a potent and agile tool to reconstruct complex metabolic pathways. Front Plant Sci 2021; 12: 719148.
[http://dx.doi.org/10.3389/fpls.2021.719148] [PMID: 34421973]

[117] Ding W, Zhang Y, Shi S. Development and application of CRISPR/Cas in microbial biotechnology. Front Bioeng Biotechnol 2020; 8: 711.
[http://dx.doi.org/10.3389/fbioe.2020.00711] [PMID: 32695770]

[118] Rainha J, Rodrigues JL, Rodrigues LR. CRISPR-Cas9: a powerful tool to efficiently engineer *saccharomyces cerevisiae*. Life (Basel) 2020; 11(1): 13.
[http://dx.doi.org/10.3390/life11010013] [PMID: 33375364]

[119] Baek S, Utomo JC, Lee JY, Dalal K, Yoon YJ, Ro D-K. The yeast platform engineered for synthetic gRNA-landing pads enables multiple gene integrations by a single gRNA/Cas9 system. Metab Eng 2021; 64: 111-21.
[http://dx.doi.org/10.1016/j.ymben.2021.01.011] [PMID: 33549837]

[120] van Wyk N, Kroukamp H, Espinosa MI, von Wallbrunn C, Wendland J, Pretorius IS. Blending wine yeast phenotypes with the aid of CRISPR DNA editing technologies. Int J Food Microbiol 2020; 324: 108615.

[http://dx.doi.org/10.1016/j.ijfoodmicro.2020.108615] [PMID: 32371236]

[121] Hao H, Wang X, Jia H, *et al.* Large fragment deletion using a CRISPR/Cas9 system in Saccharomyces cerevisiae. Anal Biochem 2016; 509: 118-23.
[http://dx.doi.org/10.1016/j.ab.2016.07.008] [PMID: 27402178]

[122] Fischer S, Procopio S, Becker T. Self-cloning brewing yeast: a new dimension in beverage production. Eur Food Res Technol 2013; 237: 851-63.
[http://dx.doi.org/10.1007/s00217-013-2092-9]

[123] Walker GM, Walker RSK. Enhancing Yeast Alcoholic Fermentations. Adv Appl Microbiol. Academic Press 2018; Vol. 105: pp. 87-129.
[http://dx.doi.org/10.1016/bs.aambs.2018.05.003]

[124] United Nations. Transforming our world: the 2030 Agenda for Sustainable Development | Department of Economic and Social Affairs. United Nations Gen Assem 2015.

[125] WHO. One Health. World Heal Organ 2017.https://www.who.int/news-room/q-a-detail/one-health

[126] Nevoigt E. Progress in metabolic engineering of Saccharomyces cerevisiae. Microbiol Mol Biol Rev 2008; 72(3): 379-412.
[http://dx.doi.org/10.1128/MMBR.00025-07] [PMID: 18772282]

[127] Mans R, Daran JG, Pronk JT. Under pressure: evolutionary engineering of yeast strains for improved performance in fuels and chemicals production. Curr Opin Biotechnol 2018; 50: 47-56.
[http://dx.doi.org/10.1016/j.copbio.2017.10.011] [PMID: 29156423]

[128] Çakar ZP, Turanli-Yildiz B, Alkim C, Yilmaz U. Evolutionary engineering of Saccharomyces cerevisiae for improved industrially important properties. FEMS Yeast Res 2012; 12(2): 171-82.
[http://dx.doi.org/10.1111/j.1567-1364.2011.00775.x] [PMID: 22136139]

[129] Kuroda K, Ueda M. Adaptive Evolution of Yeast Under Heat Stress and Genetic Reconstruction to Generate Thermotolerant Yeast. Orig Evol Biodivers. Cham: Springer International Publishing 2018; pp. 23-36.
[http://dx.doi.org/10.1007/978-3-319-95954-2_2]

[130] Sauer U. Evolutionary Engineering of Industrially Important Microbial Phenotypes. Adv Biochem Eng Biotechnol. Springer-Verlag 2001; pp. 129-69.

[131] Viña-Gonzalez J, Alcalde M. *In vivo* site-directed recombination (SDR): An efficient tool to reveal beneficial epistasis. Methods Enzymol 2020; 643: 1-13.
[http://dx.doi.org/10.1016/bs.mie.2020.04.021] [PMID: 32896276]

[132] Nespolo RF, Solano-Iguaran JJ, Paleo-López R, Quintero-Galvis JF, Cubillos FA, Bozinovic F. Performance, genomic rearrangements, and signatures of adaptive evolution: Lessons from fermentative yeasts. Ecol Evol 2020; 10(12): 5240-50.
[http://dx.doi.org/10.1002/ece3.6208] [PMID: 32607147]

[133] Garcia Sanchez R, Karhumaa K, Fonseca C, *et al.* Improved xylose and arabinose utilization by an industrial recombinant Saccharomyces cerevisiae strain using evolutionary engineering. Biotechnol Biofuels 2010; 3: 13.
[http://dx.doi.org/10.1186/1754-6834-3-13] [PMID: 20550651]

[134] Wright J, Bellissimi E, de Hulster E, Wagner A, Pronk JT, van Maris AJA. Batch and continuous culture-based selection strategies for acetic acid tolerance in xylose-fermenting Saccharomyces cerevisiae. FEMS Yeast Res 2011; 11(3): 299-306.
[http://dx.doi.org/10.1111/j.1567-1364.2011.00719.x] [PMID: 21251209]

[135] Stanley D, Fraser S, Chambers PJ, Rogers P, Stanley GA. Generation and characterisation of stable ethanol-tolerant mutants of Saccharomyces cerevisiae. J Ind Microbiol Biotechnol 2010; 37(2): 139-49.
[http://dx.doi.org/10.1007/s10295-009-0655-3] [PMID: 19902282]

[136] Basso TO, de Kok S, Dario M, *et al.* Engineering topology and kinetics of sucrose metabolism in Saccharomyces cerevisiae for improved ethanol yield. Metab Eng 2011; 13(6): 694-703.
[http://dx.doi.org/10.1016/j.ymben.2011.09.005] [PMID: 21963484]

[137] Turner TL, Kim H, Kong II, Liu J-J, Zhang G-C, Jin YS. Engineering and Evolution of Saccharomyces cerevisiae to Produce Biofuels and Chemicals. Adv Biochem Eng Biotechnol 2018; 162: 175-215.
[http://dx.doi.org/10.1007/10_2016_22] [PMID: 27913828]

[138] de Melo AHF, Lopes AMM, Dezotti N, Santos IL, Teixeira GS, Goldbeck R. Evolutionary Engineering of Two Robust Brazilian Industrial Yeast Strains for Thermotolerance and Second-Generation Biofuels. Ind Biotechnol (New Rochelle NY) 2020; 16: 91-8.
[http://dx.doi.org/10.1089/ind.2019.0031]

[139] Ko JK, Enkh-Amgalan T, Gong G, Um Y, Lee S. Improved bioconversion of lignocellulosic biomass by *Saccharomyces cerevisiae* engineered for tolerance to acetic acid. Glob Change Biol Bioenergy 2020; 12: 90-100.
[http://dx.doi.org/10.1111/gcbb.12656]

[140] Wei F, Li M, Wang M, Li H, Li Z, Qin W, *et al.* A C6/C5 co-fermenting *Saccharomyces cerevisiae* strain with the alleviation of antagonism between xylose utilization and robustness. Glob Change Biol Bioenergy 2021; 13: 83-97.
[http://dx.doi.org/10.1111/gcbb.12778]

[141] Xie C-Y, Yang B-X, Wu Y-J, Xia Z-Y, Gou M, Sun Z-Y, *et al.* Construction of industrial xylose-fermenting *Saccharomyces cerevisiae* strains through combined approaches. Process Biochem 2020; 96: 80-9.
[http://dx.doi.org/10.1016/j.procbio.2020.05.022]

[142] Passoth V. Conventional and Non-conventional Yeasts for the Production of Biofuels. Yeast Divers Hum Welf. Singapore: Springer Singapore 2017; pp. 385-416.
[http://dx.doi.org/10.1007/978-981-10-2621-8_15]

[143] Moreno AD, Tomás-Pejó E, Olsson L, Geijer C. *Candida intermedia* CBS 141442: A Novel Glucose/Xylose Co-Fermenting Isolate for Lignocellulosic Bioethanol Production. Energies 2020; 13: 5363.
[http://dx.doi.org/10.3390/en13205363]

[144] Dragone G, Kerssemakers AAJ, Driessen JLSP, Yamakawa CK, Brumano LP, Mussatto SI. Innovation and strategic orientations for the development of advanced biorefineries. Bioresour Technol 2020; 302: 122847.
[http://dx.doi.org/10.1016/j.biortech.2020.122847] [PMID: 32008863]

[145] Buschke N, Schäfer R, Becker J, Wittmann C. Metabolic engineering of industrial platform microorganisms for biorefinery applications--optimization of substrate spectrum and process robustness by rational and evolutive strategies. Bioresour Technol 2013; 135: 544-54.
[http://dx.doi.org/10.1016/j.biortech.2012.11.047] [PMID: 23260271]

[146] Goffeau A, Barrell BG, Bussey H, *et al.* Life with 6000 Genes. Science (80-) 1996; 274: 546-67.
[http://dx.doi.org/10.1126/science.274.5287.546]

[147] Botstein D. Genetics: Yeast as a model organism. Science (80-) 1997; 277: 1259-60.
[http://dx.doi.org/10.1126/science.277.5330.1259]

[148] Cheng K-K, Wu J, Lin Z-N, Zhang J-A. Aerobic and sequential anaerobic fermentation to produce xylitol and ethanol using non-detoxified acid pretreated corncob. Biotechnol Biofuels 2014; 7(1): 166.
[http://dx.doi.org/10.1186/s13068-014-0166-y] [PMID: 25431622]

[149] Madhavan A, Jose AA, Binod P, *et al.* Synthetic biology and metabolic engineering approaches and its impact on non-conventional yeast and biofuel production. Front Energy Res 2017; 5.
[http://dx.doi.org/10.3389/fenrg.2017.00008]

[150] Nandal P, Sharma S, Arora A. Bioprospecting non-conventional yeasts for ethanol production from rice straw hydrolysate and their inhibitor tolerance. Renew Energy 2020; 147: 1694-703.
[http://dx.doi.org/10.1016/j.renene.2019.09.067]

[151] Radecka D, Mukherjee V, Mateo RQ, Stojiljkovic M, Foulquié-Moreno MR, Thevelein JM. Looking beyond Saccharomyces: the potential of non-conventional yeast species for desirable traits in bioethanol fermentation. FEMS Yeast Res 2015; 15(6): fov053.
[http://dx.doi.org/10.1093/femsyr/fov053] [PMID: 26126524]

[152] Paulino de Souza J, Dias do Prado C, Eleutherio ECA, Bonatto D, Malavazi I, Ferreira da Cunha A. Improvement of Brazilian bioethanol production - Challenges and perspectives on the identification and genetic modification of new strains of Saccharomyces cerevisiae yeasts isolated during ethanol process. Fungal Biol 2018; 122(6): 583-91.
[http://dx.doi.org/10.1016/j.funbio.2017.12.006] [PMID: 29801803]

[153] Greetham D, Zaky AS, Du C. Exploring the tolerance of marine yeast to inhibitory compounds for improving bioethanol production. Sustain Energy Fuels 2019; 3: 1545-53.
[http://dx.doi.org/10.1039/C9SE00029A]

[154] Bazoti SFSF, Golunski S, Pereira Siqueira D, *et al.* Second-generation ethanol from non-detoxified sugarcane hydrolysate by a rotting wood isolated yeast strain. Bioresour Technol 2017; 244(Pt 1): 582-7.
[http://dx.doi.org/10.1016/j.biortech.2017.08.007] [PMID: 28803109]

[155] Lopes MR, Lara CA, Moura MEF, *et al.* Characterisation of the diversity and physiology of cellobiose-fermenting yeasts isolated from rotting wood in Brazilian ecosystems. Fungal Biol 2018; 122(7): 668-76.
[http://dx.doi.org/10.1016/j.funbio.2018.03.008] [PMID: 29880202]

[156] Nguyen NH, Suh SO, Marshall CJ, Blackwell M. Morphological and ecological similarities: wood-boring beetles associated with novel xylose-fermenting yeasts, Spathaspora passalidarum gen. sp. nov. and Candida jeffriesii sp. nov. Mycol Res 2006; 110(Pt 10): 1232-41.
[http://dx.doi.org/10.1016/j.mycres.2006.07.002] [PMID: 17011177]

[157] Stratford M, Steels H, Nebe-von-Caron G, Novodvorska M, Hayer K, Archer DB. Extreme resistance to weak-acid preservatives in the spoilage yeast Zygosaccharomyces bailii. Int J Food Microbiol 2013; 166(1): 126-34.
[http://dx.doi.org/10.1016/j.ijfoodmicro.2013.06.025] [PMID: 23856006]

[158] Yuangsaard N, Yongmanitchai W, Yamada M, Limtong S. Selection and characterization of a newly isolated thermotolerant Pichia kudriavzevii strain for ethanol production at high temperature from cassava starch hydrolysate. Antonie van Leeuwenhoek 2013; 103(3): 577-88.
[http://dx.doi.org/10.1007/s10482-012-9842-8] [PMID: 23132277]

[159] Khambhaty Y, Upadhyay D, Kriplani Y, Joshi N, Mody K, Gandhi MR. Bioethanol from Macroalgal Biomass: Utilization of Marine Yeast for Production of the Same. BioEnergy Res 2013; 6: 188-95.
[http://dx.doi.org/10.1007/s12155-012-9249-4]

[160] Limtong S, Sringiew C, Yongmanitchai W. Production of fuel ethanol at high temperature from sugar cane juice by a newly isolated Kluyveromyces marxianus. Bioresour Technol 2007; 98(17): 3367-74.
[http://dx.doi.org/10.1016/j.biortech.2006.10.044] [PMID: 17537627]

[161] Saucedo-Luna J, Castro-Montoya AJ, Martinez-Pacheco MM, Sosa-Aguirre CR, Campos-Garcia J. Efficient chemical and enzymatic saccharification of the lignocellulosic residue from Agave tequilana bagasse to produce ethanol by Pichia caribbica. J Ind Microbiol Biotechnol 2011; 38(6): 725-32.
[http://dx.doi.org/10.1007/s10295-010-0853-z] [PMID: 21072557]

[162] Dhaliwal SS, Oberoi HS, Sandhu SK, Nanda D, Kumar D, Uppal SK. Enhanced ethanol production from sugarcane juice by galactose adaptation of a newly isolated thermotolerant strain of *Pichia kudriavzevii*. Bioresour Technol 2011; 102(10): 5968-75.
[http://dx.doi.org/10.1016/j.biortech.2011.02.015] [PMID: 21398115]

[163]   Nakanishi SC, Soares LB, Biazi LE, *et al.* Fermentation strategy for second generation ethanol production from sugarcane bagasse hydrolyzate by *Spathaspora passalidarum* and *Scheffersomyces stipitis.* Biotechnol Bioeng 2017; 114(10): 2211-21.
[http://dx.doi.org/10.1002/bit.26357] [PMID: 28627711]

[164]   Cadete RM, Melo-Cheab MA, Dussán KJ, *et al.* Production of bioethanol in sugarcane bagasse hemicellulosic hydrolysate by Scheffersomyces parashehatae, *Scheffersomyces illinoinensis* and *Spathaspora arborariae* isolated from Brazilian ecosystems. J Appl Microbiol 2017; 123(5): 1203-13.
[http://dx.doi.org/10.1111/jam.13559] [PMID: 28799253]

[165]   Bonatto C, Scapini T, Zanivan J, *et al.* Utilization of seawater and wastewater from shrimp production in the fermentation of papaya residues to ethanol. Bioresour Technol 2021; 321: 124501.
[http://dx.doi.org/10.1016/j.biortech.2020.124501] [PMID: 33310410]

[166]   Bonatto C, Venturin B, Mayer DA, *et al.* Experimental data and modelling of 2G ethanol production by Wickerhamomyces sp. UFFS-CE-3.1.2. Renew Energy 2019; 145: 2445-50.
[http://dx.doi.org/10.1016/j.renene.2019.08.010]

[167]   Sehnem NT, Hickert LR, da Cunha-Pereira F, de Morais MA, Ayub MAZ. Bioconversion of soybean and rice hull hydrolysates into ethanol and xylitol by furaldehyde-tolerant strains of *Saccharomyces cerevisiae*, Wickerhamomyces anomalus, and their cofermentations. Biomass Convers Biorefin 2017; 7: 199-206.
[http://dx.doi.org/10.1007/s13399-016-0224-8]

[168]   Greetham D, Adams JM, Du C. The utilization of seawater for the hydrolysis of macroalgae and subsequent bioethanol fermentation. Sci Rep 2020; 10(1): 9728.
[http://dx.doi.org/10.1038/s41598-020-66610-9] [PMID: 32546695]

[169]   Atitallah I, Ntaikou I, Antonopoulou G, *et al.* Evaluation of the non-conventional yeast strain Wickerhamomyces anomalus (Pichia anomala) X19 for enhanced bioethanol production using date palm sap as renewable feedstock. Renew Energy 2020; 154: 71-81.
[http://dx.doi.org/10.1016/j.renene.2020.03.010]

[170]   Jayabalan R, Marimuthu S, Thangaraj P, *et al.* Preservation of kombucha tea-effect of temperature on tea components and free radical scavenging properties. J Agric Food Chem 2008; 56(19): 9064-71.
[http://dx.doi.org/10.1021/jf8020893] [PMID: 18781766]

[171]   Indira D, Das B, Bhawsar H, *et al.* Investigation on the production of bioethanol from black tea waste biomass in the seawater-based system. Bioresour Technol Rep 2018; 4: 209-13.
[http://dx.doi.org/10.1016/j.biteb.2018.11.003]

[172]   Martorell P, Stratford M, Steels H, Fernández-Espinar MT, Querol A. Physiological characterization of spoilage strains of *Zygosaccharomyces bailii* and *Zygosaccharomyces rouxii* isolated from high sugar environments. Int J Food Microbiol 2007; 114(2): 234-42.
[http://dx.doi.org/10.1016/j.ijfoodmicro.2006.09.014] [PMID: 17239464]

[173]   Hohmann S. Osmotic stress signaling and osmoadaptation in yeasts. Microbiol Mol Biol Rev 2002; 66(2): 300-72.
[http://dx.doi.org/10.1128/MMBR.66.2.300-372.2002] [PMID: 12040128]

[174]   Pribylova L, de Montigny J, Sychrova H. Osmoresistant yeast *Zygosaccharomyces rouxii*: the two most studied wild-type strains (ATCC 2623 and ATCC 42981) differ in osmotolerance and glycerol metabolism. Yeast 2007; 24(3): 171-80.
[http://dx.doi.org/10.1002/yea.1470] [PMID: 17351908]

[175]   Leandro MJ, Sychrová H, Prista C, Loureiro-Dias MC. The osmotolerant fructophilic yeast *Zygosaccharomyces rouxii* employs two plasma-membrane fructose uptake systems belonging to a new family of yeast sugar transporters. Microbiology 2011; 157(Pt 2): 601-8.
[http://dx.doi.org/10.1099/mic.0.044446-0] [PMID: 21051487]

[176]   Mukherjee V, Radecka D, Aerts G, Verstrepen KJ, Lievens B, Thevelein JM. Phenotypic landscape of

non-conventional yeast species for different stress tolerance traits desirable in bioethanol fermentation. Biotechnol Biofuels 2017; 10: 216.
[http://dx.doi.org/10.1186/s13068-017-0899-5] [PMID: 28924451]

[177] Greetham D, Zaky A, Makanjuola O, Du C. A brief review on bioethanol production using marine biomass, marine microorganism and seawater. Curr Opin Green Sustain Chem 2018; 14: 53-9.
[http://dx.doi.org/10.1016/j.cogsc.2018.06.008]

[178] Camargo AF, Scapini T, Stefanski FS, *et al.* Reducing the Water Footprint of Bioethanol. Ethanol as a Green Altern Fuel Insight Perspect. 1st ed.. Nova Science Publishers 2019; pp. 199-220.

[179] Zaky AS, Greetham D, Tucker GA, Du C. The establishment of a marine focused biorefinery for bioethanol production using seawater and a novel marine yeast strain. Sci Rep 2018; 8(1): 12127.
[http://dx.doi.org/10.1038/s41598-018-30660-x] [PMID: 30108287]

[180] Ben Atitallah I, Antonopoulou G, Ntaikou I, *et al.* On the evaluation of different saccharification schemes for enhanced bioethanol production from potato peels waste *via* a newly isolated yeast strain of Wickerhamomyces anomalus. Bioresour Technol 2019; 289: 121614.
[http://dx.doi.org/10.1016/j.biortech.2019.121614] [PMID: 31203181]

[181] Basso RF, Alcarde AR, Portugal CB. Could non-Saccharomyces yeasts contribute on innovative brewing fermentations? Food Res Int 2016; 86: 112-20.
[http://dx.doi.org/10.1016/j.foodres.2016.06.002]

[182] da Cunha AC, Gomes LS, Godoy-Santos F, *et al.* High-affinity transport, cyanide-resistant respiration, and ethanol production under aerobiosis underlying efficient high glycerol consumption by Wickerhamomyces anomalus. J Ind Microbiol Biotechnol 2019; 46(5): 709-23.
[http://dx.doi.org/10.1007/s10295-018-02119-5] [PMID: 30680472]

[183] Kumar S, Singh SP, Mishra IM, Adhikari DK. Ethanol and xylitol production from glucose and xylose at high temperature by Kluyveromyces sp. IIPE453. J Ind Microbiol Biotechnol 2009; 36(12): 1483-9.
[http://dx.doi.org/10.1007/s10295-009-0636-6] [PMID: 19768475]

[184] Das B, Sarkar S, Maiti S, Bhattacharjee S. Studies on production of ethanol from cheese whey using *Kluyveromyces marxianus*. Mater Today Proc 2016; 3: 3253-7.
[http://dx.doi.org/10.1016/j.matpr.2016.10.006]

[185] Yanase S, Hasunuma T, Yamada R, *et al.* Direct ethanol production from cellulosic materials at high temperature using the thermotolerant yeast *Kluyveromyces marxianus* displaying cellulolytic enzymes. Appl Microbiol Biotechnol 2010; 88(1): 381-8.
[http://dx.doi.org/10.1007/s00253-010-2784-z] [PMID: 20676628]

[186] Sandoval-Nuñez D, Arellano-Plaza M, Gschaedler A, Arrizon J, Amaya-Delgado L. A comparative study of lignocellulosic ethanol productivities by *Kluyveromyces marxianus* and *Saccharomyces cerevisiae*. Clean Technol Environ Policy 2018; 20: 1491-9.
[http://dx.doi.org/10.1007/s10098-017-1470-6]

[187] Lopes MR, Morais CG, Kominek J, *et al.* Genomic analysis and D-xylose fermentation of three novel Spathaspora species: Spathaspora girioi sp. nov., Spathaspora hagerdaliae f. a., sp. nov. and Spathaspora gorwiae f. a., sp. nov. FEMS Yeast Res 2016; 16(4): 1-12.
[http://dx.doi.org/10.1093/femsyr/fow044] [PMID: 27188884]

[188] Cadete RM, Melo MA, Dussán KJ, *et al.* Diversity and physiological characterization of D-xylos--fermenting yeasts isolated from the Brazilian Amazonian Forest. PLoS One 2012; 7(8): e43135.
[http://dx.doi.org/10.1371/journal.pone.0043135] [PMID: 22912807]

[189] Agbogbo FK, Coward-Kelly G, Torry-Smith M, Wenger KS. Fermentation of glucose/xylose mixtures using Pichia stipitis. Process Biochem 2006; 41: 2333-6.
[http://dx.doi.org/10.1016/j.procbio.2006.05.004]

[190] Dussán KJ, Silva DDV, Perez VH, da Silva SS. Evaluation of oxygen availability on ethanol production from sugarcane bagasse hydrolysate in a batch bioreactor using two strains of xylose-

fermenting yeast. Renew Energy 2016; 87: 703-10.
[http://dx.doi.org/10.1016/j.renene.2015.10.065]

[191]   Antunes FAF, Thomé LC, Santos JC, *et al.* Multi-scale study of the integrated use of the carbohydrate fractions of sugarcane bagasse for ethanol and xylitol production. Renew Energy 2021; 163: 1343-55.
[http://dx.doi.org/10.1016/j.renene.2020.08.020]

[192]   Toivola A, Yarrow D, van den Bosch E, van Dijken JP, Scheffers WA. Alcoholic fermentation of D-xylose by yeasts. Appl Environ Microbiol 1984; 47(6): 1221-3.
[http://dx.doi.org/10.1128/aem.47.6.1221-1223.1984] [PMID: 16346558]

[193]   Lara CA, Santos RO, Cadete RM, *et al.* Identification and characterisation of xylanolytic yeasts isolated from decaying wood and sugarcane bagasse in Brazil. Antonie van Leeuwenhoek 2014; 105(6): 1107-19.
[http://dx.doi.org/10.1007/s10482-014-0172-x] [PMID: 24748334]

[194]   Yuvadetkun P, Leksawasdi N, Boonmee M. Kinetic modeling of *Candida shehatae* ATCC 22984 on xylose and glucose for ethanol production. Prep Biochem Biotechnol 2017; 47(3): 268-75.
[http://dx.doi.org/10.1080/10826068.2016.1224244] [PMID: 27552485]

[195]   Sreenath HK, Jeffries TW. Production of ethanol from wood hydrolyzate by yeasts. Bioresour Technol 2000; 72: 253-60.
[http://dx.doi.org/10.1016/S0960-8524(99)00113-3]

[196]   Abbi M, Kuhad RC, Singh A. Fermentation of xylose and rice straw hydrolysate to ethanol by Candida shehatae NCL-3501. J Ind Microbiol 1996; 17(1): 20-3.
[http://dx.doi.org/10.1007/BF01570143] [PMID: 8987687]

[197]   Ryabova OB, Chmil OM, Sibirny AA. Xylose and cellobiose fermentation to ethanol by the thermotolerant methylotrophic yeast *Hansenula polymorpha.* FEMS Yeast Res 2003; 4(2): 157-64.
[http://dx.doi.org/10.1016/S1567-1356(03)00146-6] [PMID: 14613880]

[198]   Morais CG, Sena LMF, Lopes MR, *et al.* Production of ethanol and xylanolytic enzymes by yeasts inhabiting rotting wood isolated in sugarcane bagasse hydrolysate. Fungal Biol 2020; 124(7): 639-47.
[http://dx.doi.org/10.1016/j.funbio.2020.03.005] [PMID: 32540187]

[199]   Raj K, Krishnan C. Improved co-production of ethanol and xylitol from low-temperature aqueous ammonia pretreated sugarcane bagasse using two-stage high solids enzymatic hydrolysis and Candida tropicalis. Renew Energy 2020; 153: 392-403.
[http://dx.doi.org/10.1016/j.renene.2020.02.042]

[200]   Antunes FAF, Chandel AK, Brumano LP, *et al.* A novel process intensification strategy for second-generation ethanol production from sugarcane bagasse in fluidized bed reactor. Renew Energy 2018; 124: 189-96.
[http://dx.doi.org/10.1016/j.renene.2017.06.004]

[201]   Morais CG, Cadete RM, Uetanabaro APT, Rosa LH, Lachance MA, Rosa CA. D-xylose-fermenting and xylanase-producing yeast species from rotting wood of two Atlantic Rainforest habitats in Brazil. Fungal Genet Biol 2013; 60: 19-28.
[http://dx.doi.org/10.1016/j.fgb.2013.07.003] [PMID: 23872280]

[202]   Faria NT, Marques S, Ferreira FC, Fonseca C. Production of xylanolytic enzymes by Moesziomyces spp. using xylose, xylan and brewery's spent grain as substrates. N Biotechnol 2019; 49: 137-43.
[http://dx.doi.org/10.1016/j.nbt.2018.11.001] [PMID: 30423436]

[203]   Tiwari S, Avchar R, Arora R, *et al.* Xylanolytic and ethanologenic potential of gut associated yeasts from different species of termites from india. Mycobiology 2020; 48(6): 501-11.
[http://dx.doi.org/10.1080/12298093.2020.1830742] [PMID: 33312017]

[204]   Carrasco M, Rozas JM, Barahona S, Alcaíno J, Cifuentes V, Baeza M. Diversity and extracellular enzymatic activities of yeasts isolated from King George Island, the sub-Antarctic region. BMC Microbiol 2012; 12: 251.

[http://dx.doi.org/10.1186/1471-2180-12-251] [PMID: 23131126]

[205] Laitila A, Wilhelmson A, Kotaviita E, Olkku J, Home S, Juvonen R. Yeasts in an industrial malting ecosystem. J Ind Microbiol Biotechnol 2006; 33(11): 953-66.
[http://dx.doi.org/10.1007/s10295-006-0150-z] [PMID: 16758169]

[206] Suh SO, White MM, Nguyen NH, Blackwell M. The status and characterization of *Enteroramus dimorphus*: a xylose-fermenting yeast attached to the gut of beetles. Mycologia 2004; 96(4): 756-60.
[http://dx.doi.org/10.2307/3762109] [PMID: 21148896]

[207] Stefanini I. Yeast-insect associations: It takes guts. Yeast 2018; 35(4): 315-30.
[http://dx.doi.org/10.1002/yea.3309] [PMID: 29363168]

[208] Alves SL, Müller C, Bonatto C, Scapini T, Camargo AF, Fongaro G, *et al.* Bioprospection of enzymes and microorganisms in insects to improve second-generation ethanol production. Ind Biotechnol (New Rochelle NY) 2019; 15: 336-49.
[http://dx.doi.org/10.1089/ind.2019.0019]

[209] Urbina H, Frank R, Blackwell M. *Scheffersomyces cryptocercus:* a new xylose-fermenting yeast associated with the gut of wood roaches and new combinations in the Sugiyamaella yeast clade. Mycologia 2013; 105(3): 650-60.
[http://dx.doi.org/10.3852/12-094] [PMID: 23233509]

[210] Urbina H, Schuster J, Blackwell M. The gut of Guatemalan passalid beetles: a habitat colonized by cellobiose- and xylose-fermenting yeasts. Fungal Ecol 2013; 6: 339-55.
[http://dx.doi.org/10.1016/j.funeco.2013.06.005]

[211] Suh SO, Marshall CJ, McHugh JV, Blackwell M. Wood ingestion by passalid beetles in the presence of xylose-fermenting gut yeasts. Mol Ecol 2003; 12(11): 3137-45.
[http://dx.doi.org/10.1046/j.1365-294X.2003.01973.x] [PMID: 14629392]

[212] Meriggi N, Di Paola M, Cavalieri D, Stefanini I. *Saccharomyces cerevisiae* - Insects Association: Impacts, Biogeography, and Extent. Front Microbiol 2020; 11: 1629.
[http://dx.doi.org/10.3389/fmicb.2020.01629] [PMID: 32760380]

[213] Blackwell M. Made for each other: ascomycete yeasts and insects. Microbiol Spectr 2017; 5(3): 945-62.
[http://dx.doi.org/10.1128/microbiolspec.FUNK-0081-2016] [PMID: 28597823]

CHAPTER 10

Yeasts in the Beverage Industry: Patagonia Gets Wild

Melisa Gonzalez Flores[1], **María C. Bruzone**[2], **Andrea Origone**[1], **Julieta A. Burini**[2], **María E. Rodríguez**[1], **Christian A. Lopes**[1] and **Diego Libkind**[2,*]

[1] *Instituto de Investigación y Desarrollo en Ingeniería de Procesos, Biotecnología y Energías Alternativas (PROBIEN, CONICET-UNCo), Neuquén, Argentina*

[2] *Centro de Referencia en Levaduras y Tecnología Cervecera (CRELTEC), Instituto Andino Patagónico de Tecnologías Biológicas y Geoambientales (IPATEC), CONICET / Universidad Nacional del Comahue, Quintral 1250 (8400), Bariloche, Rio Negro, Argentina*

Abstract: Yeasts are intimately involved in the production of fermented alcoholic beverages being the most popular examples of beer, cider and wine. The present chapter reviews the impact of yeasts in the production of these three fermented beverages and focuses on recent innovation trends regarding the use of non-conventional yeasts for the increase of flavour complexity and/or the development of novel special products that better meet current customer's demands. The granting of regional identity by using locally sourced yeast strains is also revised, and the experience gathered in the region of Andean Patagonia (Argentina) related to the isolation, screening, selection, improvement (in some cases) and all the way to the industrial application is described. North-western Patagonia natural forests harbour yeasts species of great scientific and fundamental relevance, among which the cryotolerant species *Saccharomyces uvarum* and *Saccharomyces eubayanus* are the most important for this chapter. The successful cases reviewed here of the study and application of Patagonian cold-adapted wild *Saccharomyces* yeasts for beer, cider, and wine innovation demonstrate that the laborious journey from nature to industry application is feasible and advantageous.

Keywords: Andes, Argentina, Beer, Cider, Craft industry, Euby, Hybridization, Microbe domestication, Native starter, Native yeast, Natural environments, Patagonia, *Saccharomyces cerevisiae*, *Saccharomyces eubayanus*, *Saccharomyces uvarum*, Selective isolation, Wild yeast, Wine, Yeast bioprospection, Yeast isolation.

---

* **Corresponding author Diego Libkind:** Centro de Referencia en Levaduras y Tecnología Cervecera (CRELTEC), Instituto Andino Patagónico de Tecnologías Biológicas y Geoambientales (IPATEC), CONICET / Universidad Nacional del Comahue, Quintral 1250 (8400), Bariloche, Rio Negro, Argentina; Tel: +54 9 2944 623911; E-mail: libkindfd@comahue-conicet.gob.ar

Sérgio Luiz Alves Júnior, Helen Treichel, Thiago Olitta Basso and Boris Ugarte Stambuk (Eds.)
All rights reserved-© 2022 Bentham Science Publishers

INTRODUCTION

Yeast, alone or in consortia with other microorganisms, has a profound role in the industrial and traditional production of many beverages. These are generally recognized as fermented and usually contain alcohol. Humans have consumed fermented beverages since the Neolithic period (*c.*10 000 BC [1];), however, it is still unclear whether, in ancient times, our ancestors accidentally stumbled across fermented beverages like wine or beer, or was it a product intended as such. Undoubtedly, alcoholic beverages have been part of the diet and culture of many of the civilizations that have preceded us and are among the most popular products consumed today. Fermented alcoholic beverages are complex solutions of thousands of chemical compounds that originate from the metabolism of yeasts and other microorganisms from a sugar substrate during fermentation, and from later stages that include secondary fermentations and / or chemical reactions during aging. The most popular non-distilled fermented beverages are obtained from cereal starches (by enzymatic pre-hydrolysis) in the case of beer (barley and wheat) and / or from fruits (which do not require pre-hydrolysis) in the case of wine (grapes) and cider (apples and pears). Yeasts of the genus *Saccharomyces* are the most prevalent microorganisms in the production of these fermented beverages [2]. The genus is composed of eight natural species, namely *Saccharomyces cerevisiae, S. paradoxus, S. uvarum, S. mikatae, S. kudriavzevii, S. arboricola, S. eubayanus* and *S. jurei* [3] (Fig. **1**). *S. cerevisiae* is by far the most recognized and ubiquitous species in the production of fermented foods and beverages. Nevertheless, the cryophilic species *S. uvarum* and *S. kudriavzevii,* and the hybrid species *S. bayanus* and *S. pastorianus*, play a fundamental role in the production of beverages such as beer and wine [2]. Furthermore, *S. paradoxus* and the latest additions to the genus *S. eubayanus* and *S. jurei*, are also being studied for their application in the fermentation industry [4 - 6].

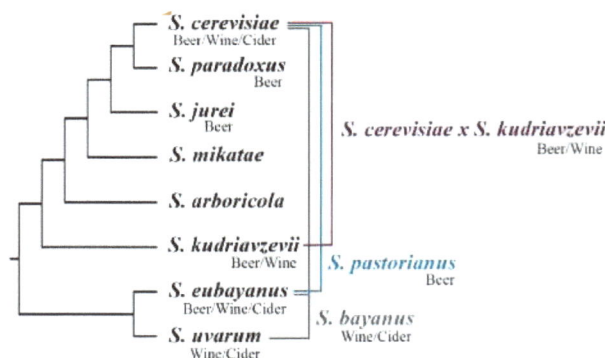

Fig. (1). Phylogenetic relationships of biological recognized species *Saccharomyces* species (black font), along with the most industrial relevant hybrids (non-black font), and their participation in the fermented beverages reviewed in this chapter.

During fermentation, yeasts produce alcohol, carbon dioxide and a range of secondary metabolites, such as esters, volatile fatty acids, higher alcohols, organic acids, volatile sulphur compounds and volatile phenols, that contribute significantly to the flavour and aroma of the final product [7, 8]. Certain strains of these yeasts, like many other microorganisms associated with man-made niches, have gone through a domestication process referred to as the artificial selection and breeding of wild specimens to obtain more fitted cultivated variants that better meet human or industrial requirements. Domesticated strains show improved adaptation to sugar-rich, oxygen-limited environments and high tolerance to ethanol, as well as other novel phenotypes, which can be specified for each fermentation environment. For instance, some beer yeasts can metabolize maltotriose (a beer-specific sugar), while wine yeasts can withstand the predominant sterilization agents in the winery (sulfite) and vineyard (copper sulphate) (for a review, see Steensels *et al.* [9]). These new domesticated traits result from the accumulation of defined genetic and genomic changes that only recently began to be elucidated and that may include inter-species gene introgressions (*i.e. S. uvarum* [10];) as well as inter-species hybridizations (*i.e. S. pastorianus* and *S. bayanus* [11];). Several non-*Saccharomyces* yeasts that are typically associated with early stages of spontaneous man-made fermentations are also relevant in the production of industrial and traditional fermented beverages. These include species of the genera *Brettanomyces*, *Torulaspora*, *Schizosaccharomyces*, *Metschnikowia, Hanseniaspora, Pichia, Lachancea* and *Kluyveromyces,* among others [12]. Because of their ability to modify the sensory quality of the final products, they are normally considered contaminants. However, due to new market trends in favor of products with differential and more complex organoleptic characteristics, they began to gain relevance in the beverage industry, contributing positively to the sensory quality of wine [13] and for bioflavouring purposes in brewing [14]. Another source of non-conventional yeasts with productive potential are natural environments and, even though these non-domesticated (wild) yeasts typically show inadequate fermentative traits for their implementation in the industry, exceptions exist mostly within the genus *Saccharomyces*. This approach has the advantage of providing additional regional identity and exclusivity to the beverages produced with local wild yeasts. Thus, the study and application of wild *Saccharomyces* for alcoholic beverage innovation have gained special attention in the last few years. In some cases, genetic improvement of wild yeasts was achieved using either genetic engineering or editing, mutagenesis or using non-GMO producing techniques such as directed experimental evolution and/or hybridization [4, 15, 16]. The whole process of isolating a wild yeast and inserting it into the market of fermented beverages is long and involves different research and development stages; initiating with yeast search and isolation and ending in scale-up trials and the technology transfer to the industry (Fig. **2**). As result, there are hitherto not many successful examples of complete journeys "from nature to bioprocess" that can be cited. Recently, significant advances related to the discovery in Andean Patagonia (Argentina and Chile) of novel populations of known species of *Saccharomyces* (namely *S. uvarum*) [10, 17, 18] and even the discovery of *S. eubayanus* [11], lead to the assessment of their potential for alcoholic beverage production. In this chapter we review the role of yeasts in the production of beer, wine and cider and describe the Patagonian experiences in the use of autochthonous *Saccharomyces* strains for the development of commercial products of these fermented beverages.

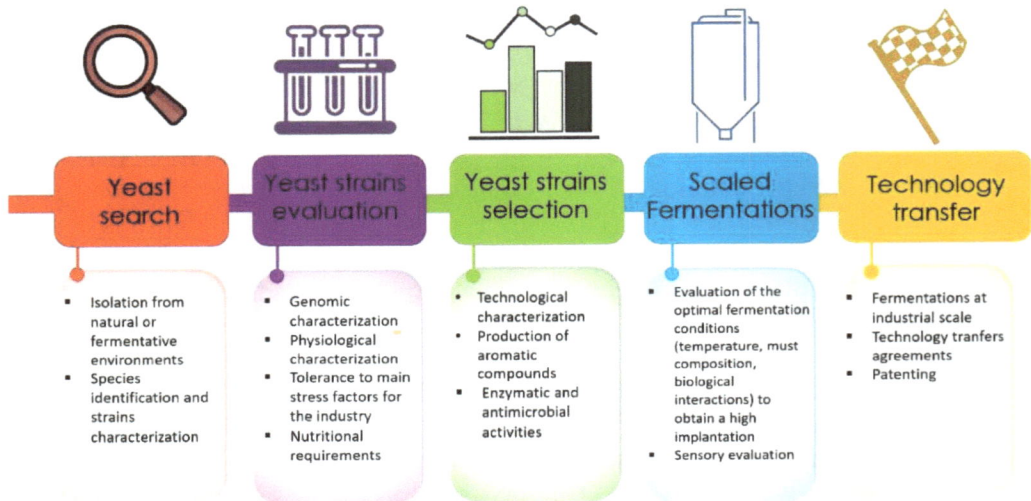

Fig. (2). Primary steps for yeast starters selection and evaluation for their use in the fermented beverage industry.

Natural Environments for Yeast Bioprospection: The Case of Andean Patagonia

In the last years, several yeast biodiversity studies have been carried out in the Patagonia region of Argentina and Chile, which have resulted in the isolation and characterization of several interesting species, being the highlight *S. eubayanus*. Patagonia is a geographical region that is located in the southern end of South America. This region has a great diversity of natural environments with very different environmental characteristics (high mountains, lakes, glaciers, steppes). It harbors a great diversity of yeasts and, over the past decades, different studies have been carried out with the aim of finding yeasts that can be used in different biotechnological processes (e.g. production of carotenoid pigments, mycosporines (UV sunscreens) and cold-active enzymes) [19 - 22]. The search for new flavours and aromas for innovation in the beverage industry is also leading to the bioprospection of yeasts from these natural environments; and finding new *Saccharomyces* strains, or other wild species, has been a goal for many researchers around the world [11, 14, 23, 24].

With the aim of finding wild *Saccharomyces* yeasts that have the potential to be used in the fermented beverage industry, different strategies have been implemented. Tolerance to ethanol (generally levels higher than 4% *v/v*) and the ability to metabolize sugars, are usually two critical characteristics for the selection of these yeasts [24, 25]. One of the most efficient strategies for *Saccharomyces* strains isolation, and other wild yeasts is the method described by Sampaio *et al*. [23], which uses an enrichment strategy with the selective media YNB (yeast nitrogen base; Difco), supplemented with 8% *v/v* of ethanol and 1% w/v of raffinose as the sole source of carbon. Once the pure yeast cultures are obtained, the species identification is generally carried out through the implementation of molecular biology techniques. Formerly, rapid identification could be achieved with Mini/Microsatellite PCR-fingerprinting method (MSP-PCR) using the (GAC)5 primer [11, 26], and those strains with atypical or ambiguous MSP-PCR results were corroborated by sequencing the ITS region [27]. Currently, in *Saccharomyces* genus, the approach is to look for strategies of multilocus or complete genome sequencing [28, 29].

For the past decade, the use of these methods allowed the isolation of yeasts from the *Saccharomyces* genus, and also a number of representatives of different yeast genera that are considered of interest to the beverage industry. For example, Eizaguirre *et al*. [30] analyzed 400 samples from different sources (bark, *Cyttaria* spp., leaves and soil) associated with diverse tree species (mainly of the endemic genus *Nothofagus*), and obtained 563 yeast isolates. The species with the highest abundance was *Saccharomyces eubayanus*, but also *S. uvarum* was present. Regarding non-*Saccharomyces* yeasts, the isolations belonged mainly to the genera *Lachancea*, *Hanseniaspora,* and *Torulaspora*; these yeasts have been defined of interest for the development of differential beverages based on global innovation trends [14, 24].

BEER INNOVATION AND YEAST BIOTECHNOLOGY

Beer is the alcoholic beverage with the highest consumption worldwide and is produced through a biotechnological process using four main ingredients: water, malt, hops, and yeast. The process basically comprises the production of wort by maceration of malted grains (generally barley) in water at a defined temperature and pH (favoring the action of amylase enzymes), followed by a boiling stage, where the microbial load present is lowered, and hop alpha and beta acids are isomerized providing the bitterness to the beer. Once cold, the wort is inoculated with a brewing yeast, the fermentative agent that converts simple sugars (mainly glucose, maltose and, in some cases, maltotriose) into carbon dioxide and ethanol. Besides these two products, yeasts produce a considerable number of organic compounds, which are essential for beer flavour and aroma [31, 32]. Nowadays, beers are classified into styles, which define certain organoleptic attributes that allow consumers to select the type of beer they want to drink. One of the main criteria for classifying the hundreds of different styles of beer is based on the type of fermentation, and therefore the type of yeast involved. Brewing yeasts are classified mainly into two groups, those of high fermentation (ale or top-fermenting yeasts) and those of low fermentation (lager or bottom-fermenting yeasts). The former is used to make ale beers, which are produced with domesticated strains of the species *Saccharomyces cerevisiae*, in a process that generally occurs between 18 and 24 °C and is usually characterized by a complexity of aromas and flavours of fruit, flowers or spices. On the other hand, lager beers are produced using the hybrid species *Saccharomyces pastorianus* (*S. cerevisiae* x *S. eubayanus* hybrid) at low temperatures (between 5 and 15 °C) and are characterized by a more neutral and crispy organoleptic profile; *S. pastorianus* exhibits high resistance to various stress factors, which makes it very useful for industrial beer production [33, 34]. *S. cerevisiae* and *S. pastorianus* have undergone a domestication process. The constant exposure to very specific conditions has resulted in a continuous selection imposed by the brewing environment, making these yeasts specially adapted to beer production [11, 35 - 37]. Through centuries the human practice of harvesting and re-pitching yeasts based on their fermentative performance and organoleptic characteristics, has resulted in the evolution of mechanisms that efficiently ferment wort sugars such as maltose and maltotriose, in most cases eliminating the production of phenolic off-flavours (POF) and improved flocculation, among others [15, 36 - 39].

In most modern beer fermentations, the inoculation of a single yeast strain is a distinctive practice compared with other fermentative beverages. The use of pure yeast cultures was initiated by Christian Hansen with the isolation of the first pure strain of a lager yeast at the Carlsberg brewery. In November 1883, the Old Carlsberg Brewery started using one of Hansen's isolates, Unterhefe Nr. 1 (*S. carlsbergensis*, now reclassified as *S. pastorianus*), to produce their lager beers. This was the first reported use of a yeast starter culture consisting of one well-defined and characterized microbial strain [9, 15]. This practice had a significant impact on the consistency and quality of beer production, but probably also was one of the reasons for the limited number of strains and phenotypic divergence of commercial brewing yeasts for over a hundred years [24, 40]. Despite the widespread use of traditional brewing yeasts, non-conventional microorganisms have begun to gain a place in brewing, as the growth of the craft beer segment and the growing number of specialized consumers have driven the search for product innovation. So, for the past years, the role of different yeasts and bacteria as protagonists in fermentations with distinctive characteristics has received special attention [4, 14, 24, 41]. Natural environments are a source of non-conventional yeasts with the potential to be used in the brewing industry. However, these yeasts in their wild state often lack adequate fermentation characteristics to be implemented directly in the industry. Generally, these yeasts have low fermentation yields and are more sensitive to ethanol stress, but they open a range of possibilities by providing distinctive aromas and flavours, as well as new approaches and characteristics that impact the organoleptic profile of beer [42, 43].

Search for Brewing Potential in Wild Yeasts

For yeast to be transferred to the beer industry, a series of tests and experiments have to be performed before a certain strain is selected (Fig. **2**). Studies aimed at the selection and application of non-conventional yeasts for brewing generally study characteristics such as use of wort sugars (glucose, maltose and maltotriose), fermentative performance (fermentation rate, attenuation, lag phase), osmotolerance, tolerance to ethanol, performance at low temperatures, tolerance to hop compounds, amino acids metabolism, production of volatile phenols (POF phenotype), production of differential flavour and aroma compounds, and also the synthesis of enzymes that influence organoleptic properties of beer (such as beta-glucosidases or glucoamylases) [24, 43, 44]. Even once a strain has been selected at a laboratory scale, it must be evaluated at larger scales before a product can be delivered to the market. The potential of wild yeasts to impart new aromas and flavours to beer is overshadowed by their lack of adaptation to the brewing environment. These yeasts, since they have not been subjected to a domestication process, present characteristics that are often detrimental when making beer, such as poor flocculation (which makes clarification difficult) and inability to use all the available sugars present in beer wort (limiting the efficiency of fermentation) [4]. This is the reason why transferring a wild species to the productive sector is not an easy task. Currently, the cases of wild yeast species for the production of beer are limited, both in the industrial and craft segments. As some commercial examples we can mention the yeast WildBrew™ Philly Sour (Lallemand), a unique species of *Lachancea* selected from nature by the University of the Sciences in Philadelphia; *Metschnikowia reukaufii* (The Yeast Bay) a wild yeast isolated from flowers in the Berkeley Hills of California; and *Lachancea* yeasts from Lachancea LLC, that were isolated from wasps and bees. Probably the most emblematic case represents the use of *S. eubayanus* wild strains to produce both industrial and craft beers, which will be the focus of the following sections.

Saccharomyces eubayanus, the Mother of the Lager Brewing Yeast

A notable example is the case of *Saccharomyces eubayanus*, which since its discovery has been intensively studied for brewing applications, and also for the understanding of evolution, ecology and population genomics of the *Saccharomyces* genus. Although the hybrid nature of *S. pastorianus*, the yeast responsible for the production of lager beer, has been known for decades [45 - 48], the identification of the non-*cerevisiae* parent remained a mystery until 2011. *Saccharomyces eubayanus* was initially found in association with trees of the genus *Nothofagus* (southern beech) in Patagonia (Argentina) and was identified at the genome level with 99.56% identity as one of the parents of brewing yeast *S. pastorianus* [11]. It was subsequently isolated in East Asia, North America, New Zealand and more extensively in Patagonia (Argentina and Chile) but, surprisingly, not in Europe (site of the supposedly original hybridization event/s that gave rise to the lager yeast [48]). This cold-tolerant yeast is most abundant and diverse in the Patagonian region of South America (with the genetic characterization of more than 200 strains isolates), where there are two major populations (PopulationA/PA and PopulationB/PB) that recent genomic data suggested are further divided into five subpopulations (PA-1, PA-2, PB-1, PB-2, and PB-3) [28 - 30, 49]. The levels of diversity found within Patagonia is further underscored by the restriction of four subpopulations to this region, suggesting that Patagonia is the origin of *S. eubayanus* diversity or at least the last common ancestor of the PA and PB-Holarctic populations, the latter of which gave rise to lager-brewing hybrids [29, 50, 51]. The large number of Patagonian isolates belonging to different genetic subpopulations can harbor a phenotypic diversity of great interest for the brewing industry as the growth in craft beer market continues, and the increasing interest and demands for diverse beers of consumers is leading brewers to push the limits of their production towards new alternatives for innovation of their products [41]. The discovery of *S. eubyanus* has not only aroused interest in the study of its potential in brewing, but has also shown to be useful for the production of cider [52, 53] and wine [54], as detailed in the following sections.

Technological Features of *S. eubayanus* for Brewing

Numerous studies have been carried out to evaluate the fermentative behavior and brewing potential of *S. eubayanus*. The type strain CRUB 1568 (CBS 12357 / PYCC 6148) showed faster growth than most larger yeasts in laboratory media containing glucose or maltose as a carbon source at 10 °C [33], while at 20 °C it was still competitive, and displayed reduced growth at temperatures of 25 °C or higher compared with brewing yeast [55, 56]. Eizaguirre [57] evaluated the osmotolerance and tolerance to ethanol of *S. eubayanus* in YM-agar medium containing high concentrations of glucose (40, 50 and 60%) and ethanol (5, 10, 15 and 20%), showing that the type strain is able to tolerate these stressors, even at low temperatures (10 °C). Osmotolerance was also evaluated in high-gravity worts, confirming the ability of *S. eubayanus* to tolerate the stress produced by high concentrations of sugars (up to 22 °P), as well as to produce a high percentage of ethanol (7.6% *v/v*).

S. eubayanus is one of the cold-tolerant species within the *Saccharomyces* genus. Cryotolerance is a promising characteristic in brewing yeasts for the production of lager beers since the low temperature is the main factor determining the sensory properties of these beers. It is proposed that the mechanisms that govern the cryotolerance of *S. eubayanus* are related to differences in the composition of the membrane, as well as to differential expression of central metabolic genes in relation to other *Saccharomyces* yeasts [58, 59]. Also, the presence of α-glucoside transporters that work better at lower temperatures (such as MTT1) could influence their performance at these temperatures [60, 61]. Regarding its fermentation performance, *S. eubayanus* ferments the glucose and maltose from wort, with average attenuation values for the type strain of 65% [30, 61 - 63]. The low attenuation observed in general for the species compared with lager brewing yeasts (aprox 85%), is the result of the inability of *S. eubayanus* to ferment maltotriose [30, 33, 35, 58, 61, 62, 64 - 66]. This was observed not only for the type strain, but also for strains of the species belonging to the different subpopulations and isolated from different geographical sites, both in Chile and Argentina [30, 51, 63, 66]; indicating that this is a species specific trait. Eizaguirre *et al*. [30] performed the first large assessment of fermentation performance of multiple *S. eubayanus* isolates. Sixty natural isolates representing the five subpopulations were assessed and, based on fermentation performance, concluded that population B generally showed better traits for brewing. *S. eubayanus* also presents low flocculation evidenced by the biomass in suspension, although at the end of the fermentation there is a rapid drop in yeasts [57, 59, 62].

Regarding the production of flavour compounds, the main distinctive characteristic of *S. eubayanus* is the production of the volatile phenols 4-vinylguaiacol (4-VG) [56, 59, 67, 68] and 4-vinylphenol (4-VP) [63]. They are produced as a result of the decarboxylation of ferulic and coumaric acid from wort, involving the action of *FDC1* and *PAD1* genes [69, 70]. As a result, clove and / or smoked aromas and flavours predominate sensorially in the fermentations carried out with *S. eubayanus*. In wild yeast, the *PAD1* and *FDC1* enzymes help detoxify phenyl acrylic acids found in plant cell walls, which explains why these genes are present and functional. In most domesticated ale and lager yeasts, they have been inactivated or eliminated by multiple types of mutations that led to their loss of function [36, 37, 71]. On the other hand, the fermentations carried out with

S. eubayanus stand out for a moderate production of acetate and ethyl esters, such as 3-methylbutyl acetate, ethyl hexanoate and ethyl octanoate [56, 59, 63], which add desirable fruit flavours to the beer. It also produces moderate concentrations of higher alcohols such as 2-phenylethanol, 2-methylpropanol, 3-methylbutanol and 2-methylbutanol [56, 59, 63]; which gives complexity to beer flavour but are generally considered unpleasant at high concentrations. The production of sulfuric aromas (for example, ethanethiol and ethyl thioacetate) has also been observed during fermentations, but these are normally reduced with maturation (*lagering*) [56, 72]. Differential production of volatile compounds has been found between different strains of *S. eubayanus,* demonstrating intraspecific phenotypic variability [63, 66, 68].

Expanding the Limits for Brewing

The brewing potential of these wild yeasts can be taken further through various strategies. Adaptation to beer wort through the method of directed experimental evolution or selective hybridization between strains or species with complementary desirable characteristics, are techniques that do not generate genetically modified organisms (GMOs) and can be used to improve or simplify the application of these non-conventional yeasts in the industry [73]. The most prominent case is the hybridization of wild strains of *S. eubayanus* and domesticated strains of *S. cerevisiae* for the generation of new lager yeast variants to extend their phenotypic diversity [56, 57, 62, 64, 72]. Interspecific hybridization can be achieved mainly by three approaches: spore-to-spore mating, rare mating and mass mating [74, 75]. Hebly *et al.* [64] and Krogerus *et al.* [59] used mass mating techniques for the obtention of *S. cerevisiae* and *S. eubayanus* hybrids, and observed that they inherited beneficial properties of both parents (sugar utilization and superior performance at low temperatures), even out-performing them in terms of fermentation. Mertens *et al.* [56] carried out the most extensive work generating 31 hybrids between *S. cerevisiae* and *S. eubayanus* using a robot-assisted spore-spore technique. The resulting hybrids showed improvements in tolerance to low and high temperatures; the best of them showed high fermentation performance and desirable aromas and flavours that were significantly different from the aroma production of currently available lager yeasts, making them interesting new yeast strains for commercial lager beer production. On the other hand, simultaneous or sequential inoculation with traditional ale or lager yeasts may be another way to enhance fermentation with non-conventional yeasts in a controlled manner and improve the quality of the final products [76, 77].

There are other improvement methods that involve genetic manipulation, such as genetic engineering and synthetic biology, which are being used or investigated, but which involve the generation of GMOs, which both the industry and the consumers still do not fully accept.

From Nature to Beer Industry

Despite the apparent limitations of brewing with pure strains of *S. eubayanus*, few strains do possess many advantageous traits for lager brewing, such as efficient biomass production at low temperature (down to 4 °C), efficient maltose utilization and production of differential aroma compounds, which can be exploited for brewing innovation as a tool for productive diversification and value addition. In fact, the case of *S. eubayanus* is one of the examples of the successful transfer of a wild yeast to the industry. In 2016, Heineken launched 'H41 Wild Lager' and 'Wild Lager H71 Patagonia', which were the first commercial products that exploited pure strains of this species. Later, two other beers were launched involving *S. eubayanus* strains isolated from different locations: 'Wild Lager H32 Himalayas' and 'Wild Lager H35 Blue Ridge Mountains'. Simultaneously, work with local brewers from Bariloche (Patagonia, Argentina) on the semi-pilot scale (20-50 Lts) started in order to understand the behavior of *S. eubayanus* outside the laboratory. In 2018, the commercial licenses for beers made with *S. eubayanus* were signed by ACAB (Association of Artisanal Brewers of Bariloche and the Andean Region) under the name of Wild Patagonia Project, where the local breweries were engaged together with local scientists from IPATEC in the development of their own styles. Later that year, the International Workshop on Brewing Yeasts (IWOBY) took place in Bariloche together with the International Specialized Symposium on Yeasts (ISSY34), where the first *S. eubayanus* beer tour was organized. Six local products brewed by six different breweries were presented there: Blest's Wild Mother, Konna's EubaKonna, Berlina's Pan del Indio, Bachmann's Wild Lager, Wesley's Wild Wild Lager and Manush's Sauvage Wild Lager. In 2019, the first 100% genuine beer from Patagonia was developed employing ingredients exclusively sourced from north-western Patagonia, including *S. eubayanus*. The production of hops in the region dates to the 1900´s and has been demonstrated to possess excellent characteristics for brewing [78]. Moreover, pristine water is provided by the Andes rivers. Even though barley is only partially cultivated in the area, experimental parcels were dedicated for this purpose and were malted locally by the La Alazana distillery. The beer had excellent acceptability among customers and current ongoing efforts aim to allow its expansion and continuous production.

The brewing industry is phasing significant changes in order to meet customers and market demands for beers with more innovative ingredients and flavours, preferably from regional and sustainable sources. It was demonstrated here that Patagonian Andes represents a nice example of how autochthonous microbiological resources can be exploited in biotechnological processes such as the beer industry.

CIDER INNOVATION AND YEAST BIOTECHNOLOGY

Cider is the alcoholic beverage that results from the fermentation of apple juice obtained from the grinding and pressing of this innocuous and clean fruit [79]. The alcoholic concentrations admitted by the Argentine Food Code for cider is 4.5 ± 0.3% *v/v*. The main producers and consumers of cider worldwide are the United Kingdom, Ireland, France, Spain, and the United States [80, 81], whereas Argentina is among the top 10 countries with the largest cider making worldwide [82]. The production of this beverage in Argentina is directly related to the growing of pome fruits (apples and pears), for this reason, up to 75% of the national cider production is concentrated in North Patagonia where fruticulture is one of the most relevant agricultural activities [83].

Even though cider is a beverage with an old tradition in Argentina, associated with European immigrations, its consumption has demonstrated an important seasonality linked to the festive summer dates [84]. In recent years, there has been a great effort and investment by both governmental organisms and cider producers, in changing consumer trends implying different marketing strategies. These include the production and sale of these beverages throughout the year, the diversification of packaging (bottle, can or barrel, the last one to be consumed "spilled") and the elaboration of new varietal ciders with flavours and aromas differentiated from the "traditional" beverage. These changes intend to turn cider into a direct competitor of beer in the market of soft drink beverages [85]. These strategies are not necessarily in association with the development of technological strategies supported by the producers in these new challenges. Similar phenomena in the cider industry are observed around the world. In order to contextualize this situation, a comparison with the winemaking industry could be useful. Hundreds of commercial yeast starters have been developed for winemaking, each with the most diverse properties; however, the number of yeasts starters for cider making is still low (Table **1**). In fact, no commercial yeast starters have been developed for cider making at all in Argentina, so many producers use yeasts originally selected for winemaking.

In this context, a strategic plan aimed at identifying and selecting the appropriate starter culture for each type of regional cider was developed (Fig. **2**). The "hunting" and selection of yeasts was carried out promoting the unconventional species *S. uvarum*. This *Saccharomyces* species has been systematically isolated in environments related to apple musts fermentation in the world and particularly in North Patagonia [86 - 94]. Moreover, the great diversity of *S. uvarum* strains that coexist in Patagonia, both in natural environments, traditional beverages and ciders [11, 17, 18, 30, 93, 94], offers an ideal scenario to study both physiological and genomic characteristics of the different strains in this species in a comparative way, as well as their potential as starter cultures for the production of specific ciders with a 100% Patagonian imprint.

Screening of *S. uvarum* Strains for Cider Production

Patagonia seems to be a unique place in the world in terms of the diversity of the *S. uvarum* cryotolerant species. Strains of this species have been isolated from natural environments together with its sibling species *S. eubayanus*. These natural habitats include both *Nothofagus* (bark, *Cyttaria* stromata and soil) [11, 30, 95] and *Araucaria araucana* (bark and seeds) [18] forests in Andean Patagonia. *S. uvarum* was also isolated from traditional fermented beverages called apple *chichas* [93] and from ciders fermented at low temperatures in North Patagonia [94]. Further, the greatest genomic diversity of this yeast species was found in Patagonia [10], with representatives of three out of the four genomic populations described worldwide. Most strains isolated from natural environments belong to the population named South America B (SA-B), the most taxonomically distant after the Australasian one. On the other hand, strains belonging to the South American B (SA-B) and to the mixed population named South America A-Holarctic (SA-A/HOL) were isolated from fermentative environments including both apple *chichas* and ciders [17].

The species *S. uvarum* and *S. eubayanus,* as well as the hybrids *S. bayanus* (hybrid between *S. uvarum* and *S. eubayanus* with or without genomic portions of *S. cerevisiae*) and *S. pastorianus* (hybrid between *S. eubayanus* and *S. cerevisiae*), share unique physiological characteristics within the genus that include an active fructose transport system, high fermentative vigour and cryotolerance [96]. In particular, different strains of *S. uvarum* have been reported to be able to completely conduct the alcoholic fermentation of grape and apple musts, producing beverages with significantly lower ethanol and higher glycerol concentrations than *S. cerevisiae*, without an increase in the acetic acid levels [52, 75, 97 - 99]. *S. uvarum* also produces volatile compounds that directly improve the fermented products flavour, such as 2-phenylethanol and its respective acetate ester 2-phenyl-ethyl-acetate [52, 94, 100]. These physiological particularities of *S. uvarum* have been exploited for the development of the first commercial wine yeast starter of this species named VELLUTO BMV58YSEO® (Lallemand).

The yeast selection criteria for cider making should be similar but not the same required for a wine yeast. Tolerance to ethanol is one of the most important features of a wine yeast starter due to its association with wine stuck fermentations [101]; however, the lower alcohol concentrations in cider reduce the relevance of this parameter in starters selected for this beverage. Osmotic stress is another parameter involved in wine stuck fermentation [102, 103]. Nevertheless, apple musts exhibit a significantly lower concentration of sugars than grape musts and, therefore, no stuck fermentations due to this factor have been reported in cider making.

On the opposite, the main stress factors in a cider yeast selection protocol are: i) resistance to sulphite (sulphite, mostly in the form of metabisulphite, is normally used in both cider and wine industries in Patagonia); ii) ability to ferment at low temperature (temperature inside the cellars in Patagonia fluctuate considerably during the production months, this phenomenon is most relevant in cider making due to the possibility of elaborating ciders during winter by using fruits stored in packinghouses); iii) ability to consume fructose at low glucose:fructose ratios (this phenomenon is especially relevant in ciders due to the high concentrations of fructose in relation to glucose present in most apple varieties used in Patagonia, that include mainly acidic fruits) [104].

In order to satisfy some of these requirements, efforts were directed to the screening of *S. uvarum* strains that combine these characteristics of interest. In a first step, a total of 31 Patagonian strains isolated from natural environments (11 strains), apple *chicha* (16 strains), and ciders (4 strains) were evaluated in their tolerance to the most relevant stress conditions for cider making. These strains, representative of the three genomic populations described for the species in Patagonia, were compared with a pool of 27 European strains belonging to the HOL population, isolated from natural (6 strains) and fermentative (21 strains) environments, in order to determine the relationship between their physiological properties, genomic background and origin. Different scenarios were evaluated:

i). Sulphite Tolerance

The AFC allows the addition of up to 300 mg/L of total sulphite in ciders; however, most cider producers prefer to use lower doses of this compound in their products, due to the demonstrated impact of this chemical compound on the health and on sensory attributes of the ciders [104]. Even considering the maximum allowed, sulphite does not represent a real stress factor for commercial starters of *S. cerevisiae,* which can tolerate significantly higher concentrations of this compound (300-1500 mg/L) [105 - 108]. Contrarily, unconventional yeast species such as *S. uvarum* do not have the same ability as *S. cerevisiae* to tolerate these concentrations [109, 110]. In a comparative assay carried out with 58 *S. uvarum* strains, we evidenced that those isolated from natural environments (both Patagonian and European strains), and from apple *chicha* (Patagonian strains) were only able to tolerate very low concentrations of sulphite (0-200 mg/L). On the other hand, a large number of strains isolated from fermentative environments (mainly European strains but also some Patagonian strains) displayed an intermediate tolerance to this compound (200-300 mg/L) and only a few of them (isolated from wine in Europe and cider in Patagonia) presented a high tolerance to sulphite (300-900 mg/L). All these strains belonged to the HOL population. When analyzing the whole sequence of these strains, a reciprocal chromosomal translocation between chromosomes VII and XVI (VIItXVI) was detected. This translocation involves numerous genes, including the *SSU1* gene (encoding a plasma membrane sulphite pump), involved in sulphite resistance [111, 112]. In addition, these strains exhibiting the translocation presented a significantly higher expression of the *SSU1* gene than the wild type strains, which gives them high tolerance to sulphite [109, 110]. Similar results had been reported for *S. cerevisiae* [107]; however, our results evidenced this domestication fingerprint in *S. uvarum* for the first time. These strains have an important biotechnological potential for cider industry.

ii). Temperature

S. uvarum has been described by numerous authors as a cryotolerant yeast species [23, 113 - 116]. However, most of these studies compare *S. uvarum* strains belonging to the less diverse genomic population (HOL), with other species of the genus as *S. cerevisiae*, *S. kudriavzevii*; *S. eubayanus* or hybrids. Additionally, these studies use mostly *S. uvarum* isolates from European environments, where this species naturally coexists with *S. cerevisiae*. In Patagonia, *S. uvarum* also coexists with *S. cerevisiae* in fermentative environments [93, 94, 117]; however, in natural environments it is frequent to find *S. uvarum* in association with its cryotolerant sister species *S. eubayanus* [11, 18, 30, 51].

In a comparative assay carried out with 58 strains of *S. uvarum* representative of different genomic populations and origins, different fermentation temperatures were evaluated. Differential fermentation kinetics were observed. Strains isolated from natural environments and apple *chicha*s in Patagonia (belonging to SA-B and SA-A/HOL populations) evidenced better fermentation kinetics (lower lag phase and higher growth rate) than European strains (isolated from both natural and fermentative environments) and Patagonian strains obtained from cider (all of them belonging to the HOL genomic population) in fermentations carried out at 25 °C and 30 °C. Competition trials between three strains (each from a different genomic population) demonstrated that the European wine strain (HOL population) dominated the fermentation at 13 °C, while both SA-B and SA-A/HOL strains (isolated from natural environments in Patagonia) became the dominant yeast strain in fermentations carried out at 30 °C [94]. These results provided evidence about the differential behaviour exhibited by *S. uvarum* strains with regards to their accompanying species in the environment. Those yeasts strains that had to coexist and compete with the highly invasive and nomadic species *S. cerevisiae* in the niche (as described by Goddard [118]) could be forced to differentiate themselves in order to guarantee their subsistence. This could be the case observed in *S. uvarum* strains isolated from fermentative environments in Europe (HOL population) adapted to low temperatures, one of the few adaptations that the species *S. cerevisiae* has not acquired. On the other hand, in Patagonian natural environments the situation was different, despite the high number of sampling studies in these habitats, no natural *S. cerevisiae* strains were obtained. In this ecological niche *S. uvarum* coexists with their sister species, also cryotolerant, *S. eubayanus* [11, 18, 30, 51]. The coexistence of these two cryotolerant species, in absence of *S. cerevisiae*, could have influenced a better growth adaptation to a higher temperature range in *S. uvarum* strains isolated from the natural environment compared to the HOL strains. A similar result was described among the sympatric species *S. cerevisiae* and *S. paradoxus* in Portugal [23]. These authors observed that, even when the two species share a preference for high temperatures, *S. cerevisiae* is able to grow at temperatures slightly higher than *S. paradoxus*, which guarantees the coexistence of the two species in the same habitat.

Differential adaptations to temperatures at the intraspecific level can be a useful feature to be exploited for biotechnological purposes. As mentioned previously, the cider industries in North Patagonia elaborate their products from both freshly harvested apples (March-April) and stored apples from packinghouses (May-October), extending the production time. The use of concentrated apple juice is not a regular practice in Patagonian cider making, as in other areas worldwide [104]. In addition, the lack of temperature control during the fermentation processes cause a significant variation in the fermentation temperatures along the year; from 20-25 °C in summer to 8-15° C in winter (unpublished data). In fact, some producers have reported stuck fermentations caused by the low winter temperatures in cases of both spontaneous fermentations carried out without the addition of yeast starters or even in conducted fermentations carried out with starters belonging to *S. cerevisiae* species.

Numerous studies have been carried out on ciders produced in geographical areas with low temperatures, where the fermentation is carried out at cellar temperature of 12-15 °C [88, 89, 91, 92, 119]. A strong relationship between the temperature of fermentation and the yeast population dynamics was observed [89]. For example, the dominance of *S. cerevisiae* strains was detected in cider elaborated in warm seasons (average temperature 24°C) over ciders elaborated in cold seasons (10-12 °C). *S. bayanus* has been reported together with *S. cerevisiae* in ciders made at unspecified temperatures [87, 90, 120, 121], and its presence is detailed at the beginning [122], middle and end of fermentation [91, 92] in ciders made on an industrial scale at cellar temperature of 12-15°C. In these temperatures, *S. bayanus* even dominates over *S. cerevisiae* due to its cryophilic nature (S-v 2007).

iii). Fructose Consumption

Another biotechnological property of the cryotolerant species *S. uvarum* is the ability to consume fructose. This attribute has been related directly to the presence of the specific fructose-proton transporter Fsy1p. Fsy1p, also present in *S. eubayanus* and the hybrids *S. bayanus* and *S. pastorianus*, showing an affinity for fructose of at least an order of magnitude greater than the affinity exhibited by the Hxtp hexose transporters for the same molecule. This phenomenon gives *S. uvarum* a substantial adaptive advantage in environments with a high concentration of fructose [100, 123 - 125]. Different fructose consumption kinetics were observed in a comparative trial carried out with *S. uvarum* strains representative from each genomic population and origin. Three fermentation conditions were evaluated in synthetic must (MS300 with modifications) containing 120 g/L of glucose and fructose according to the following ratios 1:1, 1:2 and 1:4. These conditions simulate the concentrations detected in different apple musts used in the Patagonian cider industry. The European and Patagonian cider strains (HOL) presented better fermentation kinetics in the 1:1 condition, demonstrating a high preference for glucose in relation to fructose, whereas strains isolated from apple *chicha* (SA-A/HOL population) presented better fermentation kinetics in the 1:4 condition with a fructose consumption rate significantly higher in this condition in relation to the other strains. In addition, this strain presented a higher rate of fructose consumption in the 1:4 condition than in the 1:1 condition (unpublished data), indicating that the concentration of these sugars in the must directly affect the behaviour of this strain, as described by Berthels *et al.* (2004). The differential fructose consumption at the intraspecific level observed in this assay also shows the great biotechnological potential of the *S. uvarum* yeast diversity for the selection of specific starters for particular apple varieties used in the cider industry.

Selection of *S. uvarum* Cider Starter Cultures

The adaptations described previously for *S. uvarum*, were used to select strains with distinctive properties, to be used in pilot-scale fermentations combining different variables of interest in the cider industry. In this stage, it is important to evaluate the ability of these yeast strains to dominate a fermentation process characterized by several microorganisms naturally present in apple musts and cellar equipment, and their capacity to produce a combination of certain aromatic compounds that result in a complex cider with a high degree of acceptance by consumers [104, 126, 127]. In this context, and considering the need for excellent implantation of an *S. uvarum* yeast starter, certain variables involved in the cider elaboration process should be taken into account due to their strong influence on the implantation capacity of the starter and the sensory attributes of the final products. The most important variables are: i) fermentation temperature: this parameter is extremely important for the successful implantation of cryotolerant species and the development of the secondary aroma (aromatic compounds produced by the yeast during the fermentation); ii) apple variety: because of the already mentioned glucose: fructose ratios involved in each particular apple variety and its subsequent effect on yeast growth during the production of varietal ciders of great interest for producers [84]. It has been demonstrated that each apple variety presents different physicochemical properties that impact on the final product, including the pH that could even affect the implantation of the yeasts used as starter culture due to the differential effect of SO_2 on this parameter [104, 128, 129]. Taken these two variables into account, a set of experiments based on pilot-scale fermentations were designed:

i). Impact of the Fermentation Temperature on Starter Implantation and Aromatic Properties of Ciders

A strain of *S. uvarum* isolated from apple *chicha* was selected for this first trial. This strain had shown high fermentative vigour at 20-25 °C, high fructose consumption in musts with low glucose:fructose ratio, but low sulphite tolerance. Granny Smith apple juice (glucose:fructose ratio 1:3) and sublethal concentrations of sulphite (100 mg/L) were used in this assay. The starter implantation, fermentation kinetics, chemical composition and sensory analysis of the resulting beverage were evaluated and compared with the ciders fermented with a commercial strain of *S. cerevisiae,* a wine yeast used by regional cider producers. Fermentations were carried out at 25 °C. Relatively high implantation was observed in the final stage of the fermentations inoculated with *S. uvarum* (63%) while the commercial strain *S. cerevisiae* was not able to conduct the fermentation of this must, not being detected at the end of the fermentations. These results support the importance of the development of a specific yeast starter for each type of beverage; the use of wine yeast starters in cider elaboration is not recommended, at least without the corresponding implantation study. On the other hand, this could explain the high variability of the product detected year after year, a problem that producers claim to have. Ciders fermented with the *S. uvarum* strain showed a higher concentration of glycerol, citric acid, 2-

phenylethanol and 2-phenylethyl acetate, all of them are typical characteristics of *S. uvarum* [52, 97, 123, 130 - 134]. These compounds are directly related to a positive sensory impact, giving the beverages greater body, acidity and fruity aromas [135]. The ciders conducted with the *S. uvarum* strain were preferred (66% vs 34%) in a consumer preferences analysis [94]. In order to improve the implantation of *S. uvarum*, the same fermentations were conducted at 13 °C using the same Granny Smith apple juice. In this case, the implantation capacity observed for *S. uvarum* was 100%, and again no implantation was observed for the commercial *S. cerevisiae* yeast [94].

ii). Impact of the Apple Variety and the Addition of Sulphite on the Fermentative, Aromatic and Sensory Attributes of Ciders Conducted with Different S. uvarum Strains

Three different apple varieties were evaluated: Granny Smith, Pink Lady (both acidic apple varieties) and a bivarietal mix of Packam's/D'anjou pear varietals. Due to the high implantation observed for *S. uvarum* at 13 °C in previous studies [94], this temperature was maintained in these fermentations. On this occasion, two strains belonging to *S. uvarum* species were used: the strain isolated from *chicha* used in previous fermentations and the strain isolated from Patagonian ciders that showed a high sulphite tolerance, preference for glucose, and high fermentation kinetics at low temperature. Besides, all fermentations were carried out with 200 mg/L of sulphite, a concentration normally used by regional producers. Very low and no implantation of the *chicha S. uvarum* strain was observed in the fermentations carried out with Granny Smith and Pink Lady apple musts, respectively. Contrarily, the cider *S. uvarum* strain evidenced a 100% implantation in the two musts. This phenomenon could be caused by the elevated sulphite concentrations used in the assays and the differential sulphite tolerance of the inoculated yeast strain. Regarding the production of aromatic compounds, high concentrations of glycerol, 2-phenylethanol and its acetate ester were observed in the fermentations conducted with the cider strain [94]. Sensory analysis evidenced the preference of the consumers (71%) for the cider produced with Pink Lady apple must, while 40% of the total consumers preferred the cider elaborated with Granny Smith apple must [94]. Finally, in fermentations carried out using pear juice, these results were completely different. A 100% implantation was observed for the two analyzed yeast strains. This behaviour could be associated with the significantly higher pH of pear musts in relation to apple musts (4.2 vs 3.5, respectively). On this condition, sulphite exhibits a decrease in its antimicrobial effect of up to 60% [136]. Moreover, malolactic fermentation developed by lactic acid bacteria (LAB) was observed in the cider fermentations conducted with the *S. uvarum* strain from *chicha*. This additional fermentative process altered the final concentrations of various compounds such as lactic acid and ethyl lactate, inducing a great diversification among the ciders produced with the two *S. uvarum* strains. We have evidence to infer that the development of LAB in these fermentations, as well as their inhibition in fermentations conducted with the cider *S. uvarum* strain, is closely related to the interaction between the LAB and the particular strain that dominates the fermentation [137]. All these differences, influenced the final products at the sensory level. In the sensory analysis, the consumers preferably chose the ciders fermented with the *S. uvarum* strain obtained from apple *chicha*. With this complete background, it is clear that different strains must be selected for each type of must and also that the technological practices of each producer must be taken into account.

Bioprospecting of Yeast for Cider: From Nature to Industry

A long road must be travelled from the initial isolation and genomic and physiological characterization of yeast to its selection and use to conduct a fermentation, to produce ciders with aromatic complexity and sensory acceptance by the consumers (Fig. **2**). In our experience, we emphasize the respect for the producer demands coupled with the selection of specific yeast strains adapted to these needs. The last stage of these processes is the evaluation of the yeasts at industrial level as well as the transfer of the novel technology to the industry. At this moment, the mentioned technological developments are being tested by different cider companies in Argentina through the technological assistance agreements that should also allow genetic resource protection. The objective is clear, to work with both big companies that concentrate the biggest cider production in Argentina and small producers that bet for the innovation, diversification and controlled production of this beverage. As we mentioned at the beginning of this section, these advances should go hand in hand with the development of new market trends that accompany the demands of a forgotten industry that is rising significantly.

WINE INNOVATION AND YEAST BIOTECHNOLOGY

Cryotolerant Yeasts for Patagonian White Wines

The transformation of the grape must into wine, or winemaking, is a complex biotechnological process almost as old as human history, involving grapes, microorganisms and technology. This particular transformation, carried out empirically in its beginnings, evidenced a significant scientific advance during the last 150 years, revealing the intricate microbial and biochemical phenomena involved in the process. Since grape must comprise a nutritionally complex substrate ideal for the development of microorganisms, a great variety of non-*Saccharomyces* yeasts can be found along the whole winemaking process, including the genera *Debaryomyces, Dekkera, Hanseniaspora, Issatchenkia, Lachancea, Meyerozyma, Metschnikowia, Pichia, Torulaspora, Wickerhamomyces* and *Zygosaccharomyces,* among others [138]. These yeasts that inhabit the surface of the grapes are transferred to the musts during the winemaking process and become responsible for the initial stages of fermentation. Later, these species begin to decline in number due to diverse winemaking stress factors occurring during the process and the *Saccharomyces* species, especially *S. cerevisiae,* dominate the fermentation up to the end [139]. In much lesser extension, other species included in the *Saccharomyces* genus can also participate and even become dominant during the fermentation processes as it is the case of *S. paradoxus* in particular Croatian wines [140] and *S. uvarum* in wine fermentations conducted at low temperature in different winemaking regions [114, 131, 141, 142]. The winemaking process carried out by the mixed populations of yeasts naturally present on both grapes and winery equipment surfaces is known as natural or spontaneous fermentation. The main characteristic of the wines obtained by means of this methodology is great organoleptic complexity. Contrarily, the lack of reproducibility inherent to these natural fermentations represents a serious industrial problem. For this reason, the use of yeast starter cultures constituted by selected yeasts has been introduced as a common oenological practice, which allows better microbiological control of the process through the normalization of the involved biota. On the other hand, the most important disadvantage of starter cultures, typically constituted by a pure culture of a selected *S. cerevisiae* strain, is the production of wines with standard sensory quality, with flat aromatic profiles that lose the organoleptic complexity provided by the complex biota of indigenous yeasts present in natural fermentations [143].

The fierce competition in the wine market looking for the diversification of the commercial products has led to an increase in the diversity of yeast starters available in the market. These new developments included species different from *S. cerevisiae* such as *S. uvarum* [96, 133, 144], mixed cultures of *Saccharomyces* and one or more strains belonging to the big group of wine yeasts known as non-*Saccharomyces* [144, 145] and hybrid strains [96, 98, 146]. However, the number and diversity of wine yeast starters selected, produced and commercialized in Argentina is still very limited and mostly associated with specific regional scientific developments.

Faced with these demands, North Patagonian winemakers started together with scientific institutions a prospection of local yeasts to be used for the differentiation of Patagonian wines. Different approaches have been made in order to produce a set of different yeast starters for red and white wines, including the selection of strains belonging to *S. cerevisiae, Saccharomyces* non-*cerevisiae* and interspecific hybrids [54] as well as non-*Saccharomyces* species [147, 148].

In the following sections the specific advances on the selection and characterization of non-conventional (*Saccharomyces* non-*cerevisiae* and interspecific hybrids) yeasts carried out for the development of a starter culture for white wines able to perform fermentations at low temperature will be synthesized and discussed.

Screening of Non-Conventional Yeasts for White Wines Elaboration at Low Temperature

Based on the purpose of looking for specific yeasts to perform white wine fermentations characterized by a low fermentation temperature, our search was firstly oriented to *S. uvarum*. This species frequently found in wine fermentations at low temperature, produces wines with high glycerol concentrations and high concentrations of higher alcohols specially 2-phenyletahanol, which confers a floral aromatic feature to wines [133]. Several *S. uvarum* strains had been isolated in different works carried out in our laboratory from different sources including both natural environments and fermentative processes in Patagonia as mentioned in the cider section [18, 93, 94].

On the other hand, the recent discovery of *S. eubayanus* in Patagonian forests [11, 18], also turned this cryotolerant species into an object of study for winemaking. In this way, Andean Patagonia seems to be a great source of cryotolerant yeasts of the *Saccharomyces* genus, which could be selected and used for low temperature fermentations, adding new differential organoleptic features.

Low fermentation temperature enables the retention of esters and higher alcohols, related to the flavour [149 - 152]. For that reason, this technological phenomenon was incorporated into most white wine fermentations, in which, due to the short period of maceration (contact between grape juice and skins), the retention of aroma compounds became necessary [143, 153, 154]. Fermentation temperature should be strongly controlled since it has been reported that both yeasts growth and alcoholic fermentation, are enhanced by increasing temperatures. However, temperatures higher than 30°C could damage the yeasts cell membrane causing protein denaturation [153, 155]. Low temperature also affects the plasmatic membrane, diminishing its fluidity by increasing fatty acids saturation [156]. Yeasts adapted to ferment at low temperatures are not easily found on the market, with only a few representatives used in most wineries [98, 157].

Yeast Response to Winemaking Stress Factors

A plethora of factors that characterize the process are known to become stressors for the yeast cultures and must be evaluated during selection protocols. Besides the already mentioned stress factors in cider fermentations, SO_2 concentration and temperature, additional stressors became very important in winemaking such as grape musts sugar and nitrogen concentration, as well as, the ethanol level produced during the process [158]. This evaluation is particularly relevant when non-conventional yeasts isolated from other origins (different from wine) are proposed to be used, due to the fact that they should not necessarily be adapted to the specific winemaking environment. The effect of winemaking stress factors on yeasts has been evaluated using different methodologies. The most commonly used methods include the drop test in agar plates [115] and the microtiter plate assays [159]. Moreover, the use of different synthetic growth media has largely been employed at laboratory scales for understanding yeasts growth, including synthetic grape must, and its behaviour under different stress conditions [160]. Finally, stress factors can be evaluated individually (stressor by stressor in different culture media) [115] or combined (two or more stressors together in the same medium) [116, 161]. For the last strategy, the experimental design could also be employed in order to reduce the number of assays to be performed [54, 159].

A preliminary study carried out on a set of strains belonging to both *S. eubayanus* and *S. uvarum* species was made in order to elucidate their tolerance to some general stress factors typically present at the beginning of fermentation including temperature, pH, sugar and sulphur dioxide concentrations. Central composite experimental design was used to evaluate the combined effect of all these factors in a controlled situation. This study allows to predict the yeast behaviour after the inoculation of the grape must, as well as to identify the best conditions to guarantee its implantation. Most of these parameters affected the growth of all yeast strains analysed in different degrees, being the condition characterized by 20 °C, pH 4.5 and absence of SO_2, the one that evidenced the best yeast performance [54]. The SO_2 concentration was the stressor that caused the most drastic effect on yeast growth. Contrarily, sugar concentration in the must do not significantly affect the yeast growth, at least in the evaluated concentrations ranging from 40 to 360 g/L [54], which are extremely far from the normal limits present in grape musts between 120 and 250 g/L [153].

Nutrient availability in grape must is another key for fermentation progress. Nitrogen is known as the major limiting nutrient and is mainly composed of ammonium and amino acids, as the source of yeast assimilable nitrogen (YAN) [158]. In this regard, *S. eubayanus* was studied for the first time in a subset of cryotolerant strains of *S. uvarum* from different origins, two *S. kudriavzevii* and a wine *S. cerevisiae* yeast. A preliminary nitrogen requirement evaluation, conducted in synthetic grape must complemented with the compound (20 to 300 mg/L YAN), interestingly showed that *S. eubayanus* was the species less affected by the increasing concentrations of the nutrient. Thus, *S. eubayanus* and *S.*

uvarum strains from chicha, both selected for their best behaviours, were included in mixed cultures with a wine *S. cerevisiae* strain. Microfermentations in synthetic must with different nitrogen concentrations (60, 140 and 300 mg/L YAN) carried out at different temperatures (12 °C, 20 °C and 28 °C) evidenced the prevalence of the cryotolerant strains at 12 °C, independently from the nitrogen concentration. Also, *S. eubayanus* in both, pure and mixed cultures, showed the best fermentative behaviour at intermediate temperature, especially with low and intermediate nitrogen concentration [162], which could be related to the nutritional poor conditions that characterize natural environments. These results suggest, not only that cryotolerant *Saccharomyces* species could be interesting to avoid stuck fermentations by means of low nitrogen content but also, that the particular potential of *S. eubayanus* as a fermentative yeast of grape must with low nutritional requirements.

Ethanol is the main product of fermentation. However, high ethanol concentrations can cause severe damage to the yeast cells including the increase of yeast cell membrane fluidity with the consequent loss of cellular integrity as well as the inhibition and denaturation of glycolysis enzymes [163]. Yeast response to ethanol in fermentation processes has been largely studied [164 - 166]. According to recent results, both *S. eubayanus* and *S. uvarum* strains from Patagonia present a good tolerance to ethanol (up to 8%) [54], comparable with that observed for *S. uvarum* strains isolated from wine [133, 165].

Technological Features of *S. uvarum* and *S. eubayanus* for Winemaking

Some *S. eubayanus* and *S. uvarum* strains from different habitats also evidenced antagonist activity against reference sensitive yeast strains [52]. Antagonist activity is a technological trait of importance in winemaking yeast selection protocols, which allows the starter strain to compete with other *Saccharomyces* strains as well as to eliminate spoilage yeasts naturally present in grape musts [167]. In fact, for the first time, it was described and characterized the antagonistic activity of a strain belonging to *S. eubayanus* species [52] against wine spoilage yeasts species including *Brettanomyces bruxellensis*, *Meyerozyma guilliermondii*, *Pichia manshurica* and *Pichia membranifaciens* [168, 169]. Most strains of the two species were also able to produce different enzymes of oenological interest, such as β-glucosidases and β-xilosidases [52]. These enzymes are responsible for the enhancement of the varietal aroma in wines due to their ability to release volatile and aromatic compounds from conjugated odourless precursors present in grape musts [145, 170, 171].

About the production of particular compounds of relevance in winemaking, glycerol is the main product of the alcoholic fermentation following ethanol.

Glycerol imparts a slightly sweet taste and a pleasant viscosity, contributing to the smoothness, consistency and well-known body of the wine [135, 172]. Several studies have demonstrated the ability of the species *S. uvarum* to produce elevated concentrations of glycerol in wines, on several occasions, higher levels than the ones produced by *S. cerevisiae*, with the consequent reduction in the ethanol concentration [97, 116, 173, 174]. Then, high glycerol concentrations are an expected feature for this species; however, little is known about the same ability in *S. eubayanus*. The strains belonging to the two species including those obtained from both, natural and fermentative habitats, evidenced variable production of this compound when assayed in fermentation conditions in synthetic must at 20 °C. Some of these strains produced higher concentrations of glycerol and lower levels of ethanol than *S. cerevisiae,* without overproduction of acetic acid [74]. A relationship between physiological features and the origin of isolation was previously described for *S. uvarum* [10, 52]. This was observed in a study of tolerance to winemaking stress conditions, where *S. eubayanus* and *S. uvarum* strains from *A. araucana* showed homogeneous growth kinetics under increasing ethanol concentrations, while the response of *S. uvarum* strains from fermentative habitat was varied [54]. This homogeneous behaviour by some of the strains from natural habitat was also observed in a study of nutritional requirements, when the strains were submitted to increasing nitrogen concentrations in synthetic must (20 to 300 mg/L YAN) [162].

The difficulty of gathering all stress tolerance capacities and metabolic attributes that successfully allow to obtain white wines with a particular aromatic complexity in only one non-conventional strain extended the biotechnological development to the generation and selection of interspecific hybrid yeasts. Diverse studies have demonstrated that hybrid yeasts can better adapt to fluctuant stress conditions found in winemaking [175] since they acquire physiological properties from their parents [141, 150, 176]. Moreover, hybridization has been proposed to improve characteristics like the ability to grow at a particular temperature or the ethanol tolerance, features that are dependent on numerous loci distributed throughout the yeast genome [177 - 179].

The Hybridization Strategy

Representative strains belonging to *S. eubayanus* and *S. uvarum*, were selected according to their technological features and stress factors tolerance in order to generate artificial interspecific hybrid yeasts, using natural (non-GMO producing) methods such as mass mating and spore to spore mating [74, 75]. Prototrophic cryotolerant strains (either *S. eubayanus* or *S. uvarum*) were crossed with natural auxotrophic strains (*lys⁻*) of an *S. cerevisiae* wine strain [74, 75]. Because of the well-known genetically unstable nature of recently generated hybrids, mainly due to their alloploid genomes [180], a genomic stabilization procedure should be applied after hybrids generation [176]. Based on the adaptive evolution phenomenon, the genomic stabilization has been proposed to be done under the selective or "enriching" conditions typically present in the substrate where the hybrids should be used [180 - 182]; for that reason, *S. cerevisiae* x *S. eubayanus* and *S. cerevisiae* x *S. uvarum* hybrids were stabilized by means of successive fermentations on *Sauvignon blanc* grape must [74, 75].

In a first fermentation study carried out with *S. cerevisiae* x *S. uvarum* hybrids (obtained from different strains of *S. uvarum* but the same *S. cerevisiae)* in sterile *Sauvignon blanc* must, no differences were obtained in the chemical composition independently of the hybrid and fermentation temperature [74]. Nevertheless, some metabolites produced by the hybrids showed differences when compared with the ones from both parental strains. They evidenced intermediate concentrations of ethanol and higher glycerol levels than the ones produced by their *S. cerevisiae* parental strain, with total consumption of sugars in all cases. Additionally, 2-phenyethanol and its ethyl-acetate, which are related to *S. uvarum* metabolism and that mask the typical aromas of *Sauvignon blanc* [183], were only found at very low levels in the wines fermented by the hybrids. On the other hand, the higher alcohols 1-propanol and 1-butanol, previously observed in *S. cerevisiae* x *S. bayanus* [184], together with malic acid, as described for *S. uvarum* [185], were produced by the hybrids at the highest levels [74].

In a second study carried out with both *S. uvarum* x *S. cerevisiae* and *S. eubayanus* x *S. cerevisiae*, a broader temperature growth range (8 °C to 37 °C) in all the hybrids with regards to their parental species [75] were observed. These hybrids were able to grow at low (as low as *S. eubayanus* or *S. uvarum*) and high (as high as *S. cerevisiae*) temperatures [75]. These results are in accordance with previous reports which describe hybrids possessing subgenomes of *S. cerevisiae* and of one of the well-known cryotolerant species *S. uvarum*, *S. eubayanus* or *S. kudriavzevii* [115, 159].

Similar behaviour was observed with regards to the SO_2 tolerance among hybrids and parental strains. Hybrids exhibited an intermediate sulphite tolerance among the parental strains, being, in some cases, similar to that observed for the most tolerant parent *S. cerevisiae*. One of the most tolerant hybrid strains was the triploid *S. cerevisiae* x *S. uvarum;* suggesting that its tolerance might be associated with the presence of a higher copy number of the *S: cerevisiae* genes responsible for the sulphite tolerance [75]. Hybrids also showed intermediate tolerance to high external ethanol (8% *v/v*) and sugar (up to 300 mg/L) concentrations with regards to their parents [75]. Finally, the response of both hybrids and parental strains in response to two particular problematic winemaking situations were evaluated in fermentations on modified synthetic grape must [186]. In a first step, yeasts performance on a synthetic must with an unbalanced glucose/fructose ratio (80:160 g/L) was analysed at two (13 °C and 20 °C) different temperatures. This unbalanced ratio could be a consequence of climate change [187]. The differential affinity to the hexoses glucose and fructose typically found in grape musts has been studied for different yeast strains [96, 188]. In fact, the species *S. uvarum* has been associated with a more fructophilic character than *S. cerevisiae* [96]. In our study, only the hybrids were able to complete the unbalanced fermentations at 13°C, producing synthetic wines with significantly lower acetic acid concentrations than their parents. These data suggest the existence of the hybrid vigour phenomenon in the analysed conditions [75]. In a second step, a situation mimicking an incomplete or stuck fermentation was assayed. Both parental and hybrid strains, with the exception of *S. eubayanus*, were able to restart the fermentations at 13 °C; however, all hybrids showed the extra advantage of producing synthetic wines with reduced acetic acid content [75]. This repeated observation regarding low acetic acid production in hybrid strains had already been registered in *Sauvignon blanc* natural grape must [74], suggesting that it is a stable feature of these hybrids.

From the Laboratory to The Winemaking Industry

Yeasts employed as starter cultures should be able to compete with the diverse biota naturally present in natural grape musts, becoming the dominant yeast strain during the whole process. A good implantation ability, mostly higher than 80%, guarantees that most organoleptic features of the obtained wine are due to the inoculated starter and not to the native yeasts [189]. Different authors have studied the competition capacity of strains belonging to the species *S. uvarum, S. eubayanus* as well as hybrids containing subgenomes from these species together with *S. cerevisiae* in special conditions [190, 191]. However, scarce information is available about the implantation capacity of these microorganisms on natural grape must in real winemaking situations. Most data obtained in our laboratory suggested the presence of differential adaptive features in the hybrids and interesting potential for the winemaking industry. However, their implantation capacity, chemical composition as well as sensory acceptance of wines fermented with these microorganisms should be tested. For that reason, some of the hybrids generated were selected to be evaluated at pilot scale using natural *Sauvignon blanc* at 13 °C. Hybrids were compared with their respective parental strains. An intermediate implantation capacity was observed for both *S. uvarum* and its hybrid *S. uvarum x S. cerevisiae* in *Sauvignon blanc* wines (45 and 43%) [54]. On the other hand, the hybrid *S. cerevisiae* x *S. eubayanus* was not detected at the end of alcoholic fermentation (0% implantation) while the best results were obtained with the parental strain *S. eubayanus* (83% implantation). This surprising result obtained with *S. eubayanus* was accompanied by a low fermentation kinetic despite a complete sugar consumption. *S. cerevisiae* strains used as references also showed an intermediate implantation ability (70%).

With regards to the chemical and sensory attributes of the wines, fermentation products obtained with the *S. cerevisiae* parental strain were characterized by the highest concentrations of ethanol and total higher alcohols (mainly isoamyl alcohol). In general, with the exception of 2-phenylethanol, higher alcohols are considered unpleasant aromas, however, in concentrations lower than 400 mg/L they contribute to the wine complexity [151]. On the other hand, the wines fermented with the hybrids as well as with the parental *S. uvarum* evidenced significantly higher concentrations of total esters (mostly ethyl lactate). The production of esters by some hybrids was already reported by Gamero *et al.* [97]. Ethyl lactate, in particular, was also found in *Sauvignon blanc* wines and it was associated with fruity aromas when found in low levels [135]. The principal difference in wines produced by the parental strain *S. uvarum* and the hybrids was the predominant presence of 2-phenylethanol produced by the parental strain, also observed in previous reports [74]. Finally, *S. eubayanus* produced wines with an important amount of acetaldehyde, as well as high concentrations of glycerol, malic acid, 2-phenylethanol and 2-phenylethyl acetate (these last two compounds were produced in even higher concentrations than the same with *S. uvarum*).

In reference to the sensory analysis, two different assays were performed including a preference test carried out with untrained people (40 people) - untrained people tend to perceive the product in an integrated and general way [192] - and a descriptive analysis carried out with a trained panel. The analyses were performed separately, *i.e.* wines obtained with each hybrid were compared with the ones fermented by their particular set of parental strains. Independently of the comparison, all consumers detected important differences among wines. For the comparison between wines obtained with *S. cerevisiae*, *S. uvarum* and their respective hybrid strain, the preference tests evidenced similar percentages for the three products. The evaluation of the other trio (*S. cerevisiae*, *S. eubayanus* and their hybrid strain) evidenced a significantly low preference for the wine fermented with *S. eubayanus* (20%), while no difference was detected for the remaining two wines. Interestingly, besides the remarkable content of 2-phenylethanol and 2-phenylethile acetate (associated with floral and honey notes) [172] present in wines fermented with *S. eubayanus*, this product was not preferred by the untrained public. Perhaps the high levels of acetaldehyde produced by this strain could be the cause of rejection since it is associated with green apple when found at low levels, but it can be perceived as a pungent or irritating aroma, which could be reminiscent of green grass in higher concentrations [193]. In spite of the low acceptance, all wines evidenced particular aroma profiles according to the trained panel that included some regional oenologists. In fact, we are evaluating both hybrids and parental strains at an industrial scale in regional wineries in order to produce *Sauvignon blanc* wines with differential features to be used in the elaboration of complex wines by means of their combinations.

Our experience comprises one of the scarce reports available, up to the moment, related to the behaviour of both *S. eubayanus* and its hybrids in low temperature winemaking conditions, with a high potential for the regional industry.

CONCLUSION

The present chapter aimed to review the role of yeasts in the production of the three most popular non-distilled fermented alcoholic beverages: beer, cider and wine. In particular, we focused on the new trends that are influencing recent innovation strategies in these beverages regarding the use of non-conventional yeasts for boosting flavour complexity and / or developing new special products, and the incorporation of regional identity by using locally sourced strains. The experiences gathered in the region of Andean Patagonia (Argentina) related to the isolation, screening, selection and even improvement of local cold-adapted yeasts of *S. uvarum* and *S. eubayanus* were reviewed as well as the challenging aspects to scale-up trials and technology transfer to the beverage industry. It becomes clear that the use of adequate selective isolation protocols and well-planned screening and selection strategies are the basics for success in this type of studies. Later, the application of factorial design and central composite experiments aid in determining the most relevant factors influencing each bioprocess at lab and semi-pilot scales. The use of directed experimental evolution and / or hybridization techniques helps in the rapid acquisition of desirable traits in the non-domesticated strains. Finally, successful interaction with the industry requires from the scientific side a profound knowledge of the process and product characteristics in order to gain the confidence needed and to ensure a common language with the other party, especially when it involves craft or small companies. The successful cases reviewed here of already released commercial products are proof that the laborious road from nature to industry application is feasible and advantageous. We hope that sharing these experiences might inspire other colleagues to initiate similar types of research and development studies.

CONSENT FOR PUBLICATION

Not applicable.

CONFLICT OF INTEREST

The authors declare no conflict of interest, financial or otherwise.

ACKNOWLEDGEMENT

Declared none.

REFERENCES

[1]   Patrick CH. Alcohol, culture and society. Duke University Press 1952.

[2]   Walker GM, Stewart GG. *Saccharomyces cerevisiae* in the Production of Fermented Beverages. Beverages 2016; 2(4): 30.
[http://dx.doi.org/10.3390/beverages2040030]

[3]   Naseeb S, James SA, Alsammar H, *et al. Saccharomyces jurei* sp. nov., isolation and genetic identification of a novel yeast species from Quercus robur. Int J Syst Evol Microbiol 2017; 67(6): 2046-52.
[http://dx.doi.org/10.1099/ijsem.0.002013] [PMID: 28639933]

[4]   Gibson B, Geertman JA, Hittinger CT, *et al.* New yeasts-new brews: modern approaches to brewing yeast design and development. FEMS Yeast Res 2017; 17(4)
[http://dx.doi.org/10.1093/femsyr/fox038] [PMID: 28582493]

[5]   Nikulin J, Vidgren V, Krogerus K, Magalhães F, Valkeemäki S, Kangas-Heiska T, *et al.* Brewing potential of the wild yeast species *Saccharomyces paradoxus.* Eur Food Res Technol 2020; 246(11): 2283-97.
[http://dx.doi.org/10.1007/s00217-020-03572-2]

[6]   Hutzler M, Michel M, Kunz O, *et al.* Unique brewing-relevant properties of a strain of Saccharomyces jurei isolated from ash (Fraxinus excelsior). Front Microbiol 2021; 12: 645271.
[http://dx.doi.org/10.3389/fmicb.2021.645271] [PMID: 33868204]

[7]   Cordente AG, Curtin CD, Varela C, Pretorius IS. Flavour-active wine yeasts. Appl Microbiol Biotechnol 2012; 96(3): 601-18.
[http://dx.doi.org/10.1007/s00253-012-4370-z] [PMID: 22940803]

[8]   Holt S, Mukherjee V, Lievens B, Verstrepen KJ, Thevelein JM. Bioflavoring by non-conventional yeasts in sequential beer fermentations. Food Microbiol 2018; 72: 55-66.
[http://dx.doi.org/10.1016/j.fm.2017.11.008] [PMID: 29407405]

[9]   Steensels J, Gallone B, Voordeckers K, Verstrepen KJ. Domestication of industrial microbes. Curr Biol 2019; 29(10): R381-93.
[http://dx.doi.org/10.1016/j.cub.2019.04.025] [PMID: 31112692]

[10]   Almeida P, Gonçalves C, Teixeira S, *et al.* A Gondwanan imprint on global diversity and domestication of wine and cider yeast *Saccharomyces uvarum.* Nat Commun 2014; 5(1): 4044.
[http://dx.doi.org/10.1038/ncomms5044] [PMID: 24887054]

[11]   Libkind D, Hittinger CT, Valério E, *et al.* Microbe domestication and the identification of the wild genetic stock of lager-brewing yeast. Proc Natl Acad Sci USA 2011; 108(35): 14539-44.
[http://dx.doi.org/10.1073/pnas.1105430108] [PMID: 21873232]

[12]   Maicas S. The role of yeasts in fermentation processes. Microorganisms 2020; 8(8): 1142.
[http://dx.doi.org/10.3390/microorganisms8081142] [PMID: 32731589]

[13]   Benito S, Ruiz J, Belda I, *et al.* Application of non-Saccharomyces yeasts in wine production.Non-

conventional yeasts: From basic research to application. Springer 2019; pp. 75-89.
[http://dx.doi.org/10.1007/978-3-030-21110-3_3]

[14]   Burini JA, Eizaguirre JI, Loviso C, Libkind D. Levaduras no convencionales como herramientas de innovación y diferenciación en la producción de cerveza. Rev Argent Microbiol 2021.
[http://dx.doi.org/10.1016/j.ram.2021.01.003]

[15]   Steensels J, Verstrepen KJ. Taming wild yeast: potential of conventional and nonconventional yeasts in industrial fermentations. Annu Rev Microbiol 2014; 68: 61-80.
[http://dx.doi.org/10.1146/annurev-micro-091213-113025] [PMID: 24773331]

[16]   Karabín M, Jelínek L, Kotrba P, Cejnar R, Dostálek P. Enhancing the performance of brewing yeasts. Biotechnol Adv 2018; 36(3): 691-706.
[http://dx.doi.org/10.1016/j.biotechadv.2017.12.014] [PMID: 29277309]

[17]   Gonzalez Flores M, Rodríguez ME, Peris D, Querol A, Barrio E, Lopes CA. Human-associated migration of Holarctic *Saccharomyces uvarum* strains to Patagonia. Fungal Ecol 2020; 48.
[http://dx.doi.org/10.1016/j.funeco.2020.100990]

[18]   Rodríguez ME, Pérez-Través L, Sangorrín MP, Barrio E, Lopes CA. *Saccharomyces eubayanus* and *Saccharomyces uvarum* associated with the fermentation of Araucaria araucana seeds in Patagonia. FEMS Yeast Res 2014; 14(6): 948-65.
[http://dx.doi.org/10.1111/1567-1364.12183] [PMID: 25041507]

[19]   de Garcia V, Libkind D, Moliné M, Rosa CA, Giraudo MR. Cold-adapted yeasts in patagonian habitats.cold-adapted yeasts: biodiversity, adaptation strategies and biotechnological significance. Springer 2014; pp. 123-48.
[http://dx.doi.org/10.1007/978-3-662-45759-7_6]

[20]   Libkind D, Russo G, van Broock MR. Yeasts from extreme aquatic environments: hyperacidic freshwaters. In: Jones G E B, Hyde K D, Eds. Freshwater fungi and fungi-like organisms First ed: De Gruyter;. 2014; pp. 443-63.

[21]   Libkind D, Moliné M, Trochine A, Bellora N, de Garcia V. Biotechnologically relevant yeasts from patagonian natural environments.Biology and Biotechnology of Patagonian Microorganisms. Springer 2016; pp. 325-51.
[http://dx.doi.org/10.1007/978-3-319-42801-7_18]

[22]   Libkind D, Moliné M, Colabella F. Isolation and selection of new astaxanthin-producing strains of phaffia rhodozyma.Microbial Carotenoids. Springer 2018; pp. 297-310.
[http://dx.doi.org/10.1007/978-1-4939-8742-9_18]

[23]   Sampaio JP, Gonçalves P. Natural populations of *Saccharomyces kudriavzevii* in Portugal are associated with oak bark and are sympatric with *S. cerevisiae* and *S. paradoxus*. Appl Environ Microbiol 2008; 74(7): 2144-52.
[http://dx.doi.org/10.1128/AEM.02396-07] [PMID: 18281431]

[24]   Cubillos FA, Gibson B, Grijalva-Vallejos N, Krogerus K, Nikulin J. Bioprospecting for brewers: Exploiting natural diversity for naturally diverse beers. Yeast 2019; 36(6): 383-98.
[http://dx.doi.org/10.1002/yea.3380] [PMID: 30698853]

[25]   Alperstein L, Gardner JM, Sundstrom JF, Sumby KM, Jiranek V. Yeast bioprospecting versus synthetic biology-which is better for innovative beverage fermentation? Appl Microbiol Biotechnol 2020; 104(5): 1939-53.
[http://dx.doi.org/10.1007/s00253-020-10364-x] [PMID: 31953561]

[26]   Libkind D. Evaluación de la técnica de MSP-PCR para la caracterización molecular de aislamientos de *Rhodotorula mucilaginosa* provenientes de la Patagonia noroccidental. Rev Argent Microbiol 2007; 39(3): 133-7.
[PMID: 17990375]

[27]   Schoch CL, Seifert KA, Huhndorf S, *et al.* Nuclear ribosomal internal transcribed spacer (ITS) region

as a universal DNA barcode marker for Fungi. Proc Natl Acad Sci USA 2012; 109(16): 6241-6.
[http://dx.doi.org/10.1073/pnas.1117018109] [PMID: 22454494]

[28]   Peris D, Sylvester K, Libkind D, *et al.* Population structure and reticulate evolution of Saccharomyces eubayanus and its lager-brewing hybrids. Mol Ecol 2014; 23(8): 2031-45.
[http://dx.doi.org/10.1111/mec.12702] [PMID: 24612382]

[29]   Langdon QK, Peris D, Eizaguirre JI, *et al.* Postglacial migration shaped the genomic diversity and global distribution of the wild ancestor of lager-brewing hybrids. PLoS Genet 2020; 16(4): e1008680.
[http://dx.doi.org/10.1371/journal.pgen.1008680] [PMID: 32251477]

[30]   Eizaguirre JI, Peris D, Rodríguez ME, *et al.* Phylogeography of the wild Lager-brewing ancestor (*Saccharomyces eubayanus*) in Patagonia. Environ Microbiol 2018; 20(10): 3732-43.
[http://dx.doi.org/10.1111/1462-2920.14375] [PMID: 30105823]

[31]   White C, Zainasheff J. Yeast: the practical guide to beer fermentation. Brewers Publications 2010.

[32]   Boulton C, Quain D. Yeast management.Brewing Yeast and Fermentation. 2007; pp. 468-509.

[33]   Gibson BR, Storgårds E, Krogerus K, Vidgren V. Comparative physiology and fermentation performance of Saaz and Frohberg lager yeast strains and the parental species Saccharomyces eubayanus. Yeast 2013; 30(7): 255-66.
[http://dx.doi.org/10.1002/yea.2960] [PMID: 23695993]

[34]   Sannino C, Mezzasoma A, Buzzini P, Turchetti B. Non-conventional yeasts for producing alternative beers.Non-conventional Yeasts: from Basic Research to Application. Springer 2019; pp. 361-88.
[http://dx.doi.org/10.1007/978-3-030-21110-3_11]

[35]   Baker E, Wang B, Bellora N, *et al.* The genome sequence of Saccharomyces eubayanus and the domestication of lager-brewing yeasts. Mol Biol Evol 2015; 32(11): 2818-31.
[http://dx.doi.org/10.1093/molbev/msv168] [PMID: 26269586]

[36]   Gallone B, Steensels J, Prahl T, Soriaga L, Saels V, Herrera-Malaver B, *et al.* Domestication and divergence of Saccharomyces cerevisiae beer yeasts. Cell 2016; 166(6): 1397-410. e16.
[http://dx.doi.org/10.1016/j.cell.2016.08.020]

[37]   Gonçalves M, Pontes A, Almeida P, *et al.* Distinct domestication trajectories in top-fermenting beer yeasts and wine yeasts. Curr Biol 2016; 26(20): 2750-61.
[http://dx.doi.org/10.1016/j.cub.2016.08.040] [PMID: 27720622]

[38]   Brown CA, Murray AW, Verstrepen KJ. Rapid expansion and functional divergence of subtelomeric gene families in yeasts. Curr Biol 2010; 20(10): 895-903.
[http://dx.doi.org/10.1016/j.cub.2010.04.027] [PMID: 20471265]

[39]   McMurrough I, Madigan D, Donnelly D, Hurley J, Doyle AM, Hennigan G, *et al.* Control of ferulic acid and 4-vinyl guaiacol in brewing. J Inst Brew 1996; 102(5): 327-32.
[http://dx.doi.org/10.1002/j.2050-0416.1996.tb00918.x]

[40]   Preiss R, Tyrawa C, Krogerus K, Garshol LM, van der Merwe G. Traditional Norwegian Kveik are a genetically distinct group of domesticated *Saccharomyces cerevisiae* brewing yeasts. Front Microbiol 2018; 9: 2137.
[http://dx.doi.org/10.3389/fmicb.2018.02137] [PMID: 30258422]

[41]   de Souza Varize C, Christofoleti-Furlan RM. Muynarsk EdSM, de Melo Pereira GV, Lopes LD, Basso LC. Biotechnological applications of nonconventional yeasts.Yeasts in Biotechnology: IntechOpen. 2019; pp. 57-82.
[http://dx.doi.org/10.5772/intechopen.83035]

[42]   Basso RF, Alcarde AR, Portugal CB. Could non-Saccharomyces yeasts contribute on innovative brewing fermentations? Int Food Res J 2016; 86: 112-20.
[http://dx.doi.org/10.1016/j.foodres.2016.06.002]

[43]   Michel M, Meier-Dörnberg T, Jacob F, Methner FJ, Wagner RS, Hutzler M. Pure non-Saccharomyces

starter cultures for beer fermentation with a focus on secondary metabolites and practical applications. J Inst Brew 2016; 122(4): 569-87.
[http://dx.doi.org/10.1002/jib.381]

[44]   King A, Richard Dickinson J. Biotransformation of monoterpene alcohols by *Saccharomyces cerevisiae, Torulaspora delbrueckii* and *Kluyveromyces lactis.* Yeast 2000; 16(6): 499-506.
[http://dx.doi.org/10.1002/(SICI)1097-0061(200004)16:6<499::AID-YEA548>3.0.CO;2-E]   [PMID: 10790686]

[45]   Petersen J, Nilsson-Tillgren T, Kielland-Brandt M, Gjermansenl C, Holmber S. Structural heterozygosis at genes IL V2 and IL V5 in Saccharomyces carlsbergensiss. Curr Genet 1987; 12(3): 167-74.
[http://dx.doi.org/10.1007/BF00436875]

[46]   Martini AV, Martini A. Three newly delimited species of Saccharomyces sensu stricto. Antonie van Leeuwenhoek 1987; 53(2): 77-84.
[http://dx.doi.org/10.1007/BF00419503] [PMID: 3662481]

[47]   Kodama Y, Kielland-Brandt MC, Hansen J. Lager brewing yeast Comparative genomics. Springer 2006; pp. 145-64.
[http://dx.doi.org/10.1007/b106370]

[48]   Dunn B, Sherlock G. Reconstruction of the genome origins and evolution of the hybrid lager yeast *Saccharomyces pastorianus.* Genome Res 2008; 18(10): 1610-23.
[http://dx.doi.org/10.1101/gr.076075.108] [PMID: 18787083]

[49]   Peris D, Langdon QK, Moriarty RV, *et al.* Complex ancestries of lager-brewing hybrids were shaped by standing variation in the wild yeast *Saccharomyces eubayanus.* PLoS Genet 2016; 12(7): e1006155.
[http://dx.doi.org/10.1371/journal.pgen.1006155] [PMID: 27385107]

[50]   Libkind D, Peris D, Cubillos F, *et al.* Into the wild: new yeast genomes from natural environments and new tools for their analysis. FEMS Yeast Res 2020; 20(2): foaa008.
[http://dx.doi.org/10.1093/femsyr/foaa008]

[51]   Nespolo RF, Villarroel CA, Oporto CI, *et al.* An Out-of-Patagonia migration explains the worldwide diversity and distribution of *Saccharomyces eubayanus* lineages. PLoS Genet 2020; 16(5): e1008777.
[http://dx.doi.org/10.1371/journal.pgen.1008777] [PMID: 32357148]

[52]   González Flores M, Rodríguez ME, Oteiza JM, Barbagelata RJ, Lopes CA. Physiological characterization of *Saccharomyces uvarum* and *Saccharomyces eubayanus* from Patagonia and their potential for cidermaking. Int J Food Microbiol 2017; 249: 9-17.
[http://dx.doi.org/10.1016/j.ijfoodmicro.2017.02.018] [PMID: 28271856]

[53]   Magalhães F, Krogerus K, Vidgren V, Sandell M, Gibson B. Improved cider fermentation performance and quality with newly generated *Saccharomyces cerevisiae* × *Saccharomyces eubayanus* hybrids. J Ind Microbiol Biotechnol 2017; 44(8): 1203-13.
[http://dx.doi.org/10.1007/s10295-017-1947-7] [PMID: 28451838]

[54]   Origone AC, Del Mónaco SM, Ávila JR, González Flores M, Rodríguez ME, Lopes CA. Tolerance to winemaking stress conditions of Patagonian strains of *Saccharomyces eubayanus* and *Saccharomyces uvarum.* J Appl Microbiol 2017; 123(2): 450-63.
[http://dx.doi.org/10.1111/jam.13495] [PMID: 28543932]

[55]   Walther A, Hesselbart A, Wendland J. Genome sequence of Saccharomyces carlsbergensis, the world's first pure culture lager yeast. G3-Genes Genom Genet 2014; 4(5): 783-93.

[56]   Mertens S, Steensels J, Saels V, De Rouck G, Aerts G, Verstrepen KJ. A large set of newly created interspecific Saccharomyces hybrids increases aromatic diversity in lager beers. Appl Environ Microbiol 2015; 81(23): 8202-14.
[http://dx.doi.org/10.1128/AEM.02464-15] [PMID: 26407881]

[57]   Eizaguirre JI. 2019.

[58]   Baker EP, Peris D, Moriarty RV, Li XC, Fay JC, Hittinger CT. Mitochondrial DNA and temperature tolerance in lager yeasts. Sci Adv 2019; 5(1): eaav1869.
       [http://dx.doi.org/10.1126/sciadv.aav1869] [PMID: 30729163]

[59]   Krogerus K, Magalhães F, Vidgren V, Gibson B. New lager yeast strains generated by interspecific hybridization. J Ind Microbiol Biotechnol 2015; 42(5): 769-78.
       [http://dx.doi.org/10.1007/s10295-015-1597-6] [PMID: 25682107]

[60]   Vidgren V, Multanen J-P, Ruohonen L, Londesborough J. The temperature dependence of maltose transport in ale and lager strains of brewer's yeast. FEMS Yeast Res 2010; 10(4): 402-11.
       [http://dx.doi.org/10.1111/j.1567-1364.2010.00627.x] [PMID: 20402791]

[61]   Magalhães F, Vidgren V, Ruohonen L, Gibson B. Maltose and maltotriose utilisation by group I strains of the hybrid lager yeast *Saccharomyces pastorianus*. FEMS Yeast Res 2016; 16(5): fow053.
       [http://dx.doi.org/10.1093/femsyr/fow053] [PMID: 27364826]

[62]   Krogerus K, Arvas M, De Chiara M, *et al.* Ploidy influences the functional attributes of de novo lager yeast hybrids. Appl Microbiol Biotechnol 2016; 100(16): 7203-22.
       [http://dx.doi.org/10.1007/s00253-016-7588-3] [PMID: 27183995]

[63]   Burini JA, Eizaguirre JI, Loviso C, Libkind D. Selection of S. eubayanus strains from Patagonia (Argentina) with brewing potential and performance in the craft beer industry. Int Food Res J FOODRES-D-21-02553 Submitted

[64]   Hebly M, Brickwedde A, Bolat I, *et al.* S. cerevisiae × S. eubayanus interspecific hybrid, the best of both worlds and beyond. FEMS Yeast Res 2015; 15(3): fov005.
       [http://dx.doi.org/10.1093/femsyr/fov005] [PMID: 25743788]

[65]   Brouwers N, Gorter de Vries AR, van den Broek M, *et al.* In vivo recombination of *Saccharomyces eubayanus* maltose-transporter genes yields a chimeric transporter that enables maltotriose fermentation. PLoS Genet 2019; 15(4): e1007853.
       [http://dx.doi.org/10.1371/journal.pgen.1007853] [PMID: 30946741]

[66]   Mardones W, Villarroel CA, Krogerus K, *et al.* Molecular profiling of beer wort fermentation diversity across natural *Saccharomyces eubayanus* isolates. Microb Biotechnol 2020; 13(4): 1012-25.
       [http://dx.doi.org/10.1111/1751-7915.13545] [PMID: 32096913]

[67]   Diderich JA, Weening SM, van den Broek M, Pronk JT, Daran JG. Selection of Pof-*Saccharomyces eubayanus* variants for the construction of S. cerevisiae× S. eubayanus hybrids with reduced 4-vinyl guaiacol formation. Front Microbiol 2018; 9: 1640.
       [http://dx.doi.org/10.3389/fmicb.2018.01640] [PMID: 30100898]

[68]   Urbina K, Villarreal P, Nespolo RF, Salazar R, Santander R, Cubillos FA. Volatile compound screening using HS-SPME-GC/MS on *Saccharomyces eubayanus* strains under low-temperature pilsner wort fermentation. Microorganisms 2020; 8(5): 755.
       [http://dx.doi.org/10.3390/microorganisms8050755] [PMID: 32443420]

[69]   Mukai N, Masaki K, Fujii T, Iefuji H. Single nucleotide polymorphisms of PAD1 and FDC1 show a positive relationship with ferulic acid decarboxylation ability among industrial yeasts used in alcoholic beverage production. J Biosci Bioeng 2014; 118(1): 50-5.
       [http://dx.doi.org/10.1016/j.jbiosc.2013.12.017] [PMID: 24507903]

[70]   Vanbeneden N, Gils F, Delvaux F, Delvaux FR. Formation of 4-vinyl and 4-ethyl derivatives from hydroxycinnamic acids: Occurrence of volatile phenolic flavour compounds in beer and distribution of Pad1-activity among brewing yeasts. Food Chem 2008; 107(1): 221-30.
       [http://dx.doi.org/10.1016/j.foodchem.2007.08.008]

[71]   Langdon QK, Peris D, Baker EP, *et al.* Fermentation innovation through complex hybridization of wild and domesticated yeasts. Nat Ecol Evol 2019; 3(11): 1576-86.
       [http://dx.doi.org/10.1038/s41559-019-0998-8] [PMID: 31636426]

[72]   Krogerus K, Seppänen-Laakso T, Castillo S, Gibson B. Inheritance of brewing-relevant phenotypes in constructed *Saccharomyces cerevisiae* × *Saccharomyces eubayanus* hybrids. Microb Cell Fact 2017; 16(1): 66.
[http://dx.doi.org/10.1186/s12934-017-0679-8] [PMID: 28431563]

[73]   Gibson B, Dahabieh M, Krogerus K, *et al.* Adaptive laboratory evolution of ale and lager yeasts for improved brewing efficiency and beer quality. Annu Rev Food Sci Technol 2020; 11: 23-44.
[http://dx.doi.org/10.1146/annurev-food-032519-051715] [PMID: 31951488]

[74]   Origone AC, Rodríguez ME, Oteiza JM, Querol A, Lopes CA. *Saccharomyces cerevisiae* × *Saccharomyces uvarum* hybrids generated under different conditions share similar winemaking features. Yeast 2018; 35(1): 157-71.
[http://dx.doi.org/10.1002/yea.3295] [PMID: 29131448]

[75]   Origone AC, González Flores M, Rodríguez ME, Querol A, Lopes CA. Inheritance of winemaking stress factors tolerance in *Saccharomyces uvarum/S. eubayanus* × *S. cerevisiae* artificial hybrids. Int J Food Microbiol 2020; 320: 108500.
[http://dx.doi.org/10.1016/j.ijfoodmicro.2019.108500] [PMID: 32007764]

[76]   van Rijswijck IMH, Wolkers-Rooijackers JCM, Abee T, Smid EJ. Performance of non-conventional yeasts in co-culture with brewers' yeast for steering ethanol and aroma production. Microb Biotechnol 2017; 10(6): 1591-602.
[http://dx.doi.org/10.1111/1751-7915.12717] [PMID: 28834151]

[77]   Canonico L, Agarbati A, Comitini F, Ciani M. *Torulaspora delbrueckii* in the brewing process: A new approach to enhance bioflavour and to reduce ethanol content. Food Microbiol 2016; 56: 45-51.
[http://dx.doi.org/10.1016/j.fm.2015.12.005] [PMID: 26919817]

[78]   Trochine A, González SB, Burini JA, Cavallini L, Gastaldi B, Reiner G, *et al.* Chemical characterization of the two major hop varieties produced in Patagonia (Argentina) for the brewing industry. Brew Sci 2020; 73: 95-102.

[79]   Durieux A, Nicolay X, Simon J-P. Application of Immobilisation Technology to Cider Production: A Review.Applications of Cell Immobilisation Biotechnology. Dordrecht: Springer Netherlands 2005; pp. 275-84.
[http://dx.doi.org/10.1007/1-4020-3363-X_16]

[80]   Johansen BK. Cider Production in England and France – and Denmark?. Brygmesteren 2000; pp. 1-15.

[81]   Rowles K. Processed apple product marketing analysis: Hard Cider & Apple Wine. College of Agriculture and Life Scinces, Cornell University 2000.

[82]   Becerra LM. Sidra elaborada bajo el método Champenoise. Exportación a Chile. Aconcagua University 2016.

[83]   Argentine Ministry of Agriculture L, Fisheries. Indicadores de la cadena de valor de la Sidra. 2016.

[84]   Pedreschi R, Villarreal P. La sidra en el Mercado Argentino: Elementos del marketing estratégico y operacional. Universida ed Neuquén. Argentina 2021; pp. 1-48.

[85]   Hidalgo J. El regreso de la sidra: cuáles comprar (y cuáles no). Zona Norte Hoy 2012.

[86]   Cabranes C, Blanco D, Mangas JJ. Characterisation of fermented apple products using data obtained from GC analysis. Analyst (Lond) 1998; 123: 2175-9.
[http://dx.doi.org/10.1039/a801370e]

[87]   Coton E, Coton M, Levert D, Casaregola S, Sohier D. Yeast ecology in French cider and black olive natural fermentations. Int J Food Microbiol 2006; 108(1): 130-5.
[http://dx.doi.org/10.1016/j.ijfoodmicro.2005.10.016] [PMID: 16380183]

[88]   Dueñas M, Irastorza A, Fernandez K, Bilbao A, Huerta A. Microbial Populations and Malolactic Fermentation of Apple Cider using Traditional and Modified Methods. J Food Sci 1994; 59(5): 1060-4.

[http://dx.doi.org/10.1111/j.1365-2621.1994.tb08190.x]

[89] Morrissey WF, Davenport B, Querol A, Dobson ADW. The role of indigenous yeasts in traditional Irish cider fermentations. J Appl Microbiol 2004; 97(3): 647-55.
[http://dx.doi.org/10.1111/j.1365-2672.2004.02354.x] [PMID: 15281947]

[90] Naumov GI, Nguyen HV, Naumova ES, Michel A, Aigle M, Gaillardin C. Genetic identification of Saccharomyces bayanus var. uvarum, a cider-fermenting yeast. Int J Food Microbiol 2001; 65(3): 163-71.
[http://dx.doi.org/10.1016/S0168-1605(00)00515-8] [PMID: 11393685]

[91] Pando Bedriñana R, Querol Simón A, Suárez Valles B. Genetic and phenotypic diversity of autochthonous cider yeasts in a cellar from Asturias. Food Microbiol 2010; 27(4): 503-8.
[http://dx.doi.org/10.1016/j.fm.2009.11.018] [PMID: 20417399]

[92] Suárez Valles B, Pando Bedriñana R, González García A, Querol Simón A. A molecular genetic study of natural strains of Saccharomyces isolated from Asturian cider fermentations. J Appl Microbiol 2007; 103(4): 778-86.
[http://dx.doi.org/10.1111/j.1365-2672.2007.03314.x] [PMID: 17897179]

[93] Rodríguez ME, Pérez-Través L, Sangorrín MP, Barrio E, Querol A, Lopes CA. Saccharomyces uvarum is responsible for the traditional fermentation of apple chicha in Patagonia. FEMS Yeast Res 2017; 17(1): fow109.
[PMID: 28011906]

[94] González Flores M, Rodríguez ME, Origone AC, Oteiza JM, Querol A, Lopes CA. Saccharomyces uvarum isolated from patagonian ciders shows excellent fermentative performance for low temperature cidermaking. Int Food Res J 2019; 126: 108656.
[http://dx.doi.org/10.1016/j.foodres.2019.108656] [PMID: 31732032]

[95] Mestre MC, Fontenla S, Rosa CA. Ecology of cultivable yeasts in pristine forests in northern Patagonia (Argentina) influenced by different environmental factors. Can J Microbiol 2014; 60(6): 371-82.
[http://dx.doi.org/10.1139/cjm-2013-0897] [PMID: 24849380]

[96] Tronchoni J, Gamero A, Arroyo-López FN, Barrio E, Querol A. Differences in the glucose and fructose consumption profiles in diverse Saccharomyces wine species and their hybrids during grape juice fermentation. Int J Food Microbiol 2009; 134(3): 237-43.
[http://dx.doi.org/10.1016/j.ijfoodmicro.2009.07.004] [PMID: 19632733]

[97] Gamero A, Tronchoni J, Querol A, Belloch C. Production of aroma compounds by cryotolerant Saccharomyces species and hybrids at low and moderate fermentation temperatures. J Appl Microbiol 2013; 114(5): 1405-14.
[http://dx.doi.org/10.1111/jam.12126] [PMID: 23294204]

[98] González SS, Barrio E, Gafner J, Querol A. Natural hybrids from Saccharomyces cerevisiae, Saccharomyces bayanus and Saccharomyces kudriavzevii in wine fermentations. FEMS Yeast Res 2006; 6(8): 1221-34.
[http://dx.doi.org/10.1111/j.1567-1364.2006.00126.x] [PMID: 17156019]

[99] Tosi E, Azzolini M, Guzzo F, Zapparoli G. Evidence of different fermentation behaviours of two indigenous strains of Saccharomyces cerevisiae and Saccharomyces uvarum isolated from Amarone wine. J Appl Microbiol 2009; 107(1): 210-8.
[http://dx.doi.org/10.1111/j.1365-2672.2009.04196.x] [PMID: 19245401]

[100] Masneuf I, Murat ML, Naumov GI, Tominaga T, Dubourdieu D. Hybrids Saccharomyces cerevisiae x Saccharomyces bayanus var. uvarum having a high liberating ability of some sulfur varietal aromas of vitis vinifera sauvignon blanc wines. J Int Sci Vigne Vin 2002; 36(4): 205-12.

[101] Moreno-Arribas MV, Polo MC. Winemaking biochemistry and microbiology: current knowledge and future trends. Crit Rev Food Sci Nutr 2005; 45(4): 265-86.
[http://dx.doi.org/10.1080/10408690490478118] [PMID: 16047495]

[102] Orlić S, Vojvoda T, Babić KH, Arroyo-López FN, Jeromel A, Kozina B, *et al.* Diversity and oenological characterization of indigenous Saccharomycescerevisiae associated with Žilavka grapes. World J Microbiol Biotechnol 2010; 26(8): 1483-9.
[http://dx.doi.org/10.1007/s11274-010-0323-9]

[103] Bisson LF. Fermentaciones ralentizadas o detenidas. Revista de Viticulura y Enología 2006; 10(2): 1-16.

[104] Blanco Gomis D, Mangas Alonso JJ. La manzana y la sidra: bioprocesos, tecnologías de elaboración y control. SERIDA ed. 2010.

[105] Casalone E, Colella CM, Daly S, Gallori E, Moriani L, Polsinelli M. Mechanism of resistance to sulphite in *Saccharomyces cerevisiae*. Curr Genet 1992; 22(6): 435-40.
[http://dx.doi.org/10.1007/BF00326407] [PMID: 1473174]

[106] Nardi T, Corich V, Giacomini A, Blondin B. A sulphite-inducible form of the sulphite efflux gene SSU1 in a Saccharomyces cerevisiae wine yeast. Microbiology 2010; 156(Pt 6): 1686-96.
[http://dx.doi.org/10.1099/mic.0.036723-0] [PMID: 20203053]

[107] Pérez-Ortín JE, Querol A, Puig S, Barrio E. Molecular characterization of a chromosomal rearrangement involved in the adaptive evolution of yeast strains. Genome Res 2002; 12(10): 1533-9.
[http://dx.doi.org/10.1101/gr.436602] [PMID: 12368245]

[108] Yuasa N, Nakagawa Y, Hayakawa M, Iimura Y. Distribution of the sulfite resistance gene SSU1-R and the variation in its promoter region in wine yeasts. J Biosci Bioeng 2004; 98(5): 394-7.
[http://dx.doi.org/10.1016/S1389-1723(04)00303-2] [PMID: 16233727]

[109] Querol AML, Gonzalez Flores MAA, Rodríguez ME, Barrio E, Lopes CA. R. P-T, editors. Convergent adaptation of Saccharomyces uvarum to an antimicrobial preservative in human-driven fermentation. 34 International Specialized Symposium on Yeast. Argentina. 2018.

[110] Lopes CA, Guitierrez-Macias L, Rodriguez ME, Pérez-Torrado R, Gonzalez Flores M, Barrio E, Eds. Diferential sulfite resistance as a domestication signature in Saccharomyces uvarum. Cork, Ireland 2017.

[111] Avram D, Bakalinsky AT. SSU1 encodes a plasma membrane protein with a central role in a network of proteins conferring sulfite tolerance in *Saccharomyces cerevisiae*. J Bacteriol 1997; 179(18): 5971-4.
[http://dx.doi.org/10.1128/jb.179.18.5971-5974.1997] [PMID: 9294463]

[112] Park H, Lopez NI, Bakalinsky AT. Use of sulfite resistance in Saccharomyces cerevisiae as a dominant selectable marker. Curr Genet 1999; 36(6): 339-44.
[http://dx.doi.org/10.1007/s002940050508] [PMID: 10654087]

[113] Giudici P, Caggia C, Pulvirenti A, Rainieri S. Karyotyping of *Saccharomyces* strains with different temperature profiles. J Appl Microbiol 1998; 84(5): 811-9.
[http://dx.doi.org/10.1046/j.1365-2672.1998.00416.x] [PMID: 9674135]

[114] Naumov GI, Masneuf I, Naumova ES, Aigle M, Dubourdieu D. Association of Saccharomyces bayanus var. uvarum with some French wines: genetic analysis of yeast populations. Res Microbiol 2000; 151(8): 683-91.
[http://dx.doi.org/10.1016/S0923-2508(00)90131-1] [PMID: 11081582]

[115] Belloch C, Orlic S, Barrio E, Querol A. Fermentative stress adaptation of hybrids within the *Saccharomyces* sensu stricto complex. Int J Food Microbiol 2008; 122(1-2): 188-95.
[http://dx.doi.org/10.1016/j.ijfoodmicro.2007.11.083] [PMID: 18222562]

[116] Alonso-Del-Real J, Contreras-Ruiz A, Castiglioni GL, Barrio E, Querol A. The use of mixed populations of *Saccharomyces cerevisiae* and *S. kudriavzevii* to reduce ethanol content in wine: limited aeration, inoculum proportions, and sequential inoculation. Front Microbiol 2017; 8: 2087.
[http://dx.doi.org/10.3389/fmicb.2017.02087] [PMID: 29118746]

[117]   Maidana Y, Origone A, Apablaza O, Lopes CA, Eds. Levaduras asociadas a mieles neuquinas y su potencial para elaborar hidromiel. Buenos Aires 2020.

[118]   Goddard MR. Quantifying the complexities of Saccharomyces cerevisiae's ecosystem engineering via fermentation. Ecology 2008; 89(8): 2077-82.
[http://dx.doi.org/10.1890/07-2060.1] [PMID: 18724717]

[119]   Suárez Valles B, Pando Bedriñana R, Lastra Queipo A, Mangas Alonso JJ. Screening of cider yeasts for sparkling cider production (Champenoise method). Food Microbiol 2008; 25(5): 690-7.
[http://dx.doi.org/10.1016/j.fm.2008.03.004] [PMID: 18541168]

[120]   Cabranes C, Moreno J, Mangas JJ. Dynamics of Yeast Populations during Cider Fermentation in the Asturian Region of Spain. Appl Environ Microbiol 1990; 56(12): 3881-4.
[http://dx.doi.org/10.1128/aem.56.12.3881-3884.1990] [PMID: 16348385]

[121]   Tyakht A, Kopeliovich A, Klimenko N, *et al.* Characteristics of bacterial and yeast microbiomes in spontaneous and mixed-fermentation beer and cider. Food Microbiol 2021; 94: 103658.
[http://dx.doi.org/10.1016/j.fm.2020.103658] [PMID: 33279083]

[122]   Suaréz Valles B, Pando Bedriñana R, Fernández Tascón NF, Gonzalez Garcia A, Rodríguez Madrera R. Analytical differentiation of cider inoculated with yeast (*Saccharomyces cerevisiae*) isolated from Asturian (Spain) apple juice. Lebensm Wiss Technol 2005; 38: 455-61.
[http://dx.doi.org/10.1016/j.lwt.2004.07.008]

[123]   Lorenzini M, Simonato B, Slaghenaufi D, Ugliano M, Zapparoli G. Assessment of yeasts for apple juice fermentation and production of cider volatile compounds. Lebensm Wiss Technol 2019; 99: 224-30.
[http://dx.doi.org/10.1016/j.lwt.2018.09.075]

[124]   Anjos J, Rodrigues de Sousa H, Roca C, *et al.* Fsy1, the sole hexose-proton transporter characterized in Saccharomyces yeasts, exhibits a variable fructose:H(+) stoichiometry. Biochim Biophys Acta 2013; 1828(2): 201-7.
[http://dx.doi.org/10.1016/j.bbamem.2012.08.011] [PMID: 22922355]

[125]   Gonçalves P, Rodrigues de Sousa H, Spencer-Martins I. FSY1, a novel gene encoding a specific fructose/H(+) symporter in the type strain of *Saccharomyces carlsbergensis*. J Bacteriol 2000; 182(19): 5628-30.
[http://dx.doi.org/10.1128/JB.182.19.5628-5630.2000] [PMID: 10986274]

[126]   Picinelli Lobo A, Antón-Díaz MJ, Pando Bedriñana R, Fernández García O, Hortal-García R, Suárez Valles B. Chemical, olfactometric and sensory description of single-variety cider apple juices obtained by cryo-extraction. Lebensm Wiss Technol 2018; 90: 193-200.
[http://dx.doi.org/10.1016/j.lwt.2017.12.033]

[127]   Pando Bedriñana R, Picinelli Lobo A, Suárez Valles B. Influence of the method of obtaining freeze-enriched juices and year of harvest on the chemical and sensory characteristics of Asturian ice ciders. Food Chem 2019; 274: 376-83.
[http://dx.doi.org/10.1016/j.foodchem.2018.08.141] [PMID: 30372954]

[128]   Daccache MA, Koubaa M, Maroun RG, Salameh D, Louka N, Vorobiev E. Suitability of the Lebanese "Ace spur" apple variety for cider production using *Hanseniaspora* sp. Yeast. Fermentation (Basel) 2020; 6(1): 1-14.
[http://dx.doi.org/10.3390/fermentation6010032]

[129]   Jarvis B. Cider, perry, fruit wines and other alcoholic fruit beverages. 1996; pp. 97-134.
[http://dx.doi.org/10.1007/978-1-4615-2103-7_5]

[130]   Coloretti F, Zambonelli C, Tini V. Characterization of flocculent *Saccharomyces* interspecific hybrids for the production of sparkling wines. Food Microbiol 2006; 23(7): 672-6.
[http://dx.doi.org/10.1016/j.fm.2005.11.002] [PMID: 16943067]

[131]   Demuyter C, Lollier M, Legras JL, Le Jeune C. Predominance of *Saccharomyces uvarum* during

spontaneous alcoholic fermentation, for three consecutive years, in an Alsatian winery. J Appl Microbiol 2004; 97(6): 1140-8.
[http://dx.doi.org/10.1111/j.1365-2672.2004.02394.x] [PMID: 15546404]

[132] Gangl H, Batusic M, Tscheik G, Tiefenbrunner W, Hack C, Lopandic K. Exceptional fermentation characteristics of natural hybrids from *Saccharomyces cerevisiae* and *S. kudriavzevii.* N Biotechnol 2009; 25(4): 244-51.
[http://dx.doi.org/10.1016/j.nbt.2008.10.001] [PMID: 19026772]

[133] Masneuf-Pomarède I, Bely M, Marullo P, Lonvaud-Funel A, Dubourdieu D. Reassessment of phenotypic traits for *Saccharomyces bayanus* var. *uvarum* wine yeast strains. Int J Food Microbiol 2010; 139(1-2): 79-86.
[http://dx.doi.org/10.1016/j.ijfoodmicro.2010.01.038] [PMID: 20188428]

[134] Stribny J, Gamero A, Pérez-Torrado R, Querol A. *Saccharomyces kudriavzevii* and *Saccharomyces uvarum* differ from *Saccharomyces cerevisiae* during the production of aroma-active higher alcohols and acetate esters using their amino acidic precursors. Int J Food Microbiol 2015; 205: 41-6.
[http://dx.doi.org/10.1016/j.ijfoodmicro.2015.04.003] [PMID: 25886016]

[135] Swiegers JH, Pretorius IS. Yeast modulation of wine flavor. Adv Appl Microbiol 2005; 57: 131-75.
[http://dx.doi.org/10.1016/S0065-2164(05)57005-9] [PMID: 16002012]

[136] Divol B, du Toit M, Duckitt E. Surviving in the presence of sulphur dioxide: strategies developed by wine yeasts. Appl Microbiol Biotechnol 2012; 95(3): 601-13.
[http://dx.doi.org/10.1007/s00253-012-4186-x] [PMID: 22669635]

[137] González Flores M, Origone AC, Bajda L, Rodríguez ME, Lopes CA. Evaluation of cryotolerant yeasts for the elaboration of a fermented pear beverage in Patagonia: Physicochemical and sensory attributes. Int J Food Microbiol 2021; 345: 109129.
[http://dx.doi.org/10.1016/j.ijfoodmicro.2021.109129] [PMID: 33711686]

[138] Varela C. The impact of non-*Saccharomyces* yeasts in the production of alcoholic beverages. Appl Microbiol Biotechnol 2016; 100(23): 9861-74.
[http://dx.doi.org/10.1007/s00253-016-7941-6] [PMID: 27787587]

[139] Fleet GH. Yeast interactions and wine flavour. Int J Food Microbiol 2003; 86(1-2): 11-22.
[http://dx.doi.org/10.1016/S0168-1605(03)00245-9] [PMID: 12892919]

[140] Redzepović S, Orlić S, Sikora S, Majdak A, Pretorius IS. Identification and characterization of *Saccharomyces cerevisiae* and *Saccharomyces paradoxus* strains isolated from Croatian vineyards. Lett Appl Microbiol 2002; 35(4): 305-10.
[http://dx.doi.org/10.1046/j.1472-765X.2002.01181.x] [PMID: 12358693]

[141] Sipiczk M, Romano P, Lipani G, Miklos I, Antunovics Z. Analysis of yeasts derived from natural fermentation in a Tokaj winery. Antonie van Leeuwenhoek 2001; 79(1): 97-105.
[http://dx.doi.org/10.1023/A:1010249408975] [PMID: 11392490]

[142] Masneuf-Pomarede I, Bely M, Marullo P, Albertin W. The genetics of non-conventional wine yeasts: current knowledge and future challenges. Front Microbiol 2016; 6: 1563.
[http://dx.doi.org/10.3389/fmicb.2015.01563] [PMID: 26793188]

[143] Pretorius IS. Tailoring wine yeast for the new millennium: novel approaches to the ancient art of winemaking. Yeast 2000; 16(8): 675-729.
[http://dx.doi.org/10.1002/1097-0061(20000615)16:8<675::AID-YEA585>3.0.CO;2-B] [PMID: 10861899]

[144] Bely M, Renault P, da Silva T, *et al.* Non-conventional yeasts and alcohol level reduction Alc level red in wine 2013; 33-7.

[145] Jolly N, Augustyn O, Pretorius I. The role and use of non-*Saccharomyces* yeasts in wine production. S Afr J Enol Vitic 2006; 27(1): 15-39.

[146] Bradbury JE, Richards KD, Niederer HA, Lee SA, Rod Dunbar P, Gardner RC. A homozygous diploid

subset of commercial wine yeast strains. Antonie van Leeuwenhoek 2006; 89(1): 27-37.
[http://dx.doi.org/10.1007/s10482-005-9006-1] [PMID: 16328862]

[147] Rodríguez ME, Lopes CA, Barbagelata RJ, Barda NB, Caballero AC. Influence of *Candida pulcherrima* Patagonian strain on alcoholic fermentation behaviour and wine aroma. Int J Food Microbiol 2010; 138(1-2): 19-25.
[http://dx.doi.org/10.1016/j.ijfoodmicro.2009.12.025] [PMID: 20116878]

[148] Del Mónaco SM, Rodríguez ME, Lopes CA. *Pichia kudriavzevii* as a representative yeast of North Patagonian winemaking terroir. Int J Food Microbiol 2016; 230: 31-9.
[http://dx.doi.org/10.1016/j.ijfoodmicro.2016.04.017] [PMID: 27124468]

[149] Kishimoto M, Shinohara T, Soma E, Goto S. Selection and fermentation properties of cryophilic wine yeasts. J Biosci Bioeng 1993; 75(6): 451-3.

[150] Masneuf I, Hansen J, Groth C, Piskur J, Dubourdieu D. New hybrids between *Saccharomyces* sensu stricto yeast species found among wine and cider production strains. Appl Environ Microbiol 1998; 64(10): 3887-92.
[http://dx.doi.org/10.1128/AEM.64.10.3887-3892.1998] [PMID: 9758815]

[151] Lambrechts M, Pretorius I. Yeast and its importance to wine aroma-a review. S Afr J Enol Vitic 2000; 21(1): 97-129.

[152] Beltran G, Rozes N, Mas A, Guillamon JM. Effect of low-temperature fermentation on yeast nitrogen metabolism. World J Microbiol Biotechnol 2007; 23(6): 809-15.
[http://dx.doi.org/10.1007/s11274-006-9302-6]

[153] Fleet GH, Heard GM. Yeast-Growth during Fermentation. In: Fleet, GH, Ed, Harwood Academic, Wine, Microbiology and Biotechnology, Lausanne 1993; 27-54.

[154] Falqué E, Fernández E. Effects of different skin contact times on Treixadura wine composition. Am J Enol Vitic 1996; 47(3): 309-12.

[155] Piper PW. The heat shock and ethanol stress responses of yeast exhibit extensive similarity and functional overlap. FEMS Microbiol Lett 1995; 134(2-3): 121-7.
[http://dx.doi.org/10.1111/j.1574-6968.1995.tb07925.x] [PMID: 8586257]

[156] Torija MJ, Rozès N, Poblet M, Guillamón JM, Mas A. Effects of fermentation temperature on the strain population of *Saccharomyces cerevisiae.* Int J Food Microbiol 2003; 80(1): 47-53.
[http://dx.doi.org/10.1016/S0168-1605(02)00144-7] [PMID: 12430770]

[157] Gamero A, Manzanares P, Querol A, Belloch C. Monoterpene alcohols release and bioconversion by *Saccharomyces* species and hybrids. Int J Food Microbiol 2011; 145(1): 92-7.
[http://dx.doi.org/10.1016/j.ijfoodmicro.2010.11.034] [PMID: 21176987]

[158] Bauer F, Pretorius IS. Yeast stress response and fermentation efficiency: how to survive the making of wine. S Afr J Enol Vitic 2000; 21(1): 27-51.

[159] Arroyo-López FN, Orlić S, Querol A, Barrio E. Effects of temperature, pH and sugar concentration on the growth parameters of *Saccharomyces cerevisiae, S. kudriavzevii* and their interspecific hybrid. Int J Food Microbiol 2009; 131(2-3): 120-7.
[http://dx.doi.org/10.1016/j.ijfoodmicro.2009.01.035] [PMID: 19246112]

[160] Bely M, Sablayrolles J-M, Barre P. Automatic detection of assimilable nitrogen deficiencies during alcoholic fermentation in oenological conditions. J Ferment Bioeng 1990; 70(4): 246-52.
[http://dx.doi.org/10.1016/0922-338X(90)90057-4]

[161] Masneuf-Pomarède I, Mansour C, Murat M-L, Tominaga T, Dubourdieu D. Influence of fermentation temperature on volatile thiols concentrations in Sauvignon blanc wines. Int J Food Microbiol 2006; 108(3): 385-90.
[PMID: 16524635]

[162] Su Y, Origone AC, Rodríguez ME, Querol A, Guillamón JM, Lopes CA. Fermentative behaviour and

competition capacity of cryotolerant *Saccharomyces* species in different nitrogen conditions. Int J Food Microbiol 2019; 291: 111-20.
[http://dx.doi.org/10.1016/j.ijfoodmicro.2018.11.020] [PMID: 30496940]

[163]   Stanley D, Bandara A, Fraser S, Chambers PJ, Stanley GA. The ethanol stress response and ethanol tolerance of *Saccharomyces cerevisiae*. J Appl Microbiol 2010; 109(1): 13-24.
[http://dx.doi.org/10.1111/j.1365-2672.2009.04657.x] [PMID: 20070446]

[164]   Snowdon C, Schierholtz R, Poliszczuk P, Hughes S, van der Merwe G. ETP1/YHL010c is a novel gene needed for the adaptation of *Saccharomyces cerevisiae* to ethanol. FEMS Yeast Res 2009; 9(3): 372-80.
[http://dx.doi.org/10.1111/j.1567-1364.2009.00497.x] [PMID: 19416103]

[165]   Arroyo-López FN, Salvadó Z, Tronchoni J, Guillamón JM, Barrio E, Querol A. Susceptibility and resistance to ethanol in Saccharomyces strains isolated from wild and fermentative environments. Yeast 2010; 27(12): 1005-15.
[http://dx.doi.org/10.1002/yea.1809] [PMID: 20824889]

[166]   García-Ríos E, Gutiérrez A, Salvadó Z, Arroyo-López FN, Guillamon JM. The fitness advantage of commercial wine yeasts in relation to the nitrogen concentration, temperature, and ethanol content under microvinification conditions. Appl Environ Microbiol 2014; 80(2): 704-13.
[http://dx.doi.org/10.1128/AEM.03405-13] [PMID: 24242239]

[167]   Rainieri S, Pretorius I. Selection and improvement of wine yeasts. Ann Microbiol 2000; 50(1): 15-32.

[168]   Mazzucco MB, Ganga MA, Sangorrín MP. Production of a novel killer toxin from Saccharomyces eubayanus using agro-industrial waste and its application against wine spoilage yeasts. Antonie van Leeuwenhoek 2019; 112(7): 965-73.
[http://dx.doi.org/10.1007/s10482-019-01231-5] [PMID: 30671692]

[169]   Villalba ML, Mazzucco MB, Lopes CA, Ganga MA, Sangorrín MP. Purification and characterization of *Saccharomyces eubayanus* killer toxin: Biocontrol effectiveness against wine spoilage yeasts. Int J Food Microbiol 2020; 331: 108714.
[http://dx.doi.org/10.1016/j.ijfoodmicro.2020.108714] [PMID: 32544792]

[170]   Rodríguez ME, Lopes CA, van Broock M, Valles S, Ramón D, Caballero AC. Screening and typing of Patagonian wine yeasts for glycosidase activities. J Appl Microbiol 2004; 96(1): 84-95.
[http://dx.doi.org/10.1046/j.1365-2672.2003.02032.x] [PMID: 14678162]

[171]   Rodríguez ME, Lopes C, Valles S, Giraudo MR, Caballero A. Selection and preliminary characterization of β-glycosidases producer Patagonian wild yeasts. Enzyme Microb Technol 2007; 41(6-7): 812-20.
[http://dx.doi.org/10.1016/j.enzmictec.2007.07.004]

[172]   Styger G, Prior B, Bauer FF. Wine flavor and aroma. J Ind Microbiol Biotechnol 2011; 38(9): 1145-59.
[http://dx.doi.org/10.1007/s10295-011-1018-4] [PMID: 21786136]

[173]   Castellari L, Ferruzzi M, Magrini A, Giudici P, Passarelli P. Unbalanced wine fermentation by cryotolerant vs. non-cryotolerant Saccharomyces strains. Vitis 1994; 33(1): 49-52.

[174]   Giudici P, Zambonelli C, Passarelli P, Castellari L. Improvement of wine composition with cryotolerant *Saccharomyces* strains. Am J Enol Vitic 1995; 46(1): 143-7.

[175]   Greig D, Louis EJ, Borts RH, Travisano M. Hybrid speciation in experimental populations of yeast. Science 2002; 298(5599): 1773-5.
[http://dx.doi.org/10.1126/science.1076374] [PMID: 12459586]

[176]   Pérez-Través L, Lopes CA, Barrio E, Querol A. Stabilization process in Saccharomyces intra and interspecific hybrids in fermentative conditions. Int Microbiol 2014; 17(4): 213-24.
[PMID: 26421737]

[177]   García-Ríos E, Guillén A, de la Cerda R, Pérez-Través L, Querol A, Guillamón JM. Improving the

cryotolerance of wine yeast by interspecific hybridization in the genus *Saccharomyces*. Front Microbiol 2019; 9: 3232.
[http://dx.doi.org/10.3389/fmicb.2018.03232] [PMID: 30671041]

[178] Giudici P, Solieri L, Pulvirenti AM, Cassanelli S. Strategies and perspectives for genetic improvement of wine yeasts. Appl Microbiol Biotechnol 2005; 66(6): 622-8.
[http://dx.doi.org/10.1007/s00253-004-1784-2] [PMID: 15578179]

[179] Marullo P, Bely M, Masneuf-Pomarede I, Aigle M, Dubourdieu D. Inheritable nature of enological quantitative traits is demonstrated by meiotic segregation of industrial wine yeast strains. FEMS Yeast Res 2004; 4(7): 711-9.
[http://dx.doi.org/10.1016/j.femsyr.2004.01.006] [PMID: 15093774]

[180] Sipiczki M. Interspecies hybridisation and genome chimerisation in *Saccharomyces:* combining of gene pools of species and its biotechnological perspectives. Front Microbiol 2018; 9: 3071.
[http://dx.doi.org/10.3389/fmicb.2018.03071] [PMID: 30619156]

[181] Lopandic K. *Saccharomyces* interspecies hybrids as model organisms for studying yeast adaptation to stressful environments. Yeast 2018; 35(1): 21-38.
[http://dx.doi.org/10.1002/yea.3294] [PMID: 29131388]

[182] Gorter de Vries AR, Voskamp MA, van Aalst ACA, *et al.* Laboratory evolution of a *Saccharomyces cerevisiae*× *S. eubayanus* hybrid under simulated lager-brewing conditions. Front Genet 2019; 10: 242.
[http://dx.doi.org/10.3389/fgene.2019.00242] [PMID: 31001314]

[183] Serra A, Strehaiano P, Taillandier P. Influence of temperature and pH on *Saccharomyces bayanus* var. *uvarum* growth; impact of a wine yeast interspecific hybridization on these parameters. Int J Food Microbiol 2005; 104(3): 257-65.
[http://dx.doi.org/10.1016/j.ijfoodmicro.2005.03.006] [PMID: 15979182]

[184] Bellon JR, Yang F, Day MP, Inglis DL, Chambers PJ. Designing and creating *Saccharomyces* interspecific hybrids for improved, industry relevant, phenotypes. Appl Microbiol Biotechnol 2015; 99(20): 8597-609.
[http://dx.doi.org/10.1007/s00253-015-6737-4] [PMID: 26099331]

[185] Rainieri S, Zambonelli C, Giudici P, Castellari L. Characterisation of thermotolerant *Saccharomyces cerevisiae* hybrids. Biotechnol Lett 1998; 20(6): 543-7.
[http://dx.doi.org/10.1023/A:1005389309527]

[186] Riou C, Nicaud J-M, Barre P, Gaillardin C. Stationary-phase gene expression in Saccharomyces cerevisiae during wine fermentation. Yeast 1997; 13(10): 903-15.
[http://dx.doi.org/10.1002/(SICI)1097-0061(199708)13:10<903::AID-YEA145>3.0.CO;2-1] [PMID: 9271106]

[187] Jones GV, White MA, Cooper OR, Storchmann K. Climate Change and Global Wine Quality. Clim Change 2005; 73(3): 319-43.
[http://dx.doi.org/10.1007/s10584-005-4704-2]

[188] Leandro MJ, Fonseca C, Gonçalves P. Hexose and pentose transport in ascomycetous yeasts: an overview. FEMS Yeast Res 2009; 9(4): 511-25.
[http://dx.doi.org/10.1111/j.1567-1364.2009.00509.x] [PMID: 19459982]

[189] Fleet GH. Wine yeasts for the future. FEMS Yeast Res 2008; 8(7): 979-95.
[http://dx.doi.org/10.1111/j.1567-1364.2008.00427.x] [PMID: 18793201]

[190] Arroyo-López FN, Pérez-Través L, Querol A, Barrio E. Exclusion of *Saccharomyces kudriavzevii* from a wine model system mediated by *Saccharomyces cerevisiae*. Yeast 2011; 28(6): 423-35.
[http://dx.doi.org/10.1002/yea.1848] [PMID: 21381110]

[191] Williams KM, Liu P, Fay JC. Evolution of ecological dominance of yeast species in high-sugar environments. Evolution 2015; 69(8): 2079-93.

[http://dx.doi.org/10.1111/evo.12707] [PMID: 26087012]

[192]   Lawless HT, Claassen MR. Application of the central dogma in sensory evaluation. Food Technol (Chicago) 1993; 47(6): 139-46.

[193]   Liu S-Q, Pilone GJ. An overview of formation and roles of acetaldehyde in winemaking with emphasis on microbiological implications. Int J Food Sci Technol 2000; 35(1): 49-61.
        [http://dx.doi.org/10.1046/j.1365-2621.2000.00341.x]

CHAPTER 11

Yeasts and Breadmaking

Caiti Smukowski Heil[1,†]**, Kate Howell**[2,†] **and Delphine Sicard**[3,*]

[1] *Department of Biological Sciences, North Carolina State University, Raleigh, NC, United States*

[2] *School of Agriculture and Food, University of Melbourne, Victoria 3010, Australia*

[3] *SPO, Univ Montpellier, INRAE, Institut Agro, Montpellier, France*

Abstract: The earliest known evidence of leavened bread comes from Egypt and China in the second and first millennia BC, although records of unleavened breads and potential flour production date back tens of thousands of years. In the 19th century, the discovery of yeast fermentation led to the development of industrial bakeries in parallel with traditional sourdough bakeries. While strains of *Saccharomyces cerevisiae* were selected for and became the primary yeast used in industrial breadmaking, some artisanal bakeries continued to use natural sourdough. The maintenance of these two types of bakery practices led to the evolution of two genetically and phenotypically distinct clades of *Saccharomyces* bakery yeast. In addition to *S. cerevisiae*, other yeast species are regularly found in sourdoughs, in particular yeasts of the genus *Kazachstania*. In the sourdough ecosystem, these yeasts interact with each other and with bacteria in a positive or negative way, depending on the species and strains involved. In both sourdough and yeasted industrial dough, traits of interest include aroma production, efficient maltose utilization, osmotolerance, desiccation resistance, and freeze tolerance. These traits have largely been explored in *S. cerevisiae*, but there is abundant diversity in these traits even amongst strains of *S. cerevisiae*, and in the handful of yeast species that have been surveyed outside of *Saccharomyces*. The new interest in sourdough breadmaking and the societal desire to develop more sustainable and biodiversity-friendly bakeries is now leading bakers and scientists to explore the genetic and metabolic diversity of other yeast species.

Keywords: Aroma, Baking, Bread, Domestication, Evolution, Fermentation, Genomic, *Kazachstania*, Lactic acid bacteria, Maltose, Microbial community, Microbial interaction, Ploidy, *Pichia*, *Saccharomyces*, Sourdough, Stress tolerance, *Torulaspora*, *Wickerhamomyces*, Yeast.

---

[*] **Corresponding author Delphine Sicard:** SPO, Univ Montpellier, INRAE, Institut Agro, Montpellier, France; Tel: +33 (0)6 24 72 05 01; E-mail: delphine.sicard@inrae.fr
[†] These authors have equal contribution.

Sérgio Luiz Alves Júnior, Helen Treichel, Thiago Olitta Basso and Boris Ugarte Stambuk (Eds.)
All rights reserved-© 2022 Bentham Science Publishers

INTRODUCTION

Bread has long been a staple food, perhaps starting when humans still persisted as hunter-gatherers. The art of making bread developed in the Neolithic, with the emergence of agriculture and the domestication of cereal grasses. Since then, yeast has been used unconsciously and consciously to make bread rise. In this chapter, after presenting the milestones in the history of bread, we will present the knowledge on the evolution, ecology, and metabolism of yeasts associated with the making of traditional sourdough bread and industrial bread.

HISTORY OF BREADMAKING

The origins of bread remain largely unknown. Some of the earliest records of the human processing of wild cereal grasses date back to 23,500 to 22,500 years ago in what is now Israel [1], and evidence of grinding of other starches suggests that vegetal food processing and possibly flour production was occurring in what is now known as Europe and Australia 14,000-30,000 years ago [2 - 4]. The first known breads were of a flatbread-like form made from wild grasses and tubers 14,400 years ago, recently identified in present-day Jordan [5]. Bread-like finds became more common in Neolithic sites in Europe and southwest Asia as cereals like wheat and barley were domesticated around 9000 years ago, and dome-like ovens were identified at sites in Turkey beginning in the late 8th millennium [6].

Ancient writings and artistic representations document the centrality of grain to many societies. Indeed, the control of grains becomes synonymous with power and control of people, as depicted in this quotation from a Sumerian text, "Whoever has silver, whoever has jewels, whoever has cattle, whoever has sheep shall take a seat at the gate of whoever has grain, and pass his time there." Throughout the Bronze Age, Mesopotamian households were paid with barley rations, and consumed many types of beers, porridge, cakes, and breads forming the central part of their daily diet [7]. We can speculate that some of these breads were leavened, as fermentation of beer using malted barley was practiced, and exchange between brewing and baking likely passed yeast back and forth. For example, brewing in both Mesopotamia and Egypt relied on a bread-like substance, which was only partially baked, presumably to preserve yeast viability, which was used to inoculate beer fermentation with yeast [8].

Baking in Ancient Egypt has received quite a lot of attention due to the prolific documentation of baking and brewing in art, writings, and actual remains of bread and storage vessels, especially during the period of 2500-2100 BC [9]. Egyptian breads were quite varied in size and texture, although all recovered loaves were made primarily from emmer wheat and had a dense crumb, some were flavored with coriander and fig [8, 10]. Some scholars point to the origin of what we now refer to as "sourdough starter" to this time period in Egypt, in which some dough was kept and maintained with flour and water to be used in baking the next day [11].

The first direct evidence of the presence of yeast in bread comes from Egypt and China in the second and first millennia BC [10, 12]. Scanning electron microscopy of bread and brewing remains from central Egypt 1500-1300 BC identified budding yeasts [10]. In China, the first records of beer brewing using barley date between 3400-2900 BC [13], although barley did not become an important subsistence crop there until the Han dynasty (206 BC-AD 220). Proteomic analysis of food materials resembling sourdough bread from 500–300 BC identified Saccharomycetaceae yeasts and lactic acid bacteria [12].

The rise of bakeries is often attributed to the Roman empire. As the population of the city of Rome outpaced the ability of the surrounding regions to supply food to the populace, the Roman empire began importing grain and providing grain allotments at free or subsidized prices to its poor citizens [14]. Grain would have been taken to a mill and then the flour used to bake breads in home or communal ovens, or in a bakery. Pliny the Elder records that professional bakers appeared in Rome beginning in 168 BCE [15]. Grains were traditionally ground using a rounded stone pressed manually against a flat stone bed, until milling technology originated in Greece in the 5th century BCE and the Greeks and Romans advanced mill technology to include animal-driven and water-driven mills in the centuries following [16]. Starting in the 3rd century AD, grain allotments were replaced with bread [17], which persisted until the end of the Roman empire in the 6th century AD. Water-mills spread through the Roman and Byzantine empires across Eurasia in the centuries following.

It was not until the 1700s that yeast began to be produced for the purpose of use in bread and beer. At this time, yeast was recognized as "the ferment put into drink to make it work; and into bread, to lighten and swell it," but it was not recognized as a living organism [18]. Fermentation in the late eighteenth century and into the nineteenth century was studied exclusively by chemists, but with improvements in the microscope, yeast became recognized as a living organism by several scientists beginning in 1827. In 1837, a German physiologist named Theodor Schwann published his observations that yeast consume sugar and excrete ethanol, and reproduce by budding, linking fermentation to yeast. It was at this time that Schwann consulted with the mycologist Franz Julius Ferdinand Meyen who coined the term "*Saccharomyces*," based on the Greek words for "sugar" and "mushroom." Controversy over whether yeast was a living organism persisted for several decades, but as this idea became more accepted, fermentation became increasingly studied by biologists instead of chemists. In 1860, both Louis Pasteur and Johannes Hendrik van den Broek published research attributing fermentation to the growth of yeast cells, and the following years saw great progress in the understanding of fermentation, the role of microbes, and the identification of different yeast species [19].

Meanwhile, through the 19th century, patents began to be filed to improve yeast yields for use in baking by developing recipes for growing it in liquid or sourdough-based media, or to treat yeast taken from brewing [20]. "Pressed yeast" was sold by beer and alcohol manufacturers for use in baking, but for those who could not afford it, these patents were mostly intended for individuals and small-scale baking. Various recipes for sourdough starters included different formulations of cereals, but also potatoes, hops, and other plant materials. These recipes produced quite variable results and were sour due to the presence of lactic acid bacteria. This use of sourdough starter for baking was somewhat culturally divided, with some suggestion that historically, French bakers relied more on the use of sourdough, and British and other northern European countries relied on yeast skimmed from the top of fermented beer [21]. More yeast cells were present in beer foam (or "barm" as it was often referred to), so the resulting bread was lighter and less sour than sourdough-type bread, although this also imparted a bitter taste and dark color due to the presence of hops. In 1871, Pasteur had a patent issued recognizing that yeast was often a mixture of yeasts and bacteria, which can alter the properties of fermented products. However, this was largely ignored for twenty years in yeast manufacturing, until 1891 when Jörgensen and Bergh became the first to patent a technique in which sterility was maintained to create pure culture. While difficult to corroborate, *Saccharomyces cerevisiae* was likely the dominant yeast being cultivated in these efforts.

At the start of the twentieth century, yeast to be used in baking still faced many challenges. While many bakers in Europe and North America had the ability to produce consistent breads due to improvements in milling white flour and mechanical equipment for mixing [22], yeast remained expensive and of poor quality. Improvements in media, aeration, and sterility building off of science and technology of the nineteenth century finally began to take hold and commercial yeast as we now know it began in earnest in the 1920s [23]. While the reuse of yeast from brewing persisted into the 1940s, bakers' appreciation of the need for specialized baking yeast grew through the nineteenth and into the twentieth century. Altogether, this points to a quite recent origin of modern industrial baking yeast. While the application of patents for sourdough-based yeast propagation faded in the early 1900s, home, and artisanal bakers continue to utilize sourdough starters today [11, 24, 25]. While commercial yeast is exclusively *S. cerevisiae*, sourdough starters maintain a much wider diversity of yeast species, as well as lactic and acetic acid bacteria.

EVOLUTION OF BAKERY YEASTS

Different Evolutionary Roads for Industrial and Sourdough Yeasts

There are two primary ways of making bread, either using commercial yeast or using sourdough. While commercial yeasts have been consciously selected for beer or breadmaking and sold all over the world, sourdough yeasts are selected by local bakers through the propagation of sourdough. Therefore, the evolution of sourdough yeasts depends on bakers' backslopping and breadmaking practices. Usually, natural sourdoughs are initiated by simply mixing flour and water. Regular feeding of the mix with water and flour, a process called backslopping, allows the development of a specific microbial community composed of one or two dominant yeast species and one or two dominant lactic acid bacteria. The dynamics of community succession after the initiation of a new sourdough have not been thoroughly investigated, but bakers testify that 10 to 14 days of backsloppings are enough to obtain a mature sourdough for baking good bread. Functional sourdoughs can then be maintained for decades [24, 25].

A recent population genomic analysis of the baker's yeast *S. cerevisiae* revealed that industrial and sourdough populations had undergone different evolutionary trajectories (Fig. **1**) [26].

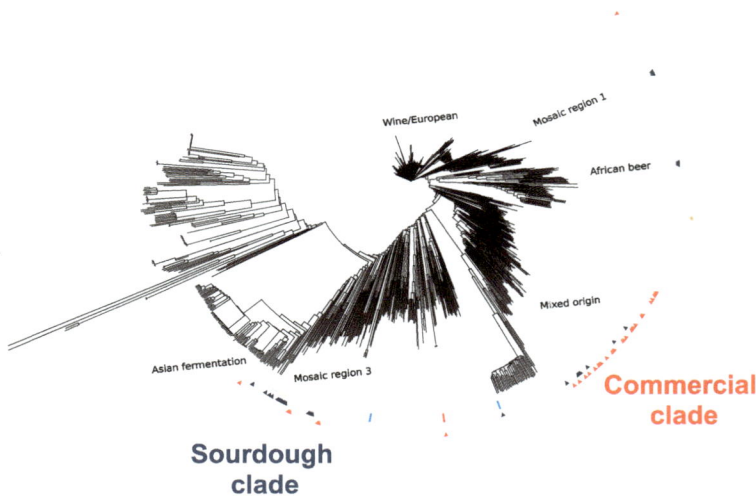

Fig. (1). Phylogenetic tree obtained from SNPs between strains of the *Saccharomyces cerevisiae* 1,011-genomes project and additional bakery strains (Figure adapted from [26]). Red triangles indicate commercial bakery strains and black triangles sourdough strains.

Baking strains of *S. cerevisiae* are clustered in two main genetic groups. The first group is mostly composed of industrial strains, which are often tetraploid and start fermentation rapidly. The second group mostly contains sourdough strains, which have higher fitness than industrial strains in a sourdough-like medium. Outside of these two main genetic groups, a handful of strains clustered with wine strains and African beer strains. Despite historical records demonstrating that bakers used beer foam or dregs as a source of yeast to make bread, there is no evidence of gene flow between beer and sourdough strains. Instead, the evolutionary history of sourdough strains seems to be linked with Asian fermentation since introgressions of Asian fermentation strains into sourdough strains have been detected. By contrast, some introgression of Ale beer strains into commercial breadmaking strains were detected, suggesting that industry may have used beer strains in their genetic crosses to select breadmaking strains.

Genetic and Phenotypic Signatures of Yeast Domestication for Breadmaking

Yeast domestication for breadmaking has only been studied in the model species *S. cerevisiae*. A recent study has shown the consequences of domestication in bakery-associated *S. cerevisiae* [26]. The use of *S. cerevisiae* for breadmaking has led to phenotypic and genotypic divergence between bakery populations and populations originating from natural environments or other fermented products. Both commercial and sourdough strains have a higher maximum CO_2 production and a higher CO_2 production rate than non-bakery strains. In addition, commercial strains have a shorter fermentation onset time than strains from other origins. Sourdough strains display better growth on maltose than commercial strains. They also have a higher population size than strains from any other origin in sourdough-like media, suggesting that they are better adapted to the sourdough environment.

The domestication of *S. cerevisiae* in bread ecosystems is associated with several genetic signatures. A majority of bakery strains are tetraploid or aneuploid [26, 27]. Polyploidization, which refers to the multiplication of a complete chromosome set, has also been found to be associated with domestication of *S. cerevisiae* for brewing. While the lager beer polyploids result from hybridization between *Saccharomyces eubayanus* and *S. cerevisiae* [28, 29], ale beer strains have been found to be autotetraploids [30]. In the case of bakery strains, tetraploids are all autotetraploids. Bakery strains display tetrasomic inheritance, leading to all possible allelic combinations between loci [27]. The null or drastically reduced fertility between diploid and tetraploid bakery strains indicates that bakery yeast populations are composed of two reproductively isolated groups of strains, isolated by post-zygotic barriers [27].

Another interesting genetic signature of domestication is the increased number of maltose and isomaltose genes in sourdough strains. In *S. cerevisiae*, the maltose gene cluster is composed of three genes, encoding the maltose transporter (permease, MAL1), maltase (MAL2) and a transcriptional regulator (MAL3). The genes involved in maltose utilization are represented in five well-described MAL loci located in subtelomeric regions. The presence of just one MAL locus is sufficient to allow for maltose fermentation. The number of copies of the *MAL12*, *MAL32*, *MAL31*, *IMA1*, *IMA3* and *IMA4* is significantly higher in sourdough strains than in non-bakery strains. This increased number of maltose gene copies is associated with increased fitness on maltose. No evidence of an increased number of maltose genes was found in commercial diploid strains.

Biodiversity, Ecology and Evolution of Sourdough Yeast Species

Beyond *S. cerevisiae*, more than 40 species of yeast have been identified in sourdough (Fig. **2**, [31], updated in Von Gastrow *et al* [162]). The vast majority of yeasts detected in sourdough are members of the family Saccharomycetaceae (phylum Ascomycota, subphylum Saccharomycotina, class Saccharomycetes, order Saccharomycetales). Among this family, two genera are frequently reported: *Saccharomyces* and *Kazachstania*. Other described yeast species belong to the more genetically distant genera *Pichia*, *Torulaspora*, and *Wickerhamomyces*.

Many yeast species found in sourdough are also detected in other human-associated environments (see reviews of [32 - 36]). For example, *Torulaspora delbrueckii* (found in cocoa, wine, cheese) [34, 37], *Pichia fermentans* (kocho, liquor, cheese, injera, beer) [38 - 41], *Pichia kudriavzevii* (soil, cacao, fruit juice, human) [42, 43], *Meyerozyma guillermondii* (coffee, cacao, wine, human) [34, 43 - 45], *Pichia membranifaciens* (kocho, olive) [40, 46], *Wickerhamomyces anomalus* (injera, doco, beer) [38, 47], *Kazachstania unispora* (Kefir, wine, corn silage) [48 - 50]. Other species appear to be more specialized and are found mostly in sourdough and not in other fermentations, including *K. bulderi*, *K. saulgeensis*, and *K. barnettii*.

The distribution of yeast species in sourdough does not appear to be structured according to geography. Most species have been found across all continents [51, 162]. At a national scale, geographical distance was not found to be correlated with sourdough yeast species diversity [24]. Some particular taxa appear to be more frequently found in specific countries or regions; however, this might be related to sampling. For example, *W. anomalus* has been frequently found in Belgium [52] and *K. bulderi* in France and Spain [25, 53]. The widespread distribution of most sourdough yeast might be explained by yeast dispersion by humans, insects, and/or birds [54 - 57].

Fig. (2). Yeast species found in sourdough. Numbers indicate the number of times a given yeast species has been detected in a sourdough. Data have been taken from Huys *et al.* [31], updated in Von Gastrow *et al.* [162]. The electronic microscopy picture of sourdough yeast has been kindly provided by Bernard Onno.

The question of the origin of yeast species in sourdough has been the subject of several studies. The bakery environment, also called "house microbiota," appears to be the primary source of yeast, with storage boxes and dough mixers sheltering *S. cerevisiae* in some bakeries [58]. A significant overlap between microbial taxa present in sourdough and bakers' hands has also been found [54]. Overlapping taxa between human hosts and the bread they bake include *K. humilis*, *K. unispora*, and *S. cerevisiae* [54]. Ingredients such as flour and water, may also be a source of yeasts [53]. Flour was found to contain *K. servazzii, P. fermentans, W. anomalus, T. delbrueckii* species, all of which are frequently encountered in sourdough. However, so far, the ecological reservoirs of *K. bulderi, K. saulgeensis,* and *K. barnettii* remain unknown.

The population dynamics of yeast colonization during sourdough initiation and backslopping has not been studied systematically. To examine the relative roles of migration, selection and random processes in sourdough yeast evolution, balanced sampling across many sourdough cultures is needed to estimate populations genetic parameters. Several studies have screened a large collection of *S. cerevisiae* strains isolated across sourdoughs in China [59], Italy [60], Spain [53], and France [26] with molecular markers. Most genetic variation was found between sourdoughs, suggesting that yeast migration between sourdoughs or sourdough exchanges between bakers are limited and that bottlenecks may occur during sourdough colonization. Further investigations are needed to shed light on the evolutionary dynamics of sourdough yeasts.

It is still unclear how random processes, which can change allelic/species frequency in a population due to random sampling, can influence the evolution of the sourdough yeast community. Community ecology is influenced by four classes of the process: selection, dispersal, speciation and drift. Genetic drift can lead to the loss or fixation of genetic variants in a population and/or the loss of rare species in the community, and therefore drive the population or the community towards less diversity. The smaller the population is, the higher genetic drift drives changes. In sourdough, yeast population density ranges from 10^4 cells per gram of sourdough to 10^9 cells per gram of sourdough. Usually, one or two yeast species dominate the community. During sourdough propagation, genetic drift is therefore expected to only slightly influence the evolution of the dominant yeast species but may influence the evolution and extinction of rare species. During the initiation of the sourdough population, sourdough may be colonized by a small number of fermenting yeast cells, leading to a bottleneck and low standing genetic variation in the sourdough community. Further studies are needed to shed light on the role of random processes in sourdough microbial community ecology.

The analysis of sourdough yeast species evolution will soon benefit from the increased number of available genomes for bakery yeast. Among the 45 yeast species that have been detected in sourdough, 40 have a genome sequenced. So far, 24 species have a good genome assembly with fewer than 20 scaffolds. This includes all genomes of the *Saccharomyces* genus, but also *T. delbrueckii* (Bioproject: PRJNA79345), *Pichia kudriavzevii* (PRJNA434433), *Pichia kluyveri* (PRJNA434537), *Kazachstania barnettii* (PRJEB35206), *Kazachstania saulgeensis* (PRJEB20516), *Kazachstania bulderi* (PRJEB44438), *Kazachstania humilis* (PRJEB44438), and *Millerozyma farinosa* (PRJNA369593). Ensuring these genetic resources are openly available and contributed to the international community will ensure that understanding of genetic diversity and drift and other evolution processes will continue.

MICROBIAL INTERACTIONS DURING SOURDOUGH PRODUCTION

The dough ecosystem is not as rich in diversity as some microbial ecosystems [61 - 63], but generally contains stable populations of yeast and bacteria. This section examines the combinations of yeasts and bacteria present, considers the measured and potential microbial interaction and briefly considers the consequences of these interactions for dough composition and final bread quality.

The abundance of yeasts and bacteria in a dough ecosystem has been understood by the perceived activity of the dough due to microbial activity. The yeast species present in dough fermentation have been mentioned above. There have been more than 40 different yeast species described in bread dough [51] and up to 70 different bacterial species [64], but in general, there are 1-2 yeast species and up to 5 bacterial species per fermentation [65]. Numerical dominance will necessarily fluctuate during the fermentation, but in general, there are generally 100 times more bacteria than yeasts [66]. The total colony-forming units for yeast is more than 8 log CFU/g, while the number of bacterial cells present can reach more than 9 log CFU/g [66], although some exceptions exist [67]. The microbial members of a dough ecosystem are dependent on the flour source, mechanism of culture and possibly related to the first reasons, country of origin or cultural tradition.

Investigating other environmental or food ecosystems shows that there is great potential in bread dough for significant and sustained interactions between yeasts and bacteria. Interactions could take many forms, including positive interactions, where metabolites are shared and growth is supported between community members, to negative interactions, where growth or activity of one microbial member is inhibited by the growth or activity of another member.

Interactions between yeasts in a dough fermentation have not been well described. However, there is a body of evidence to support yeast-yeast interactions in other food-related ecosystems and we can suppose that these mechanisms are also likely to exist in a bread dough ecosystem. While studies in this area have focused on the interactions between cells of *S. cerevisiae*, there is increasing understanding that other yeasts can communicate within members of a species and between genera. These interactions could be metabolic exchanges, where small molecules can be released or secreted by one cell and taken up by another. A metabolic example of this generally positive interaction is seen between species of yeasts fermenting grape juice to wine [68]. Ethanol is an example of a small molecule which has negative effects on other yeasts, and *S. cerevisiae* is well-known for niche construction and inhibition of other species by the production of this small molecule in a number of habitats [69]. While not specifically described in dough fermentations, small amounts of ethanol are produced by fermentative yeasts in dough fermentations, but the ethanol is generally not detected in the baked bread as the ethanol is lost during baking [70]. A recent study investigated the longevity of yeast species when grown together in the laboratory, and found that there were specific co-occurrence patterns that suggest robust microbial interactions, both positive and negative, which influenced yeast persistence in fermentation [24].

A specific interaction between yeasts, quorum sensing, has been observed in a variety of laboratory and food environments. This interaction is similar to that seen in bacteria, and is defined as intercellular signaling, which facilitates communication between microorganisms, where a cell-density-dependent regulation of gene expression occurs after reaching a critical concentration of signal chemicals [71]. While some studies have investigated filamentation [72, 73], flocculation [74] and growth arrest of non- *Saccharomyces* yeasts [75, 76] by *S. cerevisiae*, none of these studies have satisfied the criteria of quorum sensing as discussed by [77]. While the bulk of studies has been performed in *S. cerevisiae*, it is likely that alternative behaviours between other yeast species exist and could be relevant for bread dough fermentations.

Yeasts, and particularly *S. cerevisiae*, can specifically target and inhibit the growth of other yeasts by the production of a toxin. This system is called the killer system and is produced when a host cell contains a dsRNA virus, which produces a secreted toxin that kills susceptible cells and indeed other fungi [78]. The killer yeasts phenomenon is present in many species and has been described in a screen including *S. cerevisiae* of bakery origin [79]. While the mechanism has not been described as technologically relevant in a bread dough ecosystem, there are reports of *S. cerevisiae* strains with this capacity bred for use in bread production [80]. A survey of naturally occurring yeasts showed that *Kazachstania exigua* has killer activity and this attribute was exploited for the protection of fruits from a fungal pathogen [81]. While this mechanism allows yeast strain enrichment of a species through specific reduction of sensitive populations, it is as yet unclear whether there is relevance in bread dough fermentation.

Other mechanisms by which yeasts can interact with other yeasts in food fermentations could be relevant to the sourdough ecosystem. This includes a cell-cell contact inhibition mechanism, first described in *S. cerevisiae* to inhibit the growth of non- *Saccharomyces* yeasts in a liquid fermentation [76, 82]. Here, at high cellular concentrations, direct contact between cells was found to arrest the growth of *Kluyveromyces thermotolerans* and *T. delbrueckii*, and further analysis showed that a small peptide was released by *S. cerevisiae* to inhibit the growth of non- *Saccharomyces* yeasts [83]. The production of small antimicrobial peptides has been observed in other liquid fermentations, such as wine (reviewed by [84]), but have yet to be described in the semi-solid fermentation style seen in sourdough fermentation. In particular, antimicrobial peptides mediate many interkingdom microbial interactions, which are relevant for sourdough fermentation when there is an abundance of yeasts and bacteria.

Few studies have investigated the role of inter-kingdom interactions in dough fermentation, and there is clearly more to be learnt. It is clear that yeasts and bacteria interact metabolically (and otherwise) with one another in food fermentations. Lactic acid bacteria are a diverse classification of bacteria, with much diversity in food systems, and particularly sourdough fermentations [85]. One intercellular mechanism of interest is the production of antimicrobial peptides, named bacteriocins, which are produced by bacteria to inhibit, kill or interact with other bacteria [86]. Notably, these peptides tend not to interfere with the growth of yeasts or other fungi [87]. Many LAB isolates from sourdoughs produce bacteriocins (reviewed by [88, 89]), but the role of this activity has not been demonstrated in sourdough. It is possible that this behaviour could be used competitively in sourdough fermentation to enhance the dominance of a bacterial species.

METABOLIC FUNCTIONS OF BAKERY YEASTS

Maltose is a key carbohydrate present in sourdough fermentations, as it is naturally present in flour (along with glucose and fructose) and released by plant endogenous enzymes from starch. Much of the literature which relates to sugar utilisation of yeasts is related to maltose, as it is directly related to fermentation performance in the dough [51]. There is some diversity in the ability of naturally occurring yeasts to utilise maltose, and this varies amongst species [90, 91]. It appears that naturally started sourdoughs will have both maltose-fermenting and non-fermenting yeasts [92]. There is recent evidence that only *K. bulderi* can ferment maltose in standard laboratory tests [93] where *K. humilis* isolates have been selected for strong fermentative abilities [94, 95]. While *S. cerevisiae* strains may have variable fermentative capacities based on maltose utilization, it could be that accumulation of the MAL genes has been enriched during domestication of this yeast [26].

The ability to ferment other sugars such as glucose and fructose is well established in fermentative yeasts, and through the production of secreted invertase, many yeasts are able to use sucrose as a carbon source if it is added to the dough [96]. Indeed, degradation of a range of fructans of differing polymerisation levels is possible by *S. cerevisiae* secreted invertase, and this occurs during sourdough fermentation [97, 98]. This process is receiving increased attention due to the sensitivity of some consumers to the presence of fermentable oligo, di-, mono- and polysaccharides (FODMAPs) in bread and the potential role of microbial activity to reduce or eliminate these compounds [99]. Indeed, specific application of sourdough processes as suggested by Loponen and Ganzle [100] may reduce problematic FODMAPs while retaining the overall fibre content. For sourdough baking, there are specific interactions between yeast and bacteria that are relevant to consider when carbohydrate and nitrogen metabolism are considered.

Commercial and sourdough yeasts ferment carbohydrates and also produce metabolites that contribute to bread flavor, such as organic acids and aroma [101, 102]. The newly appreciated diversity of yeast species present in sourdoughs has encouraged scientists and yeast breeders to explore the flavors produced by yeast species other than *S. cerevisiae* [103]. Together with *S. cerevisiae*, *T. delbrueckii* and *S. bayanus* appear as good candidates to produce aroma complexity in bread. The yeast species *K. bulderi*, *K. humilis* and *W. anomalus* produce a lower diversity of aroma compounds than *Saccharomyces* species, although this might be related to the low number of strains that have been screened so far. Other factors such as breadmaking practices (including fermentation time, cooking, type of flours, type of cereals, cultivars, milling, crop terroir), and bakery parameters including temperature and humidity affect yeast growth and impact production of aroma [104, 105].

While the production of organic acids is responsible for the sour taste of bread and its shelf life, the production of volatile compounds is responsible for the diversity of bread aroma. To date, 192 small molecules which contribute to bread aroma have been reported in sourdough bread [102] and 75 have been found in bread made with *S. cerevisiae* [41, 70, 95, 106]. Most compounds are higher alcohols, aldehydes and esters (Fig. **3**).

The most common alcohols are 1-hexanol, 1-heptanol, 1-pentanol, 2-phenylethanol, 2-methyl-1-propanol and 3-methyl-1-butanol. The most common aldehydes are heptanal, hexanal, octoanal, nonanal, decanal, phenylacetaldehyde and benzaldehyde. The most frequently encountered esters are ethyl octanoate and ethyl acetate. Thiols, which are responsible for the pleasant fruity aroma of white wine, are not found in yeast bread. Terpenes, which are responsible for floral aroma in muscat and wine, can be found in yeast bread, but rarely. As has been shown for wine, the presence of these two classes of aroma in bread probably depends on the precursors present in the flour. Interactions between yeast and bacteria may also change the metabolism of aroma production [107].

Fig. (3). Heatmap of the aroma compounds detected in bread made with different yeast species. Data have been obtained from [41, 70, 77, 106].

Specific metabolic interactions between yeast and bacteria are responsible for many of the attractive transformations in sourdough fermentation. In general, carbohydrate degradation by bacteria is complemented by the nitrogen overflow ability of yeasts, and the stability of microbial populations present in the sourdough system is a reflection of these interactions [108]. Specific interactions between *K. humilis* and LAB have been described [109], where maltose utilisation occurred when maltose-negative *K. humilis* were grown with maltose-positive heterofermentative LAB, but the growth of the yeasts was suppressed. While there is considerable genotypic and phenotypic diversity in *F. sanfranciscensis* strains [110], the composition of several LAB and specific yeast in fermentation may assemble to best utilise the carbohydrates, amino acids available in the specific dough [111]. Indeed, diversity in strain metabolic activity of *F. sanfranciscensis* is likely to enable interactions and maintain community structures with *K. humilis* [112, 113] but does not explain the reported *S. cerevisiae* co-exclusion [24]. It is evident that metabolic interactions between yeasts and bacteria have positive consequences on the flavour, aroma and physical outcomes of bread [95], and these attractive characteristics are defining features of sourdough bread.

Available nitrogen often becomes limiting in food fermentations, and this is particularly seen in the case of wine fermentations. Here, fermentative yeasts such as *S. cerevisiae* struggle to access sufficient nitrogen to support cellular growth and fermentation [114]. In bread dough made with wheat flour, nitrogen is primarily in the form of macromolecules, such as gliadin and glutenin, which form a gluten matrix when water is added. After water is added to flour, and during fermentation, gluten are hydrolysed to more simpler forms such as peptides and free amino acids by proteinases which are naturally present in the flour and secreted by microbes [115, 116]. Total free amino acids tend to increase during sourdough fermentation [117], and this is dependent on both the yeast and bacterial species present [118], but bacteria tend to have a higher effect [117]. The release of nitrogenous compounds is very important, as many aromas and flavours are conferred by these compounds or react further to create flavours during further processing [119]. The importance of yeast and bacterial nitrogen processing is highlighted by the reduction of allergenic gluten epitopes for people with celiac disease [120]. Metabolic capabilities are likely to be different when dough containing flour other than wheat is used (see for example [121]). The diversity of secreted bacterial endo- and exo-peptidases is responsible for metabolic behavior in dough fermentations, but nitrogen metabolism of yeasts beyond *S. cerevisiae* has not been systematically addressed.

In other food fermentations, the stability of the microbial communities of yeasts and bacteria is dependent on nitrogen metabolism and thus it is likely that further activities are present in dough fermentation. For example, the populations of yeast

(*S. cerevisiae*) and bacteria (*Lactococcus lactis* and *Lactiplantibacillus plantarum*) in water kefir are stabilized by *S. cerevisiae* secreting nitrogen into the milieu for the bacteria to use [122]. Secretion of glucose and galactose by *L. lactis* by degradation of lactose allows this microbial community to thrive. Similarly, reconstruction of fungal and bacterial communities from cheese [61] showed that longevity of the interaction was affected by amino acid transfer between the kingdoms [123]. Nitrogen flow in food fermentations is complex, but is likely to be at the basis of many trophic interactions, including those in bread.

STRESS TOLERANCE OF BAKERY YEASTS: OSMOTOLERANCE, FREEZING, AND DESICCATION TOLERANCE

Yeast may face a variety of stressors related to baking, including freezing, desiccation, and osmotic stress related to changing moisture content and the presence of salt and/or sugar. This is particularly true of commercial baking yeast, which are subject to a variety of processes for packaging and distribution. *S. cerevisiae* is exclusively utilized for industrial baking and is the most well studied in its stress tolerance of various baking processes, although many stress response traits likely apply to other yeasts found in dough environments. Most commercial baking strains of yeast are used due to historical reasons and are thus not optimized to deal with many stressors associated with baking [124, 125]. Indeed, there is tremendous variability across baking strains in how they handle heat, freezing, osmotic, and oxidative stress [90, 126].

Managing osmotic stress is an essential property of yeast used in bread baking. The semi-solid dough matrix and high sugar recipes for sweet breads [up to 30% sucrose] create hyperosmotic and ionic stress to cells. These conditions cause cells to lose water, which causes impaired growth and gas production [127], and thus decreased fermentation capacity. Yeast manage this osmotic stress chiefly by the production of glycerol, which balances intracellular osmolarity and protects against dehydration. The amount of glycerol in the cell and dough can positively or negatively affect the gassing ability, bread aroma and taste, and shelf-life of loaves [128 - 131].

Glycerol homeostasis is regulated *via* the high-osmolarity glycerol (HOG) pathway, a conserved mitogen-activated protein kinase (MAPK) pathway important in stress responses and signaling [132, 133]. Many genes in this pathway are upregulated during dough fermentation [70], and various efforts aiming to improve fermentation efficiency and stress tolerance have thus focused on targets in the HOG pathway. *GPD1* is the rate-limiting enzyme involved in glycerol biosynthesis [134] and is upregulated upon the start of dough fermentation and at cold temperatures [70, 135]. Deletion of *GPD1* results in decreased glycerol production, resulting in impaired dough fermentation [130] and a decrease in freeze tolerance [135]. In contrast, strains with higher glycerol production can improve fermentation of high-sugar dough, improve CO_2 production, and increase dough height. Interestingly, several studies have observed that manipulating glycerol production has a more pronounced effect on bread traits in lab strains or strains from other fermentation environments like beer and wine [125, 130], suggesting that some baking strains have a higher base glycerol production and/or other traits impacting this phenotype.

Strong osmotolerance does not appear to be a fixed trait in baking strains and instead varies in both baking strains of *S. cerevisiae* and strains isolated from other environments [90]. Environmental sources can have a strong impact. For example, some strains of *S. cerevisiae* isolated from Chinese rice wine show signatures of positive selection on osmosensing genes [136], likely in response to high sugar content of rice wine. In the limited species that have been examined outside of *S. cerevisiae*, strains of the yeasts *T. delbrueckii, Wickerhamomyces subpelliculosus, Zygosaccharomyces rouxii, Hanseniospora vineae,* and *Barnettozyma californica* exhibit high osmotolerance and/or halotolerance, in many cases outperforming *S. cerevisiae* in these and other baking traits [137, 138] This suggests that bakers could harness natural variation within and outside of *S. cerevisiae* for greater fermentation efficiency, and in particular for working with high sugar and high salt doughs.

In addition to osmotic stress, modern bread production relies on the frozen dough for extending the shelf life of dough, easier and broader distribution of products, and meeting consumer demands for bread freshness [139]. Freezing and thawing dough impact yeast in many ways, including inducing cell damage, osmotic stress, and loss of viability, all of which influence proofing time, loaf volume, and bread firmness and structure. Thus, the utilization of yeast that has freeze tolerance is a key area of interest in the baking industry. Yeast typically responds to freezing conditions *via* modulation of the cryoprotective molecules trehalose, proline, and glycerol. Upon stress, yeast induces trehalose and glycerol synthesis, but proline appears to not be stress-activated [140]. Regardless, synthetically increased accumulation of proline provides yeast with increased tolerance to many stresses, including freezing, desiccation, oxidation, and ethanol [140]. Mutants with increased proline accumulation display higher cell viability under freezing conditions and in high sugar doughs [141 - 143]. In a related process, elevated nitric oxide synthesis, which is regulated *via* the proline oxidase Put1 and N-acetyltransferase Mpr1, also enhances bread leavening following stressors, in particular air-drying, but also freeze-thaw and oxidative stress [144, 145].

Trehalose is often associated with stress tolerance, but trehalose content varies across commercial baking strains [126]. Intracellular trehalose is most often increased through the deletion of enzymes involved in trehalose breakdown: neutral trehalase (*NTH1*) and acid trehalase (*ATH1*) [146, 147]. Deletion of one or both *NTH1* and *ATH1* can improve gassing ability in frozen dough [148 - 150], and also aid in desiccation tolerance [151]. Another mechanism to increase freeze protection involves the overexpression of particular genes. For example, overexpression of the alpha-glucosidase *MAL62* increases trehalose content *via* increased uridine diphosphoglucose UDPG-dependent trehalose synthesis [150, 152]. Overexpression of the calcineurin target *CRZ1* increases both salt and freeze tolerance in both *S. cerevisiae* and *T. delbrueckii*, and improves the leavening ability of baker's yeast in high-sugar dough [153]. Increased expression of *SNF1* protein kinase, which is an important regulator of the yeast stress response, increases CO_2 production, alters the metabolism of protectant molecules, changes cell membrane components, and increases resistance to freezing stress [154]. There is also evidence that suggests that the combination of elevated trehalose and proline provides both short-term and long-term freeze tolerance, and enhanced fermentation ability in frozen dough [143, 149].

Most yeast is not freeze tolerant, and efforts to create *S. cerevisiae* strains with greater cold tolerance *via* adaptive evolution have had rather limited success [155, 156]. Some of the most cold-tolerant evolved strains came from methods that employed outcrossing of different strains, and similar methods have been employed using interspecific hybridization to increase cold tolerance in the wine industry. Other efforts have come from the isolation of mutant strains from screens [157, 158] and bioprospecting from natural and other environments [159, 160]. Like osmotic stress, strains of yeast from species aside from *S. cerevisiae* may be particularly helpful here. Notably, strains of *T. delbrueckii* show no or significantly reduced loss of cell viability compared to their *S. cerevisiae* commercial baking strain counterparts during long-term freezing [137, 161]. Even other species in the *Saccharomyces* clade may be of interest, strains from *S. uvarum* and *S. eubayanus* are much colder tolerant than *S. cerevisiae*. Many of these strains are already used in the cold fermentation of beer and wine, and *S. uvarum* has even been isolated from sourdough starters [24].

CONCLUSION

Advances in genomics and metagenomics, together with increased worldwide sampling, have contributed dramatic insights into the species diversity, ecology, and evolution of yeasts used in fermentation. Yeasts used in baking have received relatively little attention compared to their relatives used in beer and wine. However, the baking yeast community of bakers and scientists are poised for many new advances. Increased desire from consumers for bread with more complex flavors, the use of alternative grains, a resurgence in sourdough baking in the home, and a desire for more biodiverse and sustainable bakery practices are all fueling the embrace of a wide array of yeast strains and species. Key areas of further research include a better understanding of interactions between yeast, bacteria, and their metabolites in sourdough; ecological succession and the role of selection and random processes in sourdough; sampling and understanding of traditional bread like injera; the identification of strains/species with increased stress tolerance for high sugar and frozen doughs; the impact of different grain sources on species diversity; and a better understanding of molecular signals of selection in the genome of *S. cerevisiae* strains.

CONSENT FOR PUBLICATION

Not applicable.

CONFLICT OF INTEREST

The authors declare no conflict of interest, financial or otherwise.

ACKNOWLEDGEMENTS

We acknowledge Lucas Von Gastrow for sharing his updated data file on sourdough yeast species and Thibault Nidelet for making figures.

REFERENCES

[1] Piperno DR, Weiss E, Holst I, Nadel D. Processing of wild cereal grains in the Upper Palaeolithic revealed by starch grain analysis. Nature 2004; 430(7000): 670-3.
[http://dx.doi.org/10.1038/nature02734] [PMID: 15295598]

[2] Revedin A, Aranguren B, Becattini R, *et al.* Thirty thousand-year-old evidence of plant food processing. Proc Natl Acad Sci USA 2010; 107(44): 18815-9.
[http://dx.doi.org/10.1073/pnas.1006993107] [PMID: 20956317]

[3] Fullagar R, Hayes E, Stephenson B, Field J, Matheson C, Stern N, *et al.* Evidence for Pleistocene seed grinding at Lake Mungo, south-eastern Australia. Archaeol Ocean 2015; 50(S1): 3-19.
[http://dx.doi.org/10.1002/arco.5053]

[4] Pascoe B. Dark Emu: Aboriginal Australia and the birth of agriculture. Scribe UK 2018.

[5] Arranz-Otaegui A, Gonzalez Carretero L, Ramsey MN, Fuller DQ, Richter T. Archaeobotanical evidence reveals the origins of bread 14,400 years ago in northeastern Jordan. Proc Natl Acad Sci USA 2018; 115(31): 7925-30.
[http://dx.doi.org/10.1073/pnas.1801071115] [PMID: 30012614]

[6] González Carretero L, Wollstonecroft M, Fuller DQ. A methodological approach to the study of archaeological cereal meals: a case study at Çatalhöyük East (Turkey). Veg Hist Archaeobot 2017; 26(4): 415-32.
[http://dx.doi.org/10.1007/s00334-017-0602-6] [PMID: 28706348]

[7]   Paulette T. Consumption and storage in the Bronze Age. Models of Mesopotamian Landscapes: How small-scale processes contributed to the growth of early civilizations. BAR Publishing 2013; pp. 102-8.

[8]   Darby WJ, Ghalioungui P, Grivetti L. Food: The Gift of Osiris. Academic Press 1977; p. 548.

[9]   Samuel D. Brewing and baking in ancient Egyptian art. In: Food in the arts Proceedings of the Oxford Symposium on Food and Cookery. 173-81.

[10]   Samuel D. Investigation of Ancient Egyptian Baking and Brewing Methods by Correlative Microscopy. Science 1996; 273(5274): 488-90.
        [http://dx.doi.org/10.1126/science.273.5274.488] [PMID: 8662535]

[11]   Cappelle S, Guylaine L, Gänzle M, Gobbetti M. History and Social Aspects of Sourdough. In: Gobbetti M, Gänzle M, Eds. Handbook on Sourdough Biotechnology. Boston, MA: Springer US 2013; pp. 1-10. [Internet]
        [http://dx.doi.org/10.1007/978-1-4614-5425-0_1]

[12]   Shevchenko A, Yang Y, Knaust A, *et al.* Proteomics identifies the composition and manufacturing recipe of the 2500-year old sourdough bread from Subeixi cemetery in China. J Proteomics 2014; 105: 363-71.
        [http://dx.doi.org/10.1016/j.jprot.2013.11.016] [PMID: 24291353]

[13]   Wang J, Liu L, Ball T, Yu L, Li Y, Xing F. Revealing a 5,000-y-old beer recipe in China. Proc Natl Acad Sci USA 2016; 113(23): 6444-8.
        [http://dx.doi.org/10.1073/pnas.1601465113] [PMID: 27217567]

[14]   Erdkamp P. The Food Supply of the Capital. The Cambridge Companion to Ancient Rome. Cambridge University Press 2013.
        [http://dx.doi.org/10.1017/CCO9781139025973.019]

[15]   The Natural History of Grain. Natural History. London: Taylor and Francis 1855.

[16]   Wikander O. The Water-Mill. Handbook of Ancient Water Technology, Technology and Change in History. Leiden: Brill 2000; pp. 371-400.
        [http://dx.doi.org/10.1163/9789004473829_019]

[17]   Watson A. Aurelian and the Third Century. London: Routledge 2004.
        [http://dx.doi.org/10.4324/9780203167809]

[18]   Barnett JA. A history of research on yeasts. 1: Work by chemists and biologists 1789-1850. Yeast 1998; 14(16): 1439-51.
        [http://dx.doi.org/10.1002/(SICI)1097-0061(199812)14:16<1439::AID-YEA339>3.0.CO;2-Z] [PMID: 9885150]

[19]   Barnett JA. A history of research on yeasts 2: Louis Pasteur and his contemporaries, 1850-1880. Yeast 2000; 16(8): 755-71.
        [http://dx.doi.org/10.1002/1097-0061(20000615)16:8<755::AID-YEA587>3.0.CO;2-4]   [PMID: 10861901]

[20]   Gélinas P. Mapping Early Patents on Baker's Yeast Manufacture. Compr Rev Food Sci Food Saf 2010; 9(5): 483-97.
        [http://dx.doi.org/10.1111/j.1541-4337.2010.00122.x] [PMID: 33467828]

[21]   Rubel W. Bread: A Global History. Reakton Books, Limited 2011.

[22]   Burnett J. The Baking Industry in the Nineteenth Century. Bus Hist 1963; 5(2): 98-108.
        [http://dx.doi.org/10.1080/00076796300000003]

[23]   Gélinas P. In Search of Perfect Growth Media for Baker's Yeast Production: Mapping Patents. Compr Rev Food Sci Food Saf 2012; 11(1): 13-33.
        [http://dx.doi.org/10.1111/j.1541-4337.2011.00168.x]

[24]   Landis EA, Oliverio AM, McKenney EA, *et al.* The diversity and function of sourdough starter microbiomes. In: Weigel D, Mitri S, Adams R, Eds. eLife. 2021; 10: p. e61644.
[http://dx.doi.org/10.7554/eLife.61644]

[25]   Urien C, Legrand J, Montalent P, Casaregola S, Sicard D. Fungal species diversity in french bread sourdoughs made of organic wheat flour. Front Microbiol 2019; 10: 201. https://www.ncbi.nlm.nih.gov/pmc/articles/PMC6387954/ [Internet].
[http://dx.doi.org/10.3389/fmicb.2019.00201] [PMID: 30833935]

[26]   Bigey F, Segond D, Friedrich A, *et al.* Evidence for Two Main Domestication Trajectories in *Saccharomyces cerevisiae* Linked to Distinct breadmaking Processes. Curr Biol 2021; 31(4): 722-732.e5.
[http://dx.doi.org/10.1016/j.cub.2020.11.016] [PMID: 33301710]

[27]   Albertin W, Marullo P, Aigle M, *et al.* Evidence for autotetraploidy associated with reproductive isolation in *Saccharomyces cerevisiae:* towards a new domesticated species. J Evol Biol 2009; 22(11): 2157-70.
[http://dx.doi.org/10.1111/j.1420-9101.2009.01828.x] [PMID: 19765175]

[28]   Walther A, Hesselbart A, Wendland J. Genome Sequence of *Saccharomyces carlsbergensis*, the World's First Pure Culture Lager Yeast. G3 Genes|Genomes|Genetics 2014; 4(5): 783-93.

[29]   Dunn B, Sherlock G. Reconstruction of the genome origins and evolution of the hybrid lager yeast *Saccharomyces pastorianus*. Genome Res 2008; 18(10): 1610-23.
[http://dx.doi.org/10.1101/gr.076075.108] [PMID: 18787083]

[30]   Fay JC, Liu P, Ong GT, *et al.* A polyploid admixed origin of beer yeasts derived from European and Asian wine populations. PLoS Biol 2019; 17(3): e3000147.
[http://dx.doi.org/10.1371/journal.pbio.3000147] [PMID: 30835725]

[31]   Huys G, Daniel H-M, De Vuyst L. Taxonomy and Biodiversity of Sourdough Yeasts and Lactic Acid Bacteria. In: Gobbetti M, Gänzle M, Eds. Handbook on Sourdough Biotechnology. Boston, MA: Springer US 2013; pp. 105-54. [Internet]
[http://dx.doi.org/10.1007/978-1-4614-5425-0_5]

[32]   Tamang JP, Watanabe K, Holzapfel WH. Review: Diversity of Microorganisms in Global Fermented Foods and Beverages. Front Microbiol 2016; 7: 377. https://www.ncbi.nlm.nih.gov/pmc/articles/PMC4805592/ [Internet].
[http://dx.doi.org/10.3389/fmicb.2016.00377] [PMID: 27047484]

[33]   Bokulich NA, Bamforth CW. The microbiology of malting and brewing. Microbiol Mol Biol Rev 2013; 77(2): 157-72.
[http://dx.doi.org/10.1128/MMBR.00060-12] [PMID: 23699253]

[34]   Benito Á, Calderón F, Benito S. The Influence of Non-Saccharomyces Species on Wine Fermentation Quality Parameters. Fermentation (Basel) 2019; 5(3): 54.
[http://dx.doi.org/10.3390/fermentation5030054]

[35]   Johansen PG, Owusu-Kwarteng J, Parkouda C, Padonou SW, Jespersen L. Occurrence and Importance of Yeasts in Indigenous Fermented Food and Beverages Produced in Sub-Saharan Africa. Front Microbiol 2019; 10: 1789.https://www.ncbi.nlm.nih.gov/pmc/articles/PMC6691171/ [Internet].
[http://dx.doi.org/10.3389/fmicb.2019.01789] [PMID: 31447811]

[36]   Voidarou C, Antoniadou M, Rozos G, *et al.* Fermentative Foods: Microbiology, Biochemistry, Potential Human Health Benefits and Public Health Issues. Foods 2020; 10(1): 69. https://www.ncbi.nlm.nih.gov/pmc/articles/PMC7823516/

[37]   Mota-Gutierrez J, Botta C, Ferrocino I, *et al.* Dynamics and Biodiversity of Bacterial and Yeast Communities during Fermentation of Cocoa Beans. Appl Environ Microbiol 2018; 84(19): 84. [19].
[http://dx.doi.org/10.1128/AEM.01164-18] [PMID: 30054357]

[38]   Laitila A, Wilhelmson A, Kotaviita E, Olkku J, Home S, Juvonen R. Yeasts in an industrial malting

ecosystem. J Ind Microbiol Biotechnol 2006; 33(11): 953-66.
[http://dx.doi.org/10.1007/s10295-006-0150-z] [PMID: 16758169]

[39]   Zheng X, Li K, Shi X, Ni Y, Li B, Zhuge B. Potential characterization of yeasts isolated from Kazak artisanal cheese to produce flavoring compounds. MicrobiologyOpen 2018; 7(1): e00533.
[http://dx.doi.org/10.1002/mbo3.533] [PMID: 29277964]

[40]   Birmeta G, Bakeeva A, Passoth V. Yeasts and bacteria associated with kocho, an Ethiopian fermented food produced from enset (Ensete ventricosum). Antonie van Leeuwenhoek 2019; 112(4): 651-9.
[http://dx.doi.org/10.1007/s10482-018-1192-8] [PMID: 30368690]

[41]   Liu T, Li Y, Sadiq FA, *et al.* Predominant yeasts in Chinese traditional sourdough and their influence on aroma formation in Chinese steamed bread. Food Chem 2018; 242: 404-11.
[http://dx.doi.org/10.1016/j.foodchem.2017.09.081] [PMID: 29037707]

[42]   Douglass AP, Offei B, Braun-Galleani S, *et al.* Population genomics shows no distinction between pathogenic Candida krusei and environmental Pichia kudriavzevii: One species, four names. PLoS Pathog 2018; 14(7): e1007138.
[http://dx.doi.org/10.1371/journal.ppat.1007138] [PMID: 30024981]

[43]   Pereira PV, Bravim DG, Grillo RP, *et al.* Microbial diversity and chemical characteristics of *Coffea canephora* grown in different environments and processed by dry method. World J Microbiol Biotechnol 2021; 37(3): 51.
[http://dx.doi.org/10.1007/s11274-021-03017-2] [PMID: 33594606]

[44]   De Vuyst L, Leroy F. Functional role of yeasts, lactic acid bacteria and acetic acid bacteria in cocoa fermentation processes. FEMS Microbiol Rev 2020; 44(4): 432-53.
[http://dx.doi.org/10.1093/femsre/fuaa014] [PMID: 32420601]

[45]   Yan W, Gao H, Qian X, *et al.* Biotechnological applications of the non-conventional yeast *Meyerozyma guilliermondii*. Biotechnol Adv 2021; 46: 107674.
[http://dx.doi.org/10.1016/j.biotechadv.2020.107674] [PMID: 33276074]

[46]   Argyri K, Doulgeraki AI, Manthou E, *et al.* Microbial Diversity of Fermented Greek Table Olives of Halkidiki and Konservolia Varieties from Different Regions as Revealed by Metagenomic Analysis. Microorganisms 2020; 8(8): E1241. https://www.ncbi.nlm.nih.gov/pmc/articles/PMC7464643/ [Internet].
[http://dx.doi.org/10.3390/microorganisms8081241] [PMID: 32824085]

[47]   Koricha AD, Han D-Y, Bacha K, Bai F-Y. Diversity and distribution of yeasts in indigenous fermented foods and beverages of Ethiopia. J Sci Food Agric 2020; 100(9): 3630-8.
[http://dx.doi.org/10.1002/jsfa.10391] [PMID: 32201947]

[48]   Wang H, Wang C, Guo M. Autogenic successions of bacteria and fungi in kefir grains from different origins when sub-cultured in goat milk Food Res Int 138(Pt B): 109784.2020;
[http://dx.doi.org/10.1016/j.foodres.2020.109784]

[49]   Jood I, Hoff JW, Setati ME. Evaluating fermentation characteristics of Kazachstania spp. and their potential influence on wine quality. World J Microbiol Biotechnol 2017; 33(7): 129.
[http://dx.doi.org/10.1007/s11274-017-2299-1] [PMID: 28585169]

[50]   Santos MC, Golt C, Joerger RD, Mechor GD, Mourão GB, Kung L Jr. Identification of the major yeasts isolated from high moisture corn and corn silages in the United States using genetic and biochemical methods. J Dairy Sci 2017; 100(2): 1151-60.
[http://dx.doi.org/10.3168/jds.2016-11450] [PMID: 27889126]

[51]   Carbonetto B, Ramsayer J, Nidelet T, Legrand J, Sicard D. Bakery yeasts, a new model for studies in ecology and evolution. Yeast 2018; 35(11): 591-603.
[http://dx.doi.org/10.1002/yea.3350] [PMID: 30070036]

[52]   Vrancken G, De Vuyst L, Van der Meulen R, Huys G, Vandamme P, Daniel H-M. Yeast species composition differs between artisan bakery and spontaneous laboratory sourdoughs. FEMS Yeast Res

2010; 10(4): 471-81.
[http://dx.doi.org/10.1111/j.1567-1364.2010.00621.x] [PMID: 20384785]

[53] Chiva R, Celador-Lera L, Uña JA, *et al.* Yeast biodiversity in fermented doughs and raw cereal matrices and the study of technological traits of selected strains isolated in Spain. Microorganisms 2020; 9(1): 47.
[http://dx.doi.org/10.3390/microorganisms9010047] [PMID: 33375367]

[54] Reese AT, Madden AA, Joossens M, Lacaze G, Dunn RR. Influences of Ingredients and Bakers on the Bacteria and Fungi in Sourdough Starters and Bread. MSphere 2020; 5(1): e00950-19.
https://msphere.asm.org/content/5/1/e00950-19 [Internet].
[http://dx.doi.org/10.1128/mSphere.00950-19] [PMID: 31941818]

[55] Madden AA, Epps MJ, Fukami T, *et al.* The ecology of insect–yeast relationships and its relevance to human industry. Proc R Soc B. 285: 20172733.
[http://dx.doi.org/10.1098/rspb.2017.2733]

[56] Stefanini I. Yeast-insect associations: It takes guts. Yeast 2018; 35(4): 315-30.
[http://dx.doi.org/10.1002/yea.3309] [PMID: 29363168]

[57] Francesca N, Canale DE, Settanni L, Moschetti G. Dissemination of wine-related yeasts by migratory birds. Environ Microbiol Rep 2012; 4(1): 105-12.
[http://dx.doi.org/10.1111/j.1758-2229.2011.00310.x] [PMID: 23757236]

[58] Minervini F, Lattanzi A, De Angelis M, Celano G, Gobbetti M. House microbiotas as sources of lactic acid bacteria and yeasts in traditional Italian sourdoughs. Food Microbiol 2015; 52: 66-76.
[http://dx.doi.org/10.1016/j.fm.2015.06.009] [PMID: 26338118]

[59] Yang H, Liu T, Zhang G, He G. Intraspecific diversity and fermentative properties of *Saccharomyces cerevisiae* from Chinese traditional sourdough. Lebensm Wiss Technol 2020; 124: 109195.
[http://dx.doi.org/10.1016/j.lwt.2020.109195]

[60] Palla M, Agnolucci M, Calzone A, *et al.* Exploitation of autochthonous Tuscan sourdough yeasts as potential starters. Int J Food Microbiol 2019; 302: 59-68.
[http://dx.doi.org/10.1016/j.ijfoodmicro.2018.08.004] [PMID: 30115373]

[61] Wolfe BE, Dutton RJ. Fermented foods as experimentally tractable microbial ecosystems. Cell 2015; 161(1): 49-55.
[http://dx.doi.org/10.1016/j.cell.2015.02.034] [PMID: 25815984]

[62] Blasche S, Kim Y, Oliveira AP, Patil KR. Model microbial communities for ecosystems biology. Curr Opin Syst Biol 2017; 6: 51-7.
[http://dx.doi.org/10.1016/j.coisb.2017.09.002]

[63] Ley RE, Peterson DA, Gordon JI. Ecological and evolutionary forces shaping microbial diversity in the human intestine. Cell 2006; 124(4): 837-48.
[http://dx.doi.org/10.1016/j.cell.2006.02.017] [PMID: 16497592]

[64] Gänzle MG, Zheng J. Lifestyles of sourdough lactobacilli - Do they matter for microbial ecology and bread quality? Int J Food Microbiol 2019; 302: 15-23.
[http://dx.doi.org/10.1016/j.ijfoodmicro.2018.08.019] [PMID: 30172443]

[65] Gänzle M, Ripari V. Composition and function of sourdough microbiota: From ecological theory to bread quality. Int J Food Microbiol 2016; 239: 19-25.
[http://dx.doi.org/10.1016/j.ijfoodmicro.2016.05.004] [PMID: 27240932]

[66] Ercolini D, Pontonio E, De Filippis F, *et al.* Microbial ecology dynamics during rye and wheat sourdough preparation. Appl Environ Microbiol 2013; 79(24): 7827-36.
[http://dx.doi.org/10.1128/AEM.02955-13] [PMID: 24096427]

[67] Venturi M, Guerrini S, Vincenzini M. Stable and non-competitive association of *Saccharomyces cerevisiae*, *Candida milleri* and *Lactobacillus sanfranciscensis* during manufacture of two traditional sourdough baked goods. Food Microbiol 2012; 31(1): 107-15.

[http://dx.doi.org/10.1016/j.fm.2012.02.011] [PMID: 22475948]

[68]   Howell KS, Cozzolino D, Bartowsky EJ, Fleet GH, Henschke PA. Metabolic profiling as a tool for revealing Saccharomyces interactions during wine fermentation. FEMS Yeast Res 2006; 6(1): 91-101.
[http://dx.doi.org/10.1111/j.1567-1364.2005.00010.x] [PMID: 16423074]

[69]   Goddard MR, Greig D. *Saccharomyces cerevisiae:* a nomadic yeast with no niche? FEMS Yeast Res 2015; 15(3): fov009.http://academic.oup.com/femsyr/article/doi/10.1093/femsyr/fov009/545611/Saccharomyces-cerevisiae-a-nomadic-yeast-with-no [Internet].
[http://dx.doi.org/10.1093/femsyr/fov009] [PMID: 25725024]

[70]   Aslankoohi E, Zhu B, Rezaei MN, *et al.* Dynamics of the *Saccharomyces cerevisiae* transcriptome during bread dough fermentation. Appl Environ Microbiol 2013; 79(23): 7325-33.
[http://dx.doi.org/10.1128/AEM.02649-13] [PMID: 24056467]

[71]   Miller MB, Bassler BL. Quorum sensing in bacteria. Annu Rev Microbiol 2001; 55(1): 165-99.
[http://dx.doi.org/10.1146/annurev.micro.55.1.165] [PMID: 11544353]

[72]   Avbelj M, Zupan J, Raspor P. Quorum-sensing in yeast and its potential in wine making. Appl Microbiol Biotechnol 2016; 100(18): 7841-52.
[http://dx.doi.org/10.1007/s00253-016-7758-3] [PMID: 27507587]

[73]   Chen H, Fink GR. Feedback control of morphogenesis in fungi by aromatic alcohols. Genes Dev 2006; 20(9): 1150-61.
[http://dx.doi.org/10.1101/gad.1411806] [PMID: 16618799]

[74]   Smukalla S, Caldara M, Pochet N, *et al.* FLO1 is a variable green beard gene that drives biofilm-like cooperation in budding yeast. Cell 2008; 135(4): 726-37.
[http://dx.doi.org/10.1016/j.cell.2008.09.037] [PMID: 19013280]

[75]   González B, Vázquez J, Cullen PJ, Mas A, Beltran G, Torija M-J. Aromatic amino acid-derived compounds induce morphological changes and modulate the cell growth of wine yeast species. Front Microbiol 2018; 9: 670.
[http://dx.doi.org/10.3389/fmicb.2018.00670] [PMID: 29696002]

[76]   Nissen P, Nielsen D, Arneborg N. Viable Saccharomyces cerevisiae cells at high concentrations cause early growth arrest of non-Saccharomyces yeasts in mixed cultures by a cell-cell contact-mediated mechanism. Yeast 2003; 20(4): 331-41.
[http://dx.doi.org/10.1002/yea.965] [PMID: 12627399]

[77]   Winters M, Arneborg N, Appels R, Howell K. Can community-based signalling behaviour in Saccharomyces cerevisiae be called quorum sensing? A critical review of the literature. FEMS Yeast Res 2019; 19(5): foz046.
[http://dx.doi.org/10.1093/femsyr/foz046] [PMID: 31271429]

[78]   Schmitt MJ, Breinig F. The viral killer system in yeast: from molecular biology to application. FEMS Microbiol Rev 2002; 26(3): 257-76.
[http://dx.doi.org/10.1111/j.1574-6976.2002.tb00614.x] [PMID: 12165427]

[79]   Pieczynska MD, de Visser JAGM, Korona R. Incidence of symbiotic dsRNA 'killer' viruses in wild and domesticated yeast. FEMS Yeast Res 2013; 13(8): 856-9.
[http://dx.doi.org/10.1111/1567-1364.12086] [PMID: 24028530]

[80]   Bortol A, Nudel C, Fraile E, *et al.* Isolation of yeast with killer activity and its breeding with an industrial baking strain by protoplast fusion. Appl Microbiol Biotechnol 1986; 24(5): 414-6.
[http://dx.doi.org/10.1007/BF00294599]

[81]   Perez MF, Contreras L, Garnica NM, *et al.* Native killer yeasts as biocontrol agents of postharvest fungal diseases in lemons. PLoS One 2016; 11(10): e0165590.
[http://dx.doi.org/10.1371/journal.pone.0165590] [PMID: 27792761]

[82]   Renault PE, Albertin W, Bely M. An innovative tool reveals interaction mechanisms among yeast populations under oenological conditions. Appl Microbiol Biotechnol 2013; 97(9): 4105-19.

[http://dx.doi.org/10.1007/s00253-012-4660-5] [PMID: 23292550]

[83]   Albergaria H, Francisco D, Gori K, Arneborg N, Gírio F. *Saccharomyces cerevisiae* CCMI 885 secretes peptides that inhibit the growth of some non-Saccharomyces wine-related strains. Appl Microbiol Biotechnol 2010; 86(3): 965-72.
[http://dx.doi.org/10.1007/s00253-009-2409-6] [PMID: 20039034]

[84]   Albergaria H, Arneborg N. Dominance of Saccharomyces cerevisiae in alcoholic fermentation processes: role of physiological fitness and microbial interactions. Appl Microbiol Biotechnol 2016; 100(5): 2035-46.
[http://dx.doi.org/10.1007/s00253-015-7255-0] [PMID: 26728020]

[85]   Arora K, Ameur H, Polo A, Di Cagno R, Rizzello CG, Gobbetti M. Thirty years of knowledge on sourdough fermentation: A systematic review. Trends Food Sci Technol 2021; 108: 71-83.
[http://dx.doi.org/10.1016/j.tifs.2020.12.008]

[86]   Field D, Ross RP, Hill C. Developing bacteriocins of lactic acid bacteria into next generation biopreservatives. Curr Opin Food Sci 2018; 20: 1-6.
[http://dx.doi.org/10.1016/j.cofs.2018.02.004]

[87]   Riley MA, Wertz JE. Bacteriocins: evolution, ecology, and application. Annu Rev Microbiol 2002; 56(1): 117-37.
[http://dx.doi.org/10.1146/annurev.micro.56.012302.161024] [PMID: 12142491]

[88]   Hammes WP, Gänzle MG. Sourdough breads and related products. In: Wood BJB, Ed. Microbiology of Fermented Foods. Boston, MA: Springer US 1998; pp. 199-216.http://link.springer.com/10.1007/978-1-4613-0309-1_8 [Internet]
[http://dx.doi.org/10.1007/978-1-4613-0309-1_8]

[89]   Messens W, De VL. Inhibitory substances produced by Lactobacilli isolated from sourdoughs--a review. Int J Food Microbiol 2002; 72(1-2): 31-43.
[http://dx.doi.org/10.1016/S0168-1605(01)00611-0] [PMID: 11843411]

[90]   Bell PJL, Higgins VJ, Attfield PV. Comparison of fermentative capacities of industrial baking and wild-type yeasts of the species Saccharomyces cerevisiae in different sugar media. Lett Appl Microbiol 2001; 32(4): 224-9.
[http://dx.doi.org/10.1046/j.1472-765X.2001.00894.x] [PMID: 11298930]

[91]   Daniel H-M, Moons M-C, Huret S, Vrancken G, De Vuyst L. Wickerhamomyces anomalus in the sourdough microbial ecosystem. Antonie van Leeuwenhoek 2011; 99(1): 63-73.
[http://dx.doi.org/10.1007/s10482-010-9517-2] [PMID: 20963492]

[92]   Johansson L, Nikulin J, Juvonen R, *et al.* Sourdough cultures as reservoirs of maltose-negative yeasts for low-alcohol beer brewing. Food Microbiol 2021; 94: 103629.
[http://dx.doi.org/10.1016/j.fm.2020.103629] [PMID: 33279061]

[93]   Jacques N, Sarilar V, Urien C, *et al.* Three novel ascomycetous yeast species of the Kazachstania clade, Kazachstania saulgeensis sp. nov., Kazachstaniaserrabonitensis sp. nov. and Kazachstania australis sp. nov. Reassignment of *Candida humilis* to Kazachstania humilis f.a. comb. nov. and Candida pseudohumilis to Kazachstania pseudohumilis f.a. comb. nov. Int J Syst Evol Microbiol 2016; 66(12): 5192-200.
[http://dx.doi.org/10.1099/ijsem.0.001495] [PMID: 27902197]

[94]   Vigentini I, Antoniani D, Roscini L, *et al. Candida milleri* species reveals intraspecific genetic and metabolic polymorphisms. Food Microbiol 2014; 42: 72-81.
[http://dx.doi.org/10.1016/j.fm.2014.02.011] [PMID: 24929720]

[95]   Winters M, Panayotides D, Bayrak M, *et al.* Defined co-cultures of yeast and bacteria modify the aroma, crumb and sensory properties of bread. J Appl Microbiol 2019; 127(3): 778-93.
[http://dx.doi.org/10.1111/jam.14349] [PMID: 31211891]

[96]   Nilsson U, Öste R, Jägerstad M. Cereal fructans: Hydrolysis by yeast invertase, *in vitro* and during

fermentation. J Cereal Sci 1987; 6(1): 53-60.
[http://dx.doi.org/10.1016/S0733-5210(87)80040-1]

[97]   Laurent J, Timmermans E, Struyf N, Verstrepen KJ, Courtin CM. Variability in yeast invertase activity determines the extent of fructan hydrolysis during wheat dough fermentation and final FODMAP levels in bread. Int J Food Microbiol 2020; 326: 108648.
[http://dx.doi.org/10.1016/j.ijfoodmicro.2020.108648] [PMID: 32387971]

[98]   Struyf N, Laurent J, Verspreet J, Verstrepen KJ, Courtin CM. Substrate-Limited Saccharomyces cerevisiae Yeast Strains Allow Control of Fermentation during breadmaking. J Agric Food Chem 2017; 65(16): 3368-77.
[http://dx.doi.org/10.1021/acs.jafc.7b00313] [PMID: 28367622]

[99]   Fraberger V, Call L-M, Domig KJ, D'Amico S. Applicability of Yeast Fermentation to Reduce Fructans and Other FODMAPs. Nutrients 2018; 10(9): 1247.
[http://dx.doi.org/10.3390/nu10091247] [PMID: 30200589]

[100]   Loponen J, Gänzle MG. Use of Sourdough in Low FODMAP Baking. Foods 2018; 7(7): 96.
[http://dx.doi.org/10.3390/foods7070096] [PMID: 29932101]

[101]   Pico J, Bernal J, Gómez M. Wheat bread aroma compounds in crumb and crust: A review. Food Res Int 2015; 75: 200-15.
[http://dx.doi.org/10.1016/j.foodres.2015.05.051] [PMID: 28454949]

[102]   Pétel C, Onno B, Prost C. Sourdough volatile compounds and their contribution to bread: A review. Trends Food Sci Technol 2017; 59: 105-23.
[http://dx.doi.org/10.1016/j.tifs.2016.10.015]

[103]   Aslankoohi E, Herrera-Malaver B, Rezaei MN, Steensels J, Courtin CM, Verstrepen KJ. Non-Conventional Yeast Strains Increase the Aroma Complexity of Bread. In: Louis EJ, Ed. PLoS ONE. 2016; 11: p. (10)e0165126.
[http://dx.doi.org/10.1371/journal.pone.0165126]

[104]   Zehentbauer G, Grosch W. Crust Aroma of Baguettes II. Dependence of the Concentrations of Key Odorants on Yeast Level and Dough Processing. J Cereal Sci 1998; 28(1): 93-6.
[http://dx.doi.org/10.1006/jcrs.1998.0183]

[105]   Poinot P, Arvisenet G, Grua-Priol J, *et al.* Influence of formulation and process on the aromatic profile and physical characteristics of bread. J Cereal Sci 2008; 48(3): 686-97.
[http://dx.doi.org/10.1016/j.jcs.2008.03.002]

[106]   Birch AN, Petersen MA, Arneborg N, Hansen ÅS. Influence of commercial baker's yeasts on bread aroma profiles. Food Res Int 2013; 52(1): 160-6.
[http://dx.doi.org/10.1016/j.foodres.2013.03.011]

[107]   Boudaoud S, Aouf C, Devillers H, Sicard D, Segond D. Sourdough yeast-bacteria interactions can change ferulic acid metabolism during fermentation. Food Microbiol 2021; 98: 103790.
[http://dx.doi.org/10.1016/j.fm.2021.103790] [PMID: 33875218]

[108]   De Vuyst L, Harth H, Van Kerrebroeck S, Leroy F. Yeast diversity of sourdoughs and associated metabolic properties and functionalities. Int J Food Microbiol 2016; 239: 26-34.
[http://dx.doi.org/10.1016/j.ijfoodmicro.2016.07.018] [PMID: 27470533]

[109]   Carbonetto B, Nidelet T, Guezenec S, Perez M, Segond D, Sicard D. Interactions between *Kazachstania humilis* Yeast Species and Lactic Acid Bacteria in Sourdough. Microorganisms 2020; 8(2): 240.
[http://dx.doi.org/10.3390/microorganisms8020240] [PMID: 32053958]

[110]   Rogalski E, Ehrmann MA, Vogel RF. Intraspecies diversity and genome-phenotype-associations in Fructilactobacillus sanfranciscensis. Microbiol Res 2021; 243: 126625.
[http://dx.doi.org/10.1016/j.micres.2020.126625] [PMID: 33129664]

[111]   De Vuyst L, Neysens P. The sourdough microflora: biodiversity and metabolic interactions. Trends

Food Sci Technol 2005; 16(1–3): 43-56.
[http://dx.doi.org/10.1016/j.tifs.2004.02.012]

[112]   Rogalski E, Ehrmann MA, Vogel RF. Strain-specific interaction of *Fructilactobacillus sanfranciscensis* with yeasts in the sourdough fermentation. Eur Food Res Technol 2021; 247: 1437-47.
[http://dx.doi.org/10.1007/s00217-021-03722-0]

[113]   Rogalski E, Ehrmann MA, Vogel RF. Role of *Kazachstania humilis* and *Saccharomyces cerevisiae* in the strain-specific assertiveness of *Fructilactobacillus sanfranciscensis* strains in rye sourdough. Eur Food Res Technol 2020; 246(9): 1817-27.
[http://dx.doi.org/10.1007/s00217-020-03535-7]

[114]   Blateyron L, Sablayrolles JM. Stuck and slow fermentations in enology: statistical study of causes and effectiveness of combined additions of oxygen and diammonium phosphate. J Biosci Bioeng 2001; 91(2): 184-9.
[http://dx.doi.org/10.1016/S1389-1723(01)80063-3] [PMID: 16232972]

[115]   Minervini F, De Angelis M, Di Cagno R, Gobbetti M. Ecological parameters influencing microbial diversity and stability of traditional sourdough. Int J Food Microbiol 2014; 171: 136-46.
[http://dx.doi.org/10.1016/j.ijfoodmicro.2013.11.021] [PMID: 24355817]

[116]   Thiele C, Grassl S, Gänzle M. Gluten hydrolysis and depolymerization during sourdough fermentation. J Agric Food Chem 2004; 52(5): 1307-14.
[http://dx.doi.org/10.1021/jf034470z] [PMID: 14995138]

[117]   Gänzle MG, Loponen J, Gobbetti M. Proteolysis in sourdough fermentations: mechanisms and potential for improved bread quality. Trends Food Sci Technol 2008; 19(10): 513-21.
[http://dx.doi.org/10.1016/j.tifs.2008.04.002]

[118]   Thiele C, Gänzle MG, Vogel RF. Contribution of sourdough lactobacilli, yeast, and cereal enzymes to the generation of amino acids in dough relevant for bread flavor. Cereal Chem 2002; 79(1): 45-51.
[http://dx.doi.org/10.1094/CCHEM.2002.79.1.45]

[119]   Zhao CJ, Schieber A, Gänzle MG. Formation of taste-active amino acids, amino acid derivatives and peptides in food fermentations - A review. Food Res Int 2016; 89(Pt 1): 39-47.
[http://dx.doi.org/10.1016/j.foodres.2016.08.042] [PMID: 28460929]

[120]   Di Cagno R, De Angelis M, Auricchio S, *et al.* Sourdough bread made from wheat and nontoxic flours and started with selected lactobacilli is tolerated in celiac sprue patients. Appl Environ Microbiol 2004; 70(2): 1088-96.
[http://dx.doi.org/10.1128/AEM.70.2.1088-1096.2004] [PMID: 14766592]

[121]   Brandt MJ, Hammes WP, Gänzle MG. Effects of process parameters on growth and metabolism of *Lactobacillus sanfranciscensis* and *Candida humilis* during rye sourdough fermentation. Eur Food Res Technol 2004; 218(4): 333-8.
[http://dx.doi.org/10.1007/s00217-003-0867-0]

[122]   Ponomarova O, Gabrielli N, Sévin DC, *et al.* Yeast Creates a Niche for Symbiotic Lactic Acid Bacteria through Nitrogen Overflow. Cell Syst 2017; 5(4): 345-357.e6.
[http://dx.doi.org/10.1016/j.cels.2017.09.002] [PMID: 28964698]

[123]   Morin M, Pierce EC, Dutton RJ. Changes in the genetic requirements for microbial interactions with increasing community complexity. eLife 2018; 7: e37072.
[http://dx.doi.org/10.7554/eLife.37072] [PMID: 30211673]

[124]   Steensels J, Snoek T, Meersman E, Picca Nicolino M, Voordeckers K, Verstrepen KJ. Improving industrial yeast strains: exploiting natural and artificial diversity. FEMS Microbiol Rev 2014; 38(5): 947-95.
[http://dx.doi.org/10.1111/1574-6976.12073] [PMID: 24724938]

[125]   Lahue C, Madden AA, Dunn RR, Smukowski Heil C. History and Domestication of *Saccharomyces*

cerevisiae in Bread Baking. Front Genet 2020; 11: 584718. https://www.frontiersin.org/articles/10.3389/fgene.2020.584718/full#B72 [Internet]. [http://dx.doi.org/10.3389/fgene.2020.584718] [PMID: 33262788]

[126]   Lewis JG, Learmonth RP, Attfield PV, Watson K. Stress co-tolerance and trehalose content in baking strains of Saccharomyces cerevisiae. J Ind Microbiol Biotechnol 1997; 18(1): 30-6.
[http://dx.doi.org/10.1038/sj.jim.2900347] [PMID: 9079286]

[127]   Hohmann S, Krantz M, Nordlander B. Chapter Two - Yeast Osmoregulation. In: Häussinger D, Sies H, Eds. Methods in Enzymology. Academic Press 2007; pp. 29-45. https://www.sciencedirect.com/science/article/pii/S0076687907280024 [Internet] [Osmosensing and Osmosignaling; vol. 428]

[128]   Baik M-Y, Chinachoti P. Effects of Glycerol and Moisture Redistribution on Mechanical Properties of White Bread. Cereal Chem 2002; 79(3): 376-82.
[http://dx.doi.org/10.1094/CCHEM.2002.79.3.376]

[129]   Barrett AH, Cardello AV, Mair L, *et al.* Textural Optimization of Shelf-Stable Bread: Effects of Glycerol Content and Dough-Forming Technique. Cereal Chem 2000; 77(2): 169-76.
[http://dx.doi.org/10.1094/CCHEM.2000.77.2.169]

[130]   Aslankoohi E, Rezaei MN, Vervoort Y, Courtin CM, Verstrepen KJ. Glycerol production by fermenting yeast cells is essential for optimal bread dough fermentation. PLoS One 2015; 10(3): e0119364. https://www.ncbi.nlm.nih.gov/pmc/articles/PMC4357469/ [Internet].
[http://dx.doi.org/10.1371/journal.pone.0119364] [PMID: 25764309]

[131]   Heitmann M, Zannini E, Arendt E. Impact of Saccharomyces cerevisiae metabolites produced during fermentation on bread quality parameters: A review. Crit Rev Food Sci Nutr 2018; 58(7): 1152-64.
[http://dx.doi.org/10.1080/10408398.2016.1244153] [PMID: 27874287]

[132]   Hohmann S. Control of high osmolarity signalling in the yeast *Saccharomyces cerevisiae*. FEBS Lett 2009; 583(24): 4025-9.
[http://dx.doi.org/10.1016/j.febslet.2009.10.069] [PMID: 19878680]

[133]   Brewster JL, Gustin MC. Hog1: 20 years of discovery and impact. Sci Signal 2014; 7(343): re7-7.
[http://dx.doi.org/10.1126/scisignal.2005458] [PMID: 25227612]

[134]   Albertyn J, Hohmann S, Thevelein JM, Prior BA. GPD1, which encodes glycerol-3-phosphate dehydrogenase, is essential for growth under osmotic stress in *Saccharomyces cerevisiae*, and its expression is regulated by the high-osmolarity glycerol response pathway. Mol Cell Biol 1994; 14(6): 4135-44.
[PMID: 8196651]

[135]   Panadero J, Pallotti C, Rodríguez-Vargas S, Randez-Gil F, Prieto JA. A downshift in temperature activates the high osmolarity glycerol (HOG) pathway, which determines freeze tolerance in Saccharomyces cerevisiae. J Biol Chem 2006; 281(8): 4638-45.
[http://dx.doi.org/10.1074/jbc.M512736200] [PMID: 16371351]

[136]   Li Y, Chen W, Shi Y, Liang X. Molecular cloning and evolutionary analysis of the HOG-signaling pathway genes from Saccharomyces cerevisiae rice wine isolates. Biochem Genet 2013; 51(3-4): 296-305.
[http://dx.doi.org/10.1007/s10528-012-9563-8] [PMID: 23338673]

[137]   Hernandez-Lopez MJ, Prieto JA, Randez-Gil F. Osmotolerance and leavening ability in sweet and frozen sweet dough. Comparative analysis between Torulaspora delbrueckii and Saccharomyces cerevisiae baker's yeast strains. Antonie van Leeuwenhoek 2003; 84(2): 125-34.
[http://dx.doi.org/10.1023/A:1025413520192] [PMID: 14533716]

[138]   Zhou N, Schifferdecker AJ, Gamero A, *et al. Kazachstania gamospora* and *Wickerhamomyces subpelliculosus*: Two alternative baker's yeasts in the modern bakery. Int J Food Microbiol 2017; 250: 45-58.
[http://dx.doi.org/10.1016/j.ijfoodmicro.2017.03.013] [PMID: 28365494]

[139] Luo W, Sun D-W, Zhu Z, Wang Q-J. Improving freeze tolerance of yeast and dough properties for enhancing frozen dough quality - A review of effective methods. Trends Food Sci Technol 2018; 72: 25-33.
[http://dx.doi.org/10.1016/j.tifs.2017.11.017]

[140] Takagi H. Proline as a stress protectant in yeast: physiological functions, metabolic regulations, and biotechnological applications. Appl Microbiol Biotechnol 2008; 81(2): 211-23.
[http://dx.doi.org/10.1007/s00253-008-1698-5] [PMID: 18802692]

[141] Tsolmonbaatar A, Hashida K, Sugimoto Y, Watanabe D, Furukawa S, Takagi H. Isolation of baker's yeast mutants with proline accumulation that showed enhanced tolerance to baking-associated stresses. Int J Food Microbiol 2016; 238: 233-40.
[http://dx.doi.org/10.1016/j.ijfoodmicro.2016.09.015] [PMID: 27672730]

[142] Sasano Y, Haitani Y, Ohtsu I, Shima J, Takagi H. Proline accumulation in baker's yeast enhances high-sucrose stress tolerance and fermentation ability in sweet dough. Int J Food Microbiol 2012; 152(1-2): 40-3.
[http://dx.doi.org/10.1016/j.ijfoodmicro.2011.10.004] [PMID: 22041027]

[143] Sasano Y, Haitani Y, Hashida K, Ohtsu I, Shima J, Takagi H. Simultaneous accumulation of proline and trehalose in industrial baker's yeast enhances fermentation ability in frozen dough. J Biosci Bioeng 2012; 113(5): 592-5.
[http://dx.doi.org/10.1016/j.jbiosc.2011.12.018] [PMID: 22280966]

[144] Sasano Y, Haitani Y, Hashida K, Ohtsu I, Shima J, Takagi H. Enhancement of the proline and nitric oxide synthetic pathway improves fermentation ability under multiple baking-associated stress conditions in industrial baker's yeast. Microb Cell Fact 2012; 11: 40.
[http://dx.doi.org/10.1186/1475-2859-11-40] [PMID: 22462683]

[145] Sasano Y, Takahashi S, Shima J, Takagi H. Antioxidant N-acetyltransferase Mpr1/2 of industrial baker's yeast enhances fermentation ability after air-drying stress in bread dough. Int J Food Microbiol 2010; 138(1-2): 181-5.
[http://dx.doi.org/10.1016/j.ijfoodmicro.2010.01.001] [PMID: 20096471]

[146] Kopp M, Müller H, Holzer H. Molecular analysis of the neutral trehalase gene from Saccharomyces cerevisiae. J Biol Chem 1993; 268(7): 4766-74.
[http://dx.doi.org/10.1016/S0021-9258(18)53463-3] [PMID: 8444853]

[147] Alizadeh P, Klionsky DJ. Purification and biochemical characterization of the ATH1 gene product, vacuolar acid trehalase, from Saccharomyces cerevisiae. FEBS Lett 1996; 391(3): 273-8.
[http://dx.doi.org/10.1016/0014-5793(96)00751-X] [PMID: 8764988]

[148] Shima J, Hino A, Yamada-Iyo C, *et al.* Stress tolerance in doughs of *Saccharomyces cerevisiae* trehalase mutants derived from commercial Baker's yeast. Appl Environ Microbiol 1999; 65(7): 2841-6.
[http://dx.doi.org/10.1128/AEM.65.7.2841-2846.1999] [PMID: 10388673]

[149] Dong J, Chen D, Wang G, *et al.* Improving freeze-tolerance of baker's yeast through seamless gene deletion of NTH1 and PUT1. J Ind Microbiol Biotechnol 2016; 43(6): 817-28.
[http://dx.doi.org/10.1007/s10295-016-1753-7] [PMID: 26965428]

[150] Sun X, Zhang C-Y, Wu M-Y, *et al.* MAL62 overexpression and NTH1 deletion enhance the freezing tolerance and fermentation capacity of the baker's yeast in lean dough. Microb Cell Fact 2016; 15(1): 54.
[http://dx.doi.org/10.1186/s12934-016-0453-3] [PMID: 27039899]

[151] Kim J, Alizadeh P, Harding T, Hefner-Gravink A, Klionsky DJ. Disruption of the yeast ATH1 gene confers better survival after dehydration, freezing, and ethanol shock: potential commercial applications. Appl Environ Microbiol 1996; 62(5): 1563-9.
[http://dx.doi.org/10.1128/aem.62.5.1563-1569.1996] [PMID: 8633854]

[152] Sun X, Zhang J, Fan Z-H, *et al.* MAL62 overexpression enhances uridine diphosphoglucose-dependent trehalose synthesis and glycerol metabolism for cryoprotection of baker's yeast in lean dough. Microb Cell Fact 2020; 19(1): 196. https://www.ncbi.nlm.nih.gov/pmc/articles/PMC7574194/ [Internet].
[http://dx.doi.org/10.1186/s12934-020-01454-6] [PMID: 33076920]

[153] Panadero J, Hernández-López MJ, Prieto JA, Randez-Gil F. Overexpression of the calcineurin target CRZ1 provides freeze tolerance and enhances the fermentative capacity of baker's yeast. Appl Environ Microbiol 2007; 73(15): 4824-31.
[http://dx.doi.org/10.1128/AEM.02651-06] [PMID: 17557846]

[154] Meng L, Yang X, Lin X, Jiang H-Y, Hu X-P, Liu S-X. Effect of overexpression of SNF1 on the transcriptional and metabolic landscape of baker's yeast under freezing stress. Microb Cell Fact 2021; 20(1): 10.
[http://dx.doi.org/10.1186/s12934-020-01503-0] [PMID: 33413411]

[155] Aguilera J, Andreu P, Randez-Gil F, Prieto JA. Adaptive evolution of baker's yeast in a dough-like environment enhances freeze and salinity tolerance. Microb Biotechnol 2010; 3(2): 210-21.
[http://dx.doi.org/10.1111/j.1751-7915.2009.00136.x] [PMID: 21255321]

[156] Kaminski Strauss S, Schirman D, Jona G, *et al.* Evolthon: A community endeavor to evolve lab evolution. PLoS Biol 2019; 17(3): e3000182.
[http://dx.doi.org/10.1371/journal.pbio.3000182] [PMID: 30925180]

[157] Teunissen A, Dumortier F, Gorwa M-F, *et al.* Isolation and characterization of a freeze-tolerant diploid derivative of an industrial baker's yeast strain and its use in frozen doughs. Appl Environ Microbiol 2002; 68(10): 4780-7.
[http://dx.doi.org/10.1128/AEM.68.10.4780-4787.2002] [PMID: 12324320]

[158] Matsutani K, Fukuda Y, Murata K, Kimura A, Nakamura I, Yajima N. Physical and biochemical properties of freeze-tolerant mutants of a yeast Saccharomyces cerevisiae. J Ferment Bioeng 1990; 70(4): 275-6.
[http://dx.doi.org/10.1016/0922-338X(90)90063-3]

[159] Hahn Y-S, Kawai H. Isolation and Characterization of Freeze-tolerant Yeasts from Nature Available for the Frozen-dough Method. Agric Biol Chem 1990; 54(3): 829-31.

[160] Hino A, Takano H, Tanaka Y. New freeze-tolerant yeast for frozen dough preparations. Cereal Chem 1987; 64(4): 269-75.

[161] Alves-Araújo C, Almeida MJ, Sousa MJ, Leão C. Freeze tolerance of the yeast *Torulaspora delbrueckii*: cellular and biochemical basis. FEMS Microbiol Lett 2004; 240(1): 7-14.
[http://dx.doi.org/10.1016/j.femsle.2004.09.008] [PMID: 15500973]

[162] Von Gastrow L, Gianottib A, Vernocchic P, Serrazanettid DI, Sicard D. Sourdough yeasts diversity, physiology and evolution. Handbook on sourdough biotechnology, second edition.

CHAPTER 12

Biotechnological Applications of Oleaginous Yeasts

Yasmi Louhasakul[1] and **Benjamas Cheirsilp**[2,*]

[1] *Faculty of Science Technology and Agriculture, Yala Rajabhat University, Yala 95000, Thailand*

[2] *Center of Excellence in Innovative Biotechnology for Sustainable Utilization of Bioresources, Faculty of Agro-Industry, Prince of Songkla University, Hat Yai 90110, Thailand*

Abstract: Oleaginous yeasts are potential renewable sources of alternative biofuels due to high lipid contents and fatty acid profiles similar to those of plant oils. To increase the biotechnological potential of oleaginous yeasts, strategic cultivation of them using a wide variety of low-cost materials as substrates has been investigated intensively. Their metabolisms toward various substrates for the synthesis of lipids through *de novo* and *ex novo* processes have been described. In addition, direct transesterification processes that combine cell disruption, lipid extraction, and biodiesel production into a single step are proposed. This chapter thoroughly reviews recent research into the broad characteristics of oleaginous yeasts, the utilization of promising low-cost materials as substrates for yeast cultivation, and direct processing for biodiesel production from yeast lipids.

Keywords: Direct transesterification, Lipid synthesis, Low-cost materials, Oleaginous yeasts.

INTRODUCTION

Oleaginous microorganisms, including bacteria, yeasts, molds and microalgae, are able to accumulate lipids to over 20% of their biomass [1-3]. Microalgae can produce large amounts of lipids and hydrocarbons, but they require sunlight and carbon dioxide from fuel gas [2], a large cultivation area, and a long cultivation period [3]. Most bacterial species are not lipid producers, and they accumulate complex lipids such as polyhydroxyalkanoates that are difficult to extract because these lipoids are generated in the outer membrane [4, 5]. Filamentous fungi can accumulate high intracellular lipid contents composed of triacylglycerols and specific polyunsaturated fatty acids [6], but fungi grow much slower than yeasts and form mycelia that cause high viscosity of the culture medium and decrease oxygen dispersion in the culture [7]. Unlike these micro-

* **Corresponding author Benjamas Cheirsilp:** Center of Excellence in Innovative Biotechnology for Sustainable Utilization of Bioresources, Faculty of Agro-Industry, Prince of Songkla University, Hat Yai 90110, Thailand; Tel: +6674286374; E-mail: benjamas.che@psu.ac.th

Sérgio Luiz Alves Júnior, Helen Treichel, Thiago Olitta Basso and Boris Ugarte Stambuk (Eds.)
All rights reserved-© 2022 Bentham Science Publishers

organisms, oleaginous yeasts (*i.e.*, *Yarrowia*, *Candida*, *Cryptococcus*, *Rhodotorula*, *Rhodosporidium*, *Trichosporon*, *Lipomyces*) have numerous advantages such as fast growth rate and high lipid content with high triacylglycerol fraction [8]. They typically contain a variety of lipids like triacylglycerol, diacylglycerol, monoacylglycerol, fatty acid, steryl ester, free sterol, glycerophospholipid, cardiolipin, sphingolipid, glycolipid, hydrocarbon, long-chain alcohol, wax, polyprenol, and isoprenoid quinone. Besides, four types of lipids can be found in the cytoplasmic membrane, namely glycerophospholipid, sphingolipid, steryl ester from ergosterol, and mono and diacylglycerols [9]. The oleaginous yeasts contain 80–90% triacylglycerol and a minor fraction of steryl esters, accumulated in special cell compartments known as lipid droplets (LDs) or lipid bodies (LBs) [10].

Yeast lipid costs are much higher than those of plant and animal oils because of the cost of nutrient media, and this is one of the major obstacles to large-scale yeast oil deployment [11, 12]. However, yeasts are able to utilize a wide variety of low-cost materials such as nutritional residues from agriculture and industry [13, 14]. The lipid production by oleaginous yeasts using wastes from agro-industry or industry as substrates has been extensively investigated for a variety of candidate substrates, including food waste, chicken tallow, durian peel, sorghum stalk, switchgrass, waste office paper, and crude glycerol, in order to reduce costs and make yeast lipids production economically viable [8, 15 - 20].

Yeast lipid-based biodiesel production through transesterification reactions has been investigated. The conversion of yeast lipids through transesterification reactions comprises numerous steps, including the drying of cells, the disruption of cells, lipids extraction, separation, and transesterification. In fact, several drawbacks directly affect conversion efficiency, such as long processing duration, need for large amounts of solvent, and high total costs [11]. To overcome these drawbacks, direct transesterification has been proposed. This approach excludes the cell drying step and combines the lipid extraction step with the transesterification step. Therefore, the overall production costs of biodiesel from yeast lipids could be reduced, making this process economically feasible [10 - 12]. Recently, many strategies for the direct transesterification of yeast lipids have been successfully demonstrated [13 - 16]. This article first introduces the characteristics of a broad range of oleaginous yeast species and their derived lipids. Recent challenges in the utilization of low-cost materials and wastes as alternative substrates by oleaginous yeasts and their biochemical conversion platform to produce lipids in yeast cells are highlighted. Finally, this chapter provides a current overview of published results on the direct transesterification of yeast lipids as a viable approach to economically viable biodiesel processing.

CHARACTERISTICS OF OLEAGINOUS YEASTS

Oleaginous yeasts are normally nonpathogenic unicellular budding organisms that can accumulate lipids to over 20% of their cell dry weight [17, 18]. Over 70 of the approximately 1,600 yeast species are well-known to be oleaginous [9], in five orders of Ascomycota and Basidiomycota, namely Saccharomycetales, Sporidiobolales, Tremellales, Trichosporonales and Cystobasidiales [18]. The order Saccharomycetales contains the two well-explored oleaginous yeast genera *Yarrowia* and *Lipomyces* and the new oleaginous species *Schwanniomyces etchellsii*, originally *Debaryomyces etchellsii*. The order Sporidiobolales contains *Rhodosporidium* and *Rhodotorula*. The orders Tremellales and Cystobasidiales contain diversified genera such as *Cryptococcus*, *Naganishia*, *Saitozyma*, and *Vishniacozyma*. The order Trichosporonales contains *Cutaneotrichosporon* genus, formerly *Trichosporon* (Table **1**). Depending on the species and the cultivation conditions, the lipid content in yeast biomass can be much improved to as high as 80% of the cell dry weight (Table **1**). In contrast, non-oleaginous yeasts (such as the baker's yeast *Saccharomyces cerevisiae* and the food yeast *Candida utilis*) usually cannot accumulate lipids to exceed 10% of their biomass [19]. Most well-known oleaginous yeasts contain fatty acids in the C16-C18 range of carbon atom counts. For example, *Y. lipolytica* contains mostly palmitic, stearic, oleic, linoleic, linoelaidic, and linolenic acids [20]. Recently, Carranba *et al.* [21] reported that *Y. lipolytica* Po1dL is a good candidate for lipid production in a glucose-based medium under nitrogen-limited conditions. Its lipid content reached up to 61% (w/w). *L. starkeyi* is a sustainable lipid producer, which has the potential to convert a wide variety of carbon sources into lipids in the form of triacylglycerols (TAG) for more than 70% of its dry cell weight [22]. According to Juanssilfero *et al.* [23], *L. starkeyi* NBRC10381 achieves lipid contents as high as 79.6% (w/w) in nitrogen-limited mineral media with glucose as the carbon source.

Table 1. Lipid composition of various oleaginous yeasts.

Yeast Strains	Lipid Components (%)				References
	TAG	DAG	MAG	FFA	
Yarrowia lipolytica	9.33	0.89	-	1.38	[36]
Cryptococcus vishniaccii	63.4	19.63	1.07	-	[37]
Cryptococcus curvatus	91.4	3.3	4.9	0.5	[38]
Rhodosporidium toruloides	92.2	2.7	4.7	0.4	[38]

TAG, DAG, MAG, and FFA are triacylglycerides, diacylglycerides, monoacylglycerides, and free fatty acids, respectively

Likewise, *R. toruloides* and *R. glutinis* naturally produce neutral lipids, mainly triglycerides (TAG) and carotenoids [24, 25]. Saran *et al.* [26] reported the scale-up of *R. toruloides* A29 cultivation using glucose in a 30 L bioreactor, and the lipid content increased up to 53.51% (w/w), leading to a 22-fold increase after scale-up. Maza *et al.* [25] reported that *R. glutinis* R4 showed a remarkably high lipid content of 48.9% (w/w), and the composition of fatty acids was similar to those of plant oils. *C. curvatus* is an excellent oleaginous yeast capable of utilizing both hydrophilic and hydrophobic substrates. It accumulates lipids *via* two alternative pathways, namely *de novo* and *ex novo* lipid synthesis. Under *de novo* lipid synthesis, this yeast accumulated a lipid content of 52.66% (w/w), while under *ex novo* lipid synthesis, the yeast accumulated a much higher lipid content of 70.13% (w/w) [27].

Lipid Synthesis by Oleaginous Yeasts

Two lipid synthesis pathways exist in oleaginous yeasts, with the choice based on the materials used that are either hydrophilic or hydrophobic [8, 28 - 34]. In a "*de novo*" lipids accumulation process, there is the fatty acid precursor synthesis such as malonyl-CoA and acetyl-CoA and their conversion into storage lipid through the Kennedy pathway, which is usually carried out on hydrophilic materials and typically requires nitrogen-limited culture conditions [35]. The rapid decline of the intracellular AMP (adenosine monophosphate) concentration is stimulated by nitrogen exhaustion. In order to maintain the concentration of nitrogen source necessary for cell material synthesis after the limitation of extracellular nitrogen, AMP is converted to IMP (inosine monophosphate) and NH_4^+ ions by AMP-desaminase. Then, the Krebs cycle function is affected by the excessive decrease of AMP concentration; isocitric is not transformed to α-ketoglutaric acid because NAD^+- (and in various cases also $NADP^+$-) isocitrate dehydrogenase that is allosterically activated by intracellular AMP, losses its activity leading to the accumulation of iso-citric acid in the mitochondria. The concentration of iso-citric acid is in balance with the citrate through a reaction catalyzed by isocitrate acotinase. If the concentration of citric acid inside mitochondria increases up to a critical value, the citrate is then released to the cytoplasm by exchanging with malate. Eventually, the citrate is split into acetyl-CoA and oxaloacetate by the ATP-citrate lyase (ATP-CL), the enzyme-key of lipid accumulation in oleaginous yeasts. Then cellular fatty acids are generated from acetyl-CoA by a quasi-inverted ß-oxidation process. Oxaloacetate is converted to malate by malate dehydrogenase and transported back into the mitochondria. NADPH, a coenzyme required for fatty acid synthesis, is produced when malate is oxidized into pyruvate by malic enzyme, which has also been considered crucial for oleaginous microorganisms [19, 35 - 39] (Fig. **1**).

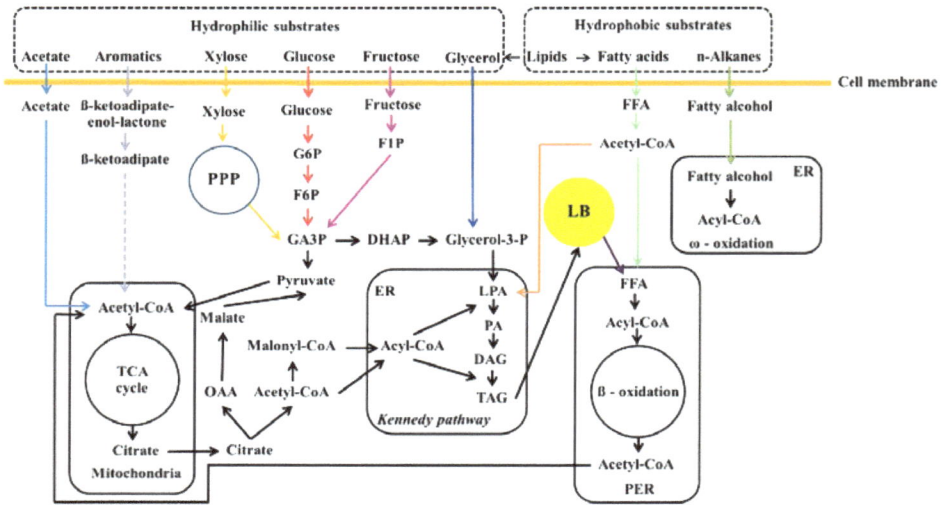

Fig. (1). Lipid metabolic pathway in yeasts. TCA cycle, tricarboxylic acid cycle; PPP, pentose phosphate pathway; ER, endoplasmic reticulum; PER, peroxisome; G6P, glucose-6-phosphate; F6P, fructose-- -phosphate; F1P, fructose-1-phosphate; GA3P, glycerol-3-phosphate; DHAP, dihydroxyacetone phosphate; OAA, oxaloacetate; FFA, free fatty acid; LPA, lysophosphatidic acid; PA, phosphatidic acid; DAG, diacylglycerol; TAG, triacylglycerol; LB, lipid body. **Source:** modified from Spagnuolo *et al.* [50] and Liu *et al.* [51].

In contrast, the "*ex novo*" lipid synthesis includes the consumption of fatty acid, oil, and triacylglycerols from the culture medium and accumulation within the cell. Still, it lacks change or modification of those components based on lipid production through hydrophobic material fermentation. The lipid synthesis is associated with cell growth and does require depletion of the nitrogen source [35]. After the hydrolysis of triacylglycerols by lipase, free fatty acids are transported into the cell and converted into shorter acyl-CoAs and acetyl-CoA by a ß-oxidation process to generate NADH and NADPH for ATP production; or are directly incorporated into triacylglycerols or steryl esters as lipid bodies for storage [38, 39] (Fig. **1**).

Lipid Composition of Oleaginous Yeasts

Oleaginous yeasts store lipids in lipid droplets (LDs) or lipid bodies (LBs). They are vital cytoplasmic organelles that appear almost throughout the cells and contribute to many cellular functions involving lipid storage, membrane synthesis, viral replication, and protein degradation [38, 39]. Yeast lipids can be mainly classified into two groups; storage lipids (neutral lipids) and structure lipids (polar lipids). Neutral lipids are mainly composed of triacylglycerols (TAG), accounting for over 90% of total lipid, and serving as primary carbon and energy storage forms in microorganisms. Other lipid components in minor quantities are free fatty acids, other types of neutral lipids (*i.e.*, monoacylglycerols, MAG; diacylglycerols, DAG; and steryl-esters), and polar lipids and sterols (*i.e.*, sphingolipids, phospholipids, and glycolipids) [40]. Further, the four types of lipids can be found in the cytoplasmic membrane, namely glycerophospholipid, sphingolipid, steryl ester from ergosterol, and mono and diacylglycerols [9]. *Y. lipolytica* also accumulates lipids primarily in the form of TAG. Similarly, the main lipid component TAG exceeded 90% in *R. toruloides* and *C. curvatus*. Neutral lipids are accumulated in lipid bodies of *C. vishniaccii* in the form of TAG (63.4%), DAG (19.63%), and MAG (1.07%) [37], as shown in Table **1**.

The composition of fatty acids in yeast lipids is mainly affected by the cultivation conditions (medium composition, temperature, and aeration rate), and they range from 12 to 22 carbons in length [14]. Fatty acids in yeast are mainly palmitic acid (C16:0), stearic acid (C18:0), oleic acid (C18:1), and linoleic acid (C18:2) [33]. Of these, the most abundant are palmitic and oleic acids. Considering saturated *versus* unsaturated fatty acids, approximately 25–45% are saturated fatty acids while 55-75% are unsaturated. Therefore, the ratio of unsaturated fatty acids to saturated ones in microbial lipids ranges from 1 to 2, similar to plant oils (such as palm oil). The wide ranges of fatty acids found in yeasts and the extents of their lipid content in a glucose-based medium under nitrogen-limited conditions are shown in Table **2**. For instance, the lipids produced by *Y. lipolytica* were mainly C16 and C18 fatty acids. Most of the fatty acids of the microbial lipid were unsaturated and corresponded primarily in oleic, palmitic, and linoleic acids [21]. Similarly, the major fatty acids produced by *L. starkeyi* were oleic acid (C18:1), palmitic acid (C16:0), stearic acid (C18:0), palmitoleic acid (C16:1), and linoleic acid (C18:2). Oleic acid accounted for about 50% of the total fatty acids [23]. Accordingly, the primary fatty acids in the lipids from *R. glutinis* are oleic, palmitic, linoleic, and stearic acids. In addition, these yeast lipids present a relatively higher concentration of unsaturated than saturated fatty acids [24]. Likewise, the major fatty acids of the lipid produced by *C. dermatis* are long-chain fatty acids with 16 and 18 carbon atoms, accounting for >90% of the total fatty acid content. Palmitic, oleic, and linoleic acids are the predominant components. Of these, oleic acid has the most significant proportion of more than 40% [35].

Table 2. Lipid content and fatty acid composition of various oleaginous yeasts.

Yeast Strains	Lipid Content (%)	Fatty Acid Composition (%)						References
		C16:0	C16:1	C18:0	C18:1	C18:2	C18:3	-
Order Saccharomycetales	-	-	-	-	-	-	-	

(Table 2) cont.....

Yeast Strains	Lipid Content (%)	Fatty Acid Composition (%)						References
		C16:0	C16:1	C18:0	C18:1	C18:2	C18:3	-
Yarrowia lipolytica	61	34.0	2.6	16.4	15.4	12.2	18.3	[31]
Lipomyces starkeyi	79.6	26.0	4.0	9.0	27.5	2.0	-	[33]
Debaryomyces etchellsii	22.4	14.6	7.7	3.6	52.6	20.4	-	[34]
Order Sporidiobolales	-	-	-	-	-	-	-	-
Rhodosporidium toruloides	53.51	22.91	1.47	11	45.02	18.1	0.64	[35]
Rhodotorula glutinis	48.93	16.78	1.81	1.35	61.60	11.64	4.23	[36]
Rhodotorula glutinis	42.80	18.17	-	4.27	49.17	19.18	2.95	[37]
Rhodotorula glacialis	68	14.4	0.5	6.3	36.3	33.1	4.8	[38]
Order Tremellales	-	-	-	-	-	-	-	-
Saitozyma podzolica	11.49	2.41	0.60	6.37	1.47	0.13	0.25	[39]
Order Cystobasidiales	-	-	-	-	-	-	-	-
Cryptococcus curvatus	52.66	18.68	-	14.03	43.66	13.02	-	[40]
Order Trichosporonales	-	-	-	-	-	-	-	-
Cutaneotrichosporon dermatis	45.88	20.07	1.36	9.38	43.01	22.21	-	[36]

All experiments used glucose as the sole substrate. C16:0, C16:1, C18:0, C18:1, C18:2, and C18:3 are palmitic, palmitoleic, strearic, oleic, linoleic and linolenic acids, respectively.

Low-cost Feedstocks and Wastes as Substrates for Oleaginous Yeasts

Valorizing low-cost substrates for lipid production by oleaginous yeasts is vital for reducing production costs and simultaneously treating organic wastes. Various low-cost materials have been evaluated as candidate feedstocks for sustainable production of lipids by oleaginous yeasts. Generally, sugar-based media (*i.e.*, glucose, fructose, xylose, sucrose, lactose), molasses, sugar-enriched wastes, whey, and methanol have been used as substrates in *de novo* lipid synthesis [39]. Hydrophobic carbon sources, such as a wide variety of plant oils like olive oil, palm oil, sunflower oil, corn oil; free fatty acid byproducts or wastes such as waste fish oils, soap-stocks, stearin-derived from tallow, hydrolyzed plant oil, free fatty acids, fatty acid esters, and n-alkanes and volatile fatty acids from agro-industrial process these have been considered as substrates for *ex novo* lipid accumulation [8].

In the last five years, low-cost hydrophilic substrates, such as glycerol-derived from biodiesel production, lignocellulosic biomass, and some low-cost hydrophobic materials involving volatile fatty acids that are byproducts or wastes from agro-industrial processes, have been mainly focused on for the production of lipids by oleaginous yeasts (Table **3**). The conversions of biodiesel-derived glycerol waste into lipids have been performed by *R. mucilagenosa, C. viswanathii, Trichosporon oleaginosus*. The glycerol is facilitated into the yeasts by active transport, and it is then phosphorylated into glycerol-3-phosphate (G3P) and converted to dihydroxyacetone phosphate (DHAP), which is used to build triacylglycerols *via* the α-glycerol phosphate acylation pathway (Fig. **1**). Bansal *et al.* [40] reported that *R. mucilagenosa* IIPL32 MTCC 25056 could produce lipids using crude glycerol as the sole source of carbon. Under nitrogen-limited conditions, a lipid concentration of 5.6 g/L was achieved with high content of monounsaturated fatty acids (MUFA) and polyunsaturated fatty acids (PUFA). Likewise, this biodiesel byproduct was used as the sole carbon source in the culture media of 13 oleaginous yeast strains, and *C. viswanathii* Y-E4 achieved a high lipid content of 51.9%, which was outstanding among the species tested [41]. The biodiesel-derived glycerol waste contains impurities like methanol, impacting cell growth and intracellular lipid production [42]. However, the study of Chen *et al.* [43] found that *T. oleaginosus* ATCC 20905 could tolerate a methanol concentration of up to 1.4% and had a high lipid production of 20.78 g/L (48.09% of its cell weight). Besides, Gao *et al.* [44] investigated the effects of four impurities in crude glycerol, including methyl oleate, sodium oleate, methanol, and NaCl; except for methanol, these impurities enhanced the lipid production by the oleaginous yeast *R. toruloide* 32489.

Table 3. Lipid production by oleaginous yeasts using low-cost feedstock and waste as substrates.

Yeast Strains	Substrates	Lipid Concentration (g/L)	Lipid Content (%)	References
Rhodococcus sp. YHY01	Food waste-derived volatile fatty acids in nitrogen-limited	2.2	69	[52]
Cryptococcus curvatus DSM 70022	Volatile fatty acids derived from waste paper	1.78	41.2	[53]
Yarrowia lipolytica MTCC 9520	Chicken tallow	4.16	-	[54]
Yarrowia lipolytica CICC 31596	70 g/L acetic acid	10.11	-	[55]
Yarrowia lipolytica CICC 31596	Food waste fermentation	3.20	21.86	[55]
Yarrowia lipolytica CICC 31596	Fruit and vegetable waste fermentation	3.08	26.02	[55]
Rhodotorula paludigena CM33	Molasses	6.1	37.1	[56]
Rhodotorula mucilagenosa IIPL32 MTCC 25056	Crude glycerol	5.6	-	[40]

(Table 3) cont.....

Yeast Strains	Substrates	Lipid Concentration (g/L)	Lipid Content (%)	References
Candida viswanathii Y-E4	Crude glycerol	13.6	51.9	[41]
Rhodotorula mucilaginosa KKUSY14	Durian peel	1.68	-	[48]
Cryptococcus curvatus ATCC 20509	Untreated waste office paper	1.39	22	[47]
Cryptococcus curvatus ATCC 20509	Treated waste office paper	5.75	37.8	[47]
Trichosporon oleaginosus ATCC 20905	Crude glycerol + 1.4% (w/v) methanol	20.78	48.09	[43]
Cryptococcus curvatus ATCC 20509	DDAP-EH corn stover hydrolysate	21.4	63.1	[61]
Trichosporon oleaginosus ATCC 20509	Sorghum stalk hydrolysate	13.1	60	[61]
Trichosporon oleaginosus ATCC 20509	Switch grass hydrolysate	12.3	58	[61]
Cryptococcus curvatus ATCC 20509	Volatile fatty acids (ratio 15:5:10)	4.93	56.85	[61]
Trichosporon oleaginosus DSM 11815	Microalgae hydrolysate	30.6	53	[62]
Cryptococcus curvatus ATCC 20509	Municipal sludge + crude glycerol + peptone	16.4	40.3	[61]

Lignocellulosic biomass and related materials consist of the three main components, cellulose, hemicellulose, and lignin, and have lesser quantities of starch, pectin, extractives, ashes, *etc* [45]. Cellulose and hemicellulose comprise polymerized sugars, such as pentose (*e.g.*, xylose) and hexose (*e.g.*, glucose) [8]. Since most oleaginous yeasts consume sugars, lignocellulosic materials are pretreated to hydrolyze cellulose and hemicellulose molecules into simple sugars, such as glucose and xylose. It is also necessary to remove inhibitors generated from lignin and sugar degradation. These include furan derivatives such as furfural derived from pentose and hydroxymethylfurfural (HMF) derived from hexose sugars, weak acids (such as formic, acetic, and levulinic acids), and phenolic compounds such as syringaldehyde, ferulic and vanillic acids [22, 45]. However, detoxification of these inhibitors is costly when preparing fermentable sugars in the hydrolysate [46].

Many research groups have been searching for oleaginous yeast species capable of converting sugars derived from lignocellulosic materials to lipids *via* the *de novo* lipid accumulation process (Table **3**). Annamalai *et al.* [47] found that *C.*

curvatus ATCC 20509 could produce lipids from enzymatically hydrolyzed waste office paper (WOP) only after lignin removal. The culture medium was supplemented with ammonium sulfate (2 g/L) and yeast extract (0.5 g/L) as nitrogen sources at a C/N ratio of 80. *R. mucilaginosa* KKUSY14 was able to tolerate acetic acid (2.39-2.6 g/L) and 5-HMF (0.27-0.86 g/L), and had good performance in converting not detoxified durian peel hydrolysate into lipids [48]. Such as tolerance is beneficial for cost-effective lipid production from lignocellulosic materials. When using sorghum stalks and switchgrass hydrolysates as substrate, *T. oleaginous* had better performance in lipid production over *L. starkeyi* and *C. albidus*. The sugar consumption rate of *T. oleaginosus* was the fastest, and this yeast achieved a high lipid content of 60%. Besides sugars, lignocellulosic hydrolysates (*i.e.*, sorghum stalks and switchgrass hydrolysates) contained acetic acid and citric acid as byproducts of enzymatic hydrolysis [49]. Interestingly, acetic acid is entirely assimilated by all yeast strains, and then enters into mitochondria to form acetyl-CoA (Fig. **1**). In contrast, citric acid enters the cytoplasm in exchange with malate, and is finally cleaved to acetyl-CoA and oxaloacetate by the ATP-citrate lyase (ATP-CL), the key enzyme of lipid accumulation in oleaginous yeasts (Fig. **1**).

Fats or hydrophobic materials including volatile fatty acids, free fatty acids, n-alkanes, waste oils, and fats are considered carbon sources for the *ex novo* lipid synthesis, in which lipid accumulation co-occurs with cell growth without nitrogen depletion [50, 51]. Free fatty acids as the primary substrates are produced from the hydrolysis of triacylglycerols/fatty acid esters and are transported into the cells by an active transport system. These fatty acids are either assimilated for cell growth requirements or used as precursors for endocellular transformations (synthesis of different fatty acids that were not previously contained in the medium). Assimilated free-fatty acids are catabolized by the ß-oxidation process into shortened chain acyl-CoAs and acetyl-CoA. The degradation is catalyzed by several acyl-CoA oxidases (Aox), resulting in: (1) energy for cell growth and the maintenance of acetyl-CoA channel inside the tricarboxylic cycle (TCA); and (2) the production of organic substances (intermediates) that create precursors for the synthesis of cellular components [50, 51] (Fig. **1**). In the case of long-chain alkanes, acetyl-CoA or propionyl-CoA is formed, enters peroxisomes to be catalyzed by the glyoxylate-cycle pathway, and interacts with the methylcitric-acid and TCA cycles in mitochondria. In addition, assimilated fatty acids could be directly built as triacylglycerols or steryl esters and stored into lipid bodies (Fig. **1**). Recently, the conversion of volatile fatty acids from sludge, food wastes, and various organic wastes, to microbial lipids has been studied (Table **3**). Food waste-derived volatile fatty acids were converted into yeast lipids by *Rhodococcus* sp. YHY01 cultivated under nitrogen-limited conditions. The yeast achieved a high lipid content of 69% (w/w) [52]. During cultivation, *Rhodococcus* sp. YHY01 presented different growth patterns towards different organic acids, including lactate, acetate, propionate, and butyrate. Lactate, propionate, and butyrate were assimilated slower than acetate [52]. Likewise, in the study of Annamalai *et al.* [53], it was found that *C. curvatus* DSM 70022 preferably consumed acetic acid over propionic and butyric acids. Acetate is directly converted into acetyl-CoA to synthesize fatty acids and various metabolic molecules, whereas lactate is converted into pyruvate and H_2O_2 through lactate oxidase or is converted into acetate, CO_2, and H_2O through lactate monooxygenase [54, 55]. In contrast, propionate is converted to propionyl CoA and then enters the tricarboxylic acid (TCA) cycle *via* methylmalonyl-CoA interconversion to succinyl-CoA. On the other hand, butyrate is metabolized through β-oxidation to produce acetoacetyl-CoA, which is further transformed into acetyl-CoA [55 - 60].

Biodiesel Production via Direct Transesterification from Yeast Lipids

The first generation biodiesel production from oleaginous yeast has been mainly performed in three steps: the first step is biomass drying followed by cell disruption and lipid extraction, and finally biodiesel production (Fig. **2**). The microbial biomass is usually dried at 60–70°C to maintain the total lipid content and its composition, and reduce the water content that would decrease the extraction efficiency of the lipids. Then, the lipids are extracted from the biomass by chemical (*i.e.*, acid, alkali, detergents) or by enzymatic and physical or mechanical (*i.e.*, sonication, bead milling, and pressure extrusion) techniques. Soxhlet extraction with hexane and the extraction method of Blight and Dyer or the extraction method of Folch using a chloroform/methanol mixture are the most performed methods in the extraction of yeast lipids. Finally, the yeast lipids are transesterified with alcohol (*i.e.*, methanol, ethanol) addition, along with an acid or base or enzymatic catalyst. Other approaches tested involve microwaves, ultrasound-assisted extraction, supercritical fluid, and heat reflux [63 - 66]. The overall process requires significant amounts of solvents and a long processing time. Therefore, direct (*in situ*) transesterification, where the disruption of cells, the extraction of lipids, and the transesterification are performed in a single step, has been proposed as an alternative approach to biodiesel production from oleaginous yeasts [10, 14] (Fig. **2**).

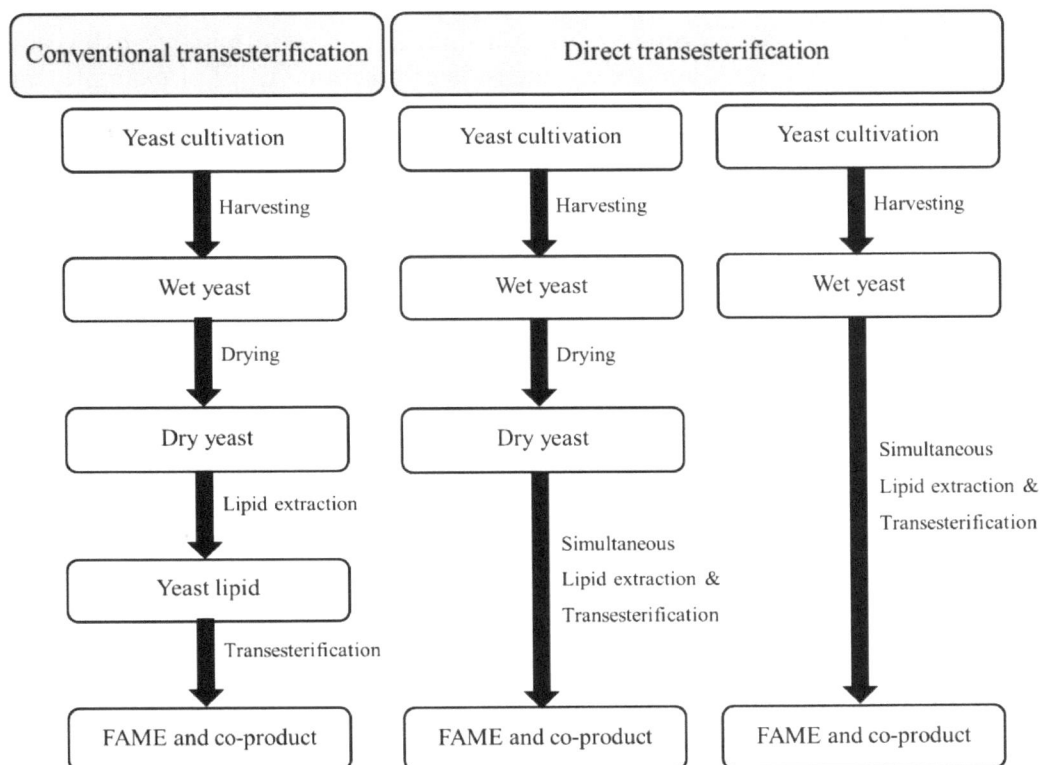

Fig. (2). Flow diagrams of yeast biodiesel production processes through conventional transesterification and direct transesterification methods. FAME stands for fatty acid methyl ester. **Source:** modified from Cheirslip and Louhasakul [14] and Kim *et al.* [68].

Direct transesterification of the yeast lipids is performed using a combination of physical or mechanical treatments (*i.e.*, bead beating/milling, microwaves, ultrasonication) and hydrophobic solvents (*i.e.*, chloroform, hexane, pentane) with the addition of alcohol and catalyst [10, 13]. However, direct transesterification of the yeast lipids can be performed without hydrophobic solvents but with an increased alcohol ratio. Cheirsilp and Louhasakul [14] evaluated the direct transesterification of *Y. lipolytica* TISTR5151 without adding hydrophobic solvent. In this case, the process gave a comparable yield of biodiesel (71.9–73% of total lipids) but either needed a longer reaction time (6 h) with a methanol/biomass ratio of 125:1 or a larger methanol/cell ratio of 209:1 to reduce the reaction period to 1 h. Likewise, Liu and Zhao [67] developed the direct transesterification of *R. toruloides* with the addition of methanol and sulfuric acid. The reaction achieved 60% fatty acid methyl ester (FAME) yield but needed a longer reaction time (20 h). Besides, the direct-transesterification efficiency can be improved by combining physical or mechanical treatments such as bead milling, ultrasonication, or microwave that can increase the mass transfer rate between immiscible phases and reduce the time of reaction [13]. For example, the direct transesterification of wet yeast cells of *Y. lipolytica* TISTR 5151 was performed without adding hydrophobic solvent but in a 2-L vigorously stirred tank reactor (VSTR)combined with bead milling. FAMEs rapidly reached their highest levels at 80.84% within a half-hour of reaction [15].

In the last five years, most direct transesterifications of yeast lipids have been performed using acid catalysts (Table **4**). The acid-catalyzed direct transesterification can prevent soap formation from free fatty acids derived from both the cells and the hydrolysis of alkyl esters, contributing water to the system. In addition, both triacylglycerol transesterification and free fatty acid esterification can be catalyzed by an acid catalyst. In a direct-transesterification process of wet *R. glutinis* BCRC 22360 cells, the highest FAME yield of 97% was achieved with the addition of methanol at 1:100 and 0.36 M H_2SO_4 as a catalyst at 90°C for 8 h [69]. Katre *et al.* [70] achieved a high FAME yield of 0.88 g/4 g biomass from direct transesterification of *Y. lipolytica* NCM 3589 for 8 h at 50 °C with the addition of methanol: chloroform (10:1) and an acid catalyst (0.2 M H_2SO_4,). In some cases, the FAME yield improved from 69% to 83% with a low methanol loading of 1:10, with some added surfactants [17]. The surfactants can improve the mass transfer rates between immiscible phases and easily bind with cell walls/membranes, subsequently bringing about cell disruptions [71].

Table 4. Direct transesterification for biodiesel production from oleaginous yeasts.

Yeast Strains	Condition	Yield	References
Yarrowia lipolytica TISTR5151	Acid [0.4 M H_2SO_4]; 1000 rpm; 50 °C	90%	[15]
Rhodotorula glutinis BCRC 22360	Acid [0.36 M H_2SO_4]; methanol ratio 1:100; 90°C; 8 h	97%	[69]
Rhodotorula glutinis BCRC 22360	Acid [0.36 M H_2SO_4]; 50 mM 3-(*N,N*-dimethylmyristylammonio) propanesul fonate (3-DMAPS, a zwitterionic surfactant); methanol ratio 1:10; 90°C; 10 h	83%	[69]
Yarrowia lipolytica NCM 3589	Acid [0.2 M H_2SO_4]; 50 °C; 8 h	0.88 g/4 g biomass	[70]
Saitozyma podzolica Zwy--3	Enzyme [2.5 g immobilized lipase dosage]; 15% water content; n-hexane/methanol molar ratio of 3:1; 35 °C	98.12%	[39]
Naganishia liquefaciens NITTS2	Enzyme [lipase 20%]; methanol/oil molar ratio of 6:4; 35 °C; 16 h. 6.4	97.13%	[72]

(Table 4) cont.....

Yeast Strains	Condition	Yield	References
Rhodosporidium diobovatum 08-225	Ionic liquid [C$_2$mim[EtSO$_4$] 2 g /g biomass ; 16.9 g methanol/g biomass; 0.056 g KOH/g biomass; 65 °C	97.1%	[73]
Rhodotorula mucilaginosa IIPL32	Acid [0.4 M HCl]; 70 °C; 20 h	97.3 mg/g	[74]
Rhodotorula glutinis BCRC 22360	Acid [0.6 M H$_2$SO$_4$]; 70 °C; 20 h	111%	[75]
Rhodotorula graminis S1/S2	Acid [0.4 M H$_2$SO$_4$]; methanol/biomass ratio of 60:1; 70°C; 3 h	123%	[76]

Besides acid-catalyzed direct transesterification, enzyme-catalyzed direct transesterification of yeast lipids has been developed to be efficient on a lab scale. Cao *et al.* [16] achieved 98.12% of FAME in direct transesterification of *S. podzolica* Zwy-2-3 without cell drying, using 2.5 g magnetic nanoparticle-immobilized lipase, 15% water content, and n-hexane/methanol molar ratio of 3:1 at 35 °C. After 10 times of reuse, the immobilized lipase still retained more than 90% of the initial enzyme activity. The direct-transesterification efficiency can be improved by combining physical or mechanical techniques such as ultrasonication and microwave [71 - 76]. These can increase the mass transfer rates between immiscible phases and reduce the reaction time. Selvakumar *et al.* [72] investigated enzyme-catalyzed direct transesterification of *N. liquefaciens* NITTS2 using ultrasound to disrupt cells and extract lipids and garbage lipase for transesterification. FAME was achieved at 97.13% at the optimal condition of 6.4 methanol/oil molar ratio and 20% enzyme dosage at 35 °C for 16 h. Ionic liquids can also be utilized in direct transesterification as a co-solvent. Gao *et al.* [73] investigated the direct transesterification of yeast biomass, *R. diobovatum* 08-225, using potassium hydroxide (KOH) as a catalyst in the presence of an ionic liquid (1-ethyl-3-methylimidazolium ethyl sulfate). The FAME yield was as high as 97.1% at the optimal conditions: methanol 6.9 g/ g-biomass, KOH 0.056 g/g-biomass, and C$_2$mim[EtSO$_4$] 2 g/g-biomass at 65 °C. An ionic liquid can drive water through hydrogen bonds and allow a more significant catalyst to stay in methoxide ion form.

CONCLUSION

Biotechnological applications of oleaginous yeasts are promising due to their beneficial characteristics in terms of high growth rate and lipid content, similar lipid composition to conventional biodiesel feedstocks, and their ability to valorize a broad range of low-cost feedstocks and industrial wastes. Understanding key biochemical pathways for lipid production from first- and second-generation feedstocks as well as developing direct process for biodiesel production from yeast cells will enable their industrialization and commercialization. Further innovations and biorefinery concepts may help overcome limitations in technical and economic feasibility and environmental footprints.

LIST OF ABBREVIATIONS

NL	Neutral Lipid
ER	Endoplasmic Reticulum
PER	Peroxisome
TCA Cycle	Tricarboxylic Acid Cycle
PPP	Pentose Phosphate Pathway
G6P	Glucose-6-phosphate
F6P	Fructose-6-phosphate
F1P	Fructose-1-phosphate
GA3P	Glycerol-3-phosphate
DHAP	Dihydroxyacetone Phosphate
OAA	Oxaloacetate
FFA	Free Fatty Acid
LPA	Lysophosphatidic Acid
PA	Phosphatidic Acid
MAG	Monoacylglycerol
DAG	Diacylglycerol
TAG	Triacylglycerol
LB	Lipid Body
LDs	Lipid Droplets
VFAs	Volatile Fatty Acids

CONSENT FOR PUBLICATION

Not applicable.

CONFLICT OF INTEREST

The authors declare no conflict of interest, financial or otherwise.

ACKNOWLEDGEMENTS

The authors were supported by Thailand Research Fund [grant no. RTA6280014].

REFERENCES

[1]   Ferreira GF, Ríos Pinto LF, Carvalho PO, *et al.* Biomass and lipid characterization of microalgae genera *Botryococcus*, *Chlorella*, and *Desmodesmus* aiming high-value fatty acid production. Biomass Convers Biorefinery 2021; 11: 1675-89.
        [http://dx.doi.org/10.1007/s13399-019-00566-3]

[2]   Ray M, Kumar N, Kumar V, Negi S, Banerjee C. Microalgae: a way forward approach towards wastewater treatment and bio-fuel production. Appl Microbiol Bioeng. Elsevier 2019; pp. 229-43.
        [http://dx.doi.org/10.1016/B978-0-12-815407-6.00012-5]

[3]   Arous F, Jaouani A, Mechichi T. Oleaginous microorganisms for simultaneous biodiesel production and wastewater treatment. Microb Wastewater Treat. Elsevier 2019; pp. 153-74.
        [http://dx.doi.org/10.1016/B978-0-12-816809-7.00008-7]

[4]   Mukhopadhyay M, Singh A, Banerjee R. Oleaginous fungi: a solution to oil crisis. Microorg Environ Manag. Dordrecht: Springer Netherlands 2012; Vol. 9789400722: pp. 403-14.
        [http://dx.doi.org/10.1007/978-94-007-2229-3_18]

[5]   El-Enshasy HA. Filamentous fungal cultures – process characteristics, products, and applications. Bioprocess Value-Added Prod from Renew Resour. Elsevier 2007; pp. 225-61.
        [http://dx.doi.org/10.1016/B978-044452114-9/50010-4]

[6]   Qin L, Liu L, Zeng A-P, Wei D. From low-cost substrates to Single Cell Oils synthesized by oleaginous yeasts. Bioresour Technol 2017; 245(Pt B): 1507-19.
        [http://dx.doi.org/10.1016/j.biortech.2017.05.163] [PMID: 28642053]

[7]   Sitepu IR, Garay LA, Sestric R, *et al.* Oleaginous yeasts for biodiesel: current and future trends in biology and production. Biotechnol Adv 2014; 32(7): 1336-60.
        [http://dx.doi.org/10.1016/j.biotechadv.2014.08.003] [PMID: 25172033]

[8]   Khot M, Raut G, Ghosh D, Alarcón-Vivero M, Contreras D, Ravikumar A. Lipid recovery from oleaginous yeasts: Perspectives and challenges for industrial applications. Fuel 2020; 259: 116292.
        [http://dx.doi.org/10.1016/j.fuel.2019.116292]

[9]   Fontanille P, Kumar V, Christophe G, Nouaille R, Larroche C. Bioconversion of volatile fatty acids into lipids by the oleaginous yeast *Yarrowia lipolytica*. Bioresour Technol 2012; 114: 443-9.
        [http://dx.doi.org/10.1016/j.biortech.2012.02.091] [PMID: 22464419]

[10]   Karamerou EE, Webb C. Cultivation modes for microbial oil production using oleaginous yeasts – A review. Biochem Eng J 2019; 151: 107322.
        [http://dx.doi.org/10.1016/j.bej.2019.107322]

[11]   Xu J, Du W, Zhao X, Zhang G, Liu D. Microbial oil production from various carbon sources and its use for biodiesel preparation. Biofuels Bioprod Biorefin 2013; 7: 65-77.
        [http://dx.doi.org/10.1002/bbb.1372]

[12]   Patel A, Arora N, Mehtani J, Pruthi V, Pruthi PA. Assessment of fuel properties on the basis of fatty acid profiles of oleaginous yeast for potential biodiesel production. Renew Sustain Energy Rev 2017; 77: 604-16.
        [http://dx.doi.org/10.1016/j.rser.2017.04.016]

[13]   Yousuf A, Khan MR, Islam MA, Wahid ZA, Pirozzi D. Technical difficulties and solutions of direct transesterification process of microbial oil for biodiesel synthesis. Biotechnol Lett 2017; 39(1): 13-23.
        [http://dx.doi.org/10.1007/s10529-016-2217-x] [PMID: 27659031]

[14]   Cheirsilp B, Louhasakul Y. Industrial wastes as a promising renewable source for production of microbial lipid and direct transesterification of the lipid into biodiesel. Bioresour Technol 2013; 142: 329-37.
        [http://dx.doi.org/10.1016/j.biortech.2013.05.012] [PMID: 23747444]

[15]   Louhasakul Y, Cheirsilp B, Maneerat S, Prasertsan P. Direct transesterification of oleaginous yeast lipids into biodiesel: Development of vigorously stirred tank reactor and process optimization.

Biochem Eng J 2018; 137: 232-8.
[http://dx.doi.org/10.1016/j.bej.2018.06.009]

[16]   Cao X, Xu H, Li F, *et al.* One-step direct transesterification of wet yeast for biodiesel production catalyzed by magnetic nanoparticle-immobilized lipase. Renew Energy 2021; 171: 11-21.
[http://dx.doi.org/10.1016/j.renene.2021.02.065]

[17]   Adrio JL. Oleaginous yeasts: Promising platforms for the production of oleochemicals and biofuels. Biotechnol Bioeng 2017; 114(9): 1915-20.
[http://dx.doi.org/10.1002/bit.26337] [PMID: 28498495]

[18]   Sreeharsha RV, Mohan SV. Obscure yet promising oleaginous yeasts for fuel and chemical production. Trends Biotechnol 2020; 38(8): 873-87.
[http://dx.doi.org/10.1016/j.tibtech.2020.02.004] [PMID: 32673589]

[19]   Papanikolaou S. Oleaginous yeasts: biochemical events related with lipid synthesis and potential biotechnological applications. Ferment Technol 2012; 01.
[http://dx.doi.org/10.4172/2167-7972.1000e103]

[20]   Darvishi F, Fathi Z, Ariana M, Moradi H. *Yarrowia lipolytica* as a workhorse for biofuel production. Biochem Eng J 2017; 127: 87-96.
[http://dx.doi.org/10.1016/j.bej.2017.08.013]

[21]   Carsanba E, Papanikolaou S, Fickers P, Erten H. Lipids by *Yarrowia lipolytica* strains cultivated on glucose in batch cultures. Microorganisms 2020; 8(7): 1054.
[http://dx.doi.org/10.3390/microorganisms8071054] [PMID: 32679918]

[22]   Takaku H, Matsuzawa T, Yaoi K, Yamazaki H. Lipid metabolism of the oleaginous yeast *Lipomyces starkeyi*. Appl Microbiol Biotechnol 2020; 104(14): 6141-8.
[http://dx.doi.org/10.1007/s00253-020-10695-9] [PMID: 32458138]

[23]   Juanssilfero AB, Kahar P, Amza RL, *et al.* Effect of inoculum size on single-cell oil production from glucose and xylose using oleaginous yeast *Lipomyces starkeyi*. J Biosci Bioeng 2018; 125(6): 695-702.
[http://dx.doi.org/10.1016/j.jbiosc.2017.12.020] [PMID: 29373308]

[24]   Wen Z, Zhang S, Odoh CK, Jin M, Zhao ZK. *Rhodosporidium toruloides* - A potential red yeast chassis for lipids and beyond. FEMS Yeast Res 2020; 20(5): foaa038.
[http://dx.doi.org/10.1093/femsyr/foaa038] [PMID: 32614407]

[25]   Maza DD, Viñarta SC, Su Y, Guillamón JM, Aybar MJ. Growth and lipid production of *Rhodotorula glutinis* R4, in comparison to other oleaginous yeasts. J Biotechnol 2020; 310: 21-31.
[http://dx.doi.org/10.1016/j.jbiotec.2020.01.012] [PMID: 32004579]

[26]   Saran S, Mathur A, Dalal J, Saxena RK. Process optimization for cultivation and oil accumulation in an oleaginous yeast *Rhodosporidium toruloides* A29. Fuel 2017; 188: 324-31.
[http://dx.doi.org/10.1016/j.fuel.2016.09.051]

[27]   Papanikolaou S, Aggelis G. Lipids of oleaginous yeasts. Part I: Biochemistry of single cell oil production. Eur J Lipid Sci Technol 2011; 113: 1031-51.
[http://dx.doi.org/10.1002/ejlt.201100014]

[28]   Patel A, Matsakas L. A comparative study on *de novo* and *ex novo* lipid fermentation by oleaginous yeast using glucose and sonicated waste cooking oil. Ultrason Sonochem 2019; 52: 364-74.
[http://dx.doi.org/10.1016/j.ultsonch.2018.12.010] [PMID: 30559080]

[29]   Beckman M. CELL BIOLOGY: Great Balls of Fat. Science (80-) 2006; 311: 1232-4.
[http://dx.doi.org/10.1126/science.311.5765.1232]

[30]   Walther TC, Farese RV Jr. Lipid droplets and cellular lipid metabolism. Annu Rev Biochem 2012; 81: 687-714.
[http://dx.doi.org/10.1146/annurev-biochem-061009-102430] [PMID: 22524315]

[31]   Deeba F, Pruthi V, Negi YS. Converting paper mill sludge into neutral lipids by oleaginous yeast

Cryptococcus vishniaccii for biodiesel production. Bioresour Technol 2016; 213: 96-102.
[http://dx.doi.org/10.1016/j.biortech.2016.02.105] [PMID: 26965670]

[32]   Do Yook S, Kim J, Woo HM, Um Y, Lee S-M. Efficient lipid extraction from the oleaginous yeast *Yarrowia lipolytica* using switchable solvents. Renew Energy 2019; 132: 61-7.
[http://dx.doi.org/10.1016/j.renene.2018.07.129]

[33]   Li Q, Kamal R, Chu Y, Wang Q, Yu X, Huang Q. Automated pressurized liquid extraction of microbial lipids from oleaginous yeasts. Appl Biochem Biotechnol 2020; 192(1): 283-95.
[http://dx.doi.org/10.1007/s12010-020-03331-9] [PMID: 32378082]

[34]   Steen EJ, Kang Y, Bokinsky G, *et al.* Microbial production of fatty-acid-derived fuels and chemicals from plant biomass. Nature 2010; 463(7280): 559-62.
[http://dx.doi.org/10.1038/nature08721] [PMID: 20111002]

[35]   Lakshmidevi R, Ramakrishnan B, Ratha SK, Bhaskar S, Chinnasamy S. Valorisation of molasses by oleaginous yeasts for single cell oil (SCO) and carotenoids production. Environ Technol Innov 2021; 21: 101281.
[http://dx.doi.org/10.1016/j.eti.2020.101281]

[36]   Wang L, Wang D, Zhang Z, *et al.* Comparative glucose and xylose coutilization efficiencies of soil-isolated yeast strains identify cutaneotrichosporon dermatis as a potential producer of lipid. ACS Omega 2020; 5(37): 23596-603.
[http://dx.doi.org/10.1021/acsomega.0c02089] [PMID: 32984679]

[37]   Arous F, Triantaphyllidou I-E, Mechichi T, Azabou S, Nasri M, Aggelis G. Lipid accumulation in the new oleaginous yeast *Debaryomyces etchellsii* correlates with ascosporogenesis. Biomass Bioenergy 2015; 80: 307-15.
[http://dx.doi.org/10.1016/j.biombioe.2015.06.019]

[38]   Amaretti A, Raimondi S, Sala M, *et al.* Single cell oils of the cold-adapted oleaginous yeast *Rhodotorula glacialis* DBVPG 4785. Microb Cell Fact 2010; 9: 73.
[http://dx.doi.org/10.1186/1475-2859-9-73] [PMID: 20863365]

[39]   Gorte O, Kugel M, Ochsenreither K. Optimization of carbon source efficiency for lipid production with the oleaginous yeast *Saitozyma podzolica* DSM 27192 applying automated continuous feeding. Biotechnol Biofuels 2020; 13(1): 181.
[http://dx.doi.org/10.1186/s13068-020-01824-7] [PMID: 33292512]

[40]   Bansal N, Dasgupta D, Hazra S, Bhaskar T, Ray A, Ghosh D. Effect of utilization of crude glycerol as substrate on fatty acid composition of an oleaginous yeast *Rhodotorula mucilagenosa* IIPL32: Assessment of nutritional indices. Bioresour Technol 2020; 309: 123330.
[http://dx.doi.org/10.1016/j.biortech.2020.123330] [PMID: 32283485]

[41]   Guerfali M, Ayadi I, Sassi H-E, Belhassen A, Gargouri A, Belghith H. Biodiesel-derived crude glycerol as alternative feedstock for single cell oil production by the oleaginous yeast *Candida viswanathii* Y-E4. Ind Crops Prod 2020; 145: 112103.
[http://dx.doi.org/10.1016/j.indcrop.2020.112103]

[42]   Kumar LR, Yellapu SK, Tyagi RD, Zhang X. A review on variation in crude glycerol composition, bio-valorization of crude and purified glycerol as carbon source for lipid production. Bioresour Technol 2019; 293: 122155.
[http://dx.doi.org/10.1016/j.biortech.2019.122155] [PMID: 31561979]

[43]   Chen J, Zhang X, Tyagi RD, Drogui P. Utilization of methanol in crude glycerol to assist lipid production in non-sterilized fermentation from *Trichosporon oleaginosus.* Bioresour Technol 2018; 253: 8-15.
[http://dx.doi.org/10.1016/j.biortech.2018.01.008] [PMID: 29328937]

[44]   Gao Z, Ma Y, Wang Q, Zhang M, Wang J, Liu Y. Effect of crude glycerol impurities on lipid preparation by *Rhodosporidium toruloides* yeast 32489. Bioresour Technol 2016; 218: 373-9.
[http://dx.doi.org/10.1016/j.biortech.2016.06.088] [PMID: 27387413]

[45] Kucharska K, Rybarczyk P, Hołowacz I, Łukajtis R, Glinka M, Kamiński M. Pretreatment of lignocellulosic materials as substrates for fermentation processes. Molecules 2018; 23(11): 2937.
[http://dx.doi.org/10.3390/molecules23112937] [PMID: 30423814]

[46] Wang J, Gao Q, Zhang H, Bao J. Inhibitor degradation and lipid accumulation potentials of oleaginous yeast Trichosporon cutaneum using lignocellulose feedstock. Bioresour Technol 2016; 218: 892-901.
[http://dx.doi.org/10.1016/j.biortech.2016.06.130] [PMID: 27441826]

[47] Annamalai N, Sivakumar N, Oleskowicz-Popiel P. Enhanced production of microbial lipids from waste office paper by the oleaginous yeast *Cryptococcus curvatus*. Fuel 2018; 217: 420-6.
[http://dx.doi.org/10.1016/j.fuel.2017.12.108]

[48] Siwina S, Leesing R. Bioconversion of durian (*Durio zibethinus* Murr.) peel hydrolysate into biodiesel by newly isolated oleaginous yeast *Rhodotorula mucilaginosa* KKUSY14. Renew Energy 2021; 163: 237-45.
[http://dx.doi.org/10.1016/j.renene.2020.08.138]

[49] Lee J-E, Vadlani PV, Min D. Sustainable production of microbial lipids from lignocellulosic biomass using oleaginous yeast cultures. J Sustain Bioenergy Syst 2017; 07: 36-50.
[http://dx.doi.org/10.4236/jsbs.2017.71004]

[50] Spagnuolo M, Yaguchi A, Blenner M. Oleaginous yeast for biofuel and oleochemical production. Curr Opin Biotechnol 2019; 57: 73-81.
[http://dx.doi.org/10.1016/j.copbio.2019.02.011] [PMID: 30875587]

[51] Liu H, Song Y, Fan X, Wang C, Lu X, Tian Y. *Yarrowia lipolytica* as an oleaginous platform for the production of value-added fatty acid-based bioproducts. Front Microbiol 2021; 11: 608662.
[http://dx.doi.org/10.3389/fmicb.2020.608662] [PMID: 33469452]

[52] Bhatia SK, Gurav R, Choi T-R, et al. Effect of synthetic and food waste-derived volatile fatty acids on lipid accumulation in *Rhodococcus* sp. YHY01 and the properties of produced biodiesel. Energy Convers Manage 2019; 192: 385-95.
[http://dx.doi.org/10.1016/j.enconman.2019.03.081]

[53] Annamalai N, Sivakumar N, Fernandez-Castane A, Oleskowicz-Popiel P. Production of microbial lipids utilizing volatile fatty acids derived from wastepaper: A biorefinery approach for biodiesel production. Fuel 2020; 276: 118087.
[http://dx.doi.org/10.1016/j.fuel.2020.118087]

[54] Jiang T, Gao C, Ma C, Xu P. Microbial lactate utilization: enzymes, pathogenesis, and regulation. Trends Microbiol 2014; 22(10): 589-99.
[http://dx.doi.org/10.1016/j.tim.2014.05.008] [PMID: 24950803]

[55] Gao R, Li Z, Zhou X, Cheng S, Zheng L. Oleaginous yeast *Yarrowia lipolytica* culture with synthetic and food waste-derived volatile fatty acids for lipid production. Biotechnol Biofuels 2017; 10: 247.
[http://dx.doi.org/10.1186/s13068-017-0942-6] [PMID: 29093751]

[56] Vajpeyi S, Chandran K. Microbial conversion of synthetic and food waste-derived volatile fatty acids to lipids. Bioresour Technol 2015; 188: 49-55.
[http://dx.doi.org/10.1016/j.biortech.2015.01.099] [PMID: 25697838]

[57] Radha P, Prabhu K, Jayakumar A. AbilashKarthik S, Ramani K. Biochemical and kinetic evaluation of lipase and biosurfactant assisted *ex novo* synthesis of microbial oil for biodiesel production by Yarrowia lipolytica utilizing chicken tallow. Process Biochem 2020; 95: 17-29.
[http://dx.doi.org/10.1016/j.procbio.2020.05.009]

[58] Gao R, Li Z, Zhou X, Bao W, Cheng S, Zheng L. Enhanced lipid production by *Yarrowia lipolytica* cultured with synthetic and waste-derived high-content volatile fatty acids under alkaline conditions. Biotechnol Biofuels 2020; 13: 3.
[http://dx.doi.org/10.1186/s13068-019-1645-y] [PMID: 31911818]

[59] Gosalawit C, Imsoonthornruksa S, Gilroyed BH, Mcnea L, Boontawan A, Ketudat-Cairns M. The

potential of the oleaginous yeast *Rhodotorula paludigena* CM33 to produce biolipids. J Biotechnol 2021; 329: 56-64.
[http://dx.doi.org/10.1016/j.jbiotec.2021.01.021] [PMID: 33549673]

[60]   Sànchez i Nogué V, Black BA, Kruger JS, *et al.* Integrated diesel production from lignocellulosic sugars *via* oleaginous yeast. Green Chem 2018; 20: 4349-65.
[http://dx.doi.org/10.1039/C8GC01905C]

[61]   Liu J, Yuan M, Liu J-N, Huang X-F. Bioconversion of mixed volatile fatty acids into microbial lipids by *Cryptococcus curvatus* ATCC 20509. Bioresour Technol 2017; 241: 645-51.
[http://dx.doi.org/10.1016/j.biortech.2017.05.085] [PMID: 28609752]

[62]   Meo A, Priebe XL, Weuster-Botz D. Lipid production with *Trichosporon oleaginosus* in a membrane bioreactor using microalgae hydrolysate. J Biotechnol 2017; 241: 1-10.
[http://dx.doi.org/10.1016/j.jbiotec.2016.10.021] [PMID: 27984117]

[63]   Zhang X, Chen J, Yan S, Tyagi RD, Surampalli RY, Li J. Lipid production for biodiesel from sludge and crude glycerol. Water Environ Res 2017; 89(5): 424-39.
[http://dx.doi.org/10.2175/106143017X14839994523424] [PMID: 28442003]

[64]   Widjaja A, Chien C-C, Ju Y-H. Study of increasing lipid production from fresh water microalgae *Chlorella vulgaris*. J Taiwan Inst Chem Eng 2009; 40: 13-20.
[http://dx.doi.org/10.1016/j.jtice.2008.07.007]

[65]   Hidalgo P, Toro C, Ciudad G, Navia R. Advances in direct transesterification of microalgal biomass for biodiesel production. Rev Environ Sci Biotechnol 2013; 12: 179-99.
[http://dx.doi.org/10.1007/s11157-013-9308-0]

[66]   Park J-Y, Park MS, Lee Y-C, Yang J-W. Advances in direct transesterification of algal oils from wet biomass. Bioresour Technol 2015; 184: 267-75.
[http://dx.doi.org/10.1016/j.biortech.2014.10.089] [PMID: 25466997]

[67]   Liu B, Zhao Z. (Kent). Biodiesel production by direct methanolysis of oleaginous microbial biomass. J Chem Technol Biotechnol 2007; 82: 775-80.
[http://dx.doi.org/10.1002/jctb.1744]

[68]   Kim B, Heo HY, Son J, *et al.* Simplifying biodiesel production from microalgae *via* wet *in situ* transesterification: A review in current research and future prospects. Algal Res 2019; 41: 101557.
[http://dx.doi.org/10.1016/j.algal.2019.101557]

[69]   Chen S-J, Kuan I-C, Tu Y-F, Lee S-L, Yu C-Y. Surfactant-assisted *in situ* transesterification of wet *Rhodotorula glutinis* biomass. J Biosci Bioeng 2020; 130(4): 397-401.
[http://dx.doi.org/10.1016/j.jbiosc.2020.05.009] [PMID: 32586661]

[70]   Katre G, Raskar S, Zinjarde S, Ravi Kumar V, Kulkarni BD. RaviKumar A. Optimization of the *in situ* transesterification step for biodiesel production using biomass of *Yarrowia lipolytica* NCIM 3589 grown on waste cooking oil. Energy 2018; 142: 944-52.
[http://dx.doi.org/10.1016/j.energy.2017.10.082]

[71]   Lee SY, Cho JM, Chang YK, Oh Y-K. Cell disruption and lipid extraction for microalgal biorefineries: A review. Bioresour Technol 2017; 244(Pt 2): 1317-28.
[http://dx.doi.org/10.1016/j.biortech.2017.06.038] [PMID: 28634124]

[72]   Selvakumar P, Sivashanmugam P. Ultrasound assisted oleaginous yeast lipid extraction and garbage lipase catalyzed transesterification for enhanced biodiesel production. Energy Convers Manage 2019; 179: 141-51.
[http://dx.doi.org/10.1016/j.enconman.2018.10.051]

[73]   Ward VCA, Munch G, Cicek N, Rehmann L. Direct conversion of the oleaginous yeast *Rhodosporidium diobovatum* to biodiesel using the ionic liquid. ACS Sustain Chem& Eng 2017; 5: 5562-70. [C 2 mim]. [EtSO 4].
[http://dx.doi.org/10.1021/acssuschemeng.7b00976]

[74]   Khot M, Ghosh D. Lipids of *Rhodotorula mucilaginosa* IIPL32 with biodiesel potential: Oil yield, fatty acid profile, fuel properties. J Basic Microbiol 2017; 57(4): 345-52.
[http://dx.doi.org/10.1002/jobm.201600618] [PMID: 28155998]

[75]   Martinez-Silveira A, Villarreal R, Garmendia G, Rufo C, Vero S. Process conditions for a rapid *in situ* transesterification for biodiesel production from oleaginous yeasts. Electron J Biotechnol 2019; 38: 1-9.
[http://dx.doi.org/10.1016/j.ejbt.2018.11.006]

[76]   Kuan I-C, Kao W-C, Chen C-L, Yu C-Y. Microbial biodiesel production by direct transesterification of *Rhodotorula glutinis* biomass. Energies 2018; 11: 1036.
[http://dx.doi.org/10.3390/en11051036]

CHAPTER 13

Improvement of Organic Agriculture with Growth-Promoting and Biocontrol Yeasts

Karen A. Achilles[1], Aline F. Camargo[2], Francisco Wilson Reichert Júnior[1], Lindomar Lerin[1], Thamarys Scapini[2], Fábio S. Stefanski[2], Caroline Dalastra[2], Helen Treichel[2] and Altemir J. Mossi[1,*]

[1] *Agroecology Laboratory, Federal University of Fronteira Sul, Erechim, Brazil*

[2] *Microbiology and Bioprocesses Laboratory, Federal University of Fronteira Sul, Erechim, Brazil*

Abstract: Organic agriculture has significantly expanded over the years, increasing population. The productive methodology adopted in globalized agricultural systems reinforced the need to develop technologies that reduce the problems caused by the excessive use of pesticides and synthetic fertilizers. Some progress is being made by applying yeasts in agriculture due to the advantages associated with their use, such as promoting plant growth, biological control, inhibition of pathogens, and production of phytohormones. This chapter discusses studies that demonstrate the potential of yeasts in agriculture for biocontrol and plant growth. Yeasts are widely disseminated in the soil, increase and promote biological control, and show positive and promising results in the management of various phytopathogens. The interactions of these organisms influence multiple processes, such as the mineralization of organic matter in the soil, nutrient cycle, disease and weed control, and ecological balance. Efforts must be made to enable the production and application of yeasts as control agents in agriculture. Considering the diversity of yeast species present in the soil, their morphological, physiological, and phenotypic properties, understanding interactions and environmental effects integrating an ecological scenario is the key to good agricultural practices in a more sustainable context.

Keywords: Biological control, Growth promotion, Organic agriculture, Sustainability, Yeasts.

INTRODUCTION

Intensive population growth and food production methods in global ecological systems under stress have aroused interest in developing productive, stable, resilient, and environmentally friendly agricultural systems that produce healthy

---

* **Corresponding author Altemir J. Mossi:** Agroecology Laboratory, Federal University of Fronteira Sul, Erechim, Brazil; Tel: +55 54 991512660; E-mail: amossiuffs@gmail.com

Sérgio Luiz Alves Júnior, Helen Treichel, Thiago Olitta Basso and Boris Ugarte Stambuk (Eds.)
All rights reserved-© 2022 Bentham Science Publishers

food and guarantee environmental integrity for future generations [1]. In conventional agriculture, weed management is carried out using pesticides, such as herbicides. In addition, the intensive and prolonged use of chemical pesticides causes environmental impacts, such as water and soil contamination, and the emergence of weed-resistant herbicides, which have increased by approximately 30% in the last ten years worldwide [2 - 4]. Resistance to pesticides has led to increased applications, higher crop losses, and farmers' mounting costs [1].

In this scenario, organic agriculture offers approaches that reduce the dependence on pesticides. Organic agriculture has expanded in many countries over the last few years; between 1999 and 2017, it has increased six-fold. In 2017, approximately 1.4 percent of the world's agricultural land was organic. Fourteen countries corresponded to more than 10% of these areas, and Australia has been a substantial contributor to exponential growth over the ten years before 2017 [5]. This increase in organic agriculture reflects the growing concern about environmental issues in intensive agriculture, mainly related to problems with the use of pesticides and commercial fertilizers. In addition, there is an increase in consumer demand for organic products, which has been supported by research and funding funds in many countries [6].

There is an expanded interest in productive and ecologically sound agriculture. Global data on organic production and markets are highly relevant for organic agriculture, considering it to be a sustainable form of production and dependent on policymakers that contribute to expanding these crops [5].

The development of new technologies that sustainably enable food security is essential in agriculture and the development of products that can reduce the excessive use of chemicals in crops to minimize environmental impacts. Significant advances are being made using microorganisms as an alternative in growth-promoting and biological control in agriculture because of their advantages over synthetic compounds, such as biodegradability, reduced half-life, and environmental safety.

The use of plant growth-promoting (PGP) microorganisms offers an alternative to reduce chemical fertilizers, resulting in increased tolerance to abiotic stresses, nutrient assimilation, pest control, plant height, root length, dry matter, *etc* [7, 8]. These benefits result from different mechanisms that help directly, for example, by phytohormone production that stimulates plant development and improves nutrient absorption by solubilization of compounds or indirectly by preventing the adverse effects of phytopathogenic microorganisms [9, 10]. PGP microorganisms have a crucial role in management systems to reduce agrochemical rates and increase the focus on biological methods for agriculture [7].

In addition to direct benefits, the potential of microorganisms for disease control in agriculture is an alternative tool in organic and conventional agriculture. It is an essential element for pest management [1]. Biological control of crops with naturally occurring microorganisms is an alternative to chemical control. It is based on a natural interaction between yeast and filamentous fungi. By different mechanisms, bacteria can protect plant crops against diseases, such as toxin and volatile compound production and specific enzyme secretion [11 - 13].

Yeast is a single-celled fungus that is abundant in the soil, with rapid growth and excellent characteristics such as PGP and biological control [14, 15]. Although research on endophytic yeasts as biocontrollers has only gained prominence in recent years, these yeasts have great potential for inhibiting phytopathogens because these microorganisms can act through several action mechanisms, thus reducing pathogen resistance in the environment [8, 16].

Yeasts can produce hormones such as indol-3-acetic acid (IAA), indol-3-pyruvic acid (IPYA), cytokinins, and several biologically active compounds that stimulate plant growth and development and increase crop productivity. Studies have identified improvements in plant growth, germination, and length after inoculation of seeds with yeast strains. When applied to biocontrol, yeasts can improve the uptake of water and nutrients, such as nitrogen and potassium, and reduce the risk of phytopathogenic infections because of the production of antimicrobial substances [17, 18].

Yeast's role in agricultural ecosystems needs advances, mainly because it is not entirely understood, and research on these microorganisms as PGP and biocontrol is scarce [10]. In this context, this chapter aims to highlight works present in the literature on the potential use of yeasts in agriculture for biocontrol and plant growth.

YEASTS IN THE NATURAL ENVIRONMENT

Living organisms can be divided into prokaryotes and eukaryotes. In the prokaryote group, we have bacteria and archaea. In eukaryotic organisms, we find cellular microorganisms, such as fungi (yeasts and molds), protozoa, algae, and higher organisms, such as plants and animals.

The Fungi Kingdom includes single-celled yeasts with approximately 680 known species distributed in two phyla (Ascomycota and Basidiomycota), multicellular molds, and macroscopic species such as mushrooms. Yeasts are non-filamentous, usually spherical or oval with a cell diameter between 1 and 10 µm, a rigid cell wall made up of the polysaccharide's glycan and mannan, which can grow in environments with low humidity, high osmotic pressure, and a wide pH range (optimum pH between 5 and 6), and many are optional anaerobic, allowing growth in different environments. Vital functions are maintained by absorbing dissolved organic matter through the plasma membrane. Yeasts are widely distributed in the environment and can be found as a white powder that covers the fruits and leaves [1].

Yeast reproduction can occur through budding or fission. During budding, the cells create a protrusion on the outer surface. The nucleus is divided with the elongation of the bud. In fission reproduction, the cell divides into two identical new cells, where the parental cell stretches and its nucleus divides; thus, two daughter cells are produced. Still, some yeasts can produce spores, and when grown in the laboratory using a solid culture medium, yeasts have colonies similar to bacteria [14].

Some yeast species can cause some pathologies in humans (*Candida albicans*, *Cryptococcus neoformans*, *Histoplasma capsulatum*, *Blastomyces dermatitidis*, and *Malassezia furfur*), animals (*Candida* spp., *C. neoformans*, *C. gattii*, *Malassezia* ssp., among others), and vegetables (*Zygosaccharomyces bailii*, *Brettanomyces* spp.). Yeasts, mainly *Saccharomyces cerevisiae*, have wide applications in industrial fermentation processes, such as in the production of alcoholic beverages, biofuel (ethanol), bakery, citric acid synthesis, and pharmaceuticals (production of B complex vitamins). Yeasts are also used in genetic engineering because they are rapidly growing, as in vaccine production against the hepatitis B virus, biosensors, and can also be used as bioremediation and biological control agents [18].

The diversity of microorganisms found in the environment is considerable, both in various species and in the abundance of individuals. Microorganisms and other living organisms do not survive in isolation and constantly interact with each other. This is essential for survival in highly variable, competitive, and oligotrophic (nutrient-poor) environments. These interactions can be divided into two large groups: positive and negative. Positive interactions allow at least one of the organisms to benefit, and none of them are harmful to the positive interactions. The following stand out Commensalism, Mutualism, and Protocooperation [19].

In the second group, there are adverse interactions, and the least one of the organisms involved is harmed, considering only the individual's point of view. Considering all the population and the ecosystem, the interaction is deemed beneficial, as it prevents population increase and acts in the natural selection, causing new adaptations. The following stand out among the negative interactions: Parasitism, antagonism, competition, and predation [20].

It should be noted that although each of the interactions has been approached in isolation, a microbial population can perform more than one type of interaction with other populations. These interactions between organisms influence various processes such as mineralization of organic matter in the soil and xenobiotics, nutrient cycling, humus formation, biological weed, disease control, and biological balance, showing the complexity of the environment. The use of yeasts as biocontrol agents has been extensively studied. Successful control of several phytopathogens has been demonstrated to be a promising strategy for controlling pathogens [16].

Yeasts found inside plant tissues are called endophytes, whereas the microorganisms that colonize plant organ surfaces are called epiphytes. These inhabit, in general, all the organs and tissues of the plants, such as the aerial parts of the plants (leaves, stems, flowers, and fruits) or in association with the underground parts (roots). Positive interactions occur between the microorganism and plant, without causing any apparent damage. These organisms can increase the compounds produced by plants, such as hormones, antimicrobials, nitrogen fixation, tooth others, and differentiating themselves from phytopathogens. However, this interaction can change depending on the organism's disposition, development, and environmental factors. Several endophytic microorganisms colonize as epiphytes and subsequently penetrate the tissues or even in the root tissues and reach other tissues through the plant's vascular system [8].

These microorganisms, especially yeasts, with their interactions with plants, can be used for biological control. Biological control is based on negative interactions using microorganisms with great adaptability and interaction with phytopathogens. Among the negative interactions, parasitism, antagonism, and competition for space or nutrients stand out. Positive interactions between the yeast and plant may promote growth or induce plant resistance to phytopathogens. It emphasizes that biological control agents can act on the pathogen using more than one action mechanism, which results in a more robust sense of the spectrum of action, preventing the appearance of resistance from phytopathogens. To select yeasts for biological control, general characteristics must be considered, such as the ability to colonize the surface of the substrates and remain viable quickly; more remarkable ability than pathogens to acquire nutrients; ability to survive under different environmental conditions and not present pathogenicity for humans [7, 8].

The microbiome of plants can be stimulated or used to improve the results in terms of agricultural productivity by applying environmentally friendly approaches based on the interaction of the soil, plant, and microorganism microbiome. Generally, in the rhizosphere, interactions occur between the roots of plants and the soil microbial community. This soil zone is influenced by the release of substrates that affect microbial activity [16].

The rhizosphere has an active environment and is subject to changes arising from temporal conditions, which results in variations in water availability, salt concentration, osmotic potential, and soil pH. These conditions make the rhizosphere an engaging environment for soil-microbial community interactions. Therefore, several yeast strains can be found in the rhizosphere, because they find a reasonable possibility of nutrients. Some yeasts that live in this environment are specialized in promoting plant growth and have some characteristics such as inhibition of pathogens, production of phytohormones, and solubilization of phosphate and nitrogen. These characteristics are relevant when considering disease control in plants and even invasive plant biocontrol [20].

Several interactions can co-occur in the soil, specifically in yeasts. They can be related to competition for nutrients, enzymatic production, mycoparasitism, nitrogen fixation, phosphate, biocontrol activity, and decomposition of organic matter, among others [20]. The soil microbiota is responsible for essential interactions within this ecosystem, balancing the microenvironment through ecological associations. In addition to biocontrol, it is possible to increase organic matter availability, degrade the lignin present in the soil, regulate photosynthesis, and synthesize bioactive substances such as phytohormones and enzymes. Thus, it is possible to associate these interactions with the promotion of plant growth.

The soil can be considered a nutrient bank, in which the growth of plants will depend on the current condition of the soil, mainly on the presence of macronutrients such as N, P, K. Nitrogen is considered fundamental to living organisms and a constituent of chlorophyll, or that is, it plays a role in photosynthesis since their soluble forms better absorb phosphorus and potassium, specialized microorganisms perform this solubilization step. Saprobic microorganisms are specialized in obtaining nutrients from decomposing organic matter, so they play an essential role in the carbon cycle, helping to maintain the levels of organic carbon in the soil; therefore, this nutrient is always available for the use of plants. Carbon in the soil also increases water retention capacity, as it keeps the soil particles closer and promotes microbial biodiversity because it is related to high levels of organic matter [9, 10].

Yeasts are abundant in the soil; however, their ecological functions in the soil and agrosystems are not entirely known, what is known is that they are found in the top 10 cm of the soil due to their ability to grow in carbon compounds and act as organic decomposition matter, phosphate solubilization, promotion of root growth, soil aggregation and biocontrol of root pathogens. Among the wide range of possibilities, the yeast strains usually found in the soil belong to the genera *Candida, Cryptococcus, Rhodotorula, Lipomyces, Sporobolomyces, Trichosporon, Pichia, Saccharomyces, Debaryomyces, Aureobasidium,* and *Williopsis* [1, 11].

The interaction between soil and plants is enhanced by the presence of microorganisms that fix atmospheric nitrogen, supplement plants with phosphorus and potassium, release enzymes and metabolites with antioxidant function, vitamins, and hormones, among which some can act as biocontrollers and maintain their tasks even in soils that experience stress due to drought or saline concentration, and the action of these microorganisms can stimulate plant growth and improve agricultural productivity. In this sense, we can understand that microbial interactions in the rhizosphere are the key to determining the dynamics and nutrient flows of plants and soil and can be used as a strategy to improve efficiency in agroecosystems [21 - 25].

GROWTH-PROMOTING YEASTS APPLIED TO ORGANIC AGRICULTURE

In organic agriculture, microorganisms are used to promote plant growth. Among the microorganisms commonly used, yeasts can be used as growth-promoting agents, but research related to this has been little investigated. To date, the role of yeasts in agricultural systems is still not well understood, and the number of studies addressing this is still scarce [10]. For this reason, this section aims to bring to light scientific knowledge, examples, current uses, and applications of yeasts as agents that promote growth and biocontrol in agriculture.

Plant Growth Promoting

Plant growth-promoting (PGP) microorganisms offer several benefits to agricultural plants and promote plant growth, directly and indirectly, acting as potential biofertilizers. Phytohormones and other molecules such as amino acids, organic acids, siderophores, lytic enzymes, and volatile compounds, in addition to increased nutrient absorption, are examples of mechanisms that microorganisms, such as PGP yeasts, are used to stimulate plant development (Fig. **1**) [10, 13]. The beneficial effects of biological compounds produced by these microorganisms begin in the early stages of plant development. It is later expressed in the suppression of phytopathogens and increased productivity [17].

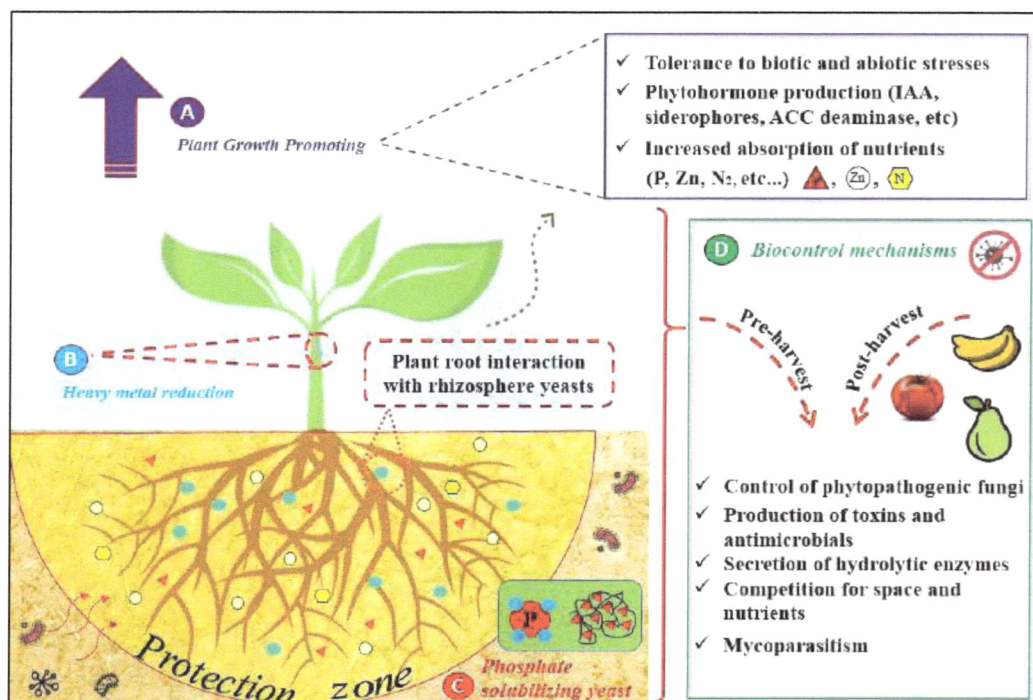

Fig. (1). Mechanisms used by yeasts to promote growth and biocontrol in plants in agriculture.

At the beginning of this century, *Saccharomyces cerevisiae* has emerged as a promising growth-promoting yeast for different crops, making it a positive alternative to chemical fertilizers. In the last two decades, new isolates have emerged as potential, not only for promoting plant growth but also for biocontrol tools in agriculture, which will be seen later [17, 26, 27]. Genres such as *Candida*, *Rhodotorula*, *Sporobolomyces*, *Trichosporon, Pseudomonas Williopsis*, and *Yarrowia* have been documented [28 - 31].

Indole-3-acetic acid (IAA) is the main phytohormone used by plants to regulate growth and is involved in several physiological processes, including lengthening and dividing plant cells, germination, vascular development, and root growth [32]. Evidence in the anatomy of sugar beet leaves has already shown that yeasts increase the thickness of the leaf blade and the middle vein, increasing the length and width of the vascular bundles. In addition to IAA production, rhizospheric soil yeast showed a relatively high tolerance to heavy metals. They can produce siderophores, aminocyclopropane-1-carboxylate (ACC) deaminase, solubilizing phosphates, zinc salts, and releasing nitrogen for plants [10, 27]. Phosphate-solubilizing microorganisms increase plant growth by solubilizing insoluble phosphates in the soil, particularly in phosphorus-deficient environments [10]. Yeasts with these characteristics have been suggested to improve growth and reduce heavy metal accumulation within plants [27].

In studies (Table **1**), ten promising strains from 69 isolates showed multifaceted characteristics providing partial information on the yeast-plant interaction, opening a new door for future agronomic developments in this segment. While many studies have investigated new potential isolates, others have sought to improve yeast performance in environments with low nitrogen [10]. This is what the results indicate: In addition to the rapid growth of *Candida tropicalis*.HY (Table **1**) in aminocyclopropane-1-carboxylate as the only source of nitrogen, this microorganism quickly colonized rice seedling roots maintaining it for at least three weeks and being able to synthesize polyamines and mobilize organic and inorganic phosphates [15]. The authors support the inclusion of this microorganism in commercial PGP inoculants for the development of sustainable agriculture.

Table 1. Yeasts, functions, and effects observed in plant growth and as biocontrol.

Microorganisms	Function	Observed Effect	Studied in	Refs.
Yeast strains isolated from Spanish vineyards (*e.g.*, *Pichia dianae* Pd-2 and *Meyerozyma guilliermondii* Mg-11)	Growth promoting	Solubilization or production of IAA, siderophores, NH3, catalase, and different hydrolytic enzymes	-*Nicotiana benthamiana* (*in vitro*); -lettuce; -corn seedlings	[10]
Yeast isolates from the phyllosphere and rhizosphere of the *Drosera spatulata* Lab.	Growth promoting	Were tested for indole-3-acetic acid-, ammonia-, and polyamine-producing abilities, calcium phosphate and zinc oxide solubilizing ability, and catalase activity	*Nicotiana benthamiana*	[9]
Saccharomyces cerevisiae	Growth promoting	Increases the length of the stem, the length of the root, the fresh weight of the stem, the dry weight of the stem, and the chlorophyll content, through the secretion of indolacetic acid and organic acids	Cucumber seeds (*Cucumis sativus* L. cv. Black Pearls)	[46]
Candida tropicalis HY	Growth promoting	Produces small amounts of indolacetic acid (IAA) and multiplies on aminocyclopropane-1-carboxylate (ACC) as the sole source of nitrogen	Rice	[15]
Rhodotorula sp.	Growth promoting	Showed growth promotion characteristics, tolerance to aluminum toxicity, and exopolysaccharide production capacity under abiotic stress	-	[47]

(Table 1) cont.....

Microorganisms	Function	Observed Effect	Studied in	Refs.
Candida subhashii	Biocontrol of plant pathogenic fungi	Strong competition for macro and micronutrients	*In vitro*	[40]
Rhodotorula glutinis	Biocontrol of plant pathogenic fungi	Reduced *Botrytis cinerea* growth and ochratoxin A production	Tomato	[37]
Candida tropicalis	Biocontrol of plant pathogenic fungi	Inhibition of spore germination and growth of *Colletotrichum musae*	Banana	[48]
Pichia guilliermondii	Biocontrol of plant pathogenic fungi	Inhibition of mycelial growth of *Botrytis cinerea, Alternaria alternata, Rhizopus nigricans*	Cherry tomato	[49]
Rhodotorula mucilaginosa	Biocontrol of plant pathogenic fungi	Growth inhibition *in vitro* of *Penicillium expansum*	Pear	[50]

For the success of sustainable agriculture in this segment, potential laboratory studies must reproduce the results on a full scale. Infield studies have observed that soil application of a yeast strain (*Meyerozyma guilliermondii*) increased the vigor index of corn seeds. In addition, phosphate solubilization efficiency similar to that of Lt-47 reduced the application of chemical fertilizers without affecting the excellent corn yield. Results, such as this and the increasing demands for regulatory policies in organic agriculture motivate scientists to explore new microorganisms and their mutualistic interactions with plants in the direction of more sustainable agriculture [33].

Biocontrol

In addition to their use as growth promoters, yeasts can also be used as biocontrol agents. The potential of yeast antagonists for biocontrol is still little explored, with only a few products based on yeasts for plant protection being present in the market [22]. Despite presenting themselves as an attractive and environmentally friendly alternative and focusing on studies for a relatively long time, yeast-based biocontrol is still not widely used. One of the reasons for this phenomenon is the difference between the laboratory results and experiments in agricultural environments [34, 35]. Biocontrol can be used for several organisms in agriculture (fungi, plants, and insects); however, studies with the use of yeast for biocontrol have, for the most part, focused on the control of phytopathogenic fungi. Thirty yeast isolated from three Greece regions showed that all isolates reduced the percentage of damage caused by *Botrytis cinerea* in tomato plants, with *Rhodotorula glutinis* Y-44 being the most efficient, reducing the incidence of the disease by 52% [36, 37]. The yeast strains inhibited vegetative growth of the evaluated fungi. The lineage of *Lachancea kluyveri* CCMA 0151 inhibited *Aspergillus ochraceus* by 80% and *Aspergillus carbonarius* by 41% in Petri dishes. Studies have reported the antagonistic effects of yeasts against phytopathogenic fungi [35]. The mechanisms of yeast action vary, such as toxin production, hydrolytic enzymes capable of degrading the cell wall, and toxic and volatile compounds (Fig. **1**) [11, 12, 38]. In addition to the previously mentioned mechanisms, competition for space and nutrients, mycoparasitism, and resistance induction in plants can also be related to antagonistic functions [38].

Competition for space and nutrients is considered the primary mode of action of yeasts against post-harvest pathogenic microorganisms. It is regarded as an efficient method when the yeast antagonist presents itself at the ideal time and place and sufficient quantity, limiting access to resources more efficiently than the pathogenic fungus [34]. However, competition is challenging to study as a mechanism of action, especially in natural environments, where resources are scarce and competitors are abundant. The battle for nutrients and space has been studied mainly in ecology and is associated with species diversity. In yeasts such as *Metschnikowia reukaufii*, the ability to restrict nutrient availability may be related to duplication of metabolism genes and duplication of nitrogen transporters, defining the community's composition due to the effects of its primary colonization [22, 39].

In vitro experiments demonstrate differences in the antifungal capacity of yeast, since yeast can present great or small antagonism to the same fungus species [40]. However, under field conditions, several mechanisms can contribute to the primary establishment of yeasts, giving them an initial advantage in colonizing the environment, avoiding the establishment and development of other fungi [22].

Toxic compounds are among the mechanisms involved in the antagonistic interactions between yeasts and phytopathogenic fungi. Yeasts can produce compounds with antibiotic and antimicrobial effects. Certain yeast strains produce extracellular protein toxins that are lethal to microorganisms of the same or different species. Producers of these toxins can eliminate the same species but are immune to toxins of the same class they produce. Yeast protein toxins are among the most studied compounds due to their antagonistic effects [34, 41]. In addition, compounds such as flocculosin, aureobasidin, and liamocins can contribute to the negative effects of yeasts [42, 43].

In addition, the release of hydrolytic enzymes appears to contribute significantly to yeast antagonism. The secretion of enzymes that degrade the cell wall is common in many host-pathogen interactions. Enzymes can be released into the medium to degrade cells and make nutrients available to enzyme-producing organisms. The cells "attacked" by these enzymes have their cell wall destroyed, causing disruption of the cell membrane, leading to cell death. These interactions are observed in mycoparasitism and fungivory and are also observed in interactions between yeasts and fungi. Among the enzymes involved in this interaction, lipases, glucanases, proteases, and chitinases have been reported [22].

In addition to the methods mentioned above, biofilm production, mycoparasitism, and resistance induction in plants have also been cited as methods of antagonism of yeasts against phytopathogenic fungi [44, 45]. Some studies showing the effects of yeasts on phytopathogenic fungi are shown in Table 1. Thus, yeasts present themselves as an alternative biocontrol, mainly of phytopathogenic fungi, which is an essential tool in the search for more sustainable agriculture.

CONCLUSION

Many challenges arise during the development of a biological control product, such as the improvement and increase in the efficiency of biocontrol in commercial conditions, development of economical methods that have a high quality of fermentation and formulation, maintenance of cell viability, identification of yeast antagonists capable of exhibiting a broad spectrum of activity against various pathogens and commodities, and the development of a comprehensive measure about the operation of the biological control system and how the environment affects the interaction between the host, pathogen, and the biocontrol agent [51].

Yeast species in the rhizospheric environment act as bioagents to suppress and inhibit the growth of disease-causing fungi and in the production of plant growth regulators, reflecting both the fresh plant and the dry weight and length of the plants. There is great potential for yeast species to be used as biocontrol agents. However, it is necessary to enable the production and application of these organisms as control agents in agriculture [52, 53]. Studies with yeasts collaborate to develop multifunctional biological preparations for agriculture because they have diverse biological activities. The strains studied in this chapter are promising for commercial inclusion as yeasts that promote plant growth for sustainable agriculture [4, 19, 20].

The limitations that have arisen to date suggest the need for further research that includes the objective of studying the diversity of yeast species existing in the investigated environment, addressing morphological, physiological, and phenotypic properties. Adapting a culture medium that provides a favorable environment for yeast growth is necessary for successful yeast development. Proper management of sample collection means more extraordinary richness and diversity in the occurrence and distribution of collected yeasts. However, it is essential to consider the types of soil that exist, the management used in soil preparation, the implanted culture, and pest control [53, 54]. A challenge to be verified is the relative contribution of soil yeasts to ecological processes that include a more significant number of microorganisms and the study of the fate of these strains within the natural soil ecosystem, either individually or through the species [14].

Future work should emphasize research at the laboratory and the field level, understand the microbiome's interactions and environmental stresses, apply recent technologies, and the integrated meta-organism approach from an ecological perspective. The products must be ecologically correct, economically viable, and a secure database to facilitate the registration process for commercialization. It is necessary to transmit more knowledge to farmers regarding the benefits of using these compounds [55].

Yeasts have many metabolic and physiological characteristics that give them advantages over conventional synthetic agents. With a focus on sustainability and climate change, chemical processes will be replaced by fermentative methods. The combination of knowledge of microbial and genetic diversity adjusted with synthetic biological tools and bioinformatics enables the production of new products with a wide range of applications [31].

The practice of sustainable agriculture and the adoption of organic farming systems improve the conservation of the environment, produce food free of chemical contaminants, and add commercial value to the final product. Studies comparing organic and conventional agriculture claim that the organic system can achieve a greater abundance of species, demonstrating the more significant microbial activity, biomass, and biodiversity [55]. The growing awareness of the negative impacts generated by the indiscriminate use of chemicals in the environment has provided several studies that reinforce the need to employ environmentally sustainable management. Given this fact, yeasts have great potential and can be used in organic agriculture, reducing the environmental impact caused by inadequate soil management [22].

CONSENT FOR PUBLICATION

Not applicable.

CONFLICT OF INTEREST

The authors declare no conflict of interest, financial or otherwise.

ACKNOWLEDGEMENT

Declared none.

REFERENCES

[1]   Baker BP, Green TA, Loker AJ. Biological control and integrated pest management in organic and conventional systems. Biol Control 2020; 140: 104095.
        [http://dx.doi.org/10.1016/j.biocontrol.2019.104095]

[2]   Badgley C, Moghtader J, Quintero E, *et al.* Organic agriculture and the global food supply. Renew Agric Food Syst 2007; 22: 86-108.
        [http://dx.doi.org/10.1017/S1742170507001640]

[3]   Camargo AF, Venturin B, Bordin ER, *et al.* A Low-Genotoxicity Bioherbicide Obtained from Trichoderma koningiopsis Fermentation in a Stirred-Tank Bioreactor. Ind Biotechnol (New Rochelle NY) 2020; 16: 176-81.
        [http://dx.doi.org/10.1089/ind.2019.0024]

[4]   Heap. The International Herbicide-Resistant Weed Database 2021.

[5]   FIBL; IFOAM. The world of organic agriculture : statistics and emerging trends 2019.

[6]   Nguyen HV, Nguyen N, Nguyen BK, Lobo A, Vu PA. Organic food purchases in an emerging market: the influence of consumers' personal factors and green marketing practices of food stores. Int J Environ Res Public Health 2019; 16(6): 1037.
        [http://dx.doi.org/10.3390/ijerph16061037] [PMID: 30909390]

[7]   Saharan BS, Nehra V. Plant growth promoting rhizobacteria: a critical review. Life Sci Med Res 2011; LSMR-21: 1-30.

[8]   Ling L, Tu Y, Ma W, *et al.* A potentially important resource: endophytic yeasts. World J Microbiol Biotechnol 2020; 36(8): 110.
        [http://dx.doi.org/10.1007/s11274-020-02889-0] [PMID: 32656593]

[9]   Fu S-F, Sun P-F, Lu H-Y, *et al.* Plant growth-promoting traits of yeasts isolated from the phyllosphere and rhizosphere of Drosera spatulata Lab. Fungal Biol 2016; 120(3): 433-48.
        [http://dx.doi.org/10.1016/j.funbio.2015.12.006] [PMID: 26895872]

[10]   Fernandez-San Millan A, Farran I, Larraya L, Ancin M, Arregui LM, Veramendi J. Plant growth-promoting traits of yeasts isolated from Spanish vineyards: benefits for seedling development. Microbiol Res 2020; 237: 126480.
        [http://dx.doi.org/10.1016/j.micres.2020.126480] [PMID: 32402946]

[11]   Walker GM, McLeod AH, Hodgson VJ. Interactions between killer yeasts and pathogenic fungi. FEMS Microbiol Lett 1995; 127(3): 213-22.
        [http://dx.doi.org/10.1111/j.1574-6968.1995.tb07476.x] [PMID: 7758935]

[12]   Urquhart EJ, Punja ZK. Hydrolytic enzymes and antifungal compounds produced by Tilletiopsis species, phyllosphere yeasts that are antagonists of powdery mildew fungi. Can J Microbiol 2002; 48(3): 219-29.

[http://dx.doi.org/10.1139/w02-008] [PMID: 11989766]

[13]   Santos A, Sánchez A, Marquina D. Yeasts as biological agents to control Botrytis cinerea. Microbiol Res 2004; 159(4): 331-8.
[http://dx.doi.org/10.1016/j.micres.2004.07.001] [PMID: 15646379]

[14]   Botha A. Yeasts in Soil. Biodivers Ecophysiol Yeasts. Berlin, Heidelberg: Springer-Verlag 2006; pp. 221-40.
[http://dx.doi.org/10.1007/3-540-30985-3_11]

[15]   Amprayn K, Rose MT, Kecskés M, Pereg L, Nguyen HT, Kennedy IR. Plant growth promoting characteristics of soil yeast (*Candida tropicalis* HY) and its effectiveness for promoting rice growth. Appl Soil Ecol 2012; 61: 295-9.
[http://dx.doi.org/10.1016/j.apsoil.2011.11.009]

[16]   Singh A, Kumari R, Yadav AN, Mishra S, Sachan A, Sachan SG. Tiny microbes, big yields: Microorganisms for enhancing food crop production for sustainable development. New Futur Dev Microb Biotechnol Bioeng. Elsevier 2020; pp. 1-15.
[http://dx.doi.org/10.1016/B978-0-12-820526-6.00001-4]

[17]   Ignatova LV, Brazhnikova YV, Berzhanova RZ, Mukasheva TD. Plant growth-promoting and antifungal activity of yeasts from dark chestnut soil. Microbiol Res 2015; 175: 78-83.
[http://dx.doi.org/10.1016/j.micres.2015.03.008] [PMID: 25843007]

[18]   Mukherjee A, Verma JP, Gaurav AK, Chouhan GK, Patel JS, Hesham AE-L. Yeast a potential bio-agent: future for plant growth and postharvest disease management for sustainable agriculture. Appl Microbiol Biotechnol 2020; 104(4): 1497-510.
[http://dx.doi.org/10.1007/s00253-019-10321-3] [PMID: 31915901]

[19]   Hallam SJ, McCutcheon JP. Microbes don't play solitaire: how cooperation trumps isolation in the microbial world. Environ Microbiol Rep 2015; 7(1): 26-8.
[http://dx.doi.org/10.1111/1758-2229.12248] [PMID: 25721597]

[20]   Aladdin A, Dib JR, Malek RA, El Enshasy HA. Killer Yeast, a Novel Biological Control of Soilborne Diseases for Good Agriculture Practice. Sustain Technol Manag Agric. Wastes, Singapore: Springer Singapore 2018; pp. 71-86.
[http://dx.doi.org/10.1007/978-981-10-5062-6_6]

[21]   Wilson CL, Wisniewski ME. Biological control of postharvest diseases of fruits and vegetables: an emerging technology*. Annu Rev Phytopathol 1989; 27: 425-41.
[http://dx.doi.org/10.1146/annurev.py.27.090189.002233]

[22]   Freimoser FM, Rueda-Mejia MP, Tilocca B, Migheli Q. Biocontrol yeasts: mechanisms and applications. World J Microbiol Biotechnol 2019; 35(10): 154.
[http://dx.doi.org/10.1007/s11274-019-2728-4] [PMID: 31576429]

[23]   Naik K, Mishra S, Srichandan H, Singh PK, Sarangi PK. Plant growth promoting microbes: Potential link to sustainable agriculture and environment. Biocatal Agric Biotechnol 2019; 21: 101326.
[http://dx.doi.org/10.1016/j.bcab.2019.101326]

[24]   Sarabia M, Cazares S, González-Rodríguez A, Mora F, Carreón-Abud Y, Larsen J. Plant growth promotion traits of rhizosphere yeasts and their response to soil characteristics and crop cycle in maize agroecosystems. Rhizosphere 2018; 6: 67-73.
[http://dx.doi.org/10.1016/j.rhisph.2018.04.002]

[25]   Sarabia M, Jakobsen I, Grønlund M, Carreon-Abud Y, Larsen J. Rhizosphere yeasts improve P uptake of a maize arbuscular mycorrhizal association. Appl Soil Ecol 2018; 125: 18-25.
[http://dx.doi.org/10.1016/j.apsoil.2017.12.012]

[26]   Ramos-Garza J, Bustamante-Brito R, Ángeles de Paz G, *et al.* Isolation and characterization of yeasts associated with plants growing in heavy-metal- and arsenic-contaminated soils. Can J Microbiol 2016; 62(4): 307-19.

[http://dx.doi.org/10.1139/cjm-2015-0226] [PMID: 26936448]

[27]   El-Maraghy SS, Tohamy TA, Hussein KA. Expression of SidD gene and physiological characterization of the rhizosphere plant growth-promoting yeasts. Heliyon 2020; 6(7): e04384.
[http://dx.doi.org/10.1016/j.heliyon.2020.e04384] [PMID: 32671269]

[28]   Medina A, Vassileva M, Caravaca F, Roldán A, Azcón R. Improvement of soil characteristics and growth of Dorycnium pentaphyllum by amendment with agrowastes and inoculation with AM fungi and/or the yeast Yarowia lipolytica. Chemosphere 2004; 56(5): 449-56.
[http://dx.doi.org/10.1016/j.chemosphere.2004.04.003] [PMID: 15212910]

[29]   Nassar AH, El-Tarabily KA, Sivasithamparam K. Promotion of plant growth by an auxin-producing isolate of the yeast Williopsis saturnus endophytic in maize (Zea mays L.) roots. Biol Fertil Soils 2005; 42: 97-108.
[http://dx.doi.org/10.1007/s00374-005-0008-y]

[30]   El-Tarabily KA, Sivasithamparam K. Potential of yeasts as biocontrol agents of soil-borne fungal plant pathogens and as plant growth promoters. Mycoscience 2006; 47: 25-35.
[http://dx.doi.org/10.1007/S10267-005-0268-2]

[31]   Sheth A, Borse P. Sugarcane Vinasse, Molasses, Yeast Cream. Adv Sugarcane Biorefinery. Elsevier 2018; pp. 153-61.
[http://dx.doi.org/10.1016/B978-0-12-804534-3.00007-0]

[32]   Luo J, Zhou J-J, Zhang J-Z. Aux/IAA Gene Family in Plants: Molecular Structure, Regulation, and Function. Int J Mol Sci 2018; 19(1): 259.
[http://dx.doi.org/10.3390/ijms19010259] [PMID: 29337875]

[33]   Nakayan P, Hameed A, Singh S, Young L-S, Hung M-H, Young C-C. Phosphate-solubilizing soil yeast Meyerozyma guilliermondii CC1 improves maize (Zea mays L.) productivity and minimizes requisite chemical fertilization. Plant Soil 2013; 373: 301-15.
[http://dx.doi.org/10.1007/s11104-013-1792-z]

[34]   Spadaro D, Droby S. Development of biocontrol products for postharvest diseases of fruit: The importance of elucidating the mechanisms of action of yeast antagonists. Trends Food Sci Technol 2016; 47: 39-49.
[http://dx.doi.org/10.1016/j.tifs.2015.11.003]

[35]   Gross S, Kunz L, Müller DC, Santos Kron A, Freimoser FM. Characterization of antagonistic yeasts for biocontrol applications on apples or in soil by quantitative analyses of synthetic yeast communities. Yeast 2018; 35(10): 559-66.
[http://dx.doi.org/10.1002/yea.3321] [PMID: 29752875]

[36]   Kalogiannis S, Tjamos SE, Stergiou A, Antoniou PP, Ziogas BN, Tjamos EC. Selection and evaluation of phyllosphere yeasts as biocontrol agents against grey mould of tomato. Eur J Plant Pathol 2006; 116: 69-76.
[http://dx.doi.org/10.1007/s10658-006-9040-5]

[37]   Souza ML, Passamani FRF, Ávila CL da S, Batista LR, Schwan RF, Silva CF. Use of wild yeasts as a biocontrol agent against toxigenic fungi and OTA production. Acta Sci Agron 2017; 39: 349.
[http://dx.doi.org/10.4025/actasciagron.v39i3.32659]

[38]   Rosa-Magri MM, Tauk-Tornisielo SM, Ceccato-Antonini SR. Bioprospection of yeasts as biocontrol agents against phytopathogenic molds. Braz Arch Biol Technol 2011; 54: 1-5.
[http://dx.doi.org/10.1590/S1516-89132011000100001]

[39]   Dhami MK, Hartwig T, Fukami T. Genetic basis of priority effects: insights from nectar yeast. Proc Biol Sci 2016; 283(1840): 20161455.
[http://dx.doi.org/10.1098/rspb.2016.1455] [PMID: 27708148]

[40]   Hilber-Bodmer M, Schmid M, Ahrens CH, Freimoser FM. Competition assays and physiological experiments of soil and phyllosphere yeasts identify Candida subhashii as a novel antagonist of

filamentous fungi. BMC Microbiol 2017; 17(1): 4.
[http://dx.doi.org/10.1186/s12866-016-0908-z] [PMID: 28056814]

[41]   Selitrennikoff CP. Antifungal proteins. Appl Environ Microbiol 2001; 67(7): 2883-94.
[http://dx.doi.org/10.1128/AEM.67.7.2883-2894.2001] [PMID: 11425698]

[42]   Teichmann B, Labbé C, Lefebvre F, Bölker M, Linne U, Bélanger RR. Identification of a biosynthesis gene cluster for flocculosin a cellobiose lipid produced by the biocontrol agent Pseudozyma flocculosa. Mol Microbiol 2011; 79(6): 1483-95.
[http://dx.doi.org/10.1111/j.1365-2958.2010.07533.x] [PMID: 21255122]

[43]   Price NP, Bischoff KM, Leathers TD, Cossé AA, Manitchotpisit P. Polyols, not sugars, determine the structural diversity of anti-streptococcal liamocins produced by Aureobasidium pullulans strain NRRL 50380. J Antibiot (Tokyo) 2017; 70(2): 136-41.
[http://dx.doi.org/10.1038/ja.2016.92] [PMID: 27436607]

[44]   Ianiri G, Idnurm A, Wright SAI, *et al.* Searching for genes responsible for patulin degradation in a biocontrol yeast provides insight into the basis for resistance to this mycotoxin. Appl Environ Microbiol 2013; 79(9): 3101-15.
[http://dx.doi.org/10.1128/AEM.03851-12] [PMID: 23455346]

[45]   Sun C, Fu D, Lu H, Zhang J, Zheng X, Yu T. Autoclaved yeast enhances the resistance against Penicillium expansum in postharvest pear fruit and its possible mechanisms of action. Biol Control 2018; 119: 51-8.
[http://dx.doi.org/10.1016/j.biocontrol.2018.01.010]

[46]   Kang S-M, Radhakrishnan R, You Y-H, Khan AL, Park J-M, Lee S-M, *et al.* Cucumber performance is improved by inoculation with plant growth-promoting microorganisms. Acta Agric Scand Sect B — Soil. Plant Sci 2015; 65: 36-44.
[http://dx.doi.org/10.1080/09064710.2014.960889]

[47]   Silambarasan S, Logeswari P, Cornejo P, Kannan VR. Evaluation of the production of exopolysaccharide by plant growth promoting yeast Rhodotorula sp. strain CAH2 under abiotic stress conditions. Int J Biol Macromol 2019; 121: 55-62.
[http://dx.doi.org/10.1016/j.ijbiomac.2018.10.016] [PMID: 30290257]

[48]   Zhimo VY, Dilip D, Sten J, Ravat VK, Bhutia DD, Panja B, *et al.* Antagonistic Yeasts for Biocontrol of the Banana Postharvest Anthracnose Pathogen Colletotrichum musae. J Phytopathol 2017; 165: 35-43.
[http://dx.doi.org/10.1111/jph.12533]

[49]   Zhao Y, Tu K, Su J, *et al.* Heat treatment in combination with antagonistic yeast reduces diseases and elicits the active defense responses in harvested cherry tomato fruit. J Agric Food Chem 2009; 57(16): 7565-70.
[http://dx.doi.org/10.1021/jf901437q] [PMID: 19637930]

[50]   Hu H, Xu Y, Lu HP, Xiao R, Zheng XD, Yu T. Evaluation of yeasts from Tibetan fermented products as agents for biocontrol of blue mold of Nashi pear fruits. J Zhejiang Univ B 2015; 16(4): 275-85.
[http://dx.doi.org/10.1631/jzus.B1400162] [PMID: 25845361]

[51]   Liu J, Sui Y, Wisniewski M, Droby S, Liu Y. Review: Utilization of antagonistic yeasts to manage postharvest fungal diseases of fruit. Int J Food Microbiol 2013; 167(2): 153-60.
[http://dx.doi.org/10.1016/j.ijfoodmicro.2013.09.004] [PMID: 24135671]

[52]   Fareed A, Ali SA, Hasan KA, Sultana V, Ehteshamul-Haque S. Evaluation of biocontrol and plant growth promoting potential of endophytic yeasts isolated from healthy plants. Pak J Bot 2019; 51.
[http://dx.doi.org/10.30848/PJB2019-6(44)]

[53]   Vadkertiová R, Dudášová H, Balaščáková M. Yeasts in agricultural and managed soils. Yeasts Nat Ecosyst Divers. Cham: Springer International Publishing 2017; pp. 117-44.
[http://dx.doi.org/10.1007/978-3-319-62683-3_4]

[54]   Kumari B, Mallick MA, Solanki MK, Solanki AC, Hora A, Guo W. Plant growth promoting rhizobacteria (PGPR): modern prospects for sustainable agriculture. Plant Heal Under Biot. Stress, Singapore: Springer Singapore 2019; pp. 109-27.
       [http://dx.doi.org/10.1007/978-981-13-6040-4_6]

[55]   Hole DG, Perkins AJ, Wilson JD, Alexander IH, Grice PV, Evans AD. Does organic farming benefit biodiversity? Biol Conserv 2005; 122: 113-30.
       [http://dx.doi.org/10.1016/j.biocon.2004.07.018]

CHAPTER 14

Yeasts: From the Laboratory to Bioprocesses

Barbara Dunn[1,*] and **Boris U. Stambuk**[2]

[1] *Department of Genome Sciences, University of Washington, Seattle, WA, USA*

[2] *Department of Biochemistry, Federal University of Santa Catarina, Florianopolis, SC, Brazil*

Abstract: Yeasts are important industrial platforms for the efficient production of foods, beverages, commodity chemicals, and biofuels. Although these yeasts usually have beneficial native phenotypes, it is often desirable to engineer these cell factories to increase yield, titer, and production rates, or even promote the production of new molecules. In the present chapter, we describe several classical genetic approaches to improve industrial yeast strains (mating, cell and protoplast fusion techniques, mutagenesis, genome shuffling, adaptive laboratory evolution, *etc.*), as well as methods to identify the genetic basis of phenotypic traits, including phenotypes controlled by quantitative trait loci (QTL), through bulk segregant analysis (BSA) and DNA sequencing. We then review modern technologies for industrial yeast strain improvement (genomic engineering through homologous recombination, CRISPR-Cas9, synthetic chromosomes, synthetic genomes and SCRaMbLE) both in conventional (mostly *Saccharomyces* strains) as well as non-conventional yeasts. Finally, we give several current examples (and ideas for the future) of yeast strains genetically modified in the laboratory to produce a range of commercial products and biofuels through industrial bioprocesses.

Keywords: CRISPR-Cas9, Genetic engineering, Genomic engineering, Homologous recombination, SCRaMbLE, Synthetic chromosomes, Yeast breeding.

INTRODUCTION

Yeasts have been utilized wittingly and unwittingly by humans as tiny bio-factories over many millennia, possibly starting with our ancestors (*Homo erectus* or *H. neanderthal* [1, 2]), by transforming sugars into ethyl alcohol and carbon dioxide for purposes of imbibing, and later on, baking. The processes and rituals surrounding the making of alcoholic beverages and leavened bread have played an important role in human civilization, both socially and economically [3 - 5]. In particular, the quest to create alcoholic beverages has often been an impetus for

---

* **Corresponding author Barbara Dunn:** Department of Genome Sciences, University of Washington, Seattle, WA, USA; Tel:????; E-mail: barb.dunn@gmail.com

Sérgio Luiz Alves Júnior, Helen Treichel, Thiago Olitta Basso and Boris Ugarte Stambuk (Eds.)
All rights reserved-© 2022 Bentham Science Publishers

scientific progress, especially in the areas of chemistry, biochemistry and microbiology: the brewing industry has even been cited as the foundation for the field of biotechnology [6].

But what exactly are these yeasts that have been our longtime companions? The yeasts responsible for producing much of our leavened bread, as well as almost all of the alcoholic beverages consumed in the world -including beer, wine, sake and distilled spirits- are "budding yeasts": small single-celled fungi that belong to the genus *Saccharomyces* (Latin for "sugar fungus"). These unassuming organisms are the world champions at performing the process of alcoholic fermentation. During such fermentations, *Saccharomyces* yeasts earn their Latin name by consuming the sugars present in the starting material -grapes in the case of wine and wort in the case of beer- and converting them into ethanol and carbon dioxide gas. In addition, yeasts will convert some of the more complex molecules already present in grapes or beer wort into novel characteristic flavor and aroma molecules; the whole endeavor results in delicious alcoholic beverages. In the case of leavened bread, the yeast consumes sugars created by the breakdown of the starch found in flour to produce alcohol and carbon dioxide; the latter makes bubbles in the rising dough, while the alcohol evaporates during the baking process.

Within the *Saccharomyces* genus, there are eight closely related, naturally occurring species so far known, as described in detail in a recent review ([7]; also see Chapter 4). The life cycle of *Saccharomyces* yeasts includes both asexual and sexual phases. Their genomes, like ours, are organized into linear chromosomes contained within a nucleus. Yeasts also contain cytoplasmically-located mitochondria that have their separate genomes [8]. Unlike us, however, the presence of mitochondria is not essential for *Saccharomyces* yeasts to survive, although functioning mitochondria are important for optimal yeast performance in many industrial processes. All *Saccharomyces* species possess very similar genomes, with the same number of chromosomes and with most genes (and gene order) shared among all species; additionally, the genomes are very similar at the DNA level. Likewise, all *Saccharomyces* species share the same basic life cycle and mating systems, and yeast cells can exist freely in the haploid or diploid state, where the haploid and diploid genomes are defined as containing one copy, or two copies, respectively, of each of the 16 different chromosomes. Haploid cells can exist briefly within the sexual mating cycle, derived by sporulation of the diploid cell, or they can exist indefinitely as free-living cells if they are unable to mate successfully, for example, due to physical isolation or mutations in the mating system [8].

Yeast cells undergo mitosis (in other words, they continually divide asexually, also called "clonally") when sufficient nutrients are present. This occurs by a "budding" process, where a small bulge on the side of the mother cell grows larger and larger until almost the same size as the mother. At this point, a new nucleus with its own set of chromosomes, as well as cytoplasmic organelles and mitochondria, are transported into the daughter bud, and a new cell wall grows between the daughter bud and the mother. After this, the bud separates away and starts its own mitosis process. A single mother cell produces an average of ~20-30 buds in its lifetime [8, 9]. However, when nutrients, especially nitrogen, become limiting, a diploid cell (but not haploid) can progress through meiosis to produce haploid spores, which are specialized gamete cells that can survive harsh conditions. In *Saccharomyces* yeasts, meiosis results in 4 haploid spores: two spores each of two opposite mating types, called "**a**" and "α" (see also Chapter 4). These two spore mating types can be thought of as "egg" and "sperm", where "**a**" cells can only mate with "α" cells and vice versa. When the two haploid spores mate (fuse together), they create a new diploid cell that combines the nuclear genomes of each parent spore, receiving one set of chromosomes from each parent [8].

Interestingly -and importantly for industrial applications- all eight *Saccharomyces* species are able to mate with each other: *i.e.*, haploid spores of one *Saccharomyces* species are able to mate with haploid spores of the opposite mating type of any of the other *Saccharomyces* species to form an interspecific hybrid, similar to a mule which is an interspecific hybrid between donkey and horse; such hybridization occurs both in the wild and in human-related environments (reviewed by [10, 11]; also refer to Chapter 4 of this book). These interspecific hybrids can proceed through sexual division (meiosis), although this results in mostly inviable spore progeny and thus, like the mule, they are "sterile". However, they are able to indefinitely reproduce in the mitotic asexual (clonal) manner. Other mechanisms, such as multiple rounds of spontaneous genome duplication, or aberrant mating between diploids, can lead to polyploidy (more than 2 copies of each of the basic haploid set of 16 chromosomes) within a species and similar aberrant mating of higher ploidy cells between different species can give rise to interspecific hybrids of varying ploidy levels [12, 13].

Finally, it is important to note that any diploid or polyploid yeast cell can usually tolerate not only mutations in single genes, but (amazingly) the loss or gain of a single chromosome, or even several chromosomes, leading to a state called "aneuploidy", where different chromosomes are present at different copy numbers [14, 15]. Equally important, most aneuploid and higher ploidy strains -as well as interspecific hybrids- are "sterile": *i.e.*, they either cannot mate, cannot complete meiosis, and/or cannot produce viable spores, and thus they cannot be subjected to classical genetic breeding schemes. These concepts of ploidy and aneuploidy, as well as interspecific hybridization, are very important for industrial yeasts, as discussed in detail further below.

Of the various species in the *Saccharomyces* genus, a concerted effort to collect isolates from around the world and to characterize genomic variation has been most exhaustively performed for *S. cerevisiae* because of its industrial/economic importance and because of its history as a premier laboratory model organism. The breadth and depth of these studies have allowed a population genomics approach to be taken, resulting in the ability to define the geographic origins of *S. cerevisiae*. For example, over 1000 strains of just the *S. cerevisiae* species have been collected from a diversity of environments worldwide and then whole-genome sequenced. Population analyses of the sequence data showed that this species originally emerged in Far East Asia [16, 17], where it arose as a distinct species from an ancestral yeast species. The details of exactly how *S. cerevisiae* has subsequently spread to virtually the whole world are still unknown, but distribution by animals, insects -and especially humans!- have all undoubtedly contributed to its global dissemination.

Surveying the thousands of *S. cerevisiae* strains for phenotypic characteristics reveals a huge diversity, indicating that there is a wealth of underlying traits in the population of this species that can be mined for use in industrial applications. When considering that hybrids between different *Saccharomyces* species can be made, where the non-*cerevisiae* species may have novel traits, it is apparent that there are boundless opportunities to genetically breed or perform laboratory selections -or even synthetically create- new yeasts that are tailored to specific industrial uses.

Saccharomyces yeasts have been incredibly well studied and characterized in the laboratory since the 1930's [18 - 20], and have contributed a wealth of basic biological knowledge about biochemistry, genetics, cell growth and other areas, leading to a surprisingly large number of Nobel prizes [21]. Historically, such studies generally did not utilize industrially-derived yeast strains and instead utilized specialized *S. cerevisiae* laboratory strains -some examples are S288C and related FY and BY strains; W303; A364A; SK1; Sigma1278b; Y55; and CEN.PK [8, 22, 23] that are best suited for rapid proliferation in laboratory growth conditions. In addition, industrial growth conditions -which are often harsh, including high osmolarity, ethanol, and temperature, and low pH- were typically not investigated in laboratory settings. But these important early laboratory findings and techniques are often quite easily translatable to industrial *Saccharomyces* strains. In addition, many laboratories are now directly studying industrially-derived *Saccharomyces* strains and industrial growth conditions because of their global economic and technological importance. Such research often focuses on the goal of enhancing (or even adding) desirable industrial traits; for example, better fermentation behavior, small molecule production, heat or toxin tolerance, nutrient scavenging, *etc.*

The wealth of phenotypic diversity among worldwide *S. cerevisiae* strains and sister *Saccharomyces* species, along with the existence of many robust genetic and genome modification techniques for these yeasts, has resulted in a perfect confluence of factors to allow the development of powerful and exciting new avenues to improve existing industrial strains. Traditional genetic breeding schemes, based on the methods developed for well-behaved lab yeasts, have been successfully used for industrial yeast strain improvement schemes. However, because most industrial yeast strains are "sterile", as discussed above, they are unsuitable for classic genetic breeding strategies. Furthermore, most industrial strains do not have any sort of metabolic genetic defect (auxotrophy or other type of genetic "marker"), that allows complementation of the defect when mated to other cells without the defect (a common genetic trick). But luckily, recent advances in genome sequencing and genome manipulation/modification techniques have given researchers ever-greater flexibility to develop industrial strains that are superior for desired performance traits. And most powerfully, the ability to synthetically create completely novel designer yeast strains for targeted uses is now within our capabilities.

Below we list and briefly describe a range of possible breeding and genetic manipulation methods for achieving strain improvement goals in yeasts. Also, note that there are several previously published review articles and book chapters that go into much greater detail regarding genetic and genome modification strategies to improve industrial strains, with illustrative step-by-step figures; these can be consulted for more in-depth explanations [24, 25].

CLASSICAL GENETIC APPROACHES TO IMPROVE INDUSTRIAL STRAINS

Much of what underlies strain improvement strategies involves the mixing of the genetic material of the industrial strain being improved with the genetic material from other yeast strains that display the desired improved, or even novel, trait. The mating-based and cell-fusion breeding techniques described in this section result in diploid or higher ploidy strains that can directly be selected or screened for the desired phenotype [24, 26]. These techniques represent the simplest paths to develop new strains that retain the needed qualities of the original industrial strain but which also have acquired the desired trait(s). Although mating and cell-fusion techniques may seem an "old-fashioned" way to improve strains compared to the many exciting newer avenues for directed and precise genetic modifications (see below), it should be noted that in a "competition" held between laboratories to evolve yeast strains tolerant to cold temperatures, the strategies involving mating appeared to be more successful than others [27]. This indicates that mating and cell-fusion techniques should not be overlooked when designing strain improvement strategies.

Mating/cell fusion-based techniques can be used to genetically mix *S. cerevisiae* strains; however, because so many industrially-important yeasts have been shown to be derived from interspecific hybridization events [28], it is worthwhile noting that most of these techniques can also be used to generate interspecific *Saccharomyces* hybrids (reviewed by [10, 11, 29, 30]; also refer to Chapter 4 of this book). Furthermore, protoplast fusion has even allowed the formation of hybrids between different genera [11]. Additionally, if the resulting mated/fused cells can be induced to undergo meiosis, then a large number of the resulting spores—each containing different combinations of parental genomes from meiotic recombination—can likewise be screened for desired behavior and traits.

Note that these methods are generally considered to NOT result in genetically-modified organisms (GMO's) as there is no direct modification of the genomes. Also note that with the availability of relatively inexpensive whole-genome sequencing techniques, it is always a good idea to whole-genome-sequence any new improved strains that are chosen for further use (see below), to both ensure that they are the expected combination of parent strains, and/or to gain insights into the genetic underpinnings of the improved phenotypes, which can be of use in future breeding experiments.

Direct Mating (Cell-to-cell; Spore-to-spore)

For some industrial *Saccharomyces* strains, it is possible to perform "classic" genetic manipulations to screen or select strains exhibiting desired enhanced or novel traits by utilizing the endogenous mating system of *Saccharomyces* yeasts. Most of these techniques require having two parental strains (or pools of strains) that have already been chosen for displaying one or more target phenotypes, and both strains must be able to mate with each other to produce a diploid strain that combines both parental genotypes [31, 32]. These direct mating techniques can be performed with haploid parents of the same *Saccharomyces* species, or with haploid parents or spores of different species within the *Saccharomyces* genus such that an "interspecific hybrid" strain, as described above, is created (see Fig. (**1**) in Chapter 4 of this book, plus its accompanying text).

The simplest breeding situation is where both parent strains are mating-competent stable haploids of opposite mating types. If the diploid can be selected (*i.e.*, if there is a growth condition where only the diploid will grow, but neither of the haploid parents), then a mass of cells (or spores, for unstable haploids) from each parent can be mixed together for a period of time to allow mating, after which the cell mixture is placed on selective growth conditions and the diploid obtained (*e.g.*, [33]). In general, industrial strains are not genetically suited to allow selective growth conditions where the mated/fused cells, but not the parent cells, can grow. Thus many of the methods below utilize microscopic manipulation of mating-competent cells or spores on a petri plate to be adjacent to each other so that they can mate to form the desired pairing of genomes, and the isolated colony of the mated cells can then be propagated and screened for desired traits. For some industrial strains, it is possible to create stable mating-competent diploids of opposite mating types, and mate them together to form tetraploids [34]. If the mated cells resulting from any of these techniques can be sporulated, a number of separate spores, each carrying different combinations of the parental genomes, can be screened; also, if possible, selected trait-bearing spores can be repeatedly back-crossed to one of the parent strains, if it is desired to retain the bulk of the genetic background of one of the parents.

When one or both of the parental strains are not stable haploids (*i.e.*, the strains are essentially always diploid, and exist only transiently as haploid cells), but are sporulation-competent, direct spore-to-spore mating can be used. Two parental strains are each sporulated, then the haploid spores are microscopically manipulated as described above to form a single colony of the mated diploid cells; an example is the creation of new "lager-like" beer strains with novel qualities *via* spore-to-spore mating between *S. cerevisiae* and *S. eubayanus* spores (recent review and examples [33, 35, 36]). Note that if the diploid parent strain(s) are heterozygous (*i.e.*, the parent itself is derived from 2 genetically distinct parents), then many different spores can be chosen for mating, as each will have a different combination of the parent genomes. If selection for diploids (and against haploids) is available, masses of spores from each sporulated parent can be mixed and the resulting diploids screened for desired traits. A subtype of this category is "spore-to-cell mating", where spores from a sporulated parent are microscopically manipulated adjacent to a mating-competent cell (a haploid, or a higher ploidy cell that has been induced to become mating-proficient as described below).

Rare Mating

If one of the parent strains is not able to successfully sporulate, such as for aneuploid cells, rare mating can be used to generate mated strains. Additionally, rare-mating can be utilized if it is desired that a diploid parent be combined with a haploid parent to give a triploid cell with one parent's genome at twice the copy number. Diploid (and higher ploidy) yeast cells generally do not express individual mating types and therefore do not participate in mating. However, they are able to undergo a type of chromosome recombination, although at a very low frequency, that allows them to mate to a haploid cell or spore of the opposite mating type [37, 38]. If cells of the diploid parent strain are mixed with cells of the haploid parent, both at sufficiently high numbers, this "rare mating" can occur. Note that there generally must be a selection for the mated cells so that they can be isolated out of the large background of unmated cells [26]. This method can be used for strains within a single species, but has also been successfully used to generate interspecific hybrids as potential new lager-type strains [39], and also for commercially-used wine yeasts that provide novel flavor profiles [40 - 42].

Mating within Insect Guts

An interesting idea for easily generating intra- and interspecific *Saccharomyces* hybrids comes from researchers who have observed that *Saccharomyces* spores survive passage through the intestines of social insects such as wasps and hornets, and in fact, interspecific *Saccharomyces* hybrids can be found in insect guts in the wild [43 - 45]. Recently Di Paola *et al.* [46] suggest that insects, grown in the lab (and thus with no endogenous gut microbes), could be fed mixtures of *Saccharomyces* species (or different strains of a single species), after which hybrids between the strains could be isolated from the guts; as for the other mating-based and cell-fusion techniques discussed in this section, this would represent a non-GMO method of generating new strains for industrial purposes.

Induced Mating-Type Switching

For industrial strains that are unable to mate, there now exist recently developed methods that make them transiently able to mate to other similarly treated non-mating strains; this means that even when using genetically recalcitrant industrial strains, most of the "direct mating" strategies described above can be performed. Induced mating-type switching techniques allow the facile production of intra- or interspecifc hybrid strains, within the *Saccharomyces* genus and can even allow the genetic mapping of desired traits for sterile strains, non-mating strains, or otherwise genetically intractable strains [47]. These techniques use plasmids transformed transiently into the desired parental strains, causing diploid cells to experience a "switch" in one of their two mating-type genes; thus the diploid cell can functionally behave as an "**a**" or "α" mating type (which, as described above, normally occurs only in haploids [31]). An HO-based technique [47, 48] uses the endogenous yeast HO gene, carried on a plasmid, to "switch" the mating type; a different approach uses plasmids that express mating-specific genes [49]; and a CRISPR-based technique [50] uses the bacterial Cas9 gene and a targeting RNA, to "switch". The resulting cells are "mating-competent", *i.e.*, they are able to mate to other similarly-altered diploid cells, or to normal haploid cells. The CRISPR-based technique can also be performed with aneuploid and high-ploidy cells [35], and the HO-based technique has even been used in an iterative manner to create a 2-, 3-, 4- and 6-way interspecific *Saccharomyces* hybrids [51]. Since these methods use plasmids carrying various drug resistance genes, the plasmids can be naturally lost from the cells when grown without the selective drugs, such that the resulting cells do not have any remaining genetic changes aside from a naturally-occuring change in the mating type gene, and thus could likely be considered as non-GMO.

Protoplast Fusion

In some cases, non-*Saccharomyces* yeasts may exhibit desirable characteristics for a given industrial use, and thus it may be advantageous to attempt to fuse cells from organisms that are from two different genera ("intergeneric crossing"). Because mating systems are not conserved across different genera, forced cell fusions must be utilized. "Protoplast fusion" is the most common technique, where the cell walls of each parent strain are first removed (usually by enzymatic means), exposing the cell membranes which are made of lipid bilayers. These treated cells are called "protoplasts" and the two parental protoplast populations are induced to asexually merge by mixing and incubating in the osmotically supportive medium at high concentrations, such that the cells are in close proximity and the cell membrane bilayers can fuse [52, 53]. After fusion, the cell wall regenerates and, if successful, the new hybrid cell undergoes cell divisions, sometimes with the two nuclei remaining separate but dividing and being passed to daughter cells in synchrony. Protoplast fusion has been successfully implemented for several different pairs of genera to produce strains for use in bioethanol, bioremediation and cider production [11]; it has also been used to create inter- and intraspecific *S. cerevisiae* hybrids for wine production [26, 54].

Cytoduction

Cytoduction is a specialized method that results in the transfer of subcellular organelles between cells without the transfer of nuclear genes [26, 37]; this process can be used to specifically transfer mitochondria and mitochondrial genes into a strain of interest, or to transfer the virus-like-particles that cause the "killer" phenotype [55]. However, the use of this technique is restricted to *S. cerevisiae* strains that are deficient for the *kar1* gene and is thus not of general pertinence to industrial strain breeding.

Utilizing Genomic Diversity to Select for Desired Phenotypic Traits and Identify Causal Genomic Elements

Where there are known to be many genes involved in the desired trait, or where the responsible genes or genomic regions are not known, pools of strains with a large amount of genetic diversity are needed to be able to screen or select for the phenotypes of interest. Researchers hoping to improve industrial strains can choose to sample the large amount of "standing genetic diversity" that already exists among the existing strains of *Saccharomyces* yeasts worldwide as discussed above [56], including many interspecific *Saccharomyces* hybrids found in natural and industrial environments [28]. Additionally, targeted "*de novo*" genetic diversity can be obtained by performing evolution experiments under specific growth conditions of interest. And recently-developed techniques allow entire synthetic chromosomes, or even genomes, to be designed, synthesized, and introduced into yeast cells to confer entirely novel genetic diversity, as described in a section below.

Mutagenesis

One of the earliest methods used by researchers to generate genomic diversity in populations of *Saccharomyces* yeasts was to expose yeast to chemical mutagens (methyl- or ethyl- methanesulfonate) or DNA-damaging agents such as UV light [38]. Such mutagenesis techniques are still employed in strain improvement schemes, where the mutagenized population of cells is screened for desired traits as a result of gene mutation(s) [24]. However, more recent and sophisticated techniques utilizing mixing of divergent genomes and chromosome recombination described in the following sections represent more variation to explore for improved or novel traits for industrial yeasts.

Mass-Mating and Mass Cell Fusion

Mass mating/mass cell fusion is types of direct mating or cell fusion respectively, but on a large and heterogeneous scale; in this case, genetically diverse populations of parental *Saccharomyces* yeast cells are mated or fused, such that the result is a population of extremely variable genetic diversity that can be sampled for desired behaviors. One or more of the parental starting strains should be heterogeneous, *i.e.*, consist of a genetically mixed population of cells; for example, a mutagenized population of cells, or a pool of meiotically recombined cells or spores. To perform mass matings, a very large number of cells (or spores, if the strains produce viable spores) from each starting parental population are generated. The populations are then mixed together, allowing random mating (if haploid cells or spores) or random cell fusion (if asexual cells are made into protoplasts) to occur [24, 25]. The resulting genetically mixed population can be subjected to selection or enrichment for cells carrying the desired combination of traits immediately, or several rounds of mass mating (or mass fusion) followed by selection can be performed iteratively to give further refinement or stronger expression of the desired phenotypic traits [53, 57]. The goal in this typical mass-mating or mass-fusion methods has usually been to achieve a stable hybrid line that expresses the desired traits.

Although standard mass mating can be performed only with sexually competent yeast strains, it is possible that forced hybridization of populations of homogeneous asexual yeasts, *via* transient HO gene induction or CRISPR-cutting as discussed above, could yield mating-competent cell populations that can be mixed and mate to produce hybrids of interest. It is also possible that this technique could produce sexually competent cells, *e.g.*, *via* doubling the chromosome complement of high-ploidy or aneuploid cells that are able to proceed through sporulation and produce meiotically recombined offspring, thus generating genetic mixing/diversity. Performing several rounds of this regime could serve as a potent genome shuffling technique for genetically intractable strains, and allow screening or selection of strains with novel combinations of beneficial traits [35, 47, 50].

Genome Shuffling

Genome shuffling experiments are similar to the mass-mating or mass-fusion methods but are performed iteratively over several rounds and in a manner such that genome recombination occurs at every round; this eventually results in the two starting parental genomes being mixed together into a fine-grained patchwork manner that varies from cell to cell [53]. This results in many different phenotypic combinations that can be helpful to obtain strains where the beneficial phenotypes are caused by several genes acting together.

One type of genome shuffling technique is based on a "return to growth" (RTG) strategy, which essentially interrupts the process of meiosis. A population of genetically heterogeneous diploid *Saccharomyces* cells, such as intra- or interspecific hybrids, are induced to begin meiosis in by starving the cells for certain nutrients (note that interspecific *Saccharomyces* hybrids are able to go through the process of meiosis and meiotic recombination even though they may fail to produce viable spores). However, if nutrients are added back to the culture after meiosis has been initiated—but before commitment to meiosis, and therefore before meiotic cell division, has occurred—the cells return to the mitotic growth pattern [58, 59]. However, the chromosomes of these RTG cells have already undergone meiotic recombination, which causes double-stranded breaks in the chromosomes; the repair of these breaks leads to regions of the chromosomes that have experienced a "loss of heterozygosity". Loss of heterozygosity (LOH) refers to a chromosomal region that was originally heterozygous *i.e.*, where DNA sequences differ at various sites along each parental chromosome, but has become homozygous, *i.e.* both chromosomes now share the identical DNA sequence across that region and thus the cell has lost that genetic information from one of the parents. These LOH patches occur in different places for each different cell in the RTG population, so that across the population as a whole, most or all of each parental genome is "uncovered" as an LOH patch. Successive rounds of RTG leads to progressively smaller LOH patches, but more of them, so that eventually a population of diploid cells is obtained in which the chromosomes are essentially entirely homozygous but contain a patchwork of contributions from each parent across the genome; *i.e.* the two parental genomes have been simultaneously mixed, shuffled and homozygosed [60]. This technique has been successfully utilized with a polyploid baking yeast strain [61], and represents a non-GMO and marker-free process to perform genome shuffling in sterile hybrid yeast, to screen for (and even map) improved traits.

The CRISPR/Cas9 genome editing system (described in more detail below) has also been used to introduce targeted cutting in the genome, which leads to mitotic recombination events that generate LOH patches as for meiotic recombination [62]. This method is similar to RTG but confers the ability to iteratively target the LOH regions to narrow down the region(s) of interest, and also does not require a strain to be able to go through meiosis. The authors generated a LOH panel in a heterozygous diploid *S. cerevisiae* strain, allowing fine mapping of a manganese sensitivity trait. Note that the use of the CRISPR-Cas9 system, which transiently introduces plasmids into the cell, may be considered a GM technique. Another CRISPR-based genome shuffling method allows "scarless" rearrangements, either a single targeted rearrangement, or large numbers of rearrangements, including long regions of genome duplications and other large structural changes [63]. The authors found that reshuffling the genome resulted in strains demonstrating fitness advantages in stressful environmental conditions.

Other methods to create genetic diversity and allow recombination in sterile yeast strains include using insect guts as described above, but instead feeding the insects heterogeneous populations of yeast [46]. Also, a recent report describes how to greatly increase the production of viable spores with perfectly diploid genomes from interspecific *Saccharomyces* hybrids; this is achieved by repressing two genes, *SGS1* and *MSH2*, that are responsible for anti-recombination in interspecific hybrids [64]. One of the most recent methods for genome shuffling is the SCRaMbLE technique, which depends on synthetic chromosomes in yeast designed to have many inducible recombination sites spread throughout them, as described in more detail below.

Bulk Segregant Analysis and QTL-Mapping to Identify Genes/Pathways Involved in Phenotypes

Once a population with high genomic diversity has been obtained, it can be used to identify the genetic basis of phenotypic traits, and is especially useful for phenotypes controlled by "quantitative trait loci" (QTL's): these are multiple genomic regions that vary genetically in the population and directly contribute in a quantitative manner to the phenotype. A very powerful way to map QTL's in yeast is by "bulk segregant analysis" (BSA) and is especially powerful when combined with next-generation DNA sequencing ("BSA-seq" [65 - 68]); it has become a commonly-used method to identify causal DNA variants for traits of interest in industrial strains. In the BSA protocol, hundreds or thousands of single colonies from the genetically-diverse population can be picked and assayed (robotically if possible) for the phenotype(s) of interest; those colonies displaying the two extremes of the phenotype are then separately pooled (one pool for each phenotype extreme) and each pool is whole-genome-sequenced (see below) in bulk. Any genomic variant regions that contribute to the phenotype will appear at much higher frequency in the positive group as compared to other parts of the genome [69], allowing determination of the exact mutations or gene alleles that are causing the new phenotype; these are called "QTN's" or "quantitative trait nucleotides". In fact Peltier *et al.* [70] have scoured the literature and have gathered a catalog of 284 experimentally-validated QTN's from *S. cerevisiae* that impact various traits of industrial importance. Knowledge of the genes and pathways contributing to complex traits can be used to construct novel industrial yeast strains either directly through genetic modifications or metabolic engineering/synthetic biology, or indirectly through adaptive evolution as described in the next section.

Adaptive (Directed) Evolution to Select for Enhanced Phenotypes

Adaptive evolution, also called adaptive laboratory evolution (ALE), directed or experimental evolution, or evolutionary engineering, refers to methods whereby a population of yeast cells is grown for many cell divisions (generations) under stress or nutrient conditions, called "selective pressure", that favor the beneficial phenotypes desired for the strain improvement [71 - 73]. For example, after growing an initial strain of interest in high ethanol for many generations, cells with spontaneous mutations may arise that are better able to grow (are more "fit") in that condition; these will eventually dominate the population and thus an improved strain be obtained, although often such experiments are halted before a single strain dominates and many individual cells (with varying spontaneous mutations) assayed for phenotype and behavior [73].

Such evolution experiments can be performed in serially repetitive steps, called "batch" evolution, *i.e.*, repeated cycles of growing cells in liquid or on plates for a certain amount of time, then diluting the cultures into a fresh new batch of the medium [71, 73]. Alternatively, the evolution process can be carried out in a prolonged continuous manner, using a chemostat [71] or other continuous culturing methods, where the fresh medium is slowly dripped into the culture while the culture is siphoned off at the same rate, keeping the culture continuously growing. The selective pressure (stress or nutrient condition) can be kept constant, or can gradually increase during the evolution timecourse. Evolution experiments that start with a single strain to select for a more fit phenotype is an approach often used to incrementally enhance or "fine tune" an already existing strain, but such experiments can also be performed with a starting mixture of genetically diverse strains (obtained by any of the methods above) prior to the evolution, allowing "competition" between strains to reveal the most-fit genotypes [74].

Evolutionary adaptive selection has been used to enhance industrially relevant traits such as substrate utilization, ethanol production, and stress tolerance. For example, adaptive evolution has yielded *S. cerevisiae* strains able to tolerate high temperatures and produce higher ethanol [75], or to grow on xylose [76], and even for other *Saccharomyces* species, for example, *S. eubayanus* strains better able to tolerate ethanol [77]. Other traits that have been successfully introduced into yeasts *via* adaptive evolution are reviewed by [38, 78] and new developments and future directions have been described by [73, 79]. Experimental evolution protocols are themselves evolving new twists and enhancements, such as a fully automated, *in vivo* continuous evolution setup that combines continuous mutagenesis with programmable multi-parameter selection regimes [80], or a CRISPR-based and RNA-assisted method for directed evolution on a growing cell population that continuously mutagenizes targeted genomic loci during the evolution [81].

Whole-genome and High-throughput Sequencing

To confirm that the strains resulting from any type of breeding program or genetic modification strategy have the desired genetic changes—and/or to uncover the genes and pathways involved in the enhanced phenotypes obtained after selection or screening regimes—it is best to perform whole-genome sequencing. Ever-increasing advances in DNA sequencing technology mean that it is now easy to rapidly and inexpensively sequence the whole genome of a yeast strain; note, though, that it can be difficult to interpret and assemble the complex genomes of industrial strains since they can be polyploid, aneuploid, and/or possibly interspecific hybrids with whole or partial genomes from two different species. However, it is becoming clear that combining the sequencing results from short-read (Illumina) and long-read (PacBio or Oxford Nanopore) platforms can often allow truly full-genome assembly of these strains, *i.e.*, the complete sequence of each entire chromosome in the genome, from telomere-to-telomere (for reviews see [82 - 84]). Other techniques such as proximity-based ligation sequencing [85 - 87] are also of use for generating such full genome assemblies, and may even allow "haplotype-phasing" of heterozygous genomes [88].

MODERN TECHNOLOGIES FOR STRAIN IMPROVEMENT

We turn now to strain improvement schemes that are the result of "recombinant DNA" techniques; these can mostly be described as techniques where exogenous DNA from any source is introduced into a yeast cell by various transformation methods, *i.e.*, chemical, mechanical, or electrical treatments of cells such that the cell walls and membranes allow the direct uptake of DNA [89, 90]. This is in contrast to using the "natural" mating and recombination systems of *Saccharomyces* yeasts to introduce new genes or mix genomes, as described in the previous sections. Recombinant DNA strategies for strain improvement can often be carried out more rapidly than mating-based strategies, and the resulting strains can be more genetically stable. However, most types of recombinant DNA methods require the use of markers to allow robust positive selection of yeast cells that have been transformed with the introduced DNA [91 - 93].

Two types of marker systems require the starting host strain to contain special mutations that can be complemented by gene(s) on the introduced DNA. One type is the "auxotrophic marker", where the strain has gene mutation(s) that inactivate a metabolic pathway (often in amino acid synthesis); the introduced DNA carries the functional gene (along with any other desired DNA segments), allowing selection on media lacking the metabolite. Because it is often difficult to inactivate metabolic pathways in diploid or polyploid industrial yeasts due to the need to inactivate several copies of a gene at once, this type of marker is not often used for industrial strain improvement schemes. A second marker type is the "autoselection marker" where the host strain has a set of mutated genes in a pathway such that the cells are completely inviable in almost any nutritional environment and can survive only if the gene is supplied by the introduced DNA; autoselection markers are often used for large-scale biotechnology production due to the ability to use simple and inexpensive growth media for selection [91, 93], but again they may not be suited for most industrial strain improvement programs due to the need for gene mutations.

The final category of marker, most often used for the improvement of industrial yeast strains, are the "dominant markers" [91]. These are mostly drug-resistance genes (usually conferring resistance to antibiotics such as hygromycin or G418) that are included on the introduced DNA, and the presence of the antibiotic in the growth medium selects for only the transformed cells. Recombinant DNA techniques encompass an extremely wide variety of methodologies, of which we will focus on those most commonly used for industrial yeast strains. Note that most of these recombinant DNA methods will result in organisms that would be classified as genetically modified (GM).

Genomic Engineering Through Homologous Recombination

Plasmids are circular pieces of self-replicating DNA, usually carrying both bacterial and yeast-encoded genes and regulatory regions; this allows easy amplification of the plasmids in bacteria to yield enough product for subsequent transformation into yeast. Plasmids represented the first method used in the laboratory to introduce exogenous DNA into yeast cells [94]. There are many types of plasmids that have been developed for use in *Saccharomyces* yeasts (review by [92, 93]); two main classes are replicating plasmids and integrative plasmids. Replicating plasmids have their own "origins" of DNA replication (also called "autonomously replicating sequences") that are functional in yeast, and thus remain as extrachromosomal circular pieces of DNA in the cell; although they are easily transformed into yeast, they are easily lost from cells in the absence of constant selection and thus are not often used for industrial strain improvement programs.

In contrast, integrative plasmids do not contain yeast origins-of-replication and cannot be retained in the cell unless they physically integrate into the yeast genome at a chromosomal region that is homologous to yeast DNA sequences on the plasmid. Chromosomal integration of the exogenous DNA carried on the plasmid is the desired outcome, since it is stably carried on the yeast chromosome and thus selective media are not needed to retain it. Integration occurs by the process of homologous recombination (HR), which is an extremely efficient process in yeast [95, 96]. A single HR crossover event (which can be enhanced if the plasmid is cut with a restriction enzyme within the yeast DNA) results in the integration of the entire plasmid into the chromosome, with a tandem repeat of the yeast DNA flanking either side [93]; this can be an unstable integration of the desired plasmid since it can use the flanking repeated sequences to "loop out" of the chromosome.

A better way to stably integrate desired pieces of DNA into a yeast chromosome is by a double crossover HR mechanism, as shown in Fig. (1), panel A. The desired gene to be introduced into the genome (shown as "M") is cloned into a plasmid along with small regions of yeast chromosomal sequences flanking the marker gene; restriction enzyme cutting can release the linear fragment shown in the figure. Note that this linear fragment could also be generated by polymerase chain reaction (PCR), or by *de novo* DNA synthesis [97], which is quickly becoming an affordable option for obtaining custom-designed double-stranded DNA fragments up to 3 kb in length (*e.g.*, gBlocks from IDT). The flanking yeast sequences in panel A are shown as regions upstream and downstream of an endogenous yeast gene ("*YFG*") that is to be deleted; if the flanking yeast sequences are contiguous on the yeast plasmid (not shown) then the marker gene will be inserted into the yeast chromosome without any deletion. This double crossover HR technique can be expanded to create genes with inducible promoters or fluorescent tags (Fig. **1**, panel B), or to integrate several genes of a metabolic pathway (panel C), or even an entire chromosome arm (panel D). Such techniques have been used to delete individually (and functionally analyze, *i.e.* phenotypic consequences) almost all yeast genes (96-97%) present in the genome [98, 99], as well as to also tag 97-98% of all genes in order to find the localization and expression of the proteins encoded by the *S. cerevisiae* genome [100, 101].

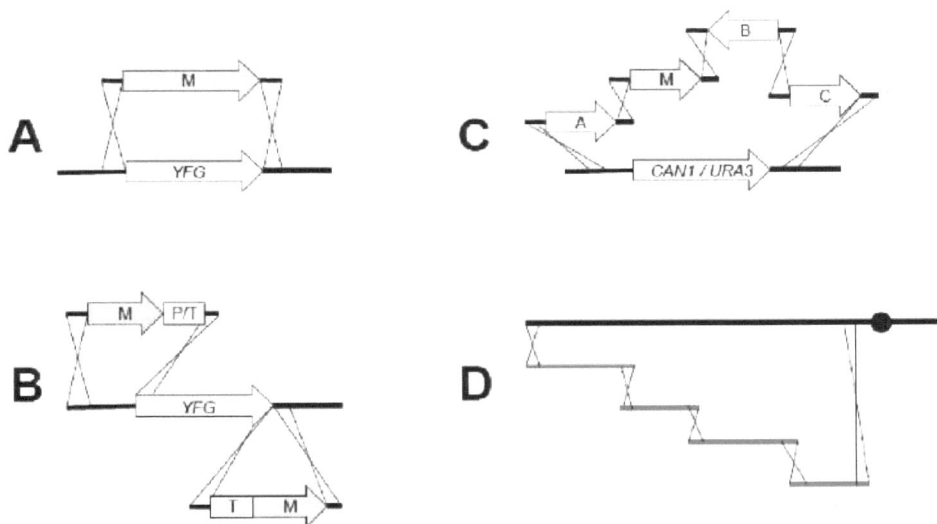

Fig. (1). Genomic engineering through homologous recombination (HR). If a marker gene (M, which restores an auxotrophy, or confers resistance to an antibiotic) contains short sequences of homology to the upstream and downstream sequence of the target locus (*YFG*), after transformation, the marker will replace *YFG* through HR (**A**). The same approach can be used to insert a different promoter (P) or a tag (T, *e.g.GFP*, His-6x) at the N- (upper part) or C-terminal (lower part) of the gene/protein (**B**). HR can also be used to insert several genes (A-C, which can be a whole metabolic pathway) at a chromosomal locus, and recombinants with the correct *in vivo* assembly can be selected by the maker gene or counter-selected with L-canavanine if they replaced the *CAN1* gene, or with 5-Fluoro-orotic acid if they replaced the *URA3* gene (**C**). A similar approach has been used to replace a chromosomal arm (black line, centromere is the black circle) with DNA synthesized (and designed) in the laboratory (gray lines), allowing the construction of synthetic chromosomes in yeasts (**D**). For simplicity, the promoters and terminator sequences of inserted genes are not shown.

CRISPR-Cas9

It is now possible to directly alter genomic regions identified as important in desired phenotypic traits and then to create customized strains with beneficial properties *via* exciting new gene- and genome-editing techniques (reviewed by [102, 103], of which the CRISPR/Cas system is currently the most versatile and most utilized [104]. Combining genome editing with long-range DNA synthesis and other advances in genetic engineering, it is now possible to create specialized yeast cell factories [105]; and in fact, the ability to synthesize entire yeast chromosomes or even whole genomes is already in hand [106, 107]. The CRISPR-Cas9 (Clustered Regularly Interspaced Short Palindromic Repeats - CRISPR-associated protein 9) genome editing system burst onto the scientific stage in 2012 [108] as a fast, easy and versatile system that enabled genomic modifications completely independent of the need for selective genetic markers (see recent reviews [104, 109]). Obviously, this was game-changing for genetic engineering of higher organisms, but even for *Saccharomyces* yeasts -despite genomes so easily manipulated due to robust HR- it became clear that CRISPR would be useful for the genetically intractable industrial strains [110 - 112].

In brief, the Cas9 protein is directed by a guide RNA (gRNA) to make a double-stranded cut in any genomic DNA that matches the sequence of the gRNA; both the Cas9 protein and the desired gRNA(s) are typically placed on a plasmid which is transformed into yeast cells, and the desired specifically-targeted genome editing occurs at a very high frequency [113, 114]. Applications of CRISPR technology in yeast now range from traditional genome editing to transcriptional modulation, to novel uses such as synthetic genome construction and creating living circuitry [115, 116], as well as applications to metabolic engineering for the production of valuable chemical products [117].

Synthetic Chromosomes, Synthetic Genomes and SCRaMbLE

One of the most ambitious, perhaps ultimate, goals of yeast strain improvement is to design and chemically synthesize from scratch an entire *S. cerevisiae* genome. Such is the aim of the Yeast 2.0 (or "Sc2.0") project [118, 119]. The hope is that by creating a completely synthetic genome, many basic biological questions can be answered, but there is also the distinct possibility that yeast strains can soon be custom-designed to carry out industrial bio-processes in ever-more efficient ways [106, 107, 120].

Among the fundamental findings that may be gleaned from this project are insights into the properties of chromosomes and genome organization, the function of RNA splicing and what role small RNAs play in yeast biology, and even finishing the task of discovering the functions of the still-remaining "function unknown" genes in the *S. cerevisiae* genome. The Sc2.0 project was initially powered by undergraduates in a "Build a Genome" course, where students synthesized the building block "chunks" of the synthetic genome while learning molecular biology and bioinformatics. Due to the success of the undergraduate class, the project expanded into an international consortium of 11 institutions across 5 countries [107]. The teams have been working in parallel fashion, each team with one to several chromosome to build in order to create the first *in silico*-designed fully synthetic eukaryotic genome [119, 121].

To reduce destabilizing genomic elements (*i.e.*, that can promote chromosome rearrangements), the design of the synthetic genome includes the removal of transposable elements and other "junk" genomic regions such as sub-telomeric elements, introns were also removed from the relatively few genes in yeast that have them, and additionally, all tRNA genes were made non-redundant and moved to a single "neochromosome" [118]. To date, the swapping out of native for synthetic chromosomes has been tolerated by the yeast strains [119, 122, 123]. One aspect of the Sc2.0 work is to define a "minimal yeast genome" [124], but additionally one could envision exploring and exploiting a more expansive synthetic yeast genome, for example, including a "pangenome" neochromosome containing all observed *S. cerevisiae* genes not found in laboratory yeasts, adding phenotypic diversity to the synthetic genome, which could be useful for industrial applications [26, 120].

One of the most ingenious and intriguing aspects of the Sc2.0 synthetic genome is the inclusion of ~4000 recombination hotspots (*lox* sites), inserted in virtually every inter-genic region, and that can be turned on or off at will [125]. When the *lox* recombination system is activated, the synthetic genome in each cell in the population experiences a "super-shuffle" of the genome, with different parts of different chromosomes joining together in new configurations and new copy numbers. This procedure is aptly known as SCRaMbLE ("Synthetic Chromosome Rearrangement and Modification by LoxP-mediated Evolution"). SCRaMbLE acts to generate genomic diversity, as do the other genomic shuffling methods discussed above, but in a rapid and controllable manner, accelerating the development of yeast strains optimized for industrial or bioprocess uses [120]. SCRaMbLE has already been utilized to create new strains showing improved xylose fermentation and production of medicinal and chemical compounds or increased stress tolerance [125 - 127], and has been shown to work in semi-synthetic industrial-background strains [128, 129]. Delving even further into the synthetic realm, recent studies have shown the feasibility of incorporation of non-natural amino acids into proteins in yeast and other microbes, and the ability to "tune" a synthetic genome to utilize such non-canonical amino acids could be an interesting tool in engineering proteins with completely novel properties and functions [130 - 132].

GENETIC MODIFICATION OF NON-CONVENTIONAL YEASTS

Non-conventional yeasts (meaning non-*Saccharomyces* yeasts) are important industrial platforms for the efficient production and secretion of proteins, lipids, and primary metabolites that have significant value as commodity chemicals [133 - 136]. Although these yeasts can have beneficial native phenotypes, it is often essential to engineer these cell factories to increase yield, titer, and production rates [137, 138]. While most non-conventional yeasts have well-established DNA-mediated transformation protocols, options for stable plasmids are significantly limited. Plasmids are initially created by combining centromeric and autonomously replicating sequences of the organism's genome with a selectable auxotrophic (or drug-resistant) marker. However, while functional plasmids are available for most non-conventional yeasts, they tend to be low copy number and show variable expression across cells in a colony, an effect due to imperfect partitioning of plasmids upon cell divisions [139].

Thus, the genetic/genomic modification of non-conventional yeasts is highly desirable, but presents some drawbacks that need to be addressed in order to perform the modifications as designed [137, 140]. The first problem is related to the fact that most non-conventional yeasts are highly inefficient at homology directed repair (homologous recombination, HR), compared with non-homologous end-joining (NHEJ, also called illegitimate recombination) [141 - 143]. In *S. cerevisiae*, HR plays a dominant role in DNA double stranded breaks repair, and homologous flanking sequences in the disruption/modification cassette (Fig. **1**, panel A) can be as short as ~35 bp is sufficient to produce nearly 100% correct target integration; these short homology sequences can be easily incorporated in the sequence of oligonucleotide primers, so that with a single PCR reaction a disruption/modification cassette can be produced [98, 144]. But to achieve high integration efficiencies in non-conventional yeasts, the homologous regions need to be significantly longer (from 100 bp up to 1-2 kb, with longer fragments increasing the efficiency [145]). To include such large homology sequences flanking the disruption/modification cassette thus may require ligation of multiple PCR products that have restriction enzyme sites, or other time-consuming strategies (*e.g.* Gibson assembly) [146 - 149].

There are several strategies to increase HR in non-conventional yeasts. Since HR is highly dependent on the cell cycle, functioning only in the S and G2 phases of the cycle [150], one strategy is to use chemical compounds that inhibit the synthesis of DNA, like hydroxyurea and nocodazoles, which significantly improves the efficiency of HR in several yeasts [151 - 153]. Another approach to enhance HR is the disruption of genes essential for the NHEJ pathway, such as DNA ligase IV or the KU70 or KU80 proteins that bind tightly to the free end of linear double-strand DNA. This approach has been used with several non-conventional yeasts, and in all cases the efficiency of HR genomic modifications increased significantly [153 - 162]. In a different approach to avoid the illegitimate recombination of a marker gene through the NHEJ pathway, a promoter-dependent disruption of genes (PRODIGE) method was developed, in which the marker expression depends on the regulatory sequences of the disrupted gene [163]. This approach has been successfully used with several *Candida* yeasts [92, 164, 165]. Another way to improve the correct genetic modification through HR is to promote a specific double strand break at the desired genomic location, through CRISPR-Cas9 or TALENT technologies [92, 137, 140, 166 - 168], and co-transforming the cells with a homologous repair/modify DNA module. Since cells that suffer a double strand break are generally non-viable, only those with the desired modification (repaired double strand break) will survive.

The last issue regarding genome engineering of non-conventional yeasts is the CUG codon; in most organisms this encodes for the amino acid leucine (CUG-Leu), but in at least 3 clades of ascomycetous yeasts this codon encodes for either serine (CUG-Ser1 and CUG-Ser2 clades) or even alanine (CUG-Ala clade) [169 - 171]. Since these amino acids changes can impact the functionality of expressed proteins, codon-optimized selectable markers or transformation modules should lack CUG codons, or those codons should be changed to other codons that exist for leucine [169, 172 - 174].

FROM THE LABORATORY TO BIOPROCESSES

An example of genetically-modified yeasts used in bioprocesses are the yeasts used in the bioethanol industry. Approximately 84% of bioethanol produced world-wide (data from 2019) is the so-called first generation ethanol (1G ethanol) which is produced from the starch present in maize in the USA, or from sucrose present in sugarcane juice in Brazil [175 - 178]. In the USA, the majority of industrial yeast strains used for 1G ethanol production are genetically modified to overexpress and secrete amylases and glucoamylases, in addition to increased ethanol production, lower glycerol as by-product, and enhanced organic acids and heat tolerance [178]. In Brazil, industrial yeast strains have been modified to improve the ethanol yield from sucrose, in addition to the same improvements described above for maize 1G ethanol production, although most distilleries use non-modified selected industrial yeast strains [176, 178 - 183].

The so-called second generation (2G) ethanol is derived from lignocellulosic biomass sources such as sugarcane bagasse, leaves or maize stalks. Recombinant *S. cerevisiae* strains for xylose fermentation, which is necessary for using these types of biomass, have been developed and are used by several countries/companies [178, 183 - 186]. Another interesting example of genetically-modified yeasts in bioprocesses is the excellent work developed by Amyris Inc. (www.amyris.com), using recombinant yeasts to produce a range of commercial products through fermentation of sugarcane juice, including artemisinic acid, a precursor of the anti-malaria artemisinin [187, 188], terpene-based advanced biofuels [189, 190], sweeteners, perfume ingredients, pharmaceuticals and other high-value chemicals that normally are extracted (in small quantities) from plants and microorganisms [191 - 193]. Chandel and co-workers [194] describe several other companies and products developed using, in most cases, bioprocesses with genetically modified yeast.

CONCLUSION

Yeasts are used in many types of bioprocesses with significant importance for humans, having been isolated from nature or genetically modified in laboratories for improved performance. With the advent of synthetic genomes and other cutting-edge genome engineering tools, the horizons appear to be opening up more and more for the development of yeast strains that will be able to produce novel medicinal compounds, create tastier and more healthful food and drink, ferment waste into eco-friendly fuels and perhaps even perform completely novel useful functions such as bioremediation, carbon capture or other helpful roles in bettering the world.

CONSENT FOR PUBLICATION

Not applicable.

CONFLICT OF INTEREST

The authors declare no conflict of interest, financial or otherwise.

ACKNOWLEDGEMENTS

BUS acknowledges the Brazilian funding agencies CNPq (grant number 308389/2019-0), CAPES (grant number 359/14), FINEP (grant number 01.09.0566.00/1421-08) and FAPESC (grant number 17293/2009-6). BD would like to acknowledge funding from the ASBC Research Council and to thank Maitreya Dunham and Gavin Sherlock for support, mentoring and enjoyable collaborations.

REFERENCES

[1] Carrigan MA, Uryasev O, Frye CB, *et al.* Hominids adapted to metabolize ethanol long before human-directed fermentation. Proc Natl Acad Sci USA 2015; 112(2): 458-63.
[http://dx.doi.org/10.1073/pnas.1404167111] [PMID: 25453080]

[2] Dunn RR, Amato KR, Archie EA, Arandjelovic M, Crittenden AN, Nichols LM. The internal, external and extended microbiomes of Hominins. Front Ecol Evol 2020; 8: 10.3389.
[http://dx.doi.org/10.3389/fevo.2020.00025]

[3] McGovern PE, Zhang J, Tang J, *et al.* Fermented beverages of pre- and proto-historic China. Proc Natl Acad Sci USA 2004; 101(51): 17593-8.
[http://dx.doi.org/10.1073/pnas.0407921102] [PMID: 15590771]

[4] Legras J-L, Merdinoglu D, Cornuet J-M, Karst F. Bread, beer and wine: *Saccharomyces cerevisiae* diversity reflects human history. Mol Ecol 2007; 16(10): 2091-102.
[http://dx.doi.org/10.1111/j.1365-294X.2007.03266.x] [PMID: 17498234]

[5] Lahue C, Madden AA, Dunn RR, Smukowski Heil C. History and domestication of *Saccharomyces cerevisiae* in bread baking. Front Genet 2020; 11: 584718.
[http://dx.doi.org/10.3389/fgene.2020.584718] [PMID: 33262788]

[6] Bud R. The Uses of Life: A History of Biotechnology. Cambridge, UK: Cambridge University Press 1994.

[7] Alsammar H, Delneri D. An update on the diversity, ecology and biogeography of the Saccharomyces genus. FEMS Yeast Res 20(3): foaa013.2020;

[8] Sherman F. Getting started with yeast. Methods Enzymol 2002; 350: 3-41.
[http://dx.doi.org/10.1016/S0076-6879(02)50954-X] [PMID: 12073320]

[9] Fabrizio P, Longo VD. The chronological life span of *Saccharomyces cerevisiae.* Aging Cell 2003; 2(2): 73-81.
[http://dx.doi.org/10.1046/j.1474-9728.2003.00033.x] [PMID: 12882320]

[10] Morales L, Dujon B. Evolutionary role of interspecies hybridization and genetic exchanges in yeasts. Microbiol Mol Biol Rev 2012; 76(4): 721-39.
[http://dx.doi.org/10.1128/MMBR.00022-12] [PMID: 23204364]

[11] Steensels J, Gallone B, Verstrepen KJ. Interspecific hybridization as a driver of fungal evolution and adaptation. Nat Rev Microbiol 2021; 19(8): 485-500.
[http://dx.doi.org/10.1038/s41579-021-00537-4] [PMID: 33767366]

[12] Hittinger CT. *Saccharomyces* diversity and evolution: a budding model genus. Trends Genet 2013; 29(5): 309-17.
[http://dx.doi.org/10.1016/j.tig.2013.01.002] [PMID: 23395329]

[13] Borneman AR, Pretorius IS. Genomic insights into the *Saccharomyces sensu stricto* complex. Genetics 2015; 199(2): 281-91.
[http://dx.doi.org/10.1534/genetics.114.173633] [PMID: 25657346]

[14] Barrio E, González S, Arias A, Belloch C, Querol A. Molecular mechanisms involved in the adaptive evolution of industrial yeasts. In: Querol A, Fleet GH, Eds. Yeasts in Food and Beverages (The Yeast Handbook). Berlin, Heidelberg: Springer-Verlag 2006; pp. 153-74.
[http://dx.doi.org/10.1007/978-3-540-28398-0_6]

[15] Storchová Z. Ploidy changes and genome stability in yeast. Yeast 2014; 31(11): 421-30.
[http://dx.doi.org/10.1002/yea.3037] [PMID: 25155743]

[16] Peter J, De Chiara M, Friedrich A, *et al.* Genome evolution across 1,011 *Saccharomyces cerevisiae* isolates. Nature 2018; 556(7701): 339-44.
[http://dx.doi.org/10.1038/s41586-018-0030-5] [PMID: 29643504]

[17]   Bendixsen DP, Gettle N, Gilchrist C, Zhang Z, Stelkens R. Genomic evidence of an ancient East Asian divergence event in wild Saccharomyces cerevisiae. Genome Biol Evol 2021; 13: evab001.

[18]   Tuite MF. Strategies for the genetic manipulation of *Saccharomyces cerevisiae*. Crit Rev Biotechnol 1992; 12(1-2): 157-88.
        [http://dx.doi.org/10.3109/07388559209069191] [PMID: 1733520]

[19]   Barnett JA. A history of research on yeasts 10: foundations of yeast genetics. Yeast 2007; 24(10): 799-845.
        [http://dx.doi.org/10.1002/yea.1513] [PMID: 17638318]

[20]   Chambers PJ, Pretorius IS. Fermenting knowledge: the history of winemaking, science and yeast research. EMBO Rep 2010; 11(12): 914-20.
        [http://dx.doi.org/10.1038/embor.2010.179] [PMID: 21072064]

[21]   Hohmann S. Nobel yeast research. FEMS Yeast Res 2016; 16(8): fow094.
        [http://dx.doi.org/10.1093/femsyr/fow094] [PMID: 27770011]

[22]   Cubillos FA, Louis EJ, Liti G. Generation of a large set of genetically tractable haploid and diploid *Saccharomyces* strains. FEMS Yeast Res 2009; 9(8): 1217-25.
        [http://dx.doi.org/10.1111/j.1567-1364.2009.00583.x] [PMID: 19840116]

[23]   Louis EJ. Historical evolution of laboratory strains of *Saccharomyces cerevisiae*. Cold Spring Harb Protoc 2016; 2016: 7.
        [http://dx.doi.org/10.1101/pdb.top077750] [PMID: 27371602]

[24]   Steensels J, Snoek T, Meersman E, Picca Nicolino M, Voordeckers K, Verstrepen KJ. Improving industrial yeast strains: exploiting natural and artificial diversity. FEMS Microbiol Rev 2014; 38(5): 947-95.
        [http://dx.doi.org/10.1111/1574-6976.12073] [PMID: 24724938]

[25]   Dunn B, Kvitek DJ, Sherlock G. Genetic manipulation of brewing yeasts: Challenges and opportunities. In: Bokulich NA, Bamforth CW, Eds. Brewing Microbiology: Current Research, Omics and Microbial Ecology. Norfolk, UK: Caister Academic Press 2017; pp. 119-44.
        [http://dx.doi.org/10.21775/9781910190616.05]

[26]   Pretorius IS. Tailoring wine yeast for the new millennium: novel approaches to the ancient art of winemaking. Yeast 2000; 16(8): 675-729.
        [http://dx.doi.org/10.1002/1097-0061(20000615)16:8<675::AID-YEA585>3.0.CO;2-B]   [PMID: 10861899]

[27]   Kaminski Strauss S, Schirman D, Jona G, *et al.* Evolthon: A community endeavor to evolve lab evolution. PLoS Biol 2019; 17(3): e3000182.
        [http://dx.doi.org/10.1371/journal.pbio.3000182] [PMID: 30925180]

[28]   Langdon QK, Peris D, Baker EP, *et al.* Fermentation innovation through complex hybridization of wild and domesticated yeasts. Nat Ecol Evol 2019; 3(11): 1576-86.
        [http://dx.doi.org/10.1038/s41559-019-0998-8] [PMID: 31636426]

[29]   Hovhannisyan H, Saus E, Ksiezopolska E, Hinks Roberts AJ, Louis EJ, Gabaldón T. Integrative Omics analysis reveals a limited transcriptional shock after yeast interspecies hybridization. Front Genet 2020; 11: 404.
        [http://dx.doi.org/10.3389/fgene.2020.00404] [PMID: 32457798]

[30]   Sipiczki M. Interspecies hybridisation and genome chimerisation in *Saccharomyces*: Combining of gene pools of species and its biotechnological perspectives. Front Microbiol 2018; 9: 3071.
        [http://dx.doi.org/10.3389/fmicb.2018.03071] [PMID: 30619156]

[31]   Haber JE. Mating-type genes and MAT switching in *Saccharomyces cerevisiae*. Genetics 2012; 191(1): 33-64.
        [http://dx.doi.org/10.1534/genetics.111.134577] [PMID: 22555442]

[32]   Fukuda N. Crossbreeding of yeasts domesticated for fermentation: Infertility challenges. Int J Mol Sci 2020; 21(21): 7985.
[http://dx.doi.org/10.3390/ijms21217985] [PMID: 33121129]

[33]   Krogerus K, Magalhães F, Vidgren V, Gibson B. New lager yeast strains generated by interspecific hybridization. J Ind Microbiol Biotechnol 2015; 42(5): 769-78.
[http://dx.doi.org/10.1007/s10295-015-1597-6] [PMID: 25682107]

[34]   Ota T, Kanai K, Nishimura H, *et al.* Generation of new hybrids by crossbreeding between bottom-fermenting yeast strains. J Biosci Bioeng 2021; 131(1): 61-7.
[http://dx.doi.org/10.1016/j.jbiosc.2020.08.009] [PMID: 33190800]

[35]   Krogerus K, Magalhães F, Vidgren V, Gibson B. Novel brewing yeast hybrids: creation and application. Appl Microbiol Biotechnol 2017; 101(1): 65-78.
[http://dx.doi.org/10.1007/s00253-016-8007-5] [PMID: 27885413]

[36]   Mertens S, Steensels J, Saels V, De Rouck G, Aerts G, Verstrepen KJ. A large set of newly created interspecific *Saccharomyces* hybrids increases aromatic diversity in lager beers. Appl Environ Microbiol 2015; 81(23): 8202-14.
[http://dx.doi.org/10.1128/AEM.02464-15] [PMID: 26407881]

[37]   Spencer JF, Spencer DM. Rare-mating and cytoduction in *Saccharomyces cerevisiae*. Methods Mol Biol 1996; 53: 39-44.
[PMID: 8925000]

[38]   Chambers PJ, Bellon JR, Schmidt SA, Varela C, Pretorius IS. Non-genetic engineering approaches for isolating and generating novel yeasts for industrial applications. In: Satyanarayana T, Kunze G, Eds. Yeast Biotechnology: Diversity and Applications. Netherlands: Springer 2009; pp. 433-57.
[http://dx.doi.org/10.1007/978-1-4020-8292-4_20]

[39]   Turgeon Z, Sierocinski T, Brimacombe CA, *et al.* Industrially applicable *de novo* Lager yeast hybrids with a unique genomic architecture: Creation and characterization. Appl Environ Microbiol 2021; 87(3): e02434-20.
[http://dx.doi.org/10.1128/AEM.02434-20] [PMID: 33188002]

[40]   de Barros Lopes M, Bellon JR, Shirley NJ, Ganter PF. Evidence for multiple interspecific hybridization in *Saccharomyces sensu stricto* species. FEMS Yeast Res 2002; 1(4): 323-31.
[http://dx.doi.org/10.1111/j.1567-1364.2002.tb00051.x] [PMID: 12702336]

[41]   Bellon JR, Eglinton JM, Siebert TE, *et al.* Newly generated interspecific wine yeast hybrids introduce flavour and aroma diversity to wines. Appl Microbiol Biotechnol 2011; 91(3): 603-12.
[http://dx.doi.org/10.1007/s00253-011-3294-3] [PMID: 21538112]

[42]   Bellon JR, Schmid F, Capone DL, Dunn BL, Chambers PJ. Introducing a new breed of wine yeast: interspecific hybridisation between a commercial *Saccharomyces cerevisiae* wine yeast and *Saccharomyces mikatae*. PLoS One 2013; 8(4): e62053.
[http://dx.doi.org/10.1371/journal.pone.0062053] [PMID: 23614011]

[43]   Reuter M, Bell G, Greig D. Increased outbreeding in yeast in response to dispersal by an insect vector. Curr Biol 2007; 17(3): R81-3.
[http://dx.doi.org/10.1016/j.cub.2006.11.059] [PMID: 17276903]

[44]   Stefanini I, Dapporto L, Legras J-L, *et al.* Role of social wasps in *Saccharomyces cerevisiae* ecology and evolution. Proc Natl Acad Sci USA 2012; 109(33): 13398-403.
[http://dx.doi.org/10.1073/pnas.1208362109] [PMID: 22847440]

[45]   Stefanini I, Dapporto L, Berná L, Polsinelli M, Turillazzi S, Cavalieri D. Social wasps are a *Saccharomyces* mating nest. Proc Natl Acad Sci USA 2016; 113(8): 2247-51.
[http://dx.doi.org/10.1073/pnas.1516453113] [PMID: 26787874]

[46]   Di Paola M, Meriggi N, Cavalieri D. Applications of wild isolates of *Saccharomyces* yeast for industrial fermentation: The gut of social insects as niche for yeast hybrids' production. Front

Microbiol 2020; 11: 578425.
[http://dx.doi.org/10.3389/fmicb.2020.578425] [PMID: 33193200]

[47]   Schwartz K, Wenger JW, Dunn B, Sherlock G. *APJ1* and *GRE3* homologs work in concert to allow growth in xylose in a natural *Saccharomyces sensu stricto* hybrid yeast. Genetics 2012; 191(2): 621-32.
[http://dx.doi.org/10.1534/genetics.112.140053] [PMID: 22426884]

[48]   Alexander WG, Peris D, Pfannenstiel BT, Opulente DA, Kuang M, Hittinger CT. Efficient engineering of marker-free synthetic allotetraploids of *Saccharomyces*. Fungal Genet Biol 2016; 89: 10-7.
[http://dx.doi.org/10.1016/j.fgb.2015.11.002] [PMID: 26555931]

[49]   Fukuda N, Kaishima M, Ishii J, Kondo A, Honda S. Continuous crossbreeding of sake yeasts using growth selection systems for a-type and α-type cells. AMB Express 2016; 6(1): 45.
[http://dx.doi.org/10.1186/s13568-016-0216-x] [PMID: 27392493]

[50]   Xie Z-X, Mitchell LA, Liu H-M, *et al.* Rapid and efficient CRISPR/Cas9-based mating-type switching of *Saccharomyces cerevisiae*. G3. Gen Genom Genet 2017; 8: 173-83.

[51]   Peris D, Alexander WG, Fisher KJ, *et al.* Synthetic hybrids of six yeast species. Nat Commun 2020; 11(1): 2085.
[http://dx.doi.org/10.1038/s41467-020-15559-4] [PMID: 32350251]

[52]   van Solingen P, van der Plaat JB. Fusion of yeast spheroplasts. J Bacteriol 1977; 130(2): 946-7.
[http://dx.doi.org/10.1128/jb.130.2.946-947.1977] [PMID: 400800]

[53]   Gong J, Zheng H, Wu Z, Chen T, Zhao X. Genome shuffling: Progress and applications for phenotype improvement. Biotechnol Adv 2009; 27(6): 996-1005.
[http://dx.doi.org/10.1016/j.biotechadv.2009.05.016] [PMID: 19463940]

[54]   Pérez-Través L, Lopes CA, Barrio E, Querol A. Evaluation of different genetic procedures for the generation of artificial hybrids in *Saccharomyces* genus for winemaking. Int J Food Microbiol 2012; 156(2): 102-11.
[http://dx.doi.org/10.1016/j.ijfoodmicro.2012.03.008] [PMID: 22503711]

[55]   Wickner RB, Edskes HK. Yeast killer elements hold their hosts hostage. PLoS Genet 2015; 11(5): e1005139.
[http://dx.doi.org/10.1371/journal.pgen.1005139] [PMID: 25973796]

[56]   Molinet J, Cubillos FA. Wild yeast for the future: Exploring the use of wild strains for wine and beer fermentation. Front Genet 2020; 11: 589350.
[http://dx.doi.org/10.3389/fgene.2020.589350] [PMID: 33240332]

[57]   Yin H, Ma Y, Deng Y, *et al.* Genome shuffling of *Saccharomyces cerevisiae* for enhanced glutathione yield and relative gene expression analysis using fluorescent quantitation reverse transcription polymerase chain reaction. J Microbiol Methods 2016; 127: 188-92.
[http://dx.doi.org/10.1016/j.mimet.2016.06.012] [PMID: 27302037]

[58]   Simchen G. Commitment to meiosis: what determines the mode of division in budding yeast? BioEssays 2009; 31(2): 169-77.
[http://dx.doi.org/10.1002/bies.200800124] [PMID: 19204989]

[59]   Dayani Y, Simchen G, Lichten M. Meiotic recombination intermediates are resolved with minimal crossover formation during return-to-growth, an analogue of the mitotic cell cycle. PLoS Genet 2011; 7(5): e1002083.
[http://dx.doi.org/10.1371/journal.pgen.1002083] [PMID: 21637791]

[60]   Laureau R, Loeillet S, Salinas F, *et al.* Extensive recombination of a yeast diploid hybrid through meiotic reversion. PLoS Genet 2016; 12(2): e1005781.
[http://dx.doi.org/10.1371/journal.pgen.1005781] [PMID: 26828862]

[61]   Serero A, Bedrat A, Baulande S, *et al.* Recombination in a sterile polyploid hybrid yeast upon meiotic

Return-To-Growth. Microbiol Res 2021; 250: 126789.
[http://dx.doi.org/10.1016/j.micres.2021.126789] [PMID: 34062341]

[62] Sadhu MJ, Bloom JS, Day L, Kruglyak L. CRISPR-directed mitotic recombination enables genetic mapping without crosses. Science 2016; 352(6289): 1113-6.
[http://dx.doi.org/10.1126/science.aaf5124] [PMID: 27230379]

[63] Fleiss A, O'Donnell S, Fournier T, *et al.* Reshuffling yeast chromosomes with CRISPR/Cas9. PLoS Genet 2019; 15(8): e1008332.
[http://dx.doi.org/10.1371/journal.pgen.1008332] [PMID: 31465441]

[64] Bozdag GO, Ono J, Denton JA, *et al.* Breaking a species barrier by enabling hybrid recombination. Curr Biol 2021; 31(4): R180-1.
[http://dx.doi.org/10.1016/j.cub.2020.12.038] [PMID: 33621502]

[65] Wenger JW, Schwartz K, Sherlock G. Bulk segregant analysis by high-throughput sequencing reveals a novel xylose utilization gene from *Saccharomyces cerevisiae*. PLoS Genet 2010; 6(5): e1000942.
[http://dx.doi.org/10.1371/journal.pgen.1000942] [PMID: 20485559]

[66] Ehrenreich IM, Torabi N, Jia Y, *et al.* Dissection of genetically complex traits with extremely large pools of yeast segregants. Nature 2010; 464(7291): 1039-42.
[http://dx.doi.org/10.1038/nature08923] [PMID: 20393561]

[67] Parts L, Cubillos FA, Warringer J, *et al.* Revealing the genetic structure of a trait by sequencing a population under selection. Genome Res 2011; 21(7): 1131-8.
[http://dx.doi.org/10.1101/gr.116731.110] [PMID: 21422276]

[68] Duveau F, Metzger BP, Gruber JD, *et al.* Mapping small effect mutations in Saccharomyces cerevisiae: impacts of experimental design and mutational properties. G3 (Bethesda) 2014; 4(7): 1205-16.
[http://dx.doi.org/10.1534/g3.114.011783] [PMID: 24789747]

[69] Cubillos FA, Brice C, Molinet J, *et al.* Identification of nitrogen consumption genetic variants in yeast through QTL mapping and bulk segregant RNA-Seq analyses. G3. G3 (Bethesda) 2017; 7(6): 1693-705.
[http://dx.doi.org/10.1534/g3.117.042127] [PMID: 28592651]

[70] Peltier E, Friedrich A, Schacherer J, Marullo P. Quantitative trait nucleotides impacting the technological performances of industrial *Saccharomyces cerevisiae* strains. Front Genet 2019; 10: 683.
[http://dx.doi.org/10.3389/fgene.2019.00683] [PMID: 31396264]

[71] Dunham MJ. Experimental evolution in yeast: a practical guide. Methods Enzymol 2010; 470: 487-507.
[http://dx.doi.org/10.1016/S0076-6879(10)70019-7] [PMID: 20946822]

[72] Marsit S, Leducq J-B, Durand É, Marchant A, Filteau M, Landry CR. Evolutionary biology through the lens of budding yeast comparative genomics. Nat Rev Genet 2017; 18(10): 581-98.
[http://dx.doi.org/10.1038/nrg.2017.49] [PMID: 28714481]

[73] Gibson B, Dahabieh M, Krogerus K, *et al.* Adaptive laboratory evolution of Ale and Lager yeasts for improved brewing efficiency and beer quality. Annu Rev Food Sci Technol 2020; 11: 23-44.
[http://dx.doi.org/10.1146/annurev-food-032519-051715] [PMID: 31951488]

[74] Delneri D. Barcode technology in yeast: application to pharmacogenomics. FEMS Yeast Res 2010; 10(8): 1083-9.
[http://dx.doi.org/10.1111/j.1567-1364.2010.00676.x] [PMID: 20846145]

[75] Caspeta L, Chen Y, Ghiaci P, *et al.* Biofuels. Altered sterol composition renders yeast thermotolerant. Science 2014; 346(6205): 75-8.
[http://dx.doi.org/10.1126/science.1258137] [PMID: 25278608]

[76] Demeke MM, Foulquié-Moreno MR, Dumortier F, Thevelein JM. Rapid evolution of recombinant *Saccharomyces cerevisiae* for Xylose fermentation through formation of extra-chromosomal circular

DNA. PLoS Genet 2015; 11(3): e1005010.
[http://dx.doi.org/10.1371/journal.pgen.1005010] [PMID: 25738959]

[77]   Mardones W, Villarroel CA, Abarca V, *et al.* Rapid selection response to ethanol in *Saccharomyces eubayanus* emulates the domestication process under brewing conditions. Microb Biotechnol 2021; 0(0): 1-18.
[http://dx.doi.org/10.1111/1751-7915.13803] [PMID: 33755311]

[78]   Deparis Q, Claes A, Foulquié-Moreno MR, Thevelein JM. Engineering tolerance to industrially relevant stress factors in yeast cell factories. FEMS Yeast Res 2017; 17(4): fox036.
[http://dx.doi.org/10.1093/femsyr/fox036] [PMID: 28586408]

[79]   Fisher KJ, Lang GI. Experimental evolution in fungi: An untapped resource. Fungal Genet Biol 2016; 94: 88-94.
[http://dx.doi.org/10.1016/j.fgb.2016.06.007] [PMID: 27375178]

[80]   Zhong Z, Wong BG, Ravikumar A, Arzumanyan GA, Khalil AS, Liu CC. Automated continuous evolution of proteins *in vivo.* ACS Synth Biol 2020; 9(6): 1270-6.
[http://dx.doi.org/10.1021/acssynbio.0c00135] [PMID: 32374988]

[81]   Jensen ED, Laloux M, Lehka BJ, *et al.* A synthetic RNA-mediated evolution system in yeast. Nucleic Acids Res 2021; 49: gkab472.

[82]   Reuter JA, Spacek DV, Snyder MP. High-throughput sequencing technologies. Mol Cell 2015; 58(4): 586-97.
[http://dx.doi.org/10.1016/j.molcel.2015.05.004] [PMID: 26000844]

[83]   Libkind D, Peris D, Cubillos FA, *et al.* Into the wild: new yeast genomes from natural environments and new tools for their analysis. FEMS Yeast Res 2020; 20: foaa008.
[http://dx.doi.org/10.1093/femsyr/foaa008]

[84]   De Coster W, Weissensteiner MH, Sedlazeck FJ. Towards population-scale long-read sequencing. Nat Rev Genet 2021; 22(9): 572-87.
[http://dx.doi.org/10.1038/s41576-021-00367-3] [PMID: 34050336]

[85]   Burton JN, Liachko I, Dunham MJ, Shendure J. Species-level deconvolution of metagenome assemblies with Hi-C-based contact probability maps. G3 (Bethesda) 2014; 4(7): 1339-46.
[http://dx.doi.org/10.1534/g3.114.011825] [PMID: 24855317]

[86]   Beitel CW, Froenicke L, Lang JM, *et al.* Strain- and plasmid-level deconvolution of a synthetic metagenome by sequencing proximity ligation products. PeerJ 2014; 2: e415.
[http://dx.doi.org/10.7717/peerj.415] [PMID: 24918035]

[87]   Marbouty M, Cournac A, Flot J-F, Marie-Nelly H, Mozziconacci J, Koszul R. Metagenomic chromosome conformation capture (meta3C) unveils the diversity of chromosome organization in microorganisms. eLife 2014; 3: e03318.
[http://dx.doi.org/10.7554/eLife.03318] [PMID: 25517076]

[88]   Kronenberg ZN, Rhie A, Koren S, *et al.* Extended haplotype-phasing of long-read de novo genome assemblies using Hi-C. Nat Commun 2021; 12(1): 1935.
[http://dx.doi.org/10.1038/s41467-020-20536-y] [PMID: 33911078]

[89]   Gietz RD, Woods RA. Genetic transformation of yeast. Biotechniques 2001; 30(4): 816-31.
[http://dx.doi.org/10.2144/01304rv02] [PMID: 11314265]

[90]   Gietz RD, Schiestl RH. High-efficiency yeast transformation using the LiAc/SS carrier DNA/PEG method. Nat Protoc 2007; 2(1): 31-4.
[http://dx.doi.org/10.1038/nprot.2007.13] [PMID: 17401334]

[91]   Siewers V. An overview on selection marker genes for transformation of *Saccharomyces cerevisiae.* Methods Mol Biol 2014; 1152: 3-15.
[http://dx.doi.org/10.1007/978-1-4939-0563-8_1] [PMID: 24744024]

[92]   David F, Siewers V. Advances in yeast genome engineering. FEMS Yeast Res 2015; 15(1): 1-14.
        [PMID: 25154295]

[93]   Gnügge R, Rudolf F. *Saccharomyces cerevisiae* Shuttle vectors. Yeast 2017; 34(5): 205-21.
        [http://dx.doi.org/10.1002/yea.3228] [PMID: 28072905]

[94]   Hinnen A, Hicks JB, Fink GR. Transformation of yeast. Proc Natl Acad Sci USA 1978; 75(4): 1929-
        33.
        [http://dx.doi.org/10.1073/pnas.75.4.1929] [PMID: 347451]

[95]   Aylon Y, Kupiec M. New insights into the mechanism of homologous recombination in yeast. Mutat
        Res 2004; 566(3): 231-48.
        [http://dx.doi.org/10.1016/j.mrrev.2003.10.001] [PMID: 15082239]

[96]   Bernardi B, Wendland J. Homologous recombination: A GRAS yeast genome editing tool.
        Fermentation (Basel) 2020; 6: 57.
        [http://dx.doi.org/10.3390/fermentation6020057]

[97]   Kosuri S, Church GM. Large-scale de novo DNA synthesis: technologies and applications. Nat
        Methods 2014; 11(5): 499-507.
        [http://dx.doi.org/10.1038/nmeth.2918] [PMID: 24781323]

[98]   Giaever G, Chu AM, Ni L, *et al.* Functional profiling of the *Saccharomyces cerevisiae* genome. Nature
        2002; 418(6896): 387-91.
        [http://dx.doi.org/10.1038/nature00935] [PMID: 12140549]

[99]   Hillenmeyer ME, Fung E, Wildenhain J, *et al.* The chemical genomic portrait of yeast: uncovering a
        phenotype for all genes. Science 2008; 320(5874): 362-5.
        [http://dx.doi.org/10.1126/science.1150021] [PMID: 18420932]

[100]   Huh WK, Falvo JV, Gerke LC, *et al.* Global analysis of protein localization in budding yeast. Nature
        2003; 425(6959): 686-91.
        [http://dx.doi.org/10.1038/nature02026] [PMID: 14562095]

[101]   Ghaemmaghami S, Huh WK, Bower K, *et al.* Global analysis of protein expression in yeast. Nature
        2003; 425(6959): 737-41.
        [http://dx.doi.org/10.1038/nature02046] [PMID: 14562106]

[102]   Fraczek MG, Naseeb S, Delneri D. History of genome editing in yeast. Yeast 2018; 35(5): 361-8.
        [http://dx.doi.org/10.1002/yea.3308] [PMID: 29345746]

[103]   Alexander WG. A history of genome editing in *Saccharomyces cerevisiae.* Yeast 2018; 35(5): 355-60.
        [http://dx.doi.org/10.1002/yea.3300] [PMID: 29247562]

[104]   Rainha J, Rodrigues JL, Rodrigues LR. CRISPR-Cas9: A powerful tool to efficiently engineer
        Saccharomyces cerevisiae. Life (Basel) 2020; 11(1): 13.
        [http://dx.doi.org/10.3390/life11010013] [PMID: 33375364]

[105]   Guirimand G, Kulagina N, Papon N, Hasunuma T, Courdavault V. Innovative tools and strategies for
        optimizing yeast cell factories. Trends Biotechnol 2021; 39(5): 488-504.
        [http://dx.doi.org/10.1016/j.tibtech.2020.08.010] [PMID: 33008642]

[106]   Pretorius IS. Synthetic genome engineering forging new frontiers for wine yeast. Crit Rev Biotechnol
        2017; 37(1): 112-36.
        [http://dx.doi.org/10.1080/07388551.2016.1214945] [PMID: 27535766]

[107]   Pretorius IS, Boeke JD. Yeast 2.0-connecting the dots in the construction of the world's first functional
        synthetic eukaryotic genome. FEMS Yeast Res 2018; 18(4): 55.
        [http://dx.doi.org/10.1093/femsyr/foy032] [PMID: 29648592]

[108]   Jinek M, Chylinski K, Fonfara I, Hauer M, Doudna JA, Charpentier E. A programmable dual-RN-
        -guided DNA endonuclease in adaptive bacterial immunity. Science 2012; 337(6096): 816-21.
        [http://dx.doi.org/10.1126/science.1225829] [PMID: 22745249]

[109] Shivram H, Cress BF, Knott GJ, Doudna JA. Controlling and enhancing CRISPR systems. Nat Chem Biol 2021; 17(1): 10-9.
[http://dx.doi.org/10.1038/s41589-020-00700-7] [PMID: 33328654]

[110] DiCarlo JE, Norville JE, Mali P, Rios X, Aach J, Church GM. Genome engineering in *Saccharomyces cerevisiae* using CRISPR-Cas systems. Nucleic Acids Res 2013; 41(7): 4336-43.
[http://dx.doi.org/10.1093/nar/gkt135] [PMID: 23460208]

[111] Stovicek V, Borodina I, Forster J. CRISPR-Cas system enables fast and simple genome editing of industrial *Saccharomyces cerevisiae* strains. Metab Eng Commun 2015; 2: 13-22.
[http://dx.doi.org/10.1016/j.meteno.2015.03.001] [PMID: 34150504]

[112] Stovicek V, Holkenbrink C, Borodina I. CRISPR/Cas system for yeast genome engineering: advances and applications. FEMS Yeast Res 2017; 17(5): fox030.
[http://dx.doi.org/10.1093/femsyr/fox030] [PMID: 28505256]

[113] Sharon E, Chen S-AA, Khosla NM, Smith JD, Pritchard JK, Fraser HB. Functional genetic variants revealed by massively parallel precise genome editing. Cell 2018; 175(2): 544-557.e16.
[http://dx.doi.org/10.1016/j.cell.2018.08.057] [PMID: 30245013]

[114] Sadhu MJ, Bloom JS, Day L, Siegel JJ, Kosuri S, Kruglyak L. Highly parallel genome variant engineering with CRISPR-Cas9. Nat Genet 2018; 50(4): 510-4.
[http://dx.doi.org/10.1038/s41588-018-0087-y] [PMID: 29632376]

[115] Giersch RM, Finnigan GC. Yeast still a beast: Diverse applications of CRISPR/Cas editing technology in *S. cerevisiae*. Yale J Biol Med 2017; 90(4): 643-51.
[PMID: 29259528]

[116] Lian J. HamediRad M, Zhao H. Advancing metabolic engineering of *Saccharomyces cerevisiae* using the CRISPR/Cas system. Biotechnol J 2018; 13: 1700601.
[http://dx.doi.org/10.1002/biot.201700601]

[117] Baptista SL, Costa CE, Cunha JT, Soares PO, Domingues L. Metabolic engineering of *Saccharomyces cerevisiae* for the production of top value chemicals from biorefinery carbohydrates. Biotechnol Adv 2021; 47: 107697.
[http://dx.doi.org/10.1016/j.biotechadv.2021.107697] [PMID: 33508428]

[118] Dymond JS, Richardson SM, Coombes CE, *et al.* Synthetic chromosome arms function in yeast and generate phenotypic diversity by design. Nature 2011; 477(7365): 471-6.
[http://dx.doi.org/10.1038/nature10403] [PMID: 21918511]

[119] Annaluru N, Muller H, Mitchell LA, *et al.* Total synthesis of a functional designer eukaryotic chromosome. Science 2014; 344(6179): 55-8.
[http://dx.doi.org/10.1126/science.1249252] [PMID: 24674868]

[120] Schindler D. Genetic engineering and synthetic genomics in yeast to understand life and boost biotechnology. Bioengineering (Basel) 2020; 7(4): 137.
[http://dx.doi.org/10.3390/bioengineering7040137] [PMID: 33138080]

[121] Richardson SM, Mitchell LA, Stracquadanio G, *et al.* Design of a synthetic yeast genome. Science 2017; 355(6329): 1040-4.
[http://dx.doi.org/10.1126/science.aaf4557] [PMID: 28280199]

[122] Wu Y, Li BZ, Zhao M, *et al.* Bug mapping and fitness testing of chemically synthesized chromosome X. Science 2017; 355(6329): 1048.
[http://dx.doi.org/10.1126/science.aaf4706] [PMID: 28280152]

[123] Mitchell LA, Wang A, Stracquadanio G, *et al.* Synthesis, debugging, and effects of synthetic chromosome consolidation: synVI and beyond. Science 2017; 355(6329): eaaf4831.
[http://dx.doi.org/10.1126/science.aaf4831] [PMID: 28280154]

[124] Dymond J, Boeke J. The *Saccharomyces cerevisiae* SCRaMbLE system and genome minimization.

Bioeng Bugs 2012; 3(3): 168-71.
[PMID: 22572789]

[125] Jin J, Jia B, Yuan YJ. Yeast chromosomal engineering to improve industrially-relevant phenotypes. Curr Opin Biotechnol 2020; 66: 165-70.
[http://dx.doi.org/10.1016/j.copbio.2020.07.003] [PMID: 32818746]

[126] Blount BA, Gowers GF, Ho JCH, *et al.* Rapid host strain improvement by *in vivo* rearrangement of a synthetic yeast chromosome. Nat Commun 2018; 9(1): 1932.
[http://dx.doi.org/10.1038/s41467-018-03143-w] [PMID: 29789540]

[127] Wang J, Xie Z-X, Ma Y, *et al.* Ring synthetic chromosome V SCRaMbLE. Nat Commun 2018; 9(1): 3783.
[http://dx.doi.org/10.1038/s41467-018-06216-y] [PMID: 30224715]

[128] Wightman ELI, Kroukamp H, Pretorius IS, Paulsen IT, Nevalainen HKM. Rapid colorimetric detection of genome evolution in SCRaMbLEd synthetic *Saccharomyces cerevisiae* strains. Microorganisms 2020; 8(12): 1914.
[http://dx.doi.org/10.3390/microorganisms8121914] [PMID: 33271913]

[129] Wightman ELI, Kroukamp H, Pretorius IS, Paulsen IT, Nevalainen HKM. Rapid optimisation of cellulolytic enzymes ratios in *Saccharomyces cerevisiae* using *in vitro* SCRaMbLE. Biotechnol Biofuels 2020; 13(1): 182.
[http://dx.doi.org/10.1186/s13068-020-01823-8] [PMID: 33292481]

[130] Gohil N, Bhattacharjee G, Singh V. Expansion of the genetic code. In: Singh V, Ed. Advances in Synthetic Biology. Singapore: Springer Singapore 2020; pp. 237-49.
[http://dx.doi.org/10.1007/978-981-15-0081-7_14]

[131] Tan L, Zheng Z, Xu Y, *et al.* Efficient selection scheme for incorporating noncanonical amino acids into proteins in *Saccharomyces cerevisiae.* Front Bioeng Biotechnol 2020; 8: 569191.
[http://dx.doi.org/10.3389/fbioe.2020.569191] [PMID: 33042970]

[132] Robertson WE, Funke LFH, de la Torre D, *et al.* Sense codon reassignment enables viral resistance and encoded polymer synthesis. Science 2021; 372(6546): 1057-62.
[http://dx.doi.org/10.1126/science.abg3029] [PMID: 34083482]

[133] Radecka D, Mukherjee V, Mateo RQ, Stojiljkovic M, Foulquié-Moreno MR, Thevelein JM. Looking beyond *Saccharomyces*: the potential of non-conventional yeast species for desirable traits in bioethanol fermentation. FEMS Yeast Res 2015; 15(6): fov053.
[http://dx.doi.org/10.1093/femsyr/fov053] [PMID: 26126524]

[134] Rebello S, Abraham A, Madhavan A, *et al.* Non-conventional yeast cell factories for sustainable bioprocesses. FEMS Microbiol Lett 2018; 365(21): fny222.
[http://dx.doi.org/10.1093/femsle/fny222] [PMID: 30212856]

[135] Sun L, Alper HS. Non-conventional hosts for the production of fuels and chemicals. Curr Opin Chem Biol 2020; 59: 15-22.
[http://dx.doi.org/10.1016/j.cbpa.2020.03.004] [PMID: 32348879]

[136] Thorwall S, Schwartz C, Chartron JW, Wheeldon I. Stress-tolerant non-conventional microbes enable next-generation chemical biosynthesis. Nat Chem Biol 2020; 16(2): 113-21.
[http://dx.doi.org/10.1038/s41589-019-0452-x] [PMID: 31974527]

[137] Löbs AK, Schwartz C, Wheeldon I. Genome and metabolic engineering in non-conventional yeasts: Current advances and applications. Synth Syst Biotechnol 2017; 2(3): 198-207.
[http://dx.doi.org/10.1016/j.synbio.2017.08.002] [PMID: 29318200]

[138] Patra P, Das M, Kundu P, Ghosh A. Recent advances in systems and synthetic biology approaches for developing novel cell-factories in non-conventional yeasts. Biotechnol Adv 2021; 47: 107695.
[http://dx.doi.org/10.1016/j.biotechadv.2021.107695] [PMID: 33465474]

[139] Cao M, Gao M, Lopez-Garcia CL, *et al.* Centromeric DNA facilitates nonconventional yeast genetic

engineering. ACS Synth Biol 2017; 6(8): 1545-53.
[http://dx.doi.org/10.1021/acssynbio.7b00046] [PMID: 28391682]

[140] Ding Y, Wang KF, Wang WJ, *et al.* Increasing the homologous recombination efficiency of eukaryotic microorganisms for enhanced genome engineering. Appl Microbiol Biotechnol 2019; 103(11): 4313-24.
[http://dx.doi.org/10.1007/s00253-019-09802-2] [PMID: 31016357]

[141] Cormack BP, Falkow S. Efficient homologous and illegitimate recombination in the opportunistic yeast pathogen *Candida glabrata*. Genetics 1999; 151(3): 979-87.
[http://dx.doi.org/10.1093/genetics/151.3.979] [PMID: 10049916]

[142] Kegel A, Martinez P, Carter SD, Aström SU. Genome wide distribution of illegitimate recombination events in *Kluyveromyces lactis*. Nucleic Acids Res 2006; 34(5): 1633-45.
[http://dx.doi.org/10.1093/nar/gkl064] [PMID: 16549875]

[143] Lieber MR. The mechanism of double-strand DNA break repair by the nonhomologous DNA end-joining pathway. Annu Rev Biochem 2010; 79: 181-211.
[http://dx.doi.org/10.1146/annurev.biochem.052308.093131] [PMID: 20192759]

[144] Petracek ME, Longtine MS. PCR-based engineering of yeast genome. Methods Enzymol 2002; 350: 445-69.
[http://dx.doi.org/10.1016/S0076-6879(02)50978-2] [PMID: 12073329]

[145] Fujitani Y, Yamamoto K, Kobayashi I. Dependence of frequency of homologous recombination on the homology length. Genetics 1995; 140(2): 797-809.
[http://dx.doi.org/10.1093/genetics/140.2.797] [PMID: 7498755]

[146] Pearson BM, Hernando Y, Schweizer M. Construction of PCR-ligated long flanking homology cassettes for use in the functional analysis of six unknown open reading frames from the left and right arms of *Saccharomyces cerevisiae* chromosome XV. Yeast 1998; 14(4): 391-9.
[http://dx.doi.org/10.1002/(SICI)1097-0061(19980315)14:4<391::AID-YEA235>3.0.CO;2-O] [PMID: 9559547]

[147] Zaragoza O. Generation of disruption cassettes *in vivo* using a PCR product and *Saccharomyces cerevisiae*. J Microbiol Methods 2003; 52(1): 141-5.
[http://dx.doi.org/10.1016/S0167-7012(02)00154-9] [PMID: 12401237]

[148] Taneja V, Paul S, Ganesan K. Directional ligation of long-flanking homology regions to selection cassettes for efficient targeted gene-disruption in *Candida albicans*. FEMS Yeast Res 2004; 4(8): 841-7.
[http://dx.doi.org/10.1016/j.femsyr.2004.05.003] [PMID: 15450191]

[149] Merryman C, Gibson DG. Methods and applications for assembling large DNA constructs. Metab Eng 2012; 14(3): 196-204.
[http://dx.doi.org/10.1016/j.ymben.2012.02.005] [PMID: 22629570]

[150] Orthwein A, Fradet-Turcotte A, Noordermeer SM, *et al.* Mitosis inhibits DNA double-strand break repair to guard against telomere fusions. Science 2014; 344(6180): 189-93.
[http://dx.doi.org/10.1126/science.1248024] [PMID: 24652939]

[151] Rosebrock AP. Synchronization and arrest of the budding yeast cell cycle using chemical and genetic methods. Cold Spring Harb Protoc 2017; 2017: 1.
[http://dx.doi.org/10.1101/pdb.prot088724] [PMID: 28049774]

[152] Tsakraklides V, Brevnova E, Stephanopoulos G, Shaw AJ. Improved gene targeting through cell cycle synchronization. PLoS One 2015; 10(7): e0133434.
[http://dx.doi.org/10.1371/journal.pone.0133434] [PMID: 26192309]

[153] Jang IS, Yu BJ, Jang JY, Jegal J, Lee JY. Improving the efficiency of homologous recombination by chemical and biological approaches in *Yarrowia lipolytica*. PLoS One 2018; 13(3): e0194954.
[http://dx.doi.org/10.1371/journal.pone.0194954] [PMID: 29566071]

[154]   Kooistra R, Hooykaas PJ, Steensma HY. Efficient gene targeting in *Kluyveromyces lactis*. Yeast 2004;
        21(9): 781-92.
        [http://dx.doi.org/10.1002/yea.1131] [PMID: 15282801]

[155]   Maassen N, Freese S, Schruff B, Passoth V, Klinner U. Nonhomologous end joining and homologous
        recombination DNA repair pathways in integration mutagenesis in the xylose-fermenting yeast *Pichia
        stipitis*. FEMS Yeast Res 2008; 8(5): 735-43.
        [http://dx.doi.org/10.1111/j.1567-1364.2008.00383.x] [PMID: 18435744]

[156]   Verbeke J, Beopoulos A, Nicaud JM. Efficient homologous recombination with short length flanking
        fragments in Ku70 deficient *Yarrowia lipolytica* strains. Biotechnol Lett 2013; 35(4): 571-6.
        [http://dx.doi.org/10.1007/s10529-012-1107-0] [PMID: 23224822]

[157]   Saraya R, Krikken AM, Kiel JA, Baerends RJ, Veenhuis M, van der Klei IJ. Novel genetic tools for
        Hansenula polymorpha. FEMS Yeast Res 2012; 12(3): 271-8.
        [http://dx.doi.org/10.1111/j.1567-1364.2011.00772.x] [PMID: 22129301]

[158]   Näätsaari L, Mistlberger B, Ruth C, Hajek T, Hartner FS, Glieder A. Deletion of the *Pichia pastoris*
        KU70 homologue facilitates platform strain generation for gene expression and synthetic biology.
        PLoS One 2012; 7(6): e39720.
        [http://dx.doi.org/10.1371/journal.pone.0039720] [PMID: 22768112]

[159]   Choo JH, Han C, Kim JY, Kang HA. Deletion of a KU80 homolog enhances homologous
        recombination in the thermotolerant yeast *Kluyveromyces marxianus*. Biotechnol Lett 2014; 36(10):
        2059-67.
        [http://dx.doi.org/10.1007/s10529-014-1576-4] [PMID: 24930110]

[160]   Oguro Y, Yamazaki H, Ara S, *et al*. Efficient gene targeting in non-homologous end-joining-deficient
        Lipomyces starkeyi strains. Curr Genet 2017; 63(4): 751-63.
        [http://dx.doi.org/10.1007/s00294-017-0679-6] [PMID: 28220186]

[161]   Ueno K, Uno J, Nakayama H, Sasamoto K, Mikami Y, Chibana H. Development of a highly efficient
        gene targeting system induced by transient repression of YKU80 expression in *Candida glabrata*.
        Eukaryot Cell 2007; 6(7): 1239-47.
        [http://dx.doi.org/10.1128/EC.00414-06] [PMID: 17513567]

[162]   Cen Y, Fiori A, Van Dijck P. Deletion of the DNA ligase IV gene in *Candida glabrata* significantly
        increases gene-targeting efficiency. Eukaryot Cell 2015; 14(8): 783-91.
        [http://dx.doi.org/10.1128/EC.00281-14] [PMID: 26048009]

[163]   Edlind TD, Henry KW, Vermitsky JP, Edlind MP, Raj S, Katiyar SK. Promoter-dependent disruption
        of genes: simple, rapid, and specific PCR-based method with application to three different yeast. Curr
        Genet 2005; 48(2): 117-25.
        [http://dx.doi.org/10.1007/s00294-005-0008-3] [PMID: 16078083]

[164]   Dudiuk C, Macedo D, Leonardelli F, *et al*. Molecular confirmation of the relationship between
        Candida guilliermondii Fks1p naturally occurring amino acid substitutions and its intrinsic reduced
        echinocandin susceptibility. Antimicrob Agents Chemother 2017; 61(5): e02644-16.
        [http://dx.doi.org/10.1128/AAC.02644-16] [PMID: 28242659]

[165]   Lopes RG, Muñoz JE, Barros LM, Alves-Jr SL, Taborda CP, Stambuk BU. The secreted acid trehalase
        encoded by the *CgATH1* gene is involved in *Candida glabrata* virulence. Mem Inst Oswaldo Cruz
        2020; 115: e200401.
        [http://dx.doi.org/10.1590/0074-02760200401] [PMID: 33146242]

[166]   Wagner JM, Alper HS. Synthetic biology and molecular genetics in non-conventional yeasts: Current
        tools and future advances. Fungal Genet Biol 2016; 89: 126-36.
        [http://dx.doi.org/10.1016/j.fgb.2015.12.001] [PMID: 26701310]

[167]   Yang Z, Blenner M. Genome editing systems across yeast species. Curr Opin Biotechnol 2020; 66:
        255-66.

[http://dx.doi.org/10.1016/j.copbio.2020.08.011] [PMID: 33011454]

[168]   Rigouin C, Croux C, Dubois G, Daboussi F, Bordes F. Genome editing in *Y. lipolytica* using TALENs. Methods Mol Biol 2021; 2307: 25-39.
[http://dx.doi.org/10.1007/978-1-0716-1414-3_2] [PMID: 33847980]

[169]   Riley R, Haridas S, Wolfe KH, *et al.* Comparative genomics of biotechnologically important yeasts. Proc Natl Acad Sci USA 2016; 113(35): 9882-7.
[http://dx.doi.org/10.1073/pnas.1603941113] [PMID: 27535936]

[170]   Krassowski T, Coughlan AY, Shen XX, *et al.* Evolutionary instability of CUG-Leu in the genetic code of budding yeasts. Nat Commun 2018; 9(1): 1887.
[http://dx.doi.org/10.1038/s41467-018-04374-7] [PMID: 29760453]

[171]   Shen XX, Opulente DA, Kominek J, *et al.* Tempo and mode of genome evolution in the budding yeast subphylum. Cell 2018; 175(6): 1533-1545.e20.
[http://dx.doi.org/10.1016/j.cell.2018.10.023] [PMID: 30415838]

[172]   Papon N, Courdavault V, Clastre M, Simkin AJ, Crèche J, Giglioli-Guivarc'h N. *Deus ex Candida* genetics: overcoming the hurdles for the development of a molecular toolbox in the CTG clade. Microbiology 2012; 158(Pt 3): 585-600.
[http://dx.doi.org/10.1099/mic.0.055244-0] [PMID: 22282522]

[173]   Defosse TA, Courdavault V, Coste AT, *et al.* A standardized toolkit for genetic engineering of CTG clade yeasts. J Microbiol Methods 2018; 144: 152-6.
[http://dx.doi.org/10.1016/j.mimet.2017.11.015] [PMID: 29155237]

[174]   Gordon ZB, Soltysiak MPM, Leichthammer C, *et al.* Development of a transformation method for *Metschnikowia borealis* and other CUG-Serine yeasts. Genes (Basel) 2019; 10(2): 78.
[http://dx.doi.org/10.3390/genes10020078] [PMID: 30678093]

[175]   Favaro L, Jansen T, van Zyl WH. Exploring industrial and natural *Saccharomyces cerevisiae* strains for the bio-based economy from biomass: the case of bioethanol. Crit Rev Biotechnol 2019; 39(6): 800-16.
[http://dx.doi.org/10.1080/07388551.2019.1619157] [PMID: 31230476]

[176]   Della-Bianca BE, Basso TO, Stambuk BU, Basso LC, Gombert AK. What do we know about the yeast strains from the Brazilian fuel ethanol industry? Appl Microbiol Biotechnol 2013; 97(3): 979-91.
[http://dx.doi.org/10.1007/s00253-012-4631-x] [PMID: 23271669]

[177]   Lopes ML, Paulillo SC, Godoy A, *et al.* Ethanol production in Brazil: a bridge between science and industry. Braz J Microbiol 2016; 47 (Suppl. 1): 64-76.
[http://dx.doi.org/10.1016/j.bjm.2016.10.003] [PMID: 27818090]

[178]   Jacobus AP, Gross J, Evans JH, Ceccato-Antonini SR, Gombert AK. *Saccharomyces cerevisiae* strains used industrially for bioethanol production. Essays Biochem 2021; 65(2): 147-61.
[http://dx.doi.org/10.1042/EBC20200160] [PMID: 34156078]

[179]   Basso TO, de Kok S, Dario M, *et al.* Engineering topology and kinetics of sucrose metabolism in *Saccharomyces cerevisiae* for improved ethanol yield. Metab Eng 2011; 13(6): 694-703.
[http://dx.doi.org/10.1016/j.ymben.2011.09.005] [PMID: 21963484]

[180]   Abreu-Cavalheiro A, Monteiro G. Solving ethanol production problems with genetically modified yeast strains. Braz J Microbiol 2014; 44(3): 665-71.
[http://dx.doi.org/10.1590/S1517-83822013000300001] [PMID: 24516432]

[181]   Paulino de Souza J, Dias do Prado C, Eleutherio ECA, Bonatto D, Malavazi I, Ferreira da Cunha A. Improvement of Brazilian bioethanol production - Challenges and perspectives on the identification and genetic modification of new strains of *Saccharomyces cerevisiae* yeasts isolated during ethanol process. Fungal Biol 2018; 122(6): 583-91.
[http://dx.doi.org/10.1016/j.funbio.2017.12.006] [PMID: 29801803]

[182]   Ceccato-Antonini SR, Covre EA. From baker's yeast to genetically modified budding yeasts: the

scientific evolution of bioethanol industry from sugarcane. FEMS Yeast Res 2021; 20: foaa065.
[http://dx.doi.org/10.1093/femsyr/foaa065]

[183] Stambuk BU. Biotechnology strategies with industrial fuel ethanol *Saccharomyces cerevisiae* strains for efficient 1st and 2nd generation bioethanol production from sugarcane. BMC Proc 2014; 8: O36.
[http://dx.doi.org/10.1186/1753-6561-8-S4-O36]

[184] dos Santos LV, Grassi MCB, Gallardo JCM, *et al.* Second-generation ethanol: The need is becoming a reality. Ind Biotechnol (New Rochelle NY) 2016; 12: 40-57.
[http://dx.doi.org/10.1089/ind.2015.0017]

[185] Dos Santos LV, Carazzolle MF, Nagamatsu ST, *et al.* Unraveling the genetic basis of xylose consumption in engineered *Saccharomyces cerevisiae* strains. Sci Rep 2016; 6: 38676.
[http://dx.doi.org/10.1038/srep38676] [PMID: 28000736]

[186] Jansen MLA, Bracher JM, Papapetridis I, *et al. Saccharomyces cerevisiae* strains for second-generation ethanol production: from academic exploration to industrial implementation. FEMS Yeast Res 2017; 17(5): fox044.
[http://dx.doi.org/10.1093/femsyr/fox044] [PMID: 28899031]

[187] Ro DK, Paradise EM, Ouellet M, *et al.* Production of the antimalarial drug precursor artemisinic acid in engineered yeast. Nature 2006; 440(7086): 940-3.
[http://dx.doi.org/10.1038/nature04640] [PMID: 16612385]

[188] Paddon CJ, Westfall PJ, Pitera DJ, *et al.* High-level semi-synthetic production of the potent antimalarial artemisinin. Nature 2013; 496(7446): 528-32.
[http://dx.doi.org/10.1038/nature12051] [PMID: 23575629]

[189] Peralta-Yahya PP, Ouellet M, Chan R, Mukhopadhyay A, Keasling JD, Lee TS. Identification and microbial production of a terpene-based advanced biofuel. Nat Commun 2011; 2: 483.
[http://dx.doi.org/10.1038/ncomms1494] [PMID: 21952217]

[190] Leavell MD, McPhee DJ, Paddon CJ. Developing fermentative terpenoid production for commercial usage. Curr Opin Biotechnol 2016; 37: 114-9.
[http://dx.doi.org/10.1016/j.copbio.2015.10.007] [PMID: 26723008]

[191] Zhang Y, Nielsen J, Liu Z. Engineering yeast metabolism for production of terpenoids for use as perfume ingredients, pharmaceuticals and biofuels. FEMS Yeast Res 2017; 17(8): fox080.
[http://dx.doi.org/10.1093/femsyr/fox080] [PMID: 29096021]

[192] Carsanba E, Pintado M, Oliveira C. Fermentation strategies for production of pharmaceutical terpenoids in engineered yeast. Pharmaceuticals (Basel) 2021; 14(4): 295.
[http://dx.doi.org/10.3390/ph14040295] [PMID: 33810302]

[193] Rienzo M, Jackson SJ, Chao LK, *et al.* High-throughput screening for high-efficiency small-molecule biosynthesis. Metab Eng 2021; 63: 102-25.
[http://dx.doi.org/10.1016/j.ymben.2020.09.004] [PMID: 33017684]

[194] Chandel AK, Forte MB, Gonçalves IS, *et al.* Brazilian biorefineries from second generation biomass: critical insights from industry and future perspectives. Biofuels Bioprod Biorefin 2021; 15.
[http://dx.doi.org/10.1002/bbb.2234]

CHAPTER 15

Are Yeasts "Humanity's Best Friends"?

Sérgio L. Alves Jr[1,*], Helen Treichel[2], Thiago O. Basso[3] and **Boris U. Stambuk[4]**

[1] *Laboratory of Biochemistry and Genetics, Federal University of Fronteira Sul, Chapecó/SC, Brazil*

[2] *Laboratory of Microbiology and Bioprocesses, Federal University of Fronteira Sul, Erechim/RS, Brazil*

[3] *Department of Chemical Engineering, University of São Paulo, São Paulo/SP, Brazil*

[4] *Department of Biochemistry, Federal University of Santa Catarina, Florianópolis/SC, Brazil*

Abstract: The beginning of the relationship between humans and yeasts is commonly assigned to the Neolithic revolution. However, the role of these microorganisms as gut symbionts of humans and other animals cannot be disregarded. In this case, the timespan of this relationship should be measured in hundreds (not tens) of thousands of years. Evidently, the hypothesis that the aforementioned symbiosis began precisely with the domestication of yeasts during the Neolithic revolution period cannot be ruled out as well. In any case, the relationship between humans and yeasts has broadly developed from the moment humanity started to domesticate them to produce bread and beverages, which seems to coincide with the Neolithic revolution period. Since then, humanity has created novel bioprocesses with yeasts, even though the role of these microorganisms was only really understood in the 19th century, especially with the studies of Louis Pasteur. Today, yeasts drive a trillion-dollar global market, which most likely presents the highest value among all sectors of industrial microbiology. In this context, this book's last chapter addresses the importance of yeasts in our society, with positive impacts on the economy and the health of humans, animals, and plants. We also discuss the role of these microorganisms in maintaining the balance and diversity of species in the environment as a whole. Finally, we close the chapter by highlighting the effects of their environmental role on human well-being and outlining the potential of wild yeasts that can drift from nature to new bioprocesses.

Keywords: Beer, Beverage, Biocontrol, Biodiesel, Biotechnology, Bread, Cheese, Chocolate, Decomposition, Ethanol, Food industry, Growth-promoting, Microbial factory, Pharmaceuticals, Probiotic, *Saccharomyces*, Single-cell protein, Vaccine, VOCs, Wine.

---

[*] **Corresponding author Sérgio L. Alves Jr:** Laboratory of Biochemistry and Genetics, Federal University of Fronteira Sul, Chapecó/SC, Brazil; Tel: +55 49 999194025; E-mail: slalvesjr@uffs.edu.br

Sérgio Luiz Alves Júnior, Helen Treichel, Thiago Olitta Basso and Boris Ugarte Stambuk (Eds.)
All rights reserved-© 2022 Bentham Science Publishers

INTRODUCTION

This is not the first time yeast has been claimed as a man's best friend [1, 2]. Since the Neolithic revolution, yeasts have been employed as fermenting microorganisms and their relationship with humans facilitated the change from a gathering-hunting lifestyle to the establishment of permanent settlements. New archaeobotanical evidence reveals that the preparation of bread-like products occurred 4,000 years before the emergence of the Neolithic agricultural way of life, probably 14,400 years ago in northeastern Jordan [3]. Although at that moment breadmaking may have occurred without fermentation (the so-called flatbread), at least from 10,000 ya, yeasts and humans have become increasingly close as the millennia have passed due to the diversification of various bioprocesses.

Yeasts have a direct or indirect bond (or a potential bond) with at least 9 of the 21 industrial sectors established by the United Nations (UN) classification — International Standard Industrial Classification of All Economic Activities (ISIC) [4]. Yeasts can also be used for heterologous expression, making them an excellent choice to act as microbial factories [5 - 12], greatly expanding their biotechnological and industrial potentials. Moreover, if one considers the approaches of Yeast Synthetic Biology, Synthetic Genomics, *Saccharomyces cerevisiae* v.2.0 (Sc2.0), and Synthetic Chromosome Recombination and Modification by LoxP-mediated Evolution (SCRaMbLE), this potential becomes even more significant [13, 14].

Furthermore, as new biotech prospects have been identified in recent years, the potential use of yeasts as bio-based (greener) alternatives to the conventional techniques in the chemical and petrochemical industries has also been envisioned [15]. Indeed, these microorganisms have proven to be increasingly versatile; not withstanding the countless industrial processes in which these microorganisms are already employed, it seems that they can be applied in an even greater variety of bioprocesses in the future. This chapter will summarise the main biotechnological applications of yeasts and outline their ecological roles, which also positively impact human welfare.

THE THINGS WE LOVE THE MOST

The most traditional biotechnological products, whose processing involves the use of yeasts, are also the most profitable ones. Besides, these products are also related to the joy, happiness, sociability, and pleasure of individuals. Therefore, they are very likely the bioproducts that humans enjoy the most. Together these industrial segments of joy comprise a trillionaire market, surpassing US$ 1.3 trillion worth of value.

Alcoholic Beverages

Most alcoholic beverages are produced due to the fermentation capacity of yeast cells. In this scenario, the species *S. cerevisiae* stands out as the primary yeast used for the beverages with the highest production volume and the most extensive market sizes (Table **1**).

Table 1. Alcoholic beverages and their markets.

Dominant Yeast Species in the Processes	Beverage	Global Production[a]	Global Market Size	References/Sources
S. cerevisiae and *S. pastorianus*	Beer	194 billion L	US$ 623 billion	[17, 19]
S. cerevisiae	Wine	29.2 billion L	US$ 327 billion	[17, 18, 20]
S. cerevisiae	Whiskey	5.2 billion L	US$ 60 billion	[21 - 23]
S. cerevisiae	Vodka	3 billion L	US$ 45 billion	[24 - 26]
S. cerevisiae	Tequila	0.25 billion L	US$ 10 billion	[27 - 29]
S. cerevisiae	Sake	0.6 billion L	US$ 9 billion	[30 - 32]
S. cerevisiae	Cachaça	1.8 billion L	US$ 2 billion	[33, 34]

[a] Approximated values per year.

Beer is a non-distilled beverage obtained from the fermentation of a wort composed of malted cereals, hops, and freshwater. Besides its millenary history, beer is now the leading alcoholic product consumed in the world. Its production has increased gradually over the last decades [16], reaching 194 billion liters in 2018. This amount represented a 50% increase in the last two decades and was six times higher than the wine production in that same year (29.2 billion liters) [17].

Although wine has a significantly lower production volume than beer, it is a higher value-added product, in such a way that the global wine market size is worth half of beer's (Table **1**). As a result, wine accounts for nearly one-quarter of the global alcoholic-beverages market [18]. Together, wine and beer represent almost 90% of the total alcoholic beverages produced on earth [17].

Considering the market of some of the most conventional alcoholic beverages alone, yeasts contribute to more than US$ 1 trillion market revenue worldwide (see Table **1**). The notoriety of this sum becomes even more evident when compared to another segment of the food and beverage industry: dairy products. This is a very representative sector of the employment of bacteria (in more significant proportion) and fungi; however, the entire dairy market, having a value of US$ 489.74 billion [35], does not even correspond to half the market size of the alcoholic beverages listed in Table **1**. Such difference between dairy and alcoholic-beverage industries proves the high economic impact of yeasts among all microorganisms. For more details on yeasts and beverage production, please refer to Chapter 10 of this book.

Bread

Bread has been widely consumed as a traditional staple food in many countries around the world. The global per capita consumption of bread is 24.5 kg per year [36]. According to the 2021 report of the market research agency "360 Market Update", the global bread market was at US$ 201 billion last year. The international bread sale has surpassed 83 million tons in one year, registering a compound annual growth rate (CAGR) of 1.2%, during the forecasted period of the aforementioned agency. The bread industry is thoroughly dominated by Europe, which accounts for over 45% of the global consumption by volume [37].

Yeasts are the main CO_2 producers in dough leavening in bakeries. The most common sourdough yeast species are *S. cerevisiae*, *Kazachstania exigua*, *C. humilis*, *Torulaspora delbrueckii*, *Pichia kudriavzevii*, and *Wickerhamomyces anomalus* [38, 39]. However, over 30 yeast species have been identified in this sour-water mixture [40], and even yeasts from house microbiotas were found on it [41]. Indeed, the sourdough microbial composition is expected to be influenced by yeasts from the environment [42, 43]. Besides their naturally-found yeasts, sourdoughs can be added along with a starter culture [39].

Yeasts that remain in sourdough after several baking cycles are selected by the stress conditions of this environment, such as nutrient starvation, low pH, reactive oxygen species, high-thermal fluctuation, and high-osmotic pressure. Besides ethanol and CO_2 production, these high-adapted yeasts contribute to sourdough products' leavening, flavor, and nutritional profiles. This contribution stems from organic acids, higher alcohols, esters, and vitamins generated by yeast metabolism. Moreover, yeasts exhibit probiotic potential, aiding in the inhibition of fungi and their mycotoxin production [39].

Most large-scale baking processes use pure starter cultures, mainly with the traditional baker's yeast *S. cerevisiae*. This is due to the difficulties and costs of maintaining a live microbial culture (sourdough starter) in industries and the time-consuming sourdough fermentation process. Also, inoculation of *S. cerevisiae* renders higher productivity and more uniform products, even though the slowly-fermented bread may be considered a higher-quality good in terms of organoleptic features, digestibility, and nutritional value [44]. For more details on yeasts and breadmaking, please refer to Chapter 11 of this book.

Chocolate

Yeasts are essential for the chocolate industry. In the first step of chocolate production, a natural, seven-day microbial fermentation of the pectinaceous pulp surrounding cocoa beans occurs under temperatures of up to 50 °C. During this fermentation, a comprehensive microbial succession of yeasts and bacteria takes place. Then, the metabolites produced by them — ethanol, lactic acid, and acetic acid — cause the death of the beans and generate flavor precursors [45].

Aiming to determine the primary contribution of yeasts to chocolate quality, Ho *et al.* [46] carried out cocoa bean fermentation with the addition of 200 ppm Natamycin, an approved food additive that inhibits yeast growth. The authors compared the resultant microbial ecology, products of metabolism, bean chemistry, and chocolate quality with those of control fermentation (without Natamycin) through these fermentations. The fermentation that did not rely on yeasts showed increased shell content as well as lower production of ethanol, higher alcohols, and esters. In addition, the quality tests revealed that cocoa beans not fermented by yeasts were not entirely brown in color and rendered a final product more acidic in nature and lacking the characteristic chocolate flavor. Meanwhile, the beans that were fermented by yeasts were fully brown and rendered chocolate with the typical organoleptic characteristics, which were preferred by sensory panels [46]. Yeasts may not only be inoculated but also be naturally found in the fermentation of cocoa beans. Indeed, studies that evaluated the spontaneous fermentation of cocoa beans demonstrated that this process presents great biodiversity of yeasts, even though it is possible to point out some more prevalent species, such as *S. cerevisiae* and species of the genera *Hanseniaspora* and *Pichia* [47 - 55].

The chocolate industry is a multibillion-dollar business, with a retail market value of US$ ~107 billion a year [56]. Of this amount, US$ 73 billion correspond to the net sales of the seven biggest multinational companies. Five of them have chocolate confectionery production plants in Europe. Europe is the world's largest chocolate producer, with a market valued at an estimated US$ 63 billion in 2019. The average per capita chocolate consumption in Europe reached 5.0 kg in 2018, but it may be as high as 11 kg per year per capita in Germany [57]. The chocolate market also has a significant impact in low-income countries. In 2016, the largest exporter of cocoa beans was Côte d'Ivoire (US$ 3.9 billion), followed by Ghana (US$ 2.5 billion) and Nigeria (US$ 0.8 billion). These three countries account for ~83% of the world value of cocoa-bean exports. From 2019 to 2025, this market is expected to grow 7.3% a year, reaching US$ 16.3 billion [56].

Cheese

Cheese is probably the least remembered product when it comes to yeast activity in favor of *Homo sapiens*. On the contrary, regarding cheeses, yeasts are most likely reminded by their negative spoiling effects. However, these single-celled fungi perform an essential role in the ripening stage of different styles of cheese, especially those that are traditionally aged and surface-ripened. In the so-referred stage, maturation starts with the rapid colonization of cheese surface by acid- and halo-tolerant yeasts, right after the brining process. Yeasts catabolize lactic acid and amino acids, thus increasing cheese pH and allowing the colonization by a less acid-tolerant bacterial community. In this sense, they contribute significantly to the flavor and typical appearance of smear-ripened cheeses, such as Gruyère, Tilsit, Reblochon, and Munster, and, to a lesser extent, the mold-ripened ones, like Camembert, Stilton, and Tomme de Savoie [58]. The yeast species *Debaryomyces hansenii*, *Yarrowia lipolytica*, *Geotrichum candidum*, *Kluyveromyces lactis*, *K. marxianus*, and *T. delbrueckii* are the most prevalently reported in various cheese-making processes [59 - 63].

WHAT ELSE CAN YEASTS OFFER US?

Biopharmaceuticals

Since the development of recombinant DNA technology in the 1970s, yeasts have been tested as microbial factories, heterologously expressing proteins from representatives of the three domains of life [64 - 66]. In 1975, at the famous "Asilomar Conference on Recombinant DNA Molecules," the scientific committee of that event pointed out the use of these microorganisms as safer alternatives to bacteria [67]. Since then, yeast has undoubtedly been the best choice of eukaryotic cells for heterologous protein expression [65, 68].

In this context, the pharmaceutical industry is the one that benefits most from genetically modified yeasts. Through different genetic engineering approaches, yeasts have been empowered to produce plant-derived therapeutics, such as the antimalarial drug artemisinin; the cancer therapeutics taxol, noscapine, and taxediene; and endocannabinoid and opioid analgesics [5, 9, 69 - 73]. Production of biopharmaceuticals by yeast cells has significant advantages over plant extraction and chemical synthesis, for instance: (i) reduced environmental harm, (ii) different approaches ensuring the biosynthesis of complex molecules, and (iii) improved scalability for compounds that normally exist at low levels in nature [9, 13, 74].

Three yeast species master the biopharmaceutical market: *Pichia pastoris*, *Hansenula polymorpha,* and *S. cerevisiae*. Yeast *P. pastoris* is employed for collagen production (Fibrogen, San Francisco, CA, USA), used as dermal filler; for production of the plasma kallikrein inhibitor Kalbitor® (Dyax, Cambridge, MA, USA), indicated against hereditary angioedema; and for insulin production (Biocon, Bangalore, India), used by patients with diabetes [11]. Recombinant human insulin is also produced by *H. polymorpha* [12, 75] and *S. cerevisiae*, the latter one accounting for half the human insulin produced in the world [76, 77]. Human insulin is a US$ 21 billion global market, forecasted to be US$ 27 billion by 2026, assuming a CAGR of 3.4% [78].

Besides insulin, those three yeasts also serve as microbial factories to produce Hepatitis-B surface antigen particles (HBsAg) that are used as Hepatitis-B vaccines [10, 75, 79, 80]. Additionally, yeasts also heterologously produce a virus-like particle vaccine for the human papillomavirus (HPV) [79, 81]. In fact, about 20% of the whole biopharmaceuticals are produced by yeasts [76]. The entire biopharmaceutical market size was worth US$ 192 billion in 2020, and it is projected to reach US$ 326 billion by 2026, considering a CAGR of 9.2% [82].

Single-cell Market

The single-cell market relies on algae, bacteria, filamentous fungi, and yeast. In this scenario, yeasts stand out for their essential amino-acids-rich proteins and low titer of nucleic acids [83]. As with the other markets mentioned above, single-cell nutrition is also a billion-dollar business. In terms of proteins, extracts from single-cell organisms are a US$ 5.3 billion market, with an 8.6% CAGR in the forecasted period of 2018–2023 [84]. Although this industrial sector is also composed of algae, bacteria, and filamentous fungi, yeast is its major player. Single-cell protein (SCP) has been recognized as an environmentally-friendly alternative to common sources of protein (like livestock) [85], and it is known as a good enrichment factor for soups, baked products, and other foodstuffs for human consumption [86]. Since it is possible to use waste as substrate for SCP production or even to use microbial cells from disposals of fermentation processes, the SCP-production costs can be low [87]. Indeed, the low production cost and high nutritional value of SCP yeasts were recognized to such an extent that in the first and second World Wars, German troops replaced up to 60% of their food with SCP yeasts [88].

Inactive forms of yeast cells are commercially available in many countries. Besides its high content of protein (up to 45% of the dry biomass) [86], *S. cerevisiae*, for example, consists of 15–45% of carbohydrates, 4–8% of lipids, and a significant amount of complex B vitamins [83, 89, 90]. Interestingly, it has been demonstrated that the amplification of genes involved in the biosynthesis of vitamins B6 (pyridoxine) and B1 (thiamin) is a common feature among fuel-ethanol *S. cerevisiae* strains [91]. It is important to emphasize that producing microorganisms for industrial microbiology purposes (such as fuel-ethanol) is another segment of the single-cell yeast (SCY) market. In this scenario, the microbial cells are produced to aid in bioprocesses with their intended biocatalysts. Thereby, taking into account the 3 million tons of dry-weight yeast cells produced per year — either for SCP purposes or as SCY to supply different bioprocess — the yeast-production market size is worth more than US$ 10 billion [85].

Probiotics and Prebiotics

While probiotics are live microorganisms that promote health benefits to the host animal [92], prebiotics is indigestible or poorly digestible feed ingredients that indirectly benefit the host by stimulating the development of health-promoting bacteria located in the gut [93, 94]. Lactic acid bacteria are notoriously the most remembered probiotics. Among the yeasts, *S. cerevisiae* var. *boulardii* is, to the best of our knowledge, the unique commercial probiotic yeast for humans. The success of this species is due to its tolerance to our body temperature, resistance to stomach acidity, and ability to inhibit the growth of pathogenic microorganisms [95]. This immunological effect results from the yeast's ability to produce killer toxins and mycocins [96], and to enhance the host's immune response by inducing natural killer cell activity [97], macrophages [98], and production of immunoglobin A (IgA) [99, 100]. Besides, yeast can modulate inflammatory reactions via the specific proliferative response of T-lymphocytes [98, 101].

There is also a direct relationship of *S. boulardii* with the prevention or mitigation of severe symptoms of infectious diarrhoea caused by *Listeria monocytogenes*, *Salmonella typhimurium*, *Pseudomonas aeruginosa*, *Staphylococcus aureus*, *Enterococcus faecalis*, *Clostridium difficile*, *Vibrio cholerae*, *Shigella flexeneri*, and enteropathogenic *Escherichia coli* (EPEC) [101 - 105]. Interesting finds, such as these have also been reported for the probiotic aptitudes of other yeast species. For example, Agarbati and co-workers [106] showed that *Lachancea thermotolerans*, *Metschnikowia ziziphicola*, *S. cerevisiae*, and *T. delbrueckii* strains isolated from un-anthropized natural environments harbor important probiotic characteristics, even though these features were strictly strain-dependent.

The above-mentioned antimicrobial activity has recently been corroborated by He and co-workers [107], who found evidence for applying live yeast as an alternative to antibiotics in broilers. The authors analyzed the broiler's growth, immune function, serum biochemical parameters, and intestinal morphology. In all these analyses, live-yeast-treated animals' performance was at least as good as that showed by antibiotic-treated chickens. Additionally, autolyzed yeast cells have also been widely used as prebiotic feed additives for poultry, and they also act as potential alternatives to in-feed antibiotics. Since prebiotics stimulates the growth of normal microbiota, it is speculated that their mode of action is similar to that of probiotics [93].

Fuels

Ethanol, the biotechnological product with the largest production volume, also involves the use of yeasts, which account for 80% of the total ethanol produced [108]. Almost 140 billion liters of this biofuel are produced annually, representing a global market value of US$ 86 billion [109]. Along with the production of beverages and bread, the species *S. cerevisiae* is the most used yeast in fuel-ethanol production [108, 110 - 112]. The United States and Brazil are the two largest fuel ethanol producers globally, accounting for 56% and 28% of worldwide production, respectively. While in the US, yeasts ferment worts with corn starch hydrolysates [113], in Brazil, production is simpler and more efficient thanks to sugarcane juice and molasses used in the fermentation vats. These substrates are rich in sucrose, a disaccharide rapidly fermented by yeast cells [114]. There are also plants in the US, Brazil, Canada, Italy, Norway, and China producing 2G ethanol [115]. This second-generation production uses lignocellulosic residues as feedstocks, which requires the fermentation of a significant concentration of pentoses, especially xylose. This represents a bottleneck for 2G plants, as the largest ethanologenic yeast (*S. cerevisiae*) is incapable of naturally fermenting pentoses [116], requiring further improvement via genetic or evolutionary engineering. As a second option, there is the prospect of new non-*Saccharomyces* yeasts known to be capable of metabolizing these sugars. This bottleneck is certainly one of the culprits for the production of 2G ethanol that still does not correspond to 1% of all production of this fuel [17, 115]. Please refer to Chapters 4, 8, and 9 of this book for comprehensive reviews on ethanol production.

Yeasts have also shown the potential to act in biodiesel production, the second most-produced liquid biofuel globally [113]. Although the first vehicles to run on diesel and ethanol were produced in approximately the same period (late 19th and early 20th century, respectively) [117], the relationship of these fuels with yeasts has a very different time span. While the optimization history of ethanol production has always kept yeast as the process protagonist, biodiesel is still mostly plant-derived (such as grains and vegetable oils) [113, 118]. However, concerns related to environmental, energy, and food security (the so-called 'food *versus* fuel' dispute), and economic aspects related to the cost of vegetable raw material (which may account for 70–85% of the entire production cost), have stimulated the production of microbial lipids or single-cell oils (SCO) as an alternative feedstock for biodiesel production [118 - 121]. In this sense, like bacteria and microalgae, yeasts appear like a good choice for SCO.

Besides their fast growth rate, yeasts can use low-cost fermentation media, such as residues from agriculture and industry [119, 122 - 129], and may have an oil content of up to 72% of the cellular dry weight [118]. The genera *Candida, Cryptococcus, Lipomyces, Rhodosporidium, Rhodotorula, Trichosporon,* and *Yarrowia* are the most reported on literature as oleaginous yeast with high potential for biodiesel production [122, 123, 124 - 126, 128 - 132, 133 - 140]. For additional information on yeasts and biodiesel, please refer to Chapter 12 of this book.

Products and Services for the Textile Industry

The US$ 1.0 trillion textile industry [141] relies on yeasts both during and after the production process. In the first case, yeasts are used for the heterologous production of significant innovations, such as spider silk and leather [5, 142, 143], as well as conventional enzymes for textile desizing, scouring, bio-polishing, and finishing processes [144 - 146]. After production, these microorganisms are used, directly or indirectly, to decolorize and detoxify textile wastewater. In both stages, yeasts also contribute to the production of laccase enzymes, which have played a central role in this industrial segment [147, 148].

During the production of clothes, laccases present themselves as environmentally-friendly alternatives for textile bleaching [146]. At the end of the process, laccases are especially important due to their low substrate specificity that allows oxidation of aminophenols, polyphenols, and polyamines [149, 150], enabling their use as potent dye removers [148]. Although some yeasts naturally produce their own laccases [151, 152], industrially, heterologous expression of bacterial and filamentous fungal laccases in *K. lactis, Y. lipolytica, S. cerevisiae, P. pastoris, P. methalonica,* and *Cryptococcus* sp. has shown to be more efficient [147, 153 - 158].

Nutrient Cycling and Decomposition

The kingdom Fungi comprises the paramount terrestrial decomposers, and in these environments, especially in forests, wood-decaying microorganisms significantly contribute to nutrient cycling [159]. Yeasts implicated in this scenario generally act via the oxidation of short-chain carbon sources into CO_2. Such carbon sources may be monomers or oligomers (*e.g.*, mono-, di-, and trisaccharides), resulting from the hydrolysis of polymers (such as the polysaccharides cellulose and hemicellulose of vegetal lignocellulosic biomasses) that mostly takes place in the soil. Cellulose and hemicellulose hydrolysis is mainly carried out by filamentous fungi, but yeasts have also been associated with this task [160, 161].

The whole decomposition process includes high biodiversity of yeasts and during the different stages of wood decay, there is a succession of different yeast species [159, 162]. In a comprehensive literature review on yeasts isolated from decaying wood from cold and temperate regions, Rönnander and Wright [159] recorded 73 species of ascomycetous yeasts and 35 of basidiomycetous. Among the major clades of yeasts associated with these environments, it is worth mentioning the following genera: *Scheffersomyces*, *Spathaspora*, *Spencermartinsiella*, *Sugiyamaella*, *Apiotrichum*, *Cutaneotrichosporon*, *Cystofilobasidium*, *Naganishia*, *Saitozyma*, *Papiliotrema*, *Pseudotremella*, *Solicoccozyma*, *Tausonia*, *Trichosporon*, *Vanrija*, and *Vishniacozyma* [161]. For a more comprehensive review of yeast ecology, please refer to Chapter 2 of this book.

Insect Attraction and Pollination

Yeasts have been associated with the attraction of insects to plants, including their floral parts, which further facilitates the pollination carried out by these invertebrates. In this sense, the effects of yeasts on insect attraction and pollination exhibit a three-fold impact, including the environmental, agronomic, and social impacts [163 - 165].

The positive environmental effect of yeasts occurs by boosting the pollination of wild species, thereby ensuring plant and animal biodiversity [163, 164, 166]. Approximately 90% of all plant species benefit from animal-mediated pollination, which is largely facilitated by floral nectar [167], where yeasts are found in fermenting sugars (mainly mono- and disaccharides) and produce volatile organic compounds (VOCs) [168]. Although these VOCs may be mostly generalized as alcohols, esters, and ketones compounds, the chemicals in each of these classes vary inter- and intra-specifically among angiosperms (plants with flower), depending on the resident microbiota [169]. In this ecological niche, the yeast genus *Metschnikowia* stands out with its species producing distinctive compounds, making it the most attractive yeast clade in floral nectar worldwide [168 - 174].

There is solid evidence indicating that the release of VOCs by yeasts is a phylogenetically ancient trait. Considering that both yeasts and insects evolutionarily emerged before angiosperms, the communication between these two probably preceded the emergence of flowers. In the 300 million years of coexistence between yeasts and insects, the insect systems of chemical-signal detection co-evolved with the yeast metabolic routes responsible for producing attractive volatile substances [175]. There are even reports today that show this symbiosis working independently of floral nectar, with yeasts releasing VOCs from distinct substrates to attract the insects [176]. From the yeast point of view, insects are a means of transport and dispersion. They also present an internal environment for sexual reproduction [177, 178] and provide yeast cells with nutrients and shelter in periods when the external environmental conditions are unfavorable. From the insects' perspective, yeast is a biocatalyst that facilitates digestion, contributes to the production of essential nutrients, and enhances their immune responses [176, 179]. For a more comprehensive review of yeast ecology, please refer to Chapter 2 of this book.

The positive agronomic impact of yeasts relies on a remarkable fact: about 75% of domesticated crops also benefit from animal-mediated pollination [167]. In the agriculture context, it is worth noting that the plant-yeast-insect relationship may be looked at the other way around. In this trio, not only can the plant provide yeasts for the insects, but the insects can also provide the plants with yeasts [164]. Indeed, this inoculation performed by pollinating insects is the most likely way for yeasts to reach the flowers [180, 181], and it has been thought to play a major role in dispersing microbial biological controllers (see below) [182]. Last but not least, as the producers of VOCs, yeasts guide insects through pollination, ensuring food production and reducing the risk to food insecurity [164, 165]. Hence, there is a significant positive social impact of this kind of plant-yeast-insect relationship.

Biological Control

Either through competition, enzyme secretion, VOCs and mycotoxin-inhibitor production, mycoparasitism (fungivory), induction of resistance, or through some of these mechanisms combined, yeasts exert plant protective effect by acting as biological controllers. Since yeasts are single-celled organisms, adhesion and biofilm formation are facilitated, increasing their environmental persistence, competitiveness, and, consequently, their biocontrol success [183 - 186]. Most of these yeast characteristics are shared with bacteria; however, genetically speaking, wild biocontrol yeasts have a safety advantage: they mostly lack plasmids, lowering the risk of taking up or passing on plasmid-based antibiotic resistance, pathogenicity factors, or toxin biosynthesis genes [183]. The same holds for horizontal gene transfer, which is far more frequent in bacteria, even though trans-kingdom conjugation has been reported from *E. coli* to *S. cerevisiae* [187].

The pre- and post-harvest protection activities of some yeasts have already been registered and commercialized as biocontrol products. Some examples include Aspire™ (based on *Candida oleophila*), BoniProtect™ (*Aureobasidium pullulans*), Candifruit™ (*C. sake*), Romeo® (based on Cerevisane®, a specific fraction of *S. cerevisiae* LAS-117), Shemer™ (*M. fructicola*), and Yieldplus™ (*Cryptococcus albidus*) [183, 188 - 192]. Biocontrol agents is a US$ 3.0 billion market, forecasted to reach US$ 7.5 billion by 2025. Although macro-organisms and biochemicals may also be employed as biocontrol agents, the microbial segment is the largest one, accounting for ~60% of the market [193]. For a detailed discussion on biological control by yeasts, please refer to Chapter 13 of this book.

Plant-growth Promoting Activities

The use of yeasts in agriculture has been exponentially increasing in the last decades. Yeast cells can act as plant growth-promoting agents by increasing the concentrations of ammonia and polyamines and producing the plant auxin indole-3-acetic acid (IAA), the cytokinin Zeatine, and the enzyme 1-aminocyclopropan--1-carboxylate (ACC) deaminase. Also, yeasts may indirectly improve plant growth by stimulating beneficial processes like nitrification, S-oxidation, and P-solubilization in soils [194, 195]. Some of the reported yeast species with the desired profiles to act as plant growth-promoting microorganisms include *Sporidiobolus ruineniae*, *Pseudozyma aphidis*, *Dothideomycetes* sp., *Sporobolomyces roseus*, *Rhodotorula* sp., *R. graminis*, *R. mucilaginosa*, *Williopsis saturnus*, *C. tropicalis*, *Cryptococcus laurentii*, *Williopsis californica*, and *S. cerevisiae* [195 - 201]. While some of these yeast species are endophytic, others are isolated from the soil, rhizosphere, and phyllosphere. For a detailed discussion on plant growth promoted by yeasts, please refer to Chapter 13 of this book.

CONCLUSION AND FURTHER CONSIDERATIONS

In terms of the market, other microbial life forms are unlikely to provide global financial turnover in the same way as yeasts, whose industrial processing sector crosses the trillion-dollar barrier. Furthermore, its benefits to humanity are not restricted to the world economy. Yeasts also play essential roles in promoting human, animal, and plant health and environmental balance. Moreover, these roles can also directly or indirectly lead to economic gains.

The play on words that assigns yeasts the role of "humanity's best friends" gains even more strength when it is verified that the largest financial volume handled by these microorganisms comes from the market segments that are directly related to joy, happiness, sociability, and pleasure. With alcoholic beverages alone (entirely dependent on yeast), the volume already exceeds US$ 1.0 trillion, and in this niche market of satisfaction, bread, chocolate, and cheese are also included. In the last three segments, other microorganisms can also be found in addition to yeasts.

However, in the production of bread and chocolates, whose markets together exceed US$ 300 billion, the process is mainly dependent on and dominated by yeasts. On the other hand, yeasts play a smaller role in the production of different types of cheese, but they are still essential for the production of certain types of this dairy product.

Yeasts are also responsible for converting sugars into bioethanol, the most produced biofuel in the world, with a global market value of US$ 86 billion. Yeasts have also shown high potential to increase biodiesel production, the second biofuel in terms of volume. In addition to being profitable, both biofuels are environmentally friendlier alternatives to fossil fuels. Moreover, this desired environmental impact can occur directly as yeasts also play an effective role in decomposing biomass (especially plants), recycling nutrients, and producing VOCs that induce animal-mediated pollination as well as the consequent reproduction of vegetable species.

The environmental role of yeasts is directly linked to agricultural production, which places them again in a prominent economic position. In agriculture, 75% of plant species depend on animal pollinators attracted by the VOCs generated by yeasts. Furthermore, the organic agriculture market is increasingly relying on the marketing of yeasts that can be used as growth promoters or as biocontrollers for agricultural pests. Thus, by ensuring greater efficiency in organic food production, yeasts reduce the environmental impacts and food insecurity. In this context, yeasts play important roles in livestock health promotion, whether as probiotics and prebiotics or SCP supplements. This yeast-cell production market size is worth more than $10 billion.

In the promotion of human health, the impact of yeasts can be seen from different angles. The ecological activities of these microorganisms can be linked to human health in the context of 'unique health' as defined by the World Health Organization (WHO). However, there are more direct impacts that depend on the heterologous expression of proteins by yeast cells. Considering the ease of genetic manipulation and the biological safety of these microbes compared to others, yeasts have been successfully used as microbial factories since the 1970s. Therefore, they have been employed to produce several biopharmaceuticals, such as cancer therapeutics, analgesics, vaccines, and hormones. Among these products, human insulin stands out with a US$ 21 billion global market, of which yeast accounts for 50% of the market value.

In this chapter, we addressed the most fruitful relationships between yeasts and humans, including the most profitable industrial activities in which these microorganisms are involved. However, despite the several benefits that yeasts provide to humanity, we recognize that we have still not exhausted all the positive impacts that they provide to humans. Yeasts are also used to produce ingredients for the food industry, including citric acid and emulsifiers, combustible gases, and bioplastics. They are also used for the bioremediation of areas contaminated by organic matter and heavy metals. However, there are also some negative effects associated with them, including the occurrence of mycoses that are predominantly caused by the members of the genus *Candida* (see Chapter 5), spoilage and off-flavors in beverages, and diseases in plants and animals. Still, these negative effects do not outweigh the significant positive impact that yeasts have on humanity. After all, a dog — man's acclaimed best friend — also bites!

CONSENT FOR PUBLICATION

Not applicable.

CONFLICT OF INTEREST

The authors declare no conflict of interest, financial or otherwise.

ACKNOWLEDGEMENTS

The authors would like to acknowledge the following Brazilian funding agencies: Conselho Nacional de Desenvolvimento Científico e Tecnológico – CNPq (grant numbers 454215/2014-2, 305258/2018-4, 429029/2018-7, and 308389/2019-0), Coordenação de Aperfeiçoamento de Pessoal de Nível Superior – CAPES (grant number 359/14), Financiadora de Estudos e Projetos – FINEP (grant number 01.09.0566.00/1421-08), Fundação de Amparo à Pesquisa do Estado de São Paulo – FAPESP (grant number 2018/17172-2), Fundação de Amparo à Pesquisa do Estado do Rio Grande do Sul – FAPERGS (grant number 17/2551-0001086-8), and Fundação de Amparo à Pesquisa do Estado de Santa Catarina – FAPESC (grant numbers 17293/2009-6 and 749/2016 T.O. 2016TR2188).

REFERENCES

[1]   Tulha J, Carvalho J, Armada R, Faria-Oliveira F, Lucas C, Pais C, *et al.* Yeast, the Man's Best Friend. Sci. Heal. Soc. Asp. Food Ind., InTech 2012.
[http://dx.doi.org/10.5772/31471]

[2]   Eberlein C, Leducq J-B, Landry CR. The genomics of wild yeast populations sheds light on the domestication of man's best (micro) friend. Mol Ecol 2015; 24(21): 5309-11.
[http://dx.doi.org/10.1111/mec.13380] [PMID: 26509691]

[3]   Arranz-Otaegui A, Gonzalez Carretero L, Ramsey MN, Fuller DQ, Richter T. Archaeobotanical evidence reveals the origins of bread 14,400 years ago in northeastern Jordan. Proc Natl Acad Sci USA 2018; 115(31): 7925-30.
[http://dx.doi.org/10.1073/pnas.1801071115] [PMID: 30012614]

[4]   United Nations. Statistical papers Series M No. 4/Rev.4 2008.

[5]   Payen C, Thompson D. The renaissance of yeasts as microbial factories in the modern age of biomanufacturing. Yeast 2019; 36(12): 685-700.
        [http://dx.doi.org/10.1002/yea.3439] [PMID: 31423599]

[6]   Tsuge Y, Kawaguchi H, Sasaki K, Kondo A. Engineering cell factories for producing building block chemicals for bio-polymer synthesis. Microb Cell Fact 2016; 15: 19.
        [http://dx.doi.org/10.1186/s12934-016-0411-0] [PMID: 26794242]

[7]   Karim A, Gerliani N, Aïder M. Kluyveromyces marxianus: An emerging yeast cell factory for applications in food and biotechnology. Int J Food Microbiol 2020; 333: 108818.
        [http://dx.doi.org/10.1016/j.ijfoodmicro.2020.108818] [PMID: 32805574]

[8]   Gündüz Ergün B, Hüccetoğulları D, Öztürk S, Çelik E, Çalık P. Established and Upcoming Yeast Expression Systems Methods Mol Biol. Humana Press Inc. 2019; Vol. 1923: pp. 1-74.
        [http://dx.doi.org/10.1007/978-1-4939-9024-5_1]

[9]   Paddon CJ, Westfall PJ, Pitera DJ, *et al.* High-level semi-synthetic production of the potent antimalarial artemisinin. Nature 2013; 496(7446): 528-32.
        [http://dx.doi.org/10.1038/nature12051] [PMID: 23575629]

[10]   Kuruti K, Vittaladevaram V, Urity SV, Palaniappan P, Bhaskar RU. Evolution of Pichia pastoris as a model organism for vaccines production in healthcare industry. Gene Rep 2020; 21: 100937.
        [http://dx.doi.org/10.1016/j.genrep.2020.100937]

[11]   Ahmad M, Hirz M, Pichler H, Schwab H. Protein expression in Pichia pastoris: recent achievements and perspectives for heterologous protein production. Appl Microbiol Biotechnol 2014; 98(12): 5301-17.
        [http://dx.doi.org/10.1007/s00253-014-5732-5] [PMID: 24743983]

[12]   Löbs A-K, Schwartz C, Wheeldon I. Genome and metabolic engineering in non-conventional yeasts: Current advances and applications. Synth Syst Biotechnol 2017; 2(3): 198-207.
        [http://dx.doi.org/10.1016/j.synbio.2017.08.002] [PMID: 29318200]

[13]   Walker RSK, Pretorius IS. Applications of Yeast Synthetic Biology Geared towards the Production of Biopharmaceuticals. Genes (Basel) 2018; 9(7): 340.
        [http://dx.doi.org/10.3390/genes9070340] [PMID: 29986380]

[14]   Blazeck J, Hill A, Liu L, *et al.* Harnessing Yarrowia lipolytica lipogenesis to create a platform for lipid and biofuel production. Nat Commun 2014; 5: 3131.
        [http://dx.doi.org/10.1038/ncomms4131] [PMID: 24445655]

[15]   Li C, Ong KL, Cui Z, *et al.* Promising advancement in fermentative succinic acid production by yeast hosts. J Hazard Mater 2021; 401: 123414.
        [http://dx.doi.org/10.1016/j.jhazmat.2020.123414] [PMID: 32763704]

[16]   Bamforth CW, Cabras I. Interesting Times: Changes for Brewing Brewing, Beer Pubs. London: Palgrave Macmillan UK 2016; pp. 15-33.
        [http://dx.doi.org/10.1057/9781137466181_2]

[17]   Eliodório KP, GC de G e. Cunha, C Müller, *et al.* Advances in yeast alcoholic fermentations for the production of bioethanol, beer and wine. In: Advances in Applied Microbiology. Academic Press Inc. 2019; 109: pp. 61-119.
        [http://dx.doi.org/10.1016/bs.aambs.2019.10.002]

[18]   Pretorius I. Conducting Wine Symphonics with the Aid of Yeast Genomics. Beverages 2016; 2: 36.
        [http://dx.doi.org/10.3390/beverages2040036]

[19]   IMARC. Global Beer Market - Growth, Trends, and Forecast. Res Mark 2021.https://www.imarcgroup.com/beer-market

[20]   GlobeNewswire. Global Wine Industry 2021.   https://www.globenewswire.com/news-release/2020/09/08/2089700/0/en/Global-Wine-Industry.html

[21]   Walker G, Hill A. Saccharomyces cerevisiae in the Production of Whisk(e)y. Beverages 2016; 2: 38.
[http://dx.doi.org/10.3390/beverages2040038]

[22]   Allied Market Research. Whiskey Market Size, Share & Value | Analysis Forecast by 2019.
https://www.alliedmarketresearch.com/whiskey-market-A06652

[23]   IWSR. Worldwide Alcohol Consumption Declines -1 . 6 %. Press Release 2019.

[24]   The Spirits Business. Analysis: The performance of vodka's biggest players 2019.
https://www.thespiritsbusiness.com/2019/06/analysis-the-performance-of-vodkas-biggest-players/

[25]   GlobeNewswire. Global Vodka Market 2021. https://www.globenewswire.com/news-release/2021/04/13/2209321/0/en/Global-Vodka-Market-to-Reach-56-4-Billion-by-2027.html

[26]   Pauley M, Maskell D. Mini-Review: The Role of Saccharomyces cerevisiae in the Production of Gin and Vodka. Beverages 2017; 3: 13.
[http://dx.doi.org/10.3390/beverages3010013]

[27]   Lappe-Oliveras P, Moreno-Terrazas R, Arrizón-Gaviño J, Herrera-Suárez T, García-Mendoza A, Gschaedler-Mathis A. Yeasts associated with the production of Mexican alcoholic nondistilled and distilled Agave beverages. FEMS Yeast Res 2008; 8(7): 1037-52.
[http://dx.doi.org/10.1111/j.1567-1364.2008.00430.x] [PMID: 18759745]

[28]   Insights FB. Tequila Market Size, Share, Growth | Industry Report [2028]. 2021.
https://www.fortunebusinessinsights.com/tequila-market-104172

[29]   IMARC. Tequila Market Size, Share, Growth, Industry Trends & Forecast 2021-2026 2021.
https://www.imarcgroup.com/prefeasibility-report-tequila-manufacturing-plant

[30]   The Insight Partners. Sake Market Forecast to 2027 2021. https://www.theinsightpartners.com/reports/sake-market

[31]   Kitagaki H, Kitamoto K. Breeding research on sake yeasts in Japan: history, recent technological advances, and future perspectives. Annu Rev Food Sci Technol 2013; 4: 215-35.
[http://dx.doi.org/10.1146/annurev-food-030212-182545] [PMID: 23464572]

[32]   Alcimed. Sake at the conquest of the world, between export and production outside Japan | Alcimed 2018.   https://www.alcimed.com/en/alcim-articles/sake-at-the-conquest-of-the-world-betwe-n-export-and-production-outside-japan/

[33]   Verruma-Bernardi MR, Oliveira AL de. Cachaça Production in Brazil and its Main Contaminant (Ethyl Carbamate). Sci Agric 2020; 77: 2020.
[http://dx.doi.org/10.1590/1678-992x-2018-0135]

[34]   Barbosa R, Pontes A, Santos RO, *et al.* Multiple Rounds of Artificial Selection Promote Microbe Secondary Domestication-The Case of Cachaça Yeasts. Genome Biol Evol 2018; 10(8): 1939-55.
[http://dx.doi.org/10.1093/gbe/evy132] [PMID: 29982460]

[35]   Grand View Research. Dairy Products Market Share & Growth Report 2020-2027 2020.
https://www.grandviewresearch.com/industry-analysis/dairy-product-market

[36]   Gao J, Zhou W. Oral processing of bread: Implications of designing healthier bread products. Trends Food Sci Technol 2021; 112: 720-34.
[http://dx.doi.org/10.1016/j.tifs.2021.04.030]

[37]   360 Market Update.. Global Bread Market Report 2021 with Top Countries Data And Covid-19 Analysis Share, Scope, Stake, Trends, Industry Size, Sales & Revenue, Growth, Opportunities and Demand with Competitive Landscape and Analysis Research Report - The Express Wire. Express 2021.   https://www.theexpresswire.com/pressrelease/Global-Bread-Market-Report-2021-wi-h-Top-Countries-Data-And-Covid-19-Analysis-Share-Scope-Stake-Trends-Indus-ry-Size-Sales-Revenue-Growth-Opportunities-and-Demand-with-Competitive-Landscape-and-Analysis-Research

[38]   De Vuyst L, Van Kerrebroeck S, Harth H, Huys G, Daniel H-M, Weckx S. Microbial ecology of sourdough fermentations: diverse or uniform? Food Microbiol 2014; 37: 11-29.
[http://dx.doi.org/10.1016/j.fm.2013.06.002] [PMID: 24230469]

[39]   De Vuyst L, Harth H, Van Kerrebroeck S, Leroy F. Yeast diversity of sourdoughs and associated metabolic properties and functionalities. Int J Food Microbiol 2016; 239: 26-34.
[http://dx.doi.org/10.1016/j.ijfoodmicro.2016.07.018] [PMID: 27470533]

[40]   Chiva R, Celador-Lera L, Uña JA, *et al.* Yeast biodiversity in fermented doughs and raw cereal matrices and the study of technological traits of selected strains isolated in spain. Microorganisms 2020; 9(1): 1-44.
[http://dx.doi.org/10.3390/microorganisms9010047] [PMID: 33375367]

[41]   Minervini F, Lattanzi A, De Angelis M, Celano G, Gobbetti M. House microbiotas as sources of lactic acid bacteria and yeasts in traditional Italian sourdoughs. Food Microbiol 2015; 52: 66-76.
[http://dx.doi.org/10.1016/j.fm.2015.06.009] [PMID: 26338118]

[42]   Vrancken G, De Vuyst L, Van der Meulen R, Huys G, Vandamme P, Daniel H-M. Yeast species composition differs between artisan bakery and spontaneous laboratory sourdoughs. FEMS Yeast Res 2010; 10(4): 471-81.
[http://dx.doi.org/10.1111/j.1567-1364.2010.00621.x] [PMID: 20384785]

[43]   Minervini F, Lattanzi A, De Angelis M, Di Cagno R, Gobbetti M. Influence of artisan bakery- or laboratory-propagated sourdoughs on the diversity of lactic acid bacterium and yeast microbiotas. Appl Environ Microbiol 2012; 78(15): 5328-40.
[http://dx.doi.org/10.1128/AEM.00572-12] [PMID: 22635989]

[44]   Albagli G. Schwartz I do M, Amaral PFF, Ferreira TF, Finotelli P V. How dried sourdough starter can enable and spread the use of Sourdough bread. Lebensm Wiss Technol 2021; 111888.
[http://dx.doi.org/10.1016/j.lwt.2021.111888]

[45]   Schwan RF, Wheals AE. The microbiology of cocoa fermentation and its role in chocolate quality. Crit Rev Food Sci Nutr 2004; 44(4): 205-21.
[http://dx.doi.org/10.1080/10408690490464104] [PMID: 15462126]

[46]   Ho VTT, Zhao J, Fleet G. Yeasts are essential for cocoa bean fermentation. Int J Food Microbiol 2014; 174: 72-87.
[http://dx.doi.org/10.1016/j.ijfoodmicro.2013.12.014] [PMID: 24462702]

[47]   Nielsen DS, Teniola OD, Ban-Koffi L, Owusu M, Andersson TS, Holzapfel WH. The microbiology of Ghanaian cocoa fermentations analysed using culture-dependent and culture-independent methods. Int J Food Microbiol 2007; 114(2): 168-86.
[http://dx.doi.org/10.1016/j.ijfoodmicro.2006.09.010] [PMID: 17161485]

[48]   Daniel H-M, Vrancken G, Takrama JF, Camu N, De Vos P, De Vuyst L. Yeast diversity of Ghanaian cocoa bean heap fermentations. FEMS Yeast Res 2009; 9(5): 774-83.
[http://dx.doi.org/10.1111/j.1567-1364.2009.00520.x] [PMID: 19473277]

[49]   Papalexandratou Z, Falony G, Romanens E, *et al.* Species diversity, community dynamics, and metabolite kinetics of the microbiota associated with traditional ecuadorian spontaneous cocoa bean fermentations. Appl Environ Microbiol 2011; 77(21): 7698-714.
[http://dx.doi.org/10.1128/AEM.05523-11] [PMID: 21926224]

[50]   Illeghems K, De Vuyst L, Papalexandratou Z, Weckx S. Phylogenetic analysis of a spontaneous cocoa bean fermentation metagenome reveals new insights into its bacterial and fungal community diversity. PLoS One 2012; 7(5): e38040.
[http://dx.doi.org/10.1371/journal.pone.0038040] [PMID: 22666442]

[51]   Moreira IM da V. Miguel MG da CP, Duarte WF, Dias DR, Schwan RF. Microbial succession and the dynamics of metabolites and sugars during the fermentation of three different cocoa (Theobroma cacao L.) hybrids. Food Res Int 2013; 54: 9-17.

[http://dx.doi.org/10.1016/j.foodres.2013.06.001]

[52] Papalexandratou Z, Lefeber T, Bahrim B, Lee OS, Daniel H-M, De Vuyst L. Hanseniaspora opuntiae, Saccharomyces cerevisiae, Lactobacillus fermentum, and Acetobacter pasteurianus predominate during well-performed Malaysian cocoa bean box fermentations, underlining the importance of these microbial species for a successful cocoa bean fermentation process. Food Microbiol 2013; 35(2): 73-85.
[http://dx.doi.org/10.1016/j.fm.2013.02.015] [PMID: 23664257]

[53] Meersman E, Steensels J, Mathawan M, *et al.* Detailed analysis of the microbial population in Malaysian spontaneous cocoa pulp fermentations reveals a core and variable microbiota. PLoS One 2013; 8(12): e81559.
[http://dx.doi.org/10.1371/journal.pone.0081559] [PMID: 24358116]

[54] Hamdouche Y, Guehi T, Durand N, Kedjebo KBD, Montet D, Meile JC. Dynamics of microbial ecology during cocoa fermentation and drying: Towards the identification of molecular markers. Food Control 2015; 48: 117-22.
[http://dx.doi.org/10.1016/j.foodcont.2014.05.031]

[55] De Vuyst L, Weckx S. The cocoa bean fermentation process: from ecosystem analysis to starter culture development. J Appl Microbiol 2016; 121(1): 5-17.
[http://dx.doi.org/10.1111/jam.13045] [PMID: 26743883]

[56] Voora V, Bermúdez S, Larrea C. Global Market Report: Cocoa. International Institute for Sustainable Development 2019.

[57] Ministerio de Relaciones Exteriores CBI. What is the demand for cocoa on the European market?. 2020. https://www.cbi.eu/market-information/cocoa/trade-statistics

[58] Fröhlich-Wyder M-T, Arias-Roth E, Jakob E. Cheese yeasts. Yeast 2019; 36(3): 129-41.
[http://dx.doi.org/10.1002/yea.3368] [PMID: 30512214]

[59] Yildiz M, Turgut T, Cetin B, Kesmen Z. Microbiological characteristics and identification of yeast microbiota of traditional mouldy civil cheese. Int Dairy J 2021; 116: 104955.
[http://dx.doi.org/10.1016/j.idairyj.2020.104955]

[60] Biolcati F, Andrighetto C, Bottero MT, Dalmasso A. Microbial characterization of an artisanal production of Robiola di Roccaverano cheese. J Dairy Sci 2020; 103(5): 4056-67.
[http://dx.doi.org/10.3168/jds.2019-17451] [PMID: 32173014]

[61] Cogan TM, Goerges S, Gelsomino R, *et al.* Biodiversity of the Surface Microbial Consortia from Limburger, Reblochon, Livarot, Tilsit, and Gubbeen Cheeses. Microbiol Spectr 2014; 2(1): CM-001--2012.
[http://dx.doi.org/10.1128/microbiolspec.CM-0010-2012] [PMID: 26082119]

[62] Irlinger F, Layec S, Hélinck S, Dugat-Bony E. Cheese rind microbial communities: diversity, composition and origin. FEMS Microbiol Lett 2015; 362(2): 1-11.
[http://dx.doi.org/10.1093/femsle/fnu015] [PMID: 25670699]

[63] Boutrou R, Guéguen M. Interests in Geotrichum candidum for cheese technology. Int J Food Microbiol 2005; 102(1): 1-20.
[http://dx.doi.org/10.1016/j.ijfoodmicro.2004.12.028] [PMID: 15924999]

[64] Smith JD, Robinson AS. Overexpression of an archaeal protein in yeast: secretion bottleneck at the ER. Biotechnol Bioeng 2002; 79(7): 713-23.
[http://dx.doi.org/10.1002/bit.10367] [PMID: 12209794]

[65] Nielsen KH. Protein Expression-Yeast Methods Enzymol. Academic Press Inc. 2014; Vol. 536: pp. 133-47.
[http://dx.doi.org/10.1016/B978-0-12-420070-8.00012-X]

[66] Summpunn P, Jomrit J, Panbangred W. Improvement of extracellular bacterial protein production in Pichia pastoris by co-expression of endoplasmic reticulum residing GroEL-GroES. J Biosci Bioeng

2018; 125(3): 268-74.
[http://dx.doi.org/10.1016/j.jbiosc.2017.09.007] [PMID: 29046263]

[67]   Berg P, Baltimore D, Brenner S, Roblin RO, Singer MF. Summary statement of the Asilomar conference on recombinant DNA molecules. Proc Natl Acad Sci USA 1975; 72(6): 1981-4.
[http://dx.doi.org/10.1073/pnas.72.6.1981] [PMID: 806076]

[68]   Khan S, Ullah MW, Siddique R, *et al.* Role of Recombinant DNA Technology to Improve Life. Int J Genomics 2016; 2016: 2405954.
[http://dx.doi.org/10.1155/2016/2405954] [PMID: 28053975]

[69]   Ding MZ, Yan HF, Li LF, *et al.* Biosynthesis of Taxadiene in *Saccharomyces cerevisiae*: selection of geranylgeranyl diphosphate synthase directed by a computer-aided docking strategy. PLoS One 2014; 9(10): e109348.
[http://dx.doi.org/10.1371/journal.pone.0109348] [PMID: 25295588]

[70]   Li Y, Li S, Thodey K, Trenchard I, Cravens A, Smolke CD. Complete biosynthesis of noscapine and halogenated alkaloids in yeast. Proc Natl Acad Sci USA 2018; 115(17): E3922-31.
[http://dx.doi.org/10.1073/pnas.1721469115] [PMID: 29610307]

[71]   Galanie S, Thodey K, Trenchard IJ, Filsinger Interrante M, Smolke CD. Complete biosynthesis of opioids in yeast. Science (80-) 2015; 349: 1095-100.
[http://dx.doi.org/10.1126/science.aac9373]

[72]   Luo X, Reiter MA, d'Espaux L, *et al.* Complete biosynthesis of cannabinoids and their unnatural analogues in yeast. Nature 2019; 567(7746): 123-6.
[http://dx.doi.org/10.1038/s41586-019-0978-9] [PMID: 30814733]

[73]   Zirpel B, Degenhardt F, Martin C, Kayser O, Stehle F. Engineering yeasts as platform organisms for cannabinoid biosynthesis. J Biotechnol 2017; 259: 204-12.
[http://dx.doi.org/10.1016/j.jbiotec.2017.07.008] [PMID: 28694184]

[74]   Luo Y, Li B-Z, Liu D, *et al.* Engineered biosynthesis of natural products in heterologous hosts. Chem Soc Rev 2015; 44(15): 5265-90.
[http://dx.doi.org/10.1039/C5CS00025D] [PMID: 25960127]

[75]   Gellissen G, Kunze G, Gaillardin C, *et al.* New yeast expression platforms based on methylotrophic Hansenula polymorpha and Pichia pastoris and on dimorphic Arxula adeninivorans and Yarrowia lipolytica - a comparison. FEMS Yeast Res 2005; 5(11): 1079-96.
[http://dx.doi.org/10.1016/j.femsyr.2005.06.004] [PMID: 16144775]

[76]   Nielsen J. Production of biopharmaceutical proteins by yeast: advances through metabolic engineering. Bioengineered 2013; 4(4): 207-11.
[http://dx.doi.org/10.4161/bioe.22856] [PMID: 23147168]

[77]   PETERSEN MH. World's biggest insulin producer to ensure success for Novo's new bet. MEDWATCH 2017. https://medwatch.dk/Top_picks_in_english/article10082000.ece

[78]   Insights Fortune Business. Human Insulin Market Size, Share, Growth | Global Forecast, 2026. 2020.

[79]   Wang M, Jiang S, Wang Y. Recent advances in the production of recombinant subunit vaccines in Pichia pastoris. Bioengineered 2016; 7(3): 155-65.
[http://dx.doi.org/10.1080/21655979.2016.1191707] [PMID: 27246656]

[80]   GSK. HIGHLIGHTS OF PRESCRIBING INFORMATION These highlights do not include all the information needed to use IRA safely and effectively. See full prescribing information for 2002 2002.

[81]   MERCK. Highlights of prescribing information 2008.

[82]   Market Data Forecast. Biopharmaceuticals Market Size, Share, Growth | 2021 to 2026. 2021.https://www.marketdataforecast.com/market-reports/bio-pharmaceuticals-market

[83]   Öztürk S, Cerit İ, Mutlu S, Demirkol O. Enrichment of cookies with glutathione by inactive yeast cells (Saccharomyces cerevisiae): Physicochemical and functional properties. J Cereal Sci 2017; 78: 19-24.

[http://dx.doi.org/10.1016/j.jcs.2017.06.019]

[84] GlobeNewswire. Protein Extracts from Single Cell Protein Sources Market (2013-2023). Prescient Strateg Intell 2018. https://www.globenewswire.com/news-release/2019/05/08/1819107/0/en/Protei-
-Extracts-from-Single-Cell-Protein-Sources-Market-is-Expected-to-Advance-a-
-a-CAGR-of-8-6-in-Coming-Years-P-S-Intelligence.html

[85] Matassa S, Boon N, Pikaar I, Verstraete W. Microbial protein: future sustainable food supply route with low environmental footprint. Microb Biotechnol 2016; 9(5): 568-75.
[http://dx.doi.org/10.1111/1751-7915.12369] [PMID: 27389856]

[86] Hezarjaribi M, Ardestani F, Ghorbani HR. Single Cell Protein Production by *Saccharomyces cerevisiae* Using an Optimized Culture Medium Composition in a Batch Submerged Bioprocess. Appl Biochem Biotechnol 2016; 179(8): 1336-45.
[http://dx.doi.org/10.1007/s12010-016-2069-9] [PMID: 27090426]

[87] Spalvins K, Zihare L, Blumberga D. Single cell protein production from waste biomass: comparison of various industrial by-products. Energy Procedia 2018; 147: 409-18.
[http://dx.doi.org/10.1016/j.egypro.2018.07.111]

[88] Najafpour GD. Single-Cell Protein Biochem Eng Biotechnol. Elsevier 2007; pp. 332-41.
[http://dx.doi.org/10.1016/B978-044452845-2/50014-8]

[89] Lange HC, Heijnen JJ. Statistical reconciliation of the elemental and molecular biomass composition of Saccharomyces cerevisiae. Biotechnol Bioeng 2001; 75(3): 334-44.
[http://dx.doi.org/10.1002/bit.10054] [PMID: 11590606]

[90] Zakhartsev M, Reuss M. Cell size and morphological properties of yeast Saccharomyces cerevisiae in relation to growth temperature. FEMS Yeast Res 2018; 18(6): 52.
[http://dx.doi.org/10.1093/femsyr/foy052] [PMID: 29718340]

[91] Stambuk BU, Dunn B, Alves SL Jr, Duval EH, Sherlock G. Industrial fuel ethanol yeasts contain adaptive copy number changes in genes involved in vitamin B1 and B6 biosynthesis. Genome Res 2009; 19(12): 2271-8.
[http://dx.doi.org/10.1101/gr.094276.109] [PMID: 19897511]

[92] Salminen S, Ouwehand A, Benno Y, Lee Y. Probiotics: how should they be defined? Trends Food Sci Technol 1999; 10: 107-10.
[http://dx.doi.org/10.1016/S0924-2244(99)00027-8]

[93] Ahiwe EU, Tedeschi Dos Santos TT, Graham H, Iji PA. Can probiotic or prebiotic yeast (Saccharomyces cerevisiae) serve as alternatives to in-feed antibiotics for healthy or disease-challenged broiler chickens?: a review. J Appl Poult Res 2021; 30: 100164.
[http://dx.doi.org/10.1016/j.japr.2021.100164]

[94] Gibson GR, Probert HM, Loo JV, Rastall RA, Roberfroid MB. Dietary modulation of the human colonic microbiota: updating the concept of prebiotics. Nutr Res Rev 2004; 17(2): 259-75.
[http://dx.doi.org/10.1079/NRR200479] [PMID: 19079930]

[95] Canonico L, Zannini E, Ciani M, Comitini F. Assessment of non-conventional yeasts with potential probiotic for protein-fortified craft beer production. Lebensm Wiss Technol 2021; 145: 111361.
[http://dx.doi.org/10.1016/j.lwt.2021.111361]

[96] Hatoum R, Labrie S, Fliss I. Antimicrobial and probiotic properties of yeasts: from fundamental to novel applications. Front Microbiol 2012; 3: 421.
[http://dx.doi.org/10.3389/fmicb.2012.00421] [PMID: 23267352]

[97] Naito Y, Marotta F, Kantah MK, *et al.* Gut-targeted immunonutrition boosting natural killer cell activity using Saccharomyces boulardii lysates in immuno-compromised healthy elderly subjects. Rejuvenation Res 2014; 17(2): 184-7.
[http://dx.doi.org/10.1089/rej.2013.1500] [PMID: 24059806]

[98] Czerucka D, Rampal P. Diversity of *Saccharomyces boulardii* CNCM I-745 mechanisms of action

against intestinal infections. World J Gastroenterol 2019; 25(18): 2188-203.
[http://dx.doi.org/10.3748/wjg.v25.i18.2188] [PMID: 31143070]

[99]   Buts J-P, Bernasconi P, Vaerman J-P, Dive C. Stimulation of secretory IgA and secretory component
         of immunoglobulins in small intestine of rats treated with Saccharomyces boulardii. Dig Dis Sci 1990;
         35(2): 251-6.
         [http://dx.doi.org/10.1007/BF01536771] [PMID: 2302983]

[100]   Qamar A, Aboudola S, Warny M, *et al.* Saccharomyces boulardii stimulates intestinal immunoglobulin
         A immune response to Clostridium difficile toxin A in mice. Infect Immun 2001; 69(4): 2762-5.
         [http://dx.doi.org/10.1128/IAI.69.4.2762-2765.2001] [PMID: 11254650]

[101]   Fakruddin M, Hossain MN, Ahmed MM. Antimicrobial and antioxidant activities of Saccharomyces
         cerevisiae IFST062013, a potential probiotic. BMC Complement Altern Med 2017; 17(1): 64.
         [http://dx.doi.org/10.1186/s12906-017-1591-9] [PMID: 28109187]

[102]   Czerucka D, Rampal P. Experimental effects of Saccharomyces boulardii on diarrheal pathogens.
         Microbes Infect 2002; 4(7): 733-9.
         [http://dx.doi.org/10.1016/S1286-4579(02)01592-7] [PMID: 12067833]

[103]   Vohra A. Probiotic potential of yeasts isolated from traditional indian fermented foods. Int J Microbiol
         Res 2013; 5: 390-8.
         [http://dx.doi.org/10.9735/0975-5276.5.2.390-398]

[104]   Rajkowska K, Kunicka-Styczyńska A, Rygala A. Probiotic activity of Saccharomyces cerevisiae var.
         Boulardii against human pathogens. Food Technol Biotechnol 2012; 50: 230-6.

[105]   Roostita LB, Fleet GH, Wendry SP, Apon ZM, Gemilang LU. Determination of yeasts antimicrobial
         activity in milk and meat products. Adv J Food Sci Technol 2011; 3: 442-5.

[106]   Agarbati A, Canonico L, Marini E, Zannini E, Ciani M, Comitini F. Potential probiotic yeasts sourced
         from natural environmental and spontaneous processed foods. Foods 2020; 9(3): 287.
         [http://dx.doi.org/10.3390/foods9030287] [PMID: 32143376]

[107]   He T, Mahfuz S, Piao X, *et al.* Effects of live yeast (Saccharomyces cerevisiae) as a substitute to
         antibiotic on growth performance, immune function, serum biochemical parameters and intestinal
         morphology of broilers. J Appl Anim Res 2021; 49: 15-22.
         [http://dx.doi.org/10.1080/09712119.2021.1876705]

[108]   Walker GM, Walker RSK. Enhancing Yeast Alcoholic Fermentations. Adv Appl Microbiol 2018; 105:
         87-129.
         [http://dx.doi.org/10.1016/bs.aambs.2018.05.003] [PMID: 30342724]

[109]   Grand View Research.. Ethanol Market Size, Share, Trends Report, 2020-2027 2020:110
         https://www.grandviewresearch.com/industry-analysis/ethanol-market

[110]   Basso LC, de Amorim HV, de Oliveira AJ, Lopes ML. Yeast selection for fuel ethanol production in
         Brazil. FEMS Yeast Res 2008; 8(7): 1155-63.
         [http://dx.doi.org/10.1111/j.1567-1364.2008.00428.x] [PMID: 18752628]

[111]   Lopes ML, Paulillo SC de L, Godoy A, *et al.* Ethanol production in Brazil: a bridge between science
         and industry. Braz J Microbiol 2016; 47 (Suppl. 1): 64-76.
         [http://dx.doi.org/10.1016/j.bjm.2016.10.003] [PMID: 27818090]

[112]   Stambuk BU. Yeasts: The leading figures on bioethanol production.Ethanol as a green Altern fuel
         insight Perspect. 1st ed. Hauppauge, NY, USA: Nova Science Publishers 2019; pp. 57-91.

[113]   Bonatto C, Camargo AF, Scapini T, *et al.* Biomass to bioenergy research: current and future trends for
         biofuels Recent Dev Bioenergy Res. Elsevier 2020; pp. 1-17.
         [http://dx.doi.org/10.1016/B978-0-12-819597-0.00001-5]

[114]   Marques WL, Raghavendran V, Stambuk BU, Gombert AK. Sucrose and Saccharomyces cerevisiae: a
         relationship most sweet. FEMS Yeast Res 2016; 16(1): fov107.

[http://dx.doi.org/10.1093/femsyr/fov107] [PMID: 26658003]

[115]   Jansen MLA, Bracher JM, Papapetridis I, *et al.* Saccharomyces cerevisiae strains for second-generation ethanol production: from academic exploration to industrial implementation. FEMS Yeast Res 2017; 17(5): 44.
[http://dx.doi.org/10.1093/femsyr/fox044] [PMID: 28899031]

[116]   Patiño MA, Ortiz JP, Velásquez M, Stambuk BU. d-Xylose consumption by nonrecombinant Saccharomyces cerevisiae: A review. Yeast 2019; 36: 541-56.
[http://dx.doi.org/10.1002/yea.3429] [PMID: 31254359]

[117]   Walker GM. 125th Anniversary Review: Fuel Alcohol: Current Production and Future Challenges. J Inst Brew 2011; 117: 3-22.
[http://dx.doi.org/10.1002/j.2050-0416.2011.tb00438.x]

[118]   Meng X, Yang J, Xu X, Zhang L, Nie Q, Xian M. Biodiesel production from oleaginous microorganisms. Renew Energy 2009; 34: 1-5.
[http://dx.doi.org/10.1016/j.renene.2008.04.014]

[119]   Ngamsirisomsakul M, Reungsang A, Kongkeitkajorn MB. Assessing oleaginous yeasts for their potentials on microbial lipid production from sugarcane bagasse and the effects of physical changes on lipid production. Bioresour Technol Rep 2021; 14: 100650.
[http://dx.doi.org/10.1016/j.biteb.2021.100650]

[120]   Lim S, Teong LK. Recent trends, opportunities and challenges of biodiesel in Malaysia: An overview. Renew Sustain Energy Rev 2010; 14: 938-54.
[http://dx.doi.org/10.1016/j.rser.2009.10.027]

[121]   Bušić A, Kundas S, Morzak G, *et al.* Recent Trends in Biodiesel and Biogas Production. Food Technol Biotechnol 2018; 56(2): 152-73.
[http://dx.doi.org/10.17113/ftb.56.02.18.5547] [PMID: 30228791]

[122]   Chaiyaso T, Manowattana A, Techapun C, Watanabe M. Efficient bioconversion of enzymatic corncob hydrolysate into biomass and lipids by oleaginous yeast *Rhodosporidium paludigenum* KM281510. Prep Biochem Biotechnol 2019; 49(6): 545-56.
[http://dx.doi.org/10.1080/10826068.2019.1591985] [PMID: 30929597]

[123]   Bansal N, Dasgupta D, Hazra S, Bhaskar T, Ray A, Ghosh D. Effect of utilization of crude glycerol as substrate on fatty acid composition of an oleaginous yeast Rhodotorula mucilagenosa IIPL32: Assessment of nutritional indices. Bioresour Technol 2020; 309: 123330.
[http://dx.doi.org/10.1016/j.biortech.2020.123330] [PMID: 32283485]

[124]   Chen J, Zhang X, Tyagi RD, Drogui P. Utilization of methanol in crude glycerol to assist lipid production in non-sterilized fermentation from Trichosporon oleaginosus. Bioresour Technol 2018; 253: 8-15.
[http://dx.doi.org/10.1016/j.biortech.2018.01.008] [PMID: 29328937]

[125]   Guerfali M, Ayadi I, Sassi H-E, Belhassen A, Gargouri A, Belghith H. Biodiesel-derived crude glycerol as alternative feedstock for single cell oil production by the oleaginous yeast Candida viswanathii Y-E4. Ind Crops Prod 2020; 145: 112103.
[http://dx.doi.org/10.1016/j.indcrop.2020.112103]

[126]   Johnravindar D, Karthikeyan OP, Selvam A, Murugesan K, Wong JWC. Lipid accumulation potential of oleaginous yeasts: A comparative evaluation using food waste leachate as a substrate. Bioresour Technol 2018; 248(Pt A): 221-8.
[http://dx.doi.org/10.1016/j.biortech.2017.06.151] [PMID: 28736146]

[127]   Poontawee R, Yongmanitchai W, Limtong S. Lipid production from a mixture of sugarcane top hydrolysate and biodiesel-derived crude glycerol by the oleaginous red yeast, Rhodosporidiobolus fluvialis. Process Biochem 2018; 66: 150-61.
[http://dx.doi.org/10.1016/j.procbio.2017.11.020]

[128] Brar KK, Sarma AK, Aslam M, Polikarpov I, Chadha BS. Potential of oleaginous yeast Trichosporon sp., for conversion of sugarcane bagasse hydrolysate into biodiesel. Bioresour Technol 2017; 242: 161-8.
[http://dx.doi.org/10.1016/j.biortech.2017.03.155] [PMID: 28438358]

[129] Louhasakul Y, Cheirsilp B, Maneerat S, Prasertsan P. Potential use of flocculating oleaginous yeasts for bioconversion of industrial wastes into biodiesel feedstocks. Renew Energy 2019; 136: 1311-9.
[http://dx.doi.org/10.1016/j.renene.2018.10.002]

[130] Tsigie YA, Wang CY, Truong CT, Ju YH. Lipid production from Yarrowia lipolytica Po1g grown in sugarcane bagasse hydrolysate. Bioresour Technol 2011; 102(19): 9216-22.
[http://dx.doi.org/10.1016/j.biortech.2011.06.047] [PMID: 21757339]

[131] Tiukova IA, Brandenburg J, Blomqvist J, *et al.* Proteome analysis of xylose metabolism in *Rhodotorula toruloides* during lipid production. Biotechnol Biofuels 2019; 12: 137.
[http://dx.doi.org/10.1186/s13068-019-1478-8] [PMID: 31171938]

[132] Martinez-Silveira A, Villarreal R, Garmendia G, Rufo C, Vero S. Process conditions for a rapid in situ transesterification for biodiesel production from oleaginous yeasts. Electron J Biotechnol 2019; 38: 1-9.
[http://dx.doi.org/10.1016/j.ejbt.2018.11.006]

[133] Ryu BG, Kim J, Kim K, Choi YE, Han JI, Yang JW. High-cell-density cultivation of oleaginous yeast Cryptococcus curvatus for biodiesel production using organic waste from the brewery industry. Bioresour Technol 2013; 135: 357-64.
[http://dx.doi.org/10.1016/j.biortech.2012.09.054] [PMID: 23177209]

[134] Maza DD, Viñarta SC, Su Y, Guillamón JM, Aybar MJ. Growth and lipid production of Rhodotorula glutinis R4, in comparison to other oleaginous yeasts. J Biotechnol 2020; 310: 21-31.
[http://dx.doi.org/10.1016/j.jbiotec.2020.01.012] [PMID: 32004579]

[135] Liu L ping, Zong M hua, Hu Y, Li N, Lou W yong, Wu H. Efficient microbial oil production on crude glycerol by Lipomyces starkeyi AS 2.1560 and its kinetics. Process Biochem 2017; 58: 230-8.
[http://dx.doi.org/10.1016/j.procbio.2017.03.024]

[136] Liu Y, Wang Y, Liu H, Zhang J. Enhanced lipid production with undetoxified corncob hydrolysate by Rhodotorula glutinis using a high cell density culture strategy. Bioresour Technol 2015; 180: 32-9.
[http://dx.doi.org/10.1016/j.biortech.2014.12.093] [PMID: 25585258]

[137] Jiru TM, Groenewald M, Pohl C, Steyn L, Kiggundu N, Abate D. Optimization of cultivation conditions for biotechnological production of lipid by Rhodotorula kratochvilovae (syn, Rhodosporidium kratochvilovae) SY89 for biodiesel preparation. 3 Biotech 2017; 7: 1-11.
[http://dx.doi.org/10.1007/s13205-017-0769-7]

[138] Chen L, Zhang Y, Liu GL, Chi Z, Hu Z, Chi ZM. Cellular lipid production by the fatty acid synthase-duplicated Lipomyces kononenkoae BF1S57 strain for biodiesel making. Renew Energy 2020; 151: 707-14.
[http://dx.doi.org/10.1016/j.renene.2019.11.074]

[139] Calvey CH, Su YK, Willis LB, McGee M, Jeffries TW. Nitrogen limitation, oxygen limitation, and lipid accumulation in Lipomyces starkeyi. Bioresour Technol 2016; 200: 780-8.
[http://dx.doi.org/10.1016/j.biortech.2015.10.104] [PMID: 26580895]

[140] Abdel-Mawgoud AM, Markham KA, Palmer CM, Liu N, Stephanopoulos G, Alper HS. Metabolic engineering in the host Yarrowia lipolytica. Metab Eng 2018; 50: 192-208.
[http://dx.doi.org/10.1016/j.ymben.2018.07.016] [PMID: 30056205]

[141] Grand View Research. Textile Market Size | Industry Analysis Report, 2021-2028 2021. https://www.grandviewresearch.com/industry-analysis/textile-market

[142] Bolt Threads. Mylo - Bolt Threads. BoltthreadsCom 2020.https://boltthreads.com/technology/microsilk/

[143] Meadow M. Technology - Modern Meadow. Mod Meadow 2020. https://www.modernmeadow.com/technology

[144] Schönberger H, Schäfer T. Best Available Techniques in Textile Industry. Berlin, Germany 2003.

[145] Hasanbeigi A, Price L. A technical review of emerging technologies for energy and water efficiency and pollution reduction in the textile industry. J Clean Prod 2015; 95: 30-44.
[http://dx.doi.org/10.1016/j.jclepro.2015.02.079]

[146] Madhu A, Chakraborty JN. Developments in application of enzymes for textile processing. J Clean Prod 2017; 145: 114-33.
[http://dx.doi.org/10.1016/j.jclepro.2017.01.013]

[147] Antošová Z, Sychrová H. Yeast Hosts for the Production of Recombinant Laccases: A Review. Mol Biotechnol 2016; 58(2): 93-116.
[http://dx.doi.org/10.1007/s12033-015-9910-1] [PMID: 26698313]

[148] Samsami S, Mohamadizaniani M, Sarrafzadeh M-H, Rene ER, Firoozbahr M. Recent advances in the treatment of dye-containing wastewater from textile industries: Overview and perspectives. Process Saf Environ Prot 2020; 143: 138-63.
[http://dx.doi.org/10.1016/j.psep.2020.05.034]

[149] Giardina P, Faraco V, Pezzella C, Piscitelli A, Vanhulle S, Sannia G. Laccases: a never-ending story. Cell Mol Life Sci 2010; 67(3): 369-85.
[http://dx.doi.org/10.1007/s00018-009-0169-1] [PMID: 19844659]

[150] Viancelli A, Michelon W, Rogovski P, *et al.* A review on alternative bioprocesses for removal of emerging contaminants. Bioprocess Biosyst Eng 2020; 43(12): 2117-29.
[http://dx.doi.org/10.1007/s00449-020-02410-9] [PMID: 32681451]

[151] Jadhav JP, Parshetti GK, Kalme SD, Govindwar SP. Decolourization of azo dye methyl red by Saccharomyces cerevisiae MTCC 463. Chemosphere 2007; 68(2): 394-400.
[http://dx.doi.org/10.1016/j.chemosphere.2006.12.087] [PMID: 17292452]

[152] Lee K-M, Kalyani D, Tiwari MK, *et al.* Enhanced enzymatic hydrolysis of rice straw by removal of phenolic compounds using a novel laccase from yeast Yarrowia lipolytica. Bioresour Technol 2012; 123: 636-45.
[http://dx.doi.org/10.1016/j.biortech.2012.07.066] [PMID: 22960123]

[153] Ranieri D, Colao MC, Ruzzi M, Romagnoli G, Bianchi MM. Optimization of recombinant fungal laccase production with strains of the yeast Kluyveromyces lactis from the pyruvate decarboxylase promoter. FEMS Yeast Res 2009; 9(6): 892-902.
[http://dx.doi.org/10.1111/j.1567-1364.2009.00532.x] [PMID: 19527303]

[154] Galli C, Gentili P, Jolivalt C, Madzak C, Vadalà R. How is the reactivity of laccase affected by single-point mutations? Engineering laccase for improved activity towards sterically demanding substrates. Appl Microbiol Biotechnol 2011; 91(1): 123-31.
[http://dx.doi.org/10.1007/s00253-011-3240-4] [PMID: 21468703]

[155] Iimura Y, Sonoki T, Habe H. Heterologous expression of Trametes versicolor laccase in Saccharomyces cerevisiae. Protein Expr Purif 2018; 141: 39-43.
[http://dx.doi.org/10.1016/j.pep.2017.09.004] [PMID: 28918197]

[156] Liu N, Shen S, Jia H, *et al.* Heterologous expression of Stlac2, a laccase isozyme of Setosphearia turcica, and the ability of decolorization of malachite green. Int J Biol Macromol 2019; 138: 21-8.
[http://dx.doi.org/10.1016/j.ijbiomac.2019.07.029] [PMID: 31301394]

[157] Guo M, Lu F, Du L, Pu J, Bai D. Optimization of the expression of a laccase gene from Trametes versicolor in Pichia methanolica. Appl Microbiol Biotechnol 2006; 71(6): 848-52.
[http://dx.doi.org/10.1007/s00253-005-0210-8] [PMID: 16292528]

[158] Nishibori N, Masaki K, Tsuchioka H, Fujii T, Iefuji H. Comparison of laccase production levels in

Pichia pastoris and Cryptococcus sp. S-2. J Biosci Bioeng 2013; 115(4): 394-9.
[http://dx.doi.org/10.1016/j.jbiosc.2012.10.025] [PMID: 23200414]

[159] Rönnander J, Wright SAI. Growth of wood-inhabiting yeasts of the Faroe Islands in the presence of spent sulphite liquor. Antonie van Leeuwenhoek 2021; 114(6): 649-66.
[http://dx.doi.org/10.1007/s10482-021-01543-5] [PMID: 33851316]

[160] Yurkov AM. Yeasts of the soil - obscure but precious. Yeast 2018; 35(5): 369-78.
[http://dx.doi.org/10.1002/yea.3310] [PMID: 29365211]

[161] Cadete RM, Lopes MR, Rosa CA. Yeasts Associated with Decomposing Plant Material and Rotting Wood Yeasts Nat Ecosyst Divers. Cham: Springer International Publishing 2017; pp. 265-92.
[http://dx.doi.org/10.1007/978-3-319-62683-3_9]

[162] González AE, Martínez AT, Almendros G, Grinbergs J. A study of yeasts during the delignification and fungal transformation of wood into cattle feed in Chilean rain forest. Antonie van Leeuwenhoek 1989; 55(3): 221-36.
[http://dx.doi.org/10.1007/BF00393851] [PMID: 2757365]

[163] Pozo MI, de Vega C, Canto A, Herrera CM. Presence of yeasts in floral nectar is consistent with the hypothesis of microbial-mediated signaling in plant-pollinator interactions. Plant Signal Behav 2009; 4(11): 1102-4.
[http://dx.doi.org/10.4161/psb.4.11.9874] [PMID: 20009562]

[164] Aleklett K, Hart M, Shade A. The microbial ecology of flowers: an emerging frontier in phyllosphere research. Botany 2014; 92: 253-66.
[http://dx.doi.org/10.1139/cjb-2013-0166]

[165] Bailes EJ, Ollerton J, Pattrick JG, Glover BJ. How can an understanding of plant-pollinator interactions contribute to global food security? Curr Opin Plant Biol 2015; 26: 72-9.
[http://dx.doi.org/10.1016/j.pbi.2015.06.002] [PMID: 26116979]

[166] Klaps J, Lievens B, Álvarez-Pérez S. Towards a better understanding of the role of nectar-inhabiting yeasts in plant-animal interactions. Fungal Biol Biotechnol 2020; 7: 1.
[http://dx.doi.org/10.1186/s40694-019-0091-8] [PMID: 31921433]

[167] Roy R, Schmitt AJ, Thomas JB, Carter CJ. Review: Nectar biology: From molecules to ecosystems. Plant Sci 2017; 262: 148-64.
[http://dx.doi.org/10.1016/j.plantsci.2017.04.012] [PMID: 28716410]

[168] Jacquemyn H, Pozo MI, Álvarez-Pérez S, Lievens B, Fukami T. Yeast-nectar interactions: metacommunities and effects on pollinators. Curr Opin Insect Sci 2021; 44: 35-40.
[http://dx.doi.org/10.1016/j.cois.2020.09.014] [PMID: 33065340]

[169] Rering CC, Beck JJ, Hall GW, McCartney MM, Vannette RL. Nectar-inhabiting microorganisms influence nectar volatile composition and attractiveness to a generalist pollinator. New Phytol 2018; 220(3): 750-9.
[http://dx.doi.org/10.1111/nph.14809] [PMID: 28960308]

[170] Herrera CM, Pozo MI, Medrano M. Yeasts in nectar of an early-blooming herb: sought by bumble bees, detrimental to plant fecundity. Ecology 2013; 94(2): 273-9.
[http://dx.doi.org/10.1890/12-0595.1] [PMID: 23691645]

[171] Schaeffer RN, Phillips CR, Duryea MC, Andicoechea J, Irwin RE. Nectar yeasts in the tall Larkspur Delphinium barbeyi (Ranunculaceae) and effects on components of pollinator foraging behavior. PLoS One 2014; 9(10): e108214.
[http://dx.doi.org/10.1371/journal.pone.0108214] [PMID: 25272164]

[172] Lachance MA, Starmer WT, Rosa CA, Bowles JM, Barker JSS, Janzen DH. Biogeography of the yeasts of ephemeral flowers and their insects. FEMS Yeast Res 2001; 1(1): 1-8.
[http://dx.doi.org/10.1016/S1567-1356(00)00003-9] [PMID: 12702457]

[173] Brysch-Herzberg M. Ecology of yeasts in plant-bumblebee mutualism in Central Europe. FEMS

Microbiol Ecol 2004; 50(2): 87-100.
[http://dx.doi.org/10.1016/j.femsec.2004.06.003] [PMID: 19712367]

[174] Cullen NP, Fetters AM, Ashman T-L. Integrating microbes into pollination. Curr Opin Insect Sci 2021; 44: 48-54.
[http://dx.doi.org/10.1016/j.cois.2020.11.002] [PMID: 33248285]

[175] Becher PG, Hagman A, Verschut V, *et al.* Chemical signaling and insect attraction is a conserved trait in yeasts. Ecol Evol 2018; 8(5): 2962-74.
[http://dx.doi.org/10.1002/ece3.3905] [PMID: 29531709]

[176] Stefanini I. Yeast-insect associations: It takes guts. Yeast 2018; 35(4): 315-30.
[http://dx.doi.org/10.1002/yea.3309] [PMID: 29363168]

[177] Stefanini I, Dapporto L, Berná L, Polsinelli M, Turillazzi S, Cavalieri D. Social wasps are a Saccharomyces mating nest. Proc Natl Acad Sci USA 2016; 113(8): 2247-51.
[http://dx.doi.org/10.1073/pnas.1516453113] [PMID: 26787874]

[178] Reuter M, Bell G, Greig D. Increased outbreeding in yeast in response to dispersal by an insect vector. Curr Biol 2007; 17(3): R81-3.
[http://dx.doi.org/10.1016/j.cub.2006.11.059] [PMID: 17276903]

[179] Meriggi N, Di Paola M, Cavalieri D, Stefanini I. *Saccharomyces cerevisiae* - Insects Association: Impacts, Biogeography, and Extent. Front Microbiol 2020; 11: 1629.
[http://dx.doi.org/10.3389/fmicb.2020.01629] [PMID: 32760380]

[180] Pozo MI, Lachance M-A, Herrera CM. Nectar yeasts of two southern Spanish plants: the roles of immigration and physiological traits in community assembly. FEMS Microbiol Ecol 2012; 80(2): 281-93.
[http://dx.doi.org/10.1111/j.1574-6941.2011.01286.x] [PMID: 22224447]

[181] Belisle M, Peay KG, Fukami T. Flowers as islands: spatial distribution of nectar-inhabiting microfungi among plants of Mimulus aurantiacus, a hummingbird-pollinated shrub. Microb Ecol 2012; 63(4): 711-8.
[http://dx.doi.org/10.1007/s00248-011-9975-8] [PMID: 22080257]

[182] Mommaerts V, Put K, Vandeven J, *et al.* Development of a new dispenser for microbiological control agents and evaluation of dissemination by bumblebees in greenhouse strawberries. Pest Manag Sci 2010; 66(11): 1199-207.
[http://dx.doi.org/10.1002/ps.1995] [PMID: 20672338]

[183] Freimoser FM, Rueda-Mejia MP, Tilocca B, Migheli Q. Biocontrol yeasts: mechanisms and applications. World J Microbiol Biotechnol 2019; 35(10): 154.
[http://dx.doi.org/10.1007/s11274-019-2728-4] [PMID: 31576429]

[184] Pandin C, Le Coq D, Canette A, Aymerich S, Briandet R. Should the biofilm mode of life be taken into consideration for microbial biocontrol agents? Microb Biotechnol 2017; 10(4): 719-34.
[http://dx.doi.org/10.1111/1751-7915.12693] [PMID: 28205337]

[185] Ruiz-Moyano S, Hernández A, Galvan AI, *et al.* Selection and application of antifungal VOCs-producing yeasts as biocontrol agents of grey mould in fruits. Food Microbiol 2020; 92: 103556.
[http://dx.doi.org/10.1016/j.fm.2020.103556] [PMID: 32950150]

[186] Ul Hassan Z, Al Thani R, Atia FA, Alsafran M, Migheli Q, Jaoua S. Application of yeasts and yeast derivatives for the biological control of toxigenic fungi and their toxic metabolites. Environ Technol Innov 2021; 22: 101447.
[http://dx.doi.org/10.1016/j.eti.2021.101447]

[187] Moriguchi K, Yamamoto S, Tanaka K, Kurata N, Suzuki K. Trans-kingdom horizontal DNA transfer from bacteria to yeast is highly plastic due to natural polymorphisms in auxiliary nonessential recipient genes. PLoS One 2013; 8(9): e74590.
[http://dx.doi.org/10.1371/journal.pone.0074590] [PMID: 24058593]

[188]   Liu Y, Yao S, Deng L, Ming J, Zeng K. Metschnikowia citriensis sp. nov., a novel yeast species isolated from leaves with potential for biocontrol of postharvest fruit rot. Biol Control 2018; 125: 15-9.
[http://dx.doi.org/10.1016/j.biocontrol.2018.05.018]

[189]   Liu J, Sui Y, Wisniewski M, Droby S, Liu Y. Review: Utilization of antagonistic yeasts to manage postharvest fungal diseases of fruit. Int J Food Microbiol 2013; 167(2): 153-60.
[http://dx.doi.org/10.1016/j.ijfoodmicro.2013.09.004] [PMID: 24135671]

[190]   Spadaro D, Droby S. Development of biocontrol products for postharvest diseases of fruit: The importance of elucidating the mechanisms of action of yeast antagonists. Trends Food Sci Technol 2016; 47: 39-49.
[http://dx.doi.org/10.1016/j.tifs.2015.11.003]

[191]   Droby S, Wisniewski M, Macarisin D, Wilson C. Twenty years of postharvest biocontrol research: Is it time for a new paradigm? Postharvest Biol Technol 2009; 52: 137-45.
[http://dx.doi.org/10.1016/j.postharvbio.2008.11.009]

[192]   Janisiewicz WJ, Korsten L. Biological control of postharvest diseases of fruits. Annu Rev Phytopathol 2002; 40: 411-41.
[http://dx.doi.org/10.1146/annurev.phyto.40.120401.130158] [PMID: 12147766]

[193]   Kiran P, Hemant P. Biocontrol Agents Market Share | Global Industry Report 2019-2025 - Industry Coverage. Glob Mark Insights 1–274. 2019. https://www.gminsights.com/industry-analysis/biocontrol-agents-market

[194]   Mukherjee A, Verma JP, Gaurav AK, Chouhan GK, Patel JS, Hesham AE-L. Yeast a potential bio-agent: future for plant growth and postharvest disease management for sustainable agriculture. Appl Microbiol Biotechnol 2020; 104(4): 1497-510.
[http://dx.doi.org/10.1007/s00253-019-10321-3] [PMID: 31915901]

[195]   Fu S-F, Sun P-F, Lu H-Y, *et al.* Plant growth-promoting traits of yeasts isolated from the phyllosphere and rhizosphere of Drosera spatulata Lab. Fungal Biol 2016; 120(3): 433-48.
[http://dx.doi.org/10.1016/j.funbio.2015.12.006] [PMID: 26895872]

[196]   El-Tarabily KA, Sivasithamparam K. Potential of yeasts as biocontrol agents of soil-borne fungal plant pathogens and as plant growth promoters. Mycoscience 2006; 47: 25-35.
[http://dx.doi.org/10.1007/S10267-005-0268-2]

[197]   Nassar AH, El-Tarabily KA, Sivasithamparam K. Promotion of plant growth by an auxin-producing isolate of the yeast Williopsis saturnus endophytic in maize (Zea mays L.) roots. Biol Fertil Soils 2005; 42: 97-108.
[http://dx.doi.org/10.1007/s00374-005-0008-y]

[198]   Amprayn K, Rose MT, Kecskés M, Pereg L, Nguyen HT, Kennedy IR. Plant growth promoting characteristics of soil yeast (Candida tropicalis HY) and its effectiveness for promoting rice growth. Appl Soil Ecol 2012; 61: 295-9.
[http://dx.doi.org/10.1016/j.apsoil.2011.11.009]

[199]   Cloete KJ, Valentine AJ, Stander MA, Blomerus LM, Botha A. Evidence of symbiosis between the soil yeast *Cryptococcus laurentii* and a sclerophyllous medicinal shrub, *Agathosma betulina* (Berg.) Pillans. Microb Ecol 2009; 57(4): 624-32.
[http://dx.doi.org/10.1007/s00248-008-9457-9] [PMID: 18958514]

[200]   Falih AM, Wainwright M. Nitrification, S-oxidation and P-solubilization by the soil yeast *Williopsis californica* and by *Saccharomyces cerevisiae*. Mycol Res 1995; 99: 200-4.
[http://dx.doi.org/10.1016/S0953-7562(09)80886-1]

[201]   Xin G, Glawe D, Doty SL. Characterization of three endophytic, indole-3-acetic acid-producing yeasts occurring in Populus trees. Mycol Res 2009; 113(Pt 9): 973-80.
[http://dx.doi.org/10.1016/j.mycres.2009.06.001] [PMID: 19539760]

SUBJECT INDEX

Sérgio Luiz Alves Júnior, Helen Treichel, Thiago Olitta Basso and Boris Ugarte Stambuk (Eds.)
All rights reserved-© 2022 Bentham Science Publishers

U

V

W

www.ingramcontent.com/pod-product-compliance
Lightning Source LLC
Chambersburg PA
CBHW050758220326
41598CB00006B/48